Current Developments in Mathematics 1995

Edited by

Raoul Bott
Michael Hopkins
Arthur Jaffe
Isadore Singer
Daniel W. Stroock
Shing-Tung Yau

International Press
www.intlpress.com

Current Developments in Mathematics, 1995

ISBN 978-1-57146-232-9

Paperback reissue 2012. Originally published in 1994 under ISBN 1-57146-029-2 (casebound). The preface, table of contents, and papers herein are reproduced in facsimile from the original publication.

Printed in the United States of America.

Preface

In the fall of 1994, the departments of mathematics of Harvard University and the Massachusetts Institute of Technology joined forces to establish what they hope will be an annual seminar devoted to the exposition of current developments in mathematics. Each year the content of the seminar will be determined by a committee made up of representatives from each of the two departments. For the 1995 seminar, the members of the committee were Raoul Bott, Arthur Jaffe, and Shing-Tung Yau from Harvard, and Michael Hopkins, Isadore Singer and Daniel Stroock from M.I.T.

Although the committee attempts to choose subjects which they feel are of current interest, time constraints make it inevitable that only some important subjects can be covered. For this reason, the committee will sometimes avoid subjects which already enjoy wide exposure in seminars elsewhere. For example, this year the committee decided not to have talks about gauge theory invariants, in spite of the obvious appeal and enormous progress in this field.

Because the 1995 seminar was not even conceived until December 1994 and the speakers were not notified until the turn of the year, it is remarkable that we were able to produce the preliminary volume from manuscripts which were delivered in early April 1995, and, for this, we are all in the authors' debt. We hope that their efforts will be rewarded by the feedback which their talks and manuscripts provoke.

Finally, the organizers take this opportunity to thank several institutions without whose cooperation the seminar would not have been possible: the departments of mathematics at Harvard and M.I.T. for financial aid and the American Academy of Arts and Sciences for opening their facilities to us.

Editors

Contents

Fermat's last theorem

Henri Darmon
(DARMON@MATH.MCGILL.CA)
DEPARTMENT OF MATHEMATICS
MCGILL UNIVERSITY
MONTREAL, QC
CANADA H3A 2K6

Fred Diamond
(FDIAMOND@PMMS.CAM.AC.UK)
D.P.M.M.S.
CAMBRIDGE UNIVERSITY
CAMBRIDGE, CB2 1SB
UNITED KINGDOM

Richard Taylor
(TAYLORR@MATHS.OX.AC.UK)
MATHEMATICS INSTITUTE
OXFORD UNIVERSITY
24-29 ST. GILES
OXFORD, OX1 3LB
UNITED KINGDOM

The authors would like to give special thanks to N. Boston, K. Buzzard, and B. Conrad for providing so much valuable feedback on earlier versions of this paper. They are also grateful to A. Agboola, M. Bertolini, B. Edixhoven, J. Fearnley, B.H. Gross, R. Gross, L. Guo, F. Jarvis, H. Kisilevsky, E. Liverance, J. Manoharmayum, K. Ribet, D. Rohrlich, M. Rosen, R. Schoof, J.-P. Serre, C. Skinner, D. Thakur, J. Tilouine, J. Tunnell, A. Van der Poorten, and L.

Washington for their helpful comments.

Darmon thanks the members of CICMA and of the Quebec-Vermont Number Theory Seminar for many stimulating conversations on the topics of this paper, particularly in the Spring of 1995. For the same reason Diamond is grateful to the participants in an informal seminar at Columbia University in 1993-94, and Taylor thanks those attending the Oxford Number Theory Seminar in the Fall of 1995.

Parts of this paper were written while the authors held positions at other institutions: Darmon at Princeton University, Diamond at the Institute for Advanced Study, and Taylor at Cambridge University. During some of the period, Diamond enjoyed the hospitality of Princeton University, and Taylor that of Harvard University and MIT. The writing of this paper was also supported by research grants from NSERC (Darmon), EPSRC #GR/J94761 and NSF # DMS 9304580 (Diamond) and by an advanced fellowship from EPSRC (Taylor).

This article owes everything to the ideas of Wiles, and the arguments presented here are fundamentally his [W3], though they include both the work [TW] and several simplifications to the original arguments, most notably that of Faltings. In the hope of increasing clarity, we have not always stated theorems in the greatest known generality, concentrating instead on what is needed for the proof of the Shimura-Taniyama conjecture for semi-stable elliptic curves. This article can serve as an introduction to the fundamental papers [W3] and [TW], which the reader is encouraged to consult for a different, and often more in-depth, perspective on the topics considered. Another useful more advanced reference is the article [Di2] which strengthens the methods of [W3] and [TW] to prove that every elliptic curve that is semistable at 3 and 5 is modular.

Introduction

Fermat's Last Theorem

Fermat's Last Theorem states that the equation

$$x^n + y^n = z^n, \quad xyz \neq 0$$

has no integer solutions when n is greater than or equal to 3. Around 1630, Pierre de Fermat claimed that he had found a "truly wonderful" proof of this theorem, but that the margin of his copy of Diophantus' *Arithmetica* was too small to contain it:

> "Cubum autem in duos cubos, aut quadrato quadratum in duos quadrato quadratos, et generaliter nullam in infinitum ultra quadratum potestatem in duos ejusdem nominis fas est dividere; cujus rei demonstrationem mirabile sane detexi. Hanc marginis exiguitas non caperet."

Among the many challenges that Fermat left for posterity, this was to prove the most vexing. A tantalizingly simple problem about whole numbers, it stood unsolved for more than 350 years, until in 1994 Andrew Wiles finally laid it to rest.

Prehistory: The only case of Fermat's Last Theorem for which Fermat actually wrote down a proof is for the case $n = 4$. To do this, Fermat introduced the idea of *infinite descent* which is still one the main tools in the study of Diophantine equations, and was to play a central role in the proof of Fermat's Last Theorem 350 years later. To prove his Last Theorem for exponent 4, Fermat showed something slightly stronger, namely that the equation $x^4 + y^4 = z^2$ has no solutions in relatively prime integers with $xyz \neq 0$. Solutions to such an equation correspond to rational points on the elliptic curve $v^2 = u^3 - 4u$. Since every integer $n \geq 3$ is divisible either by an odd prime or by 4, the result of Fermat allowed one to reduce the study of Fermat's equation to the case where $n = \ell$ is an *odd prime.*

In 1753, Leonhard Euler wrote down a proof of Fermat's Last Theorem for the exponent $\ell = 3$, by performing what in modern language we would call a 3-descent on the curve $x^3 + y^3 = 1$ which is also an elliptic curve. Euler's argument (which seems to have contained a gap) is explained in [Edw], ch. 2, and [Dic1], p. 545.

It took mathematicians almost 100 years after Euler's achievement to handle the case $\ell = 5$; this was settled, more or less simultaneously, by Gustav Peter Lejeune Dirichlet [Dir] and Adrien Marie Legendre [Leg] in 1825. Their elementary arguments are quite involved. (Cf. [Edw], sec. 3.3.)

In 1839, Fermat's equation for exponent 7 also yielded to elementary methods, through the heroic efforts of Gabriel Lamé. Lamé's proof was even more intricate than the proof for exponent 5, and suggested that to go further, new theoretical insights would be needed.

The work of Sophie Germain: Around 1820, in a letter to Gauss, Sophie Germain proved that if ℓ is a prime and $q = 2\ell + 1$ is also prime, then Fermat's equation $x^\ell + y^\ell = z^\ell$ with exponent ℓ has no solutions (x, y, z) with $xyz \not\equiv 0 \pmod{\ell}$. Germain's theorem was the first really general proposition on Fermat's Last Theorem, unlike the previous results which considered the Fermat equation one exponent at a time.

The case where the solution (x, y, z) to $x^\ell + y^\ell = z^\ell$ satisfies $xyz \not\equiv 0 \pmod{\ell}$ was called the *first case* of Fermat's Last Theorem, and the case where ℓ divides xyz, the *second case*. It was realized at that time that the first case was generally easier to handle: Germain's theorem was extended, using similar ideas, to cases where $k\ell + 1$ is prime and k is small, and this led to a proof that there were no first case solutions to Fermat's equation with prime exponents $\ell \leq 100$, which in 1830 represented a significant advance. The division between first and second case remained fundamental in much of the later work on the subject. In 1977, Terjanian [Te] proved that if the equation $x^{2\ell} + y^{2\ell} = z^{2\ell}$ has a solution (x, y, z), then 2ℓ divides either x or y, i.e., "the first case of Fermat's Last Theorem is true for even exponents". His simple and elegant proof used only techniques that were available to Germain and her contemporaries.

The work of Kummer: The work of Ernst Eduard Kummer marked the beginning of a new era in the study of Fermat's Last Theorem. For the first time, sophisticated concepts of algebraic number theory and the theory of L-functions were brought to bear on a question that had until then been addressed only with elementary methods. While he fell short of providing a complete solution, Kummer made substantial progress. He showed how Fermat's Last Theorem is intimately tied to deep questions on class numbers of cyclotomic fields which are still an active subject of research. Kummer's approach relied on the factorization

$$(x + y)(x + \zeta_\ell y) \cdots (x + \zeta_\ell^{\ell-1} y) = z^\ell$$

of Fermat's equation over the ring $\mathbb{Z}[\zeta_\ell]$ generated by the ℓth roots of unity. One observes that the greatest common divisor of any two factors in the product on the left divides the element $(1 - \zeta_\ell)$, which is an element of norm ℓ. Since the product of these numbers is a perfect ℓ-th power, one is tempted to conclude that $(x + y), \ldots, (x + \zeta_\ell^{\ell-1} y)$ are each ℓ-th powers in the ring $\mathbb{Z}[\zeta_\ell]$ up to units in this ring, and up to powers of $(1 - \zeta_\ell)$. Such an inference would be valid if one were to replace $\mathbb{Z}[\zeta_\ell]$ by $\mathbb{Z}$, and is a direct consequence of ***unique factorization*** of integers into products of primes. We say that a ring R has property UF if every non-zero element of R is uniquely a product of primes, up to units. Mathematicians such as Lamé made attempts at proving Fermat's Last Theorem based on the mistaken assumption that the rings $\mathbb{Z}[\zeta_\ell]$ had property UF. Legend even has it that Kummer fell into this trap, although this story now has been discredited; see for example [Edw], sec. 4.1. In fact, property UF is far from being satisfied in general: one now knows that the rings $\mathbb{Z}[\zeta_\ell]$ have property UF only for $\ell < 23$ (cf. [Wa], ch. 1).

It turns out that the full force of property UF is not really needed in the applications to Fermat's Last Theorem. Say that a ring R has property UF_ℓ if the following inference is valid:

$$ab = z^\ell, \text{ and } \gcd(a,b) = 1 \Rightarrow a \text{ and } b \text{ are } \ell\text{th powers up to units of } R.$$

If a ring R has property UF, then it also has property UF_ℓ, but the converse need not be true. Kummer showed that Fermat's last theorem was true for exponent ℓ if $\mathbb{Z}[\zeta_\ell]$ satisfied the property UF_ℓ (cf. [Wa]). The proof is far from trivial, because of difficulties arising from the units in $\mathbb{Z}[\zeta_\ell]$ as well as from the possible failure of property UF. (A number of Kummer's contemporaries, such as Cauchy and Lamé, seem to have overlooked both of these difficulties in their attempts to prove Fermat's Last Theorem.)

Kummer then launched a systematic study of the property UF_ℓ for the rings $\mathbb{Z}[\zeta_\ell]$. He showed that even if $\mathbb{Z}[\zeta_\ell]$ failed to have unique factorization, it still possessed unique factorization into prime ***ideals***. He defined the ***ideal class group*** as the quotient of the group of fractional ideals by its subgroup consisting of principal ideals, and was able to establish the finiteness of this class group. The order of the class group of $\mathbb{Z}[\zeta_\ell]$, denoted h_ℓ, could be taken as a measure of the failure of the ring $\mathbb{Z}[\zeta_\ell]$ to satisfy UF. It was rather straightforward to show that if ℓ did not divide h_ℓ, then $\mathbb{Z}[\zeta_\ell]$ satisfied the property UF_ℓ. In this case, one called ℓ a *regular prime*. Kummer thus showed that Fermat's last theorem is true for exponent ℓ if ℓ is a regular prime.

He did not stop here. For it remained to give an efficient means of computing h_ℓ, or at least an efficient way of checking when ℓ divides h_ℓ. The class number h_ℓ can be factorized as a product

$$h_\ell = h_\ell^+ h_\ell^-,$$

where h_ℓ^+ is the class number of the real subfield $\mathbb{Q}(\zeta_\ell)^+$, and h_ℓ^- is defined as h_ℓ/h_ℓ^+. Essentially because of the units in $\mathbb{Q}(\zeta_\ell)^+$, the factor h_ℓ^+ is somewhat difficult to compute, while, because the units in $\mathbb{Q}(\zeta_\ell)^+$ generate the group of units in $\mathbb{Q}(\zeta_\ell)$ up to finite index, the term h_ℓ^- can be expressed in a simple closed form. Kummer showed that if ℓ divides h_ℓ^+, then ℓ divides h_ℓ^-. Hence, ℓ divides h_ℓ if and only if ℓ divides h_ℓ^-. This allowed one to avoid the difficulties inherent in the calculation of h_ℓ^+. Kummer then gave an elegant formula for h_ℓ^- by considering the Bernoulli numbers B_n, which are rational numbers defined by the formula

$$\frac{x}{e^x - 1} = \sum \frac{B_n}{n!} x^n.$$

He produced an explicit formula for the class number h_ℓ^-, and concluded that if ℓ does not divide the numerator of B_{2i}, for $1 \le i \le (\ell - 3)/2$, then ℓ is regular, and conversely.

The conceptual explanation for Kummer's formula for h_ℓ^- lies in the work of Dirichlet on the analytic class number formula, where it is shown that h_ℓ^- can be expressed as a product of special values of certain (abelian) L-series

$$L(s,\chi) = \sum_{n=1}^{\infty} \chi(n) n^{-s}$$

associated to odd Dirichlet characters. Such special values in turn can be expressed in terms of certain generalized Bernoulli numbers $B_{1,\chi}$, which are related to the Bernoulli numbers B_i via congruences mod ℓ. (For more details, see [Wa].)

These considerations led Kummer to initiate a deep study relating congruence properties of special values of L-functions and of class numbers, which was to emerge as a central concern of modern algebraic number theory, and was to reappear – in a surprisingly different guise – at the heart of Wiles' strategy for proving the Shimura-Taniyama conjecture.

Later developments: Kummer's work had multiple ramifications, and led to a very active line of enquiry pursued by many people. His formulae relating Bernoulli numbers to class numbers of cyclotomic fields were refined by Kenneth Ribet [R1], Barry Mazur and Andrew Wiles [MW], using new methods from the theory of modular curves which also play a central role in Wiles' more recent work. (Later Francisco Thaine [Th] reproved some of the results of Mazur and Wiles using techniques inspired directly from a reading of Kummer.) In a development more directly related to Fermat's Last Theorem, Wieferich proved that if ℓ^2 does not divide $2^{\ell-1} - 1$, then the first case of Fermat's Last Theorem is true for exponent ℓ. (Cf. [Ri], lecture VIII.)

There were many other refinements of similar criteria for Fermat's Last theorem to be true. Computer calculations based on these criteria led to a verification that Fermat's Last theorem is true for all odd prime exponents less than four million [BCEM], and that the first case is true for all $\ell \leq 8.858 \cdot 10^{20}$ [Su].

The condition that ℓ is a regular prime seems to hold heuristically for about 61% of the primes. (See the discussion on p. 63, and also p. 108, of [Wa], for example.) In spite of the convincing numerical evidence, it is still not known if there are infinitely many regular primes. Ironically, it is not too difficult to show that there are infinitely many irregular primes. (Cf. [Wa].)

Thus the methods introduced by Kummer, after leading to very strong results in the direction of Fermat's Last theorem, seemed to become mired in difficulties, and ultimately fell short of solving Fermat's conundrum[1].

Faltings' proof of the Mordell conjecture: In 1985, Gerd Faltings [Fa] proved the very general statement (which had previously been conjectured

[1] However, W. McCallum has recently introduced a technique, based on the method of Chabauty and Coleman, which suggests new directions for approaching Fermat's Last Theorem via the cyclotomic theory. An application of McCallum's method to showing the *second* case of Fermat's Last Theorem for regular primes is explained in [Mc].

by Mordell) that any equation in two variables corresponding to a curve of genus strictly greater than one had (at most) finitely many rational solutions. In the context of Fermat's Last Theorem, this led to the proof that for each exponent $n \geq 3$, the Fermat equation $x^n + y^n = z^n$ has at most finitely many integer solutions (up to the obvious rescaling). Andrew Granville [Gra] and Roger Heath-Brown [HB] remarked that Faltings' result implies Fermat's Last Theorem for a set of exponents of density one.

However, Fermat's Last Theorem was still not known to be true for an infinite set of prime exponents. In fact, the theorem of Faltings seemed ill-equipped for dealing with the finer questions raised by Fermat in his margin, namely of finding a complete list of rational points on *all* of the Fermat curves $x^n + y^n = 1$ simultaneously, and showing that there are no solutions on these curves when $n \geq 3$ except the obvious ones.

Mazur's work on Diophantine properties of modular curves: Although it was not realized at the time, the chain of ideas that was to lead to a proof of Fermat's Last theorem had already been set in motion by Barry Mazur in the mid seventies. The modular curves $X_0(\ell)$ and $X_1(\ell)$ introduced in section 1.2 and 1.5 give rise to another naturally occurring infinite family of Diophantine equations. These equations have certain systematic rational solutions corresponding to the cusps that are defined over $\mathbb{Q}$, and are analogous to the so-called "trivial solutions" of Fermat's equation. Replacing Fermat curves by modular curves, one could ask for a complete list of all the rational points on the curves $X_0(\ell)$ and $X_1(\ell)$. This problem is perhaps even more compelling than Fermat's Last Theorem: rational points on modular curves correspond to objects with natural geometric and arithmetic interest, namely, elliptic curves with cyclic subgroups or points of order ℓ. In [Maz1] and [Maz2], B. Mazur gave essentially a complete answer to the analogue of Fermat's Last Theorem for modular curves. More precisely, he showed that if $\ell \neq 2, 3, 5$ and 7, (i.e., $X_1(\ell)$ has genus > 0) then the curve $X_1(\ell)$ has no rational points other than the "trivial" ones, namely cusps. He proved analogous results for the curves $X_0(\ell)$ in [Maz2], which implied, in particular, that an elliptic curve over $\mathbb{Q}$ with square-free conductor has no rational cyclic subgroup of order ℓ over $\mathbb{Q}$ if ℓ is a prime which is strictly greater than 7. This result appeared a full ten years before Faltings' proof of the Mordell conjecture.

Frey's strategy: In 1986, Gerhard Frey had the insight that these constructions might provide a precise link between Fermat's Last Theorem and deep questions in the theory of elliptic curves, most notably the Shimura Taniyama conjecture. Given a solution $a^\ell + b^\ell = c^\ell$ to the Fermat equation of prime degree ℓ, we may assume without loss of generality that $a^\ell \equiv -1 \pmod 4$ and that $b^\ell \equiv 0 \pmod{32}$. Frey considered (following Hellegouarch, [He], p. 262; cf. also Kubert-Lang [KL], ch. 8, §2) the elliptic curve

$$E : y^2 = x(x - a^\ell)(x + b^\ell).$$

This curve is *semistable*, i.e., it has square-free conductor. Let $E[\ell]$ denote the group of points of order ℓ on E defined over some (fixed) algebraic closure $\bar{\mathbb{Q}}$ of $\mathbb{Q}$, and let L denote the smallest number field over which these points are defined. This extension appears as a natural generalization of the cyclotomic fields $\mathbb{Q}(\zeta_\ell)$ studied by Kummer. What singles out the field L for special attention is that it has *very little ramification*: using Tate's analytic description of E at the primes dividing abc, it could be shown that L was ramified only at 2 and ℓ, and that the ramification of L at these two primes was rather restricted. (See theorem 2.15 of section 2.2 for a precise statement.) Moreover, the results of Mazur on the curve $X_0(\ell)$ could be used to show that L is *large*, in the following precise sense. The space $E[\ell]$ is a vector space of dimension 2 over the finite field $\mathbb{F}_\ell$ with ℓ elements, and the absolute Galois group $G_\mathbb{Q} = \mathrm{Gal}\,(\bar{\mathbb{Q}}/\mathbb{Q})$ acts $\mathbb{F}_\ell$-linearly on $E[\ell]$. Choosing an $\mathbb{F}_\ell$-basis for $E[\ell]$, the action is described by a representation

$$\bar{\rho}_{E,\ell} : \mathrm{Gal}\,(L/\mathbb{Q}) \hookrightarrow GL_2(\mathbb{F}_\ell).$$

Mazur's results in [Maz1] and [Maz2] imply that $\bar{\rho}_{E,\ell}$ is *irreducible* if $\ell > 7$ (using the fact that E is *semi-stable*). In fact, combined with earlier results of Serre [Se6], Mazur's results imply that for $\ell > 7$, the representation $\bar{\rho}_{E,\ell}$ is surjective, so that $\mathrm{Gal}\,(L/\mathbb{Q})$ is actually isomorphic to $GL_2(\mathbb{F}_\ell)$ in this case.

Serre's conjectures: In [Se7], Jean-Pierre Serre made a careful study of mod ℓ Galois representations $\bar{\rho} : G_\mathbb{Q} \longrightarrow GL_2(\mathbb{F}_\ell)$ (and, more generally, of representations into $GL_2(k)$, where k is any finite field). He was able to make very precise conjectures (see section 3.2) relating these representations to modular forms mod ℓ. In the context of the representations $\bar{\rho}_{E,\ell}$ that occur in Frey's construction, Serre's conjecture predicted that they arose from modular forms (mod ℓ) of weight two and level two. Such modular forms, which correspond to differentials on the modular curve $X_0(2)$, do not exist because $X_0(2)$ has genus 0. Thus Serre's conjecture implied Fermat's Last Theorem. The link between fields with Galois groups contained in $GL_2(\mathbb{F}_\ell)$ and modular forms mod ℓ still appears to be very deep, and Serre's conjecture remains a tantalizing open problem.

Ribet's work: lowering the level: The conjecture of Shimura and Taniyama (cf. section 1.8) provides a direct link between elliptic curves and modular forms. It predicts that the representation $\bar{\rho}_{E,\ell}$ obtained from the ℓ-division points of the Frey curve arises from a modular form of weight 2, albeit a form whose level is quite large. (It is the product of all the primes dividing abc, where $a^\ell + b^\ell = c^\ell$ is the putative solution to Fermat's equation.) Ribet [R5] proved that, if this were the case, then $\bar{\rho}_{E,\ell}$ would *also* be associated with a modular form mod ℓ of weight 2 and level 2, in the way predicted by Serre's conjecture. This deep result allowed him to reduce Fermat's Last Theorem to the Shimura-Taniyama conjecture.

Wiles' work: proof of the Shimura-Taniyama conjecture: In [W3] Wiles proves the Shimura-Taniyama conjecture for semi-stable elliptic curves, providing the final missing step and proving Fermat's Last Theorem. After more than 350 years, the saga of Fermat's Last theorem has come to a spectacular end.

The relation between Wiles' work and Fermat's Last Theorem has been very well documented (see, for example, [R8], and the references contained therein). Hence this article will focus primarily on the breakthrough of Wiles [W3] and Taylor-Wiles [TW] which leads to the proof of the Shimura-Taniyama conjecture for semi-stable elliptic curves.

From elliptic curves to ℓ-adic representations: Wiles' opening gambit for proving the Shimura-Taniyama conjecture is to view it as part of the more general problem of relating two-dimensional Galois representations and modular forms. The Shimura-Taniyama conjecture states that if E is an elliptic curve over $\mathbb{Q}$, then E is modular. One of several equivalent definitions of modularity is that for some integer N there is an eigenform $f = \sum a_n q^n$ of weight two on $\Gamma_0(N)$ such that

$$\#E(\mathbb{F}_p) = p + 1 - a_p$$

for all but finitely primes p. (By an eigenform, here we mean a cusp form which is a normalized eigenform for the Hecke operators; see section 1 for definitions.)

This conjecture acquires a more Galois theoretic flavour when one considers the two dimensional ℓ-adic representation

$$\rho_{E,\ell} : G_{\mathbb{Q}} \longrightarrow GL_2(\mathbb{Z}_\ell)$$

obtained from the action of $G_{\mathbb{Q}}$ on the ℓ-adic Tate module of

$$E : T_\ell E = \lim_{\leftarrow} E[l^n](\bar{\mathbb{Q}}).$$

An ℓ-adic representation ρ of $G_{\mathbb{Q}}$ is said to arise from an eigenform $f = \sum a_n q^n$ with integer coefficients a_n if

$$\mathrm{tr}\,(\rho(\mathrm{Frob}\,_p)) = a_p,$$

for all but finitely many primes p at which ρ is unramified. Here $\mathrm{Frob}\,_p$ is a Frobenius element at p (see section 2), and its image under ρ is a well-defined conjugacy class.

A direct computation shows that $\#E(\mathbb{F}_p) = p + 1 - \mathrm{tr}\,(\rho_{E,\ell}(\mathrm{Frob}\,_p))$ for all primes p at which $\rho_{E,\ell}$ is unramified, so that E is modular (in the sense defined above) if and only if for some ℓ, $\rho_{E,\ell}$ arises from an eigenform. In fact the Shimura-Taniyama conjecture can be generalized to a conjecture that every ℓ-adic representation, satisfying suitable local conditions, arises from a modular form. Such a conjecture was proposed by Fontaine and Mazur [FM].

Galois groups and modular forms

Viewed in this way, the Shimura-Taniyama conjecture becomes part of a much larger picture: the emerging, partly conjectural and partly proven correspondence between certain modular forms and two dimensional representations of $G_{\mathbf{Q}}$. This correspondence, which encompasses the Serre conjectures, the Fontaine-Mazur conjecture, and the Langlands program for GL_2, represents a first step toward a higher dimensional, non-abelian generalization of class field theory.

Two-dimensional representations of $G_{\mathbf{Q}}$: In the first part of this century, class field theory gave a complete description of $G_{\mathbf{Q}}^{\mathrm{ab}}$, the maximal (continuous) abelian quotient of $G_{\mathbf{Q}}$. In fact the Kronecker-Weber theorem asserts that $G_{\mathbf{Q}}^{\mathrm{ab}} \cong \prod_p \mathbb{Z}_p^\times$, and one obtains a complete description of all one-dimensional representations of $G_{\mathbf{Q}}$. In the second half of this century much attention has focused on attempts to understand the whole group $G_{\mathbf{Q}}$, or more precisely to describe all its representations. Although there has been a fair degree of success in using modular forms to construct representations of $G_{\mathbf{Q}}$, less is known about how exhaustive these constructions are. The major results in the latter direction along these lines are the work of Langlands [Ll2] and the recent work of Wiles ([W3] completed by [TW]). Both concern two-dimensional representations of $G_{\mathbf{Q}}$ and give significant evidence that these representations are parametrised (in a very precise sense) by certain modular forms. The purpose of this article is to describe both the proven and conjectural parts of this theory, give a fairly detailed exposition of Wiles' recent contribution and explain the application to Fermat's Last theorem. To make this description somewhat more precise let us distinguish three types of representation.

Artin representations and the Langlands-Tunnell theorem: Continuous representations $\rho : G_{\mathbf{Q}} \to GL_2(\mathbb{C})$ are called (two-dimensional) Artin representations. Such representations necessarily have finite image, and are therefore semi-simple. We restrict our attention to those which are irreducible. They are conjectured to be in bijection (in a precise way) with certain newforms (a special class of eigenforms). Those ρ which are odd (i.e. the determinant of complex conjugation is -1), should correspond to weight 1 holomorphic newforms. Those which are even should correspond to certain non-holomorphic (Maass) newforms. Two partial but deep results are known.

(a) (Deligne-Serre) If f is a holomorphic weight one newform then the corresponding Artin representation can be constructed ([DS]).

(b) (Langlands-Tunnell) If ρ is a two dimensional Artin representation with soluble image then the corresponding modular form exists ([Ll2] and [Tu]).

The proof of the latter result is analytic in nature, invoking the trace formula and the theory of L-functions.

ℓ-adic representations and the Fontaine-Mazur conjecture: By an ℓ-adic representation we shall mean any continuous representation $\rho : G_{\mathbb{Q}} \to GL_2(K)$ which is unramified outside a finite set of primes and where K is a finite extension of $\mathbb{Q}_\ell$ (generalizing slightly the notion of ℓ-adic representation that was introduced before). Given a holomorphic newform f one can attach to f a system of ℓ-adic representations, following Eichler, Shimura, Deligne and Serre. These ℓ-adic representations are called modular. The Fontaine-Mazur conjecture (see [FM]) predicts if ρ is an odd, irreducible, ℓ-adic representation whose restriction to the decomposition group at ℓ is well enough behaved, then ρ is modular. (The restriction on the behaviour of the representation on the decomposition group at ℓ is essential in this conjecture; it is not true that all odd, irreducible two dimensional ℓ-adic representation are modular.) Before Wiles' work almost nothing was known about this conjecture, except that certain very special cases could be deduced from the work of Hecke, Langlands and Tunnell.

Mod ℓ representations and Serre's conjecture: A mod ℓ representation is a continuous representation $\bar{\rho} : G_{\mathbb{Q}} \longrightarrow GL_2(\bar{\mathbb{F}}_\ell)$. For example if $E/\mathbb{Q}$ is an elliptic curve then the action of $G_{\mathbb{Q}}$ on the ℓ-division points of E gives rise to a mod ℓ representation $\bar{\rho}_{E,\ell}$ which is just the reduction modulo ℓ of $\rho_{E,\ell}$. One can use the work of Eichler, Shimura, Deligne and Serre to associate to each mod ℓ eigenform a mod ℓ representation of $G_{\mathbb{Q}}$. The mod ℓ representations which arise in this way are called modular. Serre has conjectured [Se7] that every odd (absolutely) irreducible mod ℓ representation is modular and should arise from a mod ℓ eigenform with certain very specific properties. This conjecture can be thought of as having two parts.

The first asserts that every odd irreducible mod ℓ representation is modular. About this very little is known. It is known for $\bar{\rho} : G_{\mathbb{Q}} \to GL_2(\mathbb{F}_2)$ by work of Hecke. It is also known for $\bar{\rho} : G_{\mathbb{Q}} \to GL_2(\mathbb{F}_3)$. This latter result is an application of the Langlands-Tunnell theorem using the two accidents that there is a section to the homomorphism $GL_2(\mathbb{Z}[\sqrt{-2}]) \twoheadrightarrow GL_2(\mathbb{F}_3)$ and that $GL_2(\mathbb{F}_3)$ is soluble. Partial results for $\bar{\rho} : G_{\mathbb{Q}} \to GL_2(\mathbb{F}_5)$ follow from Wiles' work.

Given a mod ℓ representation arising from a mod ℓ eigenform, the second part of Serre's conjecture predicts the minimal weight and level for that mod ℓ eigenform. Here the situation is much better. There has been a lot of work over the last decade (including ideas from Mazur, Ribet, Carayol and Gross) and the problem is nearly completely resolved (see [Di1]). As was pointed out earlier, Ribet's contribution [R5] implies that, if one can show that the Galois representation $\bar{\rho}_{E,\ell}$ arising from the (semi-stable) Frey curve attached to a solution of Fermat's equation with exponent ℓ is modular, then one can show that this representation does not exist—because it would be modular of weight two and level two— and hence one can deduce Fermat's

Last Theorem.

However we have seen that to show $\bar{\rho}_{E,\ell}$ is modular it suffices to show that for some ℓ_0, the ℓ_0-adic representation ρ_{E,ℓ_0} is modular. In particular it suffices to verify that either $\rho_{E,3}$ or $\rho_{E,5}$ is modular. Hence the Shimura-Taniyama conjecture can be reduced to (part of) the Fontaine-Mazur conjecture for $\ell = 3$ and 5. We have seen that for these primes part of Serre's conjecture is known, so it turns out it suffices to prove results of the form "Serre's conjecture for ℓ implies the Fontaine-Mazur conjecture for ℓ". This is the direction of Wiles' work, although nothing quite this general has been proven yet.

Deformation theory: Thus the problem Wiles faces is to show that if ρ is an odd ℓ-adic representation which has irreducible modular reduction $\bar{\rho}$ and which is sufficiently well behaved when restricted to the decomposition group at ℓ, then ρ is modular. In fact he only proves a weakened version of such a result, but one which is sufficient to conclude that all semistable elliptic curves are modular.

Wiles approaches the problem by putting it in a more general setting. On the one hand he considers lifts of $\bar{\rho}$ to representations over complete noetherian local $\mathbb{Z}_\ell$-algebras R. For each finite set of primes Σ, one can consider lifts of type Σ; these are lifts which are well-behaved on a decomposition group at ℓ, and whose ramification at primes not in Σ is rather restricted. In particular, such a lift is unramified outside $\Sigma \cup S$ where S is the set of ramified primes of $\bar{\rho}$. A method of Mazur (see [Maz3]) can then be used to show that if $\bar{\rho}$ is absolutely irreducible, then there is a representation

$$\rho_\Sigma^{\text{univ}} : G_\mathbb{Q} \longrightarrow GL_2(R_\Sigma)$$

which is universal in the following sense. If $\rho : G_\mathbb{Q} \to GL_2(R)$ is a lift of $\bar{\rho}$ of type Σ, then there is a unique local homomorphism $R_\Sigma \longrightarrow R$ such that ρ is equivalent to the pushforward of $\rho_\Sigma^{\text{univ}}$. Thus the equivalence classes of type Σ lifts to $GL_2(R)$ can be *identified* with $\text{Hom}(R_\Sigma, R)$. The local ring R_Σ is called the *universal deformation ring* for representations of type Σ.

On the other hand Wiles constructs a candidate for a universal modular lifting of type Σ

$$\rho_\Sigma^{\text{mod}} : G_\mathbb{Q} \longrightarrow GL_2(\mathbb{T}_\Sigma).$$

The ring $\mathbb{T}_\Sigma$ is constructed from the algebra of Hecke operators acting on a certain space of modular forms. The universal property of R_Σ gives a map $R_\Sigma \to \mathbb{T}_\Sigma$. The problem thus becomes: to show that this map is an isomorphism[2]. In fact, it can be shown to be a surjection without great difficulty, and the real challenge is to prove injectivity, i.e., to show, in essence, that R_Σ is not larger than $\mathbb{T}_\Sigma$.

[2]Maps of this kind were already considered in [Maz3] and [BM], and it is conjectured in [MT] that these maps are isomorphisms in certain cases, though not in exactly the situations considered by Wiles.

By an ingenious piece of commutative algebra, Wiles found a numerical criterion for this map to be an isomorphism, and for the ring $\mathbb{T}_\Sigma$ to be a local complete intersection. This numerical criterion seems to be very close to a special case of the Bloch-Kato conjecture [BK]. Wiles further showed (by combining arguments from Galois cohomology and from the theory of congruences between modular forms) that this numerical criterion was satisfied if the minimal version $\mathbb{T}_\emptyset$ of this Hecke algebra (obtained by taking $\Sigma = \emptyset$, i.e., allowing the least possible amount of ramification in the deformations) was a complete intersection. Finally in [TW] it was proved that $\mathbb{T}_\emptyset$ is a complete intersection.

Outline of the paper

Chapter 1 recalls some basic notions from the classical theory of elliptic curves and modular forms, such as modular forms and modular curves over $\mathbb{C}$ and $\mathbb{Q}$, Hecke operators and q-expansions, and Eichler-Shimura theory. The Shimura-Taniyama conjecture is stated precisely in section 1.8.

Chapter 2 introduces the basic theory of representations of $G_{\mathbb{Q}}$. We describe Mazur's deformation theory and begin our study of the universal deformation rings using techniques from Galois cohomology and from the theory of finite flat group schemes. We also recall some basic properties of elliptic curves, both to explain Frey's argument precisely and illustrate the uses of ℓ-adic representations.

Chapter 3 explains how to associate Galois representations to modular forms. We then describe what was known and conjectured about associating modular forms to Galois representations before Wiles' work. After introducing the universal modular lifts of certain mod ℓ representations, we give the proof of Wiles' main theorems, taking for granted certain results of a more technical nature that are proved in the last two chapters.

Chapter 4 explains how to prove the necessary results concerning the structure of Hecke algebras: the generalization by Taylor and Wiles of a result of de Shalit, and the generalization by Wiles of a result of Ribet.

Chapter 5 establishes the fundamental results from commutative algebra discovered by Wiles, following modifications of the approach of Wiles and Taylor-Wiles proposed by Faltings and Lenstra.

Contents

1 Elliptic curves and modular forms

1.1 Elliptic curves

We begin with a brief review of elliptic curves. A general reference for the results discussed in this section is [Si1] and [Si2].

An elliptic curve E over a field F is a smooth projective curve over F of genus one with a distinguished F-rational point. If E/F is an elliptic curve and if ω is a non-zero holomorphic differential on E/F then E can be realised in the projective plane by an equation (called a Weierstrass equation) of the form

$$Y^2Z + a_1XYZ + a_3YZ^2 = X^3 + a_2X^2Z + a_4XZ^2 + a_6Z^3 \qquad \text{(W)}$$

such that the distinguished point is $(0:1:0)$ (sometimes denoted ∞ because it corresponds to the "point at infinity" in the affine model obtained by setting $Z = 1$) and $\omega = \frac{dx}{2y+a_1x+a_3}$. We also define the following quantities associated to (W):

$$\begin{aligned}
&b_2 = a_1^2 + 4a_2 \qquad b_4 = 2a_4 + a_1a_3 \qquad b_6 = a_3^2 + 4a_6\\
&b_8 = a_1^2a_6 + 4a_2a_6 - a_1a_3a_4 + a_2a_3^2 - a_4^2\\
&\Delta = 9b_2b_4b_6 - b_2^2b_8 - 8b_4^3 - 27b_6^2\\
&j = (b_2^2 - 24b_4)^3/\Delta.
\end{aligned}$$

One can check that the equation (W) defines an elliptic curve if and only if Δ is nonzero. One can also check that such equations define elliptic curves which are isomorphic over $\bar{F}$ if and only if they give the same quantity j. Thus j only depends on E so we will denote it j_E. The quantity Δ depends only on the pair (E,ω) so we shall denote it $\Delta(E,\omega)$. If u belongs to $F^\times$ then $u^{12}\Delta(E,u\omega) = \Delta(E,\omega)$.

An elliptic curve E/F has a natural structure of a commutative algebraic group with the distinguished F-rational point as the identity element.

An algebraic map between two elliptic curves which sends the distinguished point of one to the distinguished point of the other is automatically a morphism of algebraic groups. A map between elliptic curves which has finite kernel (and hence, is generically surjective) is called an *isogeny*.

Elliptic curves over $\mathbb{C}$: If $F = \mathbb{C}$, then the curve E is isomorphic as a complex analytic manifold to the complex torus $\mathbb{C}/\Lambda$, where Λ is a lattice in $\mathbb{C}$, i.e., a discrete $\mathbb{Z}$-submodule of $\mathbb{C}$ of rank 2. The group law on $E(\mathbb{C})$

corresponds to the usual addition in $\mathbb{C}/\Lambda$. In terms of Λ, an affine equation for E in $A^2(\mathbb{C})$ is given by

$$y^2 = 4x^3 + g_2 x + g_3,$$

where

$$g_2 = -60 \sum_{\lambda \in \Lambda - \{0\}} \frac{1}{z^4}, \qquad g_3 = -140 \sum_{\lambda \in \Lambda - \{0\}} \frac{1}{z^6}.$$

In terms of this equation, the map from $\mathbb{C}/\Lambda$ to $E(\mathbb{C})$ sends z to $(x, y) = (\wp(z), \wp'(z))$, where $\wp(z)$ is the *Weierstrass $\wp$-function* associated to the lattice Λ. (Cf. [Sil], ch. VI.) The inverse map is given by integrating the holomorphic differential ω, i.e., sending $P \in E(\mathbb{C})$ to the image of $\int_\gamma \omega$ in $\mathbb{C}/\Lambda$, where γ is any path on $E(\mathbb{C})$ from ∞ to P, and Λ is the lattice of periods $\int_\gamma \omega$, where γ ranges over the integral homology $H_1(E(\mathbb{C}), \mathbb{Z})$. Replacing ω by $u\omega$ changes Λ to $u\Lambda$, so that Λ is determined by E only up to homotheties. We scale Λ so that one of its $\mathbb{Z}$-generators is 1, and another, τ, has strictly positive imaginary part. This gives the analytic isomorphism:

$$E(\mathbb{C}) \simeq \mathbb{C}/\langle 1, \tau \rangle.$$

The complex number τ in the complex upper half plane $\mathcal{H}$ is well defined, modulo the natural action of $SL_2(\mathbb{Z})$ on $\mathcal{H}$ by Möbius transformations. (Thus the set of isomorphism classes of elliptic curves over $\mathbb{C}$ can be *identified* with the quotient $\mathcal{H}/SL_2(\mathbb{Z})$.)

The map $z \mapsto e^{2\pi i z}$ identifies $\mathbb{C}/\langle 1, \tau \rangle$ with $\mathbb{C}^\times / q^{\mathbf{Z}}$, where $q = e^{2\pi i \tau}$ is the *multiplicative Tate period*. The analytic isomorphism

$$E(\mathbb{C}) \simeq \mathbb{C}^\times / q^{\mathbf{Z}}$$

has the virtue of generalizing to the p-adic setting in certain cases, as we will see shortly.

Note that $|q| < 1$. The invariant j can be expressed in terms of q by a convergent power series with *integer* coefficients:

$$j = q^{-1} + 744 + 196884q + \cdots . \tag{1.1.1}$$

The following basic facts are a direct consequence of the analytic theory:

Proposition 1.1 *The subgroup $E[n](\mathbb{C})$ of points of order n on $E(\mathbb{C})$ is isomorphic (non-canonically) to $\mathbb{Z}/n\mathbb{Z} \times \mathbb{Z}/n\mathbb{Z}$. More generally, if F is any field of characteristic zero, the subgroup $E[n](F)$ is contained in $\mathbb{Z}/n\mathbb{Z} \times \mathbb{Z}/n\mathbb{Z}$.*

Proof. The analytic theory shows that $E(\mathbb{C})$ is isomorphic as an abstract group to a product of two circle groups, and the first statement follows. The second statement follows from the Lefschetz principle (cf. [Sil], ch. VI, §6). □

Proposition 1.2 *The endomorphism ring* $\mathrm{End}_{\mathbb{C}}(E)$ *of an elliptic curve over* $\mathbb{C}$ *is isomorphic either to* $\mathbb{Z}$ *or to an order in a quadratic imaginary field. The same is true if one replaces* $\mathbb{C}$ *by any field of characteristic* 0.

Proof: An endomorphism of $E(\mathbb{C}) \simeq \mathbb{C}/\Lambda$ induces multiplication by complex number α on the tangent space. Hence $\mathrm{End}_{\mathbb{C}}(E)$ is isomorphic to the ring of $\alpha \in \mathbb{C}$ satisfying $\alpha\Lambda \subset \Lambda$. Such a ring is isomorphic either to $\mathbb{Z}$ or to a quadratic imaginary order. The corresponding statement for fields of characteristic 0 follows as in the proof of proposition 1.1. □

If $\mathrm{End}_{\mathbb{C}}(E) \otimes \mathbb{Q}$ is a quadratic imaginary field, we say that E has *complex multiplication*.

Remark 1.3 It follows from the arithmetic theory of complex multiplication (cf. [Si2], ch. 1) that any elliptic curve E with complex multiplication is defined over an abelian extension of the quadratic imaginary field $K = \mathrm{End}_{\mathbb{C}}(E) \otimes \mathbb{Q}$. If E is defined over $\mathbb{Q}$, then K has class number one. There are only finitely many elliptic curves over $\mathbb{Q}$ with complex multiplication, up to "twists" (i.e., $\mathbb{C}$-isomorphism).

Elliptic curves over $\mathbb{Q}_p$: Now suppose that E is an elliptic curve defined over the p-adic field $\mathbb{Q}_p$. There is an equation

$$(W^{\min}) \qquad Y^2Z + a_1XYZ + a_3YZ^2 = X^3 + a_2X^2Z + a_4XZ^2 + a_6Z^3$$

for E with the property $a_i \in \mathbb{Z}_p$ for all i and $|\Delta|$ is minimal amongst all such equations for E. Although $(W^{\min})$ is not unique, the associated discriminant depends only on E and is denoted $\Delta_E^{\min}$. Moreover the reduction of $(W^{\min})$ modulo the uniformizer p defines a projective curve $\bar{E}$, which is independent of the particular minimal equation chosen. If (W) is any equation for E with coefficients in $\mathbb{Z}_p$ and with discriminant Δ, then $\Delta_E^{\min}$ divides Δ.

If $\bar{E}$ is a smooth curve we say that E has *good reduction* at p. If $\bar{E}$ has a unique singular point which is a node we say that E has *multiplicative reduction* at p. Otherwise $\bar{E}$ has a unique singular point which is a cusp and we say that E has *additive reduction* at p. If E has good or multiplicative reduction we say that it has *semi-stable* reduction at p, or simply that E is *semi-stable.*

If (W) defines a smooth curve mod p then E has good reduction at p and (W) is a minimal equation. If $\Delta \equiv 0 \bmod p$ but $b_2^2 \not\equiv 24b_4 \bmod p$, then modulo p the equation (W) defines a curve with a node. In this case E has multiplicative reduction at p and (W) is a minimal equation.

Curves with good reduction: In that case p does not divide $\Delta_E^{\min}$, and the reduction $\bar{E}$ is an elliptic curve over $\mathbb{F}_p$.

If q is any power of p, and $\mathbb{F}_q$ is the field with q elements, we define the integer N_q to be the number of solutions to the equation $(W^{\min})$ in the projective plane $\mathbb{P}^2(\mathbb{F}_q)$. Thus N_q is the order of the finite group $\bar{E}(\mathbb{F}_q)$. We define the integer a_q by the formula

$$a_q = q + 1 - N_q.$$

The integers a_q are completely determined by a_p: more precisely, we have

$$(1 - a_p p^{-s} + p^{1-2s})^{-1} = 1 + a_p p^{-s} + a_{p^2} p^{-2s} + a_{p^3} p^{-3s} + \cdots . \tag{1.1.2}$$

We call the expression on the left the (local) L-function associated to E over $\mathbb{Q}_p$, and denote it by $L(E/\mathbb{Q}_p, s)$. Concerning the size of a_p we have the following fundamental result of Hasse, whose proof can be found in [Sil], ch. V, § 1:

Theorem 1.4 $|a_p| \leq 2\sqrt{p}$.

A further division among curves of good reduction plays a significant role in our later discussion. We say that E has (good) *ordinary* reduction if p does not divide a_p, and that it has *supersingular* reduction if p divides a_p.

When E has good reduction at p, we define its local conductor at p to be $m_p(E) = 0$.

Curves of multiplicative reduction: Elliptic curves over $\mathbb{Q}_p$ which have multiplicative reduction at p can be understood by using the p-adic analytic description discovered by Tate. More precisely, we can formally invert the power series (1.1.1) expressing j in terms of q, to obtain to a power series for q in j^{-1}, having *integer* coefficients:

$$q = j^{-1} + 744j^{-2} + 750420j^{-3} + 872769632j^{-4} + \cdots . \tag{1.1.3}$$

If E has multiplicative reduction, then $j \in \mathbb{Q}_p$ is non-integral, and hence the power series (1.1.3) converges, yielding a well-defined value of q in $p\mathbb{Z}_p$. This is called Tate's p-adic period associated to E over $\mathbb{Q}_p$. Note that we have $v_p(q) = -v_p(j) = v_p(\Delta_E^{min})$.

We say that E has *split* (resp. *non-split*) multiplicative reduction at p if the two tangent lines to the node on $\bar{E}(\mathbb{F}_p)$ have slopes defined over $\mathbb{F}_p$ (resp. $\mathbb{F}_{p^2}$).

Proposition 1.5 (Tate) *There is a p-adic analytic isomorphism*

$$\Phi : \bar{\mathbb{Q}}_p^\times / q^{\mathbb{Z}} \longrightarrow E(\bar{\mathbb{Q}}_p),$$

which has the property that

$$\sigma(\Phi(x)) = \Phi(\sigma x^{\delta(\sigma)}), \quad \forall \sigma \in G_{\mathbb{Q}_p},$$

where $\delta : G_{\mathbb{Q}_p} \longrightarrow \pm 1$ is

- *the trivial character, if E has split multiplicative reduction;*
- *the unique unramified quadratic character of $G_{\mathbb{Q}_p}$, if E has non-split multiplicative reduction.*

The proof of this proposition is explained in [Si2], ch. V, for example.

We define the L-function $L(E/\mathbb{Q}_p, s)$ to be

$$L(E/\mathbb{Q}_p, s) = \begin{cases} (1-p^{-s})^{-1} & \text{if } E \text{ has split reduction,} \\ (1+p^{-s})^{-1} & \text{if } E \text{ has non-split reduction.} \end{cases} \tag{1.1.4}$$

In both cases the conductor $m_p(E)$ is defined to be 1.

Curves of additive reduction: If E has additive reduction at p, we simply define

$$L(E/\mathbb{Q}_p, s) = 1. \tag{1.1.5}$$

The conductor $m_p(E)$ is defined to be 2, if $p > 3$. When $p = 2$ or 3, it is determined by a somewhat more complicated recipe, given in [Ta].

Elliptic curves over $\mathbb{Q}$: Let E be an elliptic curve defined over $\mathbb{Q}$. In particular E may be viewed as a curve over $\mathbb{Q}_p$ for every p, and we define its (global) conductor by

$$N_E = \prod_p p^{m_p(E)}.$$

The curve E is said to be *semi-stable* if it is semi-stable over all p-adic fields $\mathbb{Q}_p$. Note that E is semi-stable if and only if its conductor N_E is square-free.

Using the fact that $\mathbb{Q}$ has class number 1, one can show that E has a *global minimal Weierstrass model* ($W^{\min}$) which gives the equation of a minimal Weierstrass model over each $\mathbb{Q}_p$. The associated discriminant, denoted $\Delta_E^{\min}$, depends only on E. The associated differential, denoted ω_E^{Neron}, is called the *Néron differential.* It is well-defined up to sign.

The following, known as the Mordell-Weil theorem, is the fundamental result about the structure of the group of rational points $E(\mathbb{Q})$. (Cf. For example [Si1].)

Theorem 1.6 *The group $E(\mathbb{Q})$ is a finitely generated abelian group. Hence*

$$E(\mathbb{Q}) \simeq T \oplus \mathbb{Z}^r,$$

where T is the (finite) torsion subgroup of $E(\mathbb{Q})$, and $r \geq 0$ is the rank *of E over $\mathbb{Q}$.*

Concerning the possible structure of T, there is the following deep result of Mazur, a variant of which also plays a crucial role in the proof of Fermat's Last Theorem:

Theorem 1.7 *If $E/\mathbb{Q}$ is an elliptic curve, then its torsion subgroup is isomorphic to one of the following possibilities:*

$$\mathbb{Z}/n\mathbb{Z}, \quad 1 \le n \le 10, \quad n = 12, \qquad \mathbb{Z}/2n\mathbb{Z} \times \mathbb{Z}/2\mathbb{Z}, \quad 1 \le n \le 4.$$

The proof is given in [Maz1] (see also [Maz2]). Thanks to this result, the structure of the torsion subgroup T is well understood. (Recently, the techniques of Mazur have been extended by Kamienny [Kam] and Merel [Mer] to prove uniform boundedness results on the torsion of elliptic curves over general number fields.)

Much more mysterious is the behaviour of the rank r. It is not known if r can be arbitrarily large, although results of Mestre [Mes] and Nagao [Na] show that it is greater or equal to 13 for infinitely many elliptic curves over $\mathbb{Q}$. It turns out that many of the deep results on $E(\mathbb{Q})$ and on r are based on the relation with L-functions.

We define the global L-function of the complex variable s by:

$$L(E/\mathbb{Q}, s) = \prod_p L(E/\mathbb{Q}_p, s). \tag{1.1.6}$$

Exercise 1.8 Using theorem 1.4, show that the infinite product defining the L-function $L(E/\mathbb{Q}, s)$ converges absolutely on the right half plane $\mathrm{Real}(s) > 3/2$.

Conjecture 1.9 (Birch-Swinnerton-Dyer) *The L-function $L(E/\mathbb{Q}, s)$ has an analytic continuation to the entire complex plane,and in particular is analytic at $s = 1$. Furthermore:*

$$\mathrm{ord}_{s=1} L(E/\mathbb{Q}, s) = r.$$

There is also a more precise form of this conjecture, which expresses the leading coefficient of $L(E/\mathbb{Q}, s)$ at $s = 1$ in terms of certain arithmetic invariants of $E/\mathbb{Q}$. For more details, see [Sil], conj. 16.5.

As we will explain in more detail in section 1.8, the analytic continuation of $L(E/\mathbb{Q}, s)$ now follows from the work of Wiles and Taylor-Wiles and a strengthening by Diamond [Di2], for a very large class of elliptic curves over $\mathbb{Q}$, which includes all the semi-stable ones.

Abelian varieties: Elliptic curves admit higher-dimensional analogues, called *abelian varieties*, which also play a role in our discussion. Analytically, the set of complex points on an abelian variety is isomorphic to a quotient $\mathbb{C}^g/\Lambda$, where Λ is a lattice in $\mathbb{C}^g$ of rank $2g$, satisfying the so-called Riemann period relations. A good introduction to the basic theory of abelian varieties can be found in [CS] and [Wel].

1.2 Modular curves and modular forms over $\mathbb{C}$

Modular curves: The group $SL_2(\mathbb{Z})$ of two by two integer matrices of determinant one acts by fractional linear (Möbius) transformations on the complex upper half plane

$$\mathcal{H} = \{z \in \mathbb{C} \mid \operatorname{Im}(z) > 0\},$$

equipped with its standard complex analytic structure. The ***principal congruence group*** $\Gamma(N)$ of level N is the subgroup of matrices in $SL_2(\mathbb{Z})$ which reduce to the identity matrix modulo the positive integer N. A subgroup Γ of $SL_2(\mathbb{Z})$ is called a ***congruence group*** if it contains $\Gamma(N)$ for some N. The level of Γ is the smallest N for which this is true. The most important examples of congruence groups are:

- The group $\Gamma_0(N)$ consisting of all matrices that reduce modulo N to an upper triangular matrix.
- The group $\Gamma_1(N)$ consisting of all matrices that reduce modulo N to a matrix of the form $\begin{pmatrix} 1 & * \\ 0 & 1 \end{pmatrix}$.
- The principal congruence group $\Gamma(N)$ of level N consisting of all matrices that reduce modulo N to the identity.

Notice the natural inclusions of normal subgroups $\Gamma(N) \subset \Gamma_1(N) \subset \Gamma_0(N)$. The quotient $\Gamma_0(N)/\Gamma_1(N)$ is canonically isomorphic to $(\mathbb{Z}/N\mathbb{Z})^\times$ via

$$\begin{pmatrix} a & b \\ c & d \end{pmatrix} \mapsto d \bmod N.$$

For any subgroup H of $(\mathbb{Z}/N\mathbb{Z})^\times$, we denote by $\Gamma_H(N)$ the group of matrices in $\Gamma_0(N)$ whose image in $\Gamma_0(N)/\Gamma_1(N)$ belongs to H.

If Γ is a congruence subgroup of $SL_2(\mathbb{Z})$, define Y_Γ to be the quotient of the upper half plane $\mathcal{H}$ by the action of Γ. One equips Y_Γ with the analytic structure coming from the projection map $\pi : \mathcal{H} \longrightarrow Y_\Gamma$. (More precisely, if $y = \pi(\tau)$, and $G_\tau \subset \Gamma$ is the stabilizer of τ in Γ, then the local ring $\mathcal{O}_{Y_\Gamma,y}$ is identified with the local ring of germs of holomorphic functions at τ which are invariant under the action of G_τ.) This makes Y_Γ into a connected complex analytic manifold of dimension one, i.e., a Riemann surface. If Γ is $\Gamma_0(N)$ (resp. $\Gamma_1(N)$, or $\Gamma(N)$), we will also denote Y_Γ by $Y_0(N)$ (resp. $Y_1(N)$, or $Y(N)$). One compactifies Y_Γ by adjoining a finite set of ***cusps*** which correspond to orbits of $\mathbb{P}^1(\mathbb{Q}) = \mathbb{Q} \cup \{\infty\}$ under Γ. Call X_Γ the corresponding compact Riemann surface. (For more details, notably on the definition of the analytic structure on X_Γ at the cusps, see for example [Kn], p. 311, or [Shi2], ch. 1.) It follows from the definition of this analytic structure that the field K_Γ of meromorphic functions on X_Γ is equal to the set of meromorphic functions on $\mathcal{H}$ satisfying

- (Transformation property): $f(\gamma\tau) = f(\tau)$, for all $\gamma \in \Gamma$;
- (Behaviour at the cusps): For all $\gamma \in SL_2(\mathbb{Z})$, the function $f(\gamma\tau)$ has a Puiseux series expansion $\sum_{-m}^{\infty} a_n q^{n/h}$ in fractional powers of $q = e^{2\pi i\tau}$.

Riemann's existence theorem (cf. for example [For], ch. 2) asserts that the analytic structure on X_Γ comes from an algebraic one, i.e., the field K_Γ is a finitely generated extension of $\mathbb{C}$ of transcendence degree 1. Thus we can, and will, view X_Γ as a complex algebraic curve over $\mathbb{C}$. If Γ is $\Gamma_0(N)$ (resp. $\Gamma_1(N)$, or $\Gamma(N)$), we will also denote X_Γ by $X_0(N)$ (resp. $X_1(N)$, or $X(N)$).

Examples and exercises:
1. For $N = 1$, the curve $X_0(N) = X_1(N) = X(N)$ is a curve of genus 0, and its field of functions is the ring $\mathbb{C}(j)$, where j is the classical modular function,

$$j(\tau) = q^{-1} + 744 + 196884q + \cdots, \quad q = e^{2\pi i\tau}.$$

(Cf., for example, [Se4], ch. 7.)

2. For $\tau \in \mathcal{H} \cup \mathbb{P}^1(\mathbb{Q}) = \bar{\mathcal{H}}$, define G_τ to be the stabilizer of τ in $PSL_2(\mathbb{Z})$, and let $e_\tau = \#(G_\tau/(G_\tau \cap \Gamma))$. Show that e_τ depends only on the Γ-orbit of τ in $\bar{\mathcal{H}}$, and that $e_\tau = 1$ for all but finitely many τ in $\bar{\mathcal{H}}/\Gamma$. Using the Riemann-Hurwitz formula (cf. [Ki], sec. 4.3) show that the genus of X_Γ is given by

$$g(\Gamma) = 1 - [PSL_2(\mathbb{Z}) : \Gamma] + \frac{1}{2} \sum_{\tau \in \bar{\mathcal{H}}/\Gamma} (e_\tau - 1).$$

Use this to compute the genus of $X_0(p)$, $X_1(p)$, and $X(p)$ for p prime. For details, see [Shi2], sec. 1.6 or [Ogg].

3. For $\Gamma = \Gamma(2)$, show that X_Γ is isomorphic to $\mathbb{P}^1$, and that Y_Γ is isomorphic to $\mathbb{P}^1 - \{0, 1, \infty\}$. Show that $\Gamma/\langle\pm 1\rangle$ is the free group on the two generators $g_1 = \begin{pmatrix} 1 & 2 \\ 0 & 1 \end{pmatrix}$ and $g_2 = \begin{pmatrix} 1 & 0 \\ 2 & 1 \end{pmatrix}$.

4. Define a homomorphism $\Gamma(2) \longrightarrow \mathbb{Z}/n\mathbb{Z} \times \mathbb{Z}/n\mathbb{Z}$, by sending g_1 to $(1, 0)$ and g_2 to $(0, 1)$, and let Γ denote its kernel. Show that Γ is not in general a congruence subgroup and that the curve $Y_\Gamma := \mathcal{H}/\Gamma$ is birationally isomorphic to the Fermat curve of degree n with affine equation $x^n + y^n = 1$.

Moduli interpretations: The points in $Y_\Gamma = \mathcal{H}/\Gamma$ can be interpreted as elliptic curves over $\mathbb{C}$ with some extra "level N" structure. More precisely,

- If $\Gamma = \Gamma_0(N)$, then the Γ-orbit of $\tau \in \mathcal{H}$ corresponds to the complex torus $E = \mathbb{C}/\langle 1, \tau\rangle$ with the distinguished cyclic subgroup of order N generated by $\frac{1}{N}$. Thus, points on $Y_0(N)$ parametrize isomorphism classes of pairs (E, C) where E is an elliptic curve over $\mathbb{C}$ and C is a cyclic subgroup of E of order N.

- If $\Gamma = \Gamma_1(N)$, then the Γ-orbit of τ corresponds to the complex torus $E = \mathbb{C}/\langle 1, \tau\rangle$ with the distinguished point of order N given by $\frac{1}{N}$. Hence, points on $Y_1(N)$ parametrize isomorphism classes of pairs (E, P) where now P is a point on E of exact order N.

Remark 1.10 One checks that the above rules set up a bijection between points on Y_Γ and elliptic curves with the appropriate structures, and that the projection $Y_1(N) \longrightarrow Y_0(N)$ sending $\Gamma_1(N)\tau$ to $\Gamma_0(N)\tau$ becomes the "forgetful" map sending (E, P) to $(E, \langle P\rangle)$.

Remark 1.11 (This remark will be used in section 1.3 when discussing Hecke operators.) Define an n-isogeny of Γ-structures to be an n-isogeny of the underlying elliptic curves which sends one Γ-structure to the other. If p is a prime not dividing N, then there are exactly $p+1$ distinct p-isogenies from $(\mathbb{C}/\langle\tau, 1\rangle, \frac{1}{N})$, whose images are the pairs:

$$\left(\mathbb{C}/\left\langle \frac{\tau+i}{p}, 1\right\rangle, \frac{1}{N}\right) \quad (i = 0, \ldots, p-1), \qquad \left(\mathbb{C}/\langle p\tau, 1\rangle, \frac{p}{N}\right).$$

If p divides N, then there are only p distinct p-isogenies from $(\mathbb{C}/\langle\tau, 1\rangle, \frac{1}{N})$, since $(\mathbb{C}/\langle p\tau, 1\rangle, \frac{p}{N})$ is not a $\Gamma_1(N)$-structure (the point p/N not being of exact order N on the complex torus $\mathbb{C}/\langle p\tau, 1\rangle$).

Modular forms: Let k be an even positive integer. A *modular form* of weight k on Γ is a holomorphic function f on $\mathcal{H}$ satisfying:

- (Transformation property): $f(\gamma\tau) = (c\tau + d)^k f(\tau)$, for all $\gamma = \begin{pmatrix} a & b \\ c & d \end{pmatrix} \in \Gamma$.
- (Behaviour at the cusps): For all $\gamma \in PSL_2(\mathbb{Z})$, the function $(c\tau + d)^{-k} f(\gamma\tau)$ has a Puiseux series expansion $\sum_0^\infty a_n q^{n/h}$ in fractional powers of $q = e^{2\pi i\tau}$. We call $\sum a_n q^{n/h}$ the Fourier expansion of f at the cusp $\gamma^{-1}(i\infty)$.

A modular form which satisfies the stronger property that the constant coefficient of its Fourier expansion at each cusp vanishes is called a *cusp form*. We denote by $M_k(\Gamma)$ the complex vector space of modular forms of weight k on Γ, and by $S_k(\Gamma)$ the space of cusp forms on Γ. (For examples, see [DI], sec. 2.2 and the references therein, especially, [Shi2], ch. 2.)

This article is mainly concerned with modular forms of weight 2, and hence we will focus our attention on these from now on. A pleasant feature of the case $k = 2$ is that the cusp forms in $S_2(\Gamma)$ admit a direct geometric interpretation as holomorphic differentials on the curve X_Γ.

Lemma 1.12 *The map $f(\tau) \mapsto \omega_f := 2\pi i f(\tau) d\tau$ is an isomorphism between the space $S_2(\Gamma)$ and the space $\Omega^1(X_\Gamma)$ of holomorphic differentials on the curve X_Γ.*

Sketch of proof: One checks that the transformation property satisfied by $f(\tau)$ under Γ causes the expression $f(\tau)d\tau$ to be Γ-invariant, and that the condition of vanishing at the cusps translates into holomorphicity of $f(\tau)d\tau$. (Note, for example, that $2\pi i d\tau = dq/q$, so that ω_f is holomorphic at $i\infty$ precisely when $f(q)$ vanishes at $q = 0$.) □

As a corollary, we find:

Corollary 1.13 *The space $S_2(\Gamma)$ is finite-dimensional, and its dimension is equal to the genus g of X_Γ.*

Proof: This follows directly from the Riemann-Roch theorem, cf. [Ki], sec. 6.3. □

To narrow still further the focus of our interest, we will be mostly concerned with the cases $\Gamma = \Gamma_0(N)$ and $\Gamma_1(N)$. A slightly more general framework is sometimes convenient, so we suppose from now on that Γ satisfies

$$\Gamma_1(N) \subset \Gamma \subset \Gamma_0(N).$$

Such a group Γ is necessarily of the form $\Gamma_H(N)$ for some subgroup H of $(\mathbb{Z}/N\mathbb{Z})^\times$. Because the transformation $\tau \mapsto \tau + 1$ belongs to Γ the forms in $S_2(\Gamma)$ are periodic functions on $\mathcal{H}$ of period 1, and hence their Fourier expansions at $i\infty$ are of the form

$$f(\tau) = \sum_{n>0} a_n q^n, \quad q = e^{2\pi i\tau}, a_n \in \mathbb{C}.$$

The Petersson inner product: The spaces $S_2(\Gamma)$ are also equipped with a natural Hermitian inner product given by

$$\langle f, g\rangle = \frac{i}{8\pi^2}\int_{X_\Gamma} \omega_f \wedge \bar{\omega}_g = \int_{\mathcal{H}/\Gamma} f(\tau)\bar{g}(\tau)dxdy,$$

where $\tau = x + iy$. This is called the Petersson inner product.

The diamond operators: Suppose now that $\Gamma = \Gamma_1(N)$ and let d be an element of $(\mathbb{Z}/N\mathbb{Z})^\times$. The map $\langle d\rangle$ which sends an elliptic curve with Γ-structure (E, P) to the pair (E, dP) gives an automorphism of Y_Γ which extends to X_Γ. It is called the *diamond operator*. For τ in $\mathcal{H}$ and $\gamma = \begin{pmatrix} a & b \\ c & d \end{pmatrix}$ in $\Gamma_0(N)$, we have

$$\langle d\rangle(\Gamma\tau) = \Gamma(\gamma\tau).$$

Hence $\langle d\rangle$ acts on $S_2(\Gamma)$, identified with the holomorphic differentials on X_Γ, by the rule

$$\langle d\rangle f(\tau) = (c\tau + d)^{-2} f\left(\frac{a\tau + b}{c\tau + d}\right).$$

In geometric terms, the diamond operators are the Galois automorphisms of the natural (branched) covering $X_1(N) \longrightarrow X_0(N)$ whose Galois group is isomorphic to $\Gamma_0(N)/\langle\pm\Gamma_1(N)\rangle = (\mathbb{Z}/N\mathbb{Z})^\times/\langle\pm1\rangle$. Given an even Dirichlet character $\chi : (\mathbb{Z}/N\mathbb{Z})^\times \longrightarrow \mathbb{C}^\times$, say that f is a modular form of level N and character χ if it belongs to the χ-eigenspace in $S_2(\Gamma_1(N))$ under this action. Let $S_2(N,\chi)$ denote the space of all such forms. Thus a function f in $S_2(N,\chi)$ is a cusp form on $\Gamma_1(N)$ which satisfies the stronger transformation property:

$$f\left(\frac{a\tau+b}{c\tau+d}\right) = \chi(d)(c\tau+d)^2 f(\tau), \text{ for all } \left(\begin{smallmatrix} a & b \\ c & d \end{smallmatrix}\right) \in \Gamma_0(N).$$

Note that if $\mathbf{1}$ is the trivial character, then $S_2(N,\mathbf{1})$ is canonically identified with $S_2(\Gamma_0(N))$, which we will also denote by $S_2(N)$. Finally note the direct sum decomposition:

$$S_2(\Gamma_1(N)) = \oplus_\chi S_2(N,\chi),$$

where the sum ranges over all the even Dirichlet characters modulo N.

Exercise 1.14 Show that if $f(\tau)$ belongs to $S_2(N)$, then $f(a\tau)$ belongs to $S_2(mN)$, for each integer a dividing m.

Jacobians of modular curves: Let V be the dual space

$$V = S_2(\Gamma)^\vee := \mathrm{Hom}(S_2(\Gamma),\mathbb{C}).$$

It is a complex vector space of dimension $g = \mathrm{genus}(X_\Gamma)$. The integral homology $\Lambda = H_1(X_\Gamma,\mathbb{Z})$ maps naturally to V by sending a homology cycle c to the functional ϕ_c defined by $\phi_c(f) = \int_c \omega_f$. The image of Λ is a lattice in V, i.e., a $\mathbb{Z}$-module of rank $2g$ which is discrete (cf. [Mu1], cor. 3.8). Fix a base point $\tau_0 \in \mathcal{H}$, and define the Abel-Jacobi map $\Phi_{\mathrm{AJ}} : X_\Gamma(\mathbb{C}) \longrightarrow V/\Lambda$ by $\Phi_{\mathrm{AJ}}(P)(f) = \int_{\tau_0}^{P} \omega_f$. Note that this is well-defined, i.e., it does not depend on the choice of path on X_Γ from τ_0 to P, up to elements in Λ.

We extend the map Φ_{AJ} by linearity to the group $\mathrm{Div}\,(X_\Gamma)$ of divisors on X_Γ, and observe that the restriction of Φ_{AJ} to the group $\mathrm{Div}^0(X_\Gamma)$ of degree 0 divisors does not depend on the choice of base-point τ_0. Moreover we have the Abel-Jacobi theorem:

Theorem 1.15 *The map*

$$\Phi_{\mathrm{AJ}} : \mathrm{Div}^0(X_\Gamma) \longrightarrow V/\Lambda$$

has a kernel consisting precisely of the group $P(X_\Gamma)$ of principal divisors on X_Γ. Hence Φ_{AJ} induces an isomorphism from $\mathrm{Pic}^0(X_\Gamma) := \mathrm{Div}^0(X_\Gamma)/P(X_\Gamma)$ to V/Λ.

For the proof, see [Mu1], ch. 3. The quotient V/Λ is a complex torus, and is equal to the group of complex points of an abelian variety. We denote this abelian variety by J_Γ, the *Jacobian variety* of X_Γ over $\mathbb{C}$. If $\Gamma = \Gamma_0(N)$ or $\Gamma_1(N)$, we will also write $J_0(N)$ or $J_1(N)$ respectively for the Jacobian J_Γ.

1.3 Hecke operators and Hecke theory

We maintain our running assumption that Γ satisfies

$$\Gamma_1(N) \subset \Gamma \subset \Gamma_0(N).$$

If p is a prime not dividing the level N, we define the Hecke operator T_p on $S_2(\Gamma)$ by the formula

$$T_p(f) = \frac{1}{p}\sum_{i=0}^{p-1} f\left(\frac{\tau+i}{p}\right) + p\langle p\rangle f(p\tau).$$

We give a more conceptual description of T_p in terms of remark 1.11 in the case $\Gamma = \Gamma_1(N)$. We have

$$\omega_{T_p(f)} = \sum \phi_i^*(\omega_f),$$

where $\phi_i(\tau) = \frac{\tau+i}{p}$, and $\phi_\infty(\tau) = \langle p\rangle p\tau$ represent the $p+1$ curves with Γ-structure that are images of $(\mathbb{C}/\langle\tau, 1\rangle, \frac{1}{N})$ by a p-isogeny, and the ϕ_i^* are the pull-back maps on differential forms on $\mathcal{H}$. (An isogeny of elliptic curves with Γ-structure is simply an isogeny between the underlying curves which sends one Γ-structure to the other.) Such a description makes it evident that $T_p(f)$ belongs to $S_2(\Gamma)$, if f does. In terms of the Fourier expansion of $f = \sum a_n q^n$, the formula for the operator T_p on $S_2(N, \chi)$ is given by:

$$T_p(f) = \sum_{p|n} a_n q^{n/p} + p\chi(p)\sum a_n q^{pn}.$$

If p divides N, then we define the Hecke operator U_p analogously, by summing again over all the cyclic p-isogenies of Γ-structures. Since there are only p of them, the formula becomes simpler:

$$U_p(f) = \frac{1}{p}\sum_{i=0}^{p-1} f\left(\frac{\tau+i}{p}\right) = \sum_{p|n} a_n q^{n/p}.$$

The reader is invited to check that the Hecke operators of the form T_p or U_q commute with each other, and also that they commute with the diamond operators introduced in the previous section.

We extend the definition of the Hecke operators to operators T_{p^n}, with $n > 1$, by the inductive formulae

$$T_{p^{n+1}} = T_p T_{p^n} - \langle p\rangle p T_{p^{n-1}}, \quad \text{if } (p, N) = 1,$$

and $T_{p^n} = U_p^n$ otherwise. We then define the operator T_n, where $n = \prod p_i^{e_i}$ is written as a product of powers of distinct primes p_i, by

$$T_n = \prod_i T_{p_i^{e_i}}.$$

This definition makes the Hecke operators multiplicative, i.e., $T_m T_n = T_{mn}$ if $(m,n) = 1$. (A more conceptual definition of the Hecke operator T_n is that $T_n(f)$ is obtained by summing the pullback of ω_f over the maps describing all the cyclic n-isogenies of Γ-structures.) The relations among the different Hecke operators can be stated succinctly by saying that they obey the following (formal) identity:

$$\prod_{p \nmid N} (1 - T_p p^{-s} + \langle p \rangle p^{1-2s})^{-1} \prod_{p \mid N} (1 - U_p p^{-s})^{-1} = \sum_n T_n n^{-s}. \qquad (1.3.1)$$

The reader can consult [DI], sec. 3 and the references therein (especially [Shi2], ch. 3 or [Kn]) for more details and different points of view on Hecke operators. Let $\mathbb{T}$ be the subring of $\mathrm{End}_{\mathbb{C}}(S_2(\Gamma))$ generated over $\mathbb{C}$ by all the Hecke operators T_p for $p \nmid N$, U_q for $q \mid N$, and $\langle d \rangle$ acting on $S_2(\Gamma)$.

Definition 1.16 A modular form f is an *eigenform* if it is a simultaneous eigenvector for *all* the Hecke operators in $\mathbb{T}$, i.e., if there exists a $\mathbb{C}$-algebra homomorphism $\lambda : \mathbb{T} \longrightarrow \mathbb{C}$ such that $Tf = \lambda(T)f$, for all $T \in \mathbb{T}$.

A direct calculation shows that the coefficients a_n of an eigenform f can be recovered from the homomorphism λ by the formula:

$$a_n(f) = a_1(f)\lambda(T_n).$$

It follows that the first Fourier coefficient a_1 of a non-zero eigenform is always non-zero, and that the non-trivial eigenspaces for $\mathbb{T}$ are all one-dimensional:

Proposition 1.17 *Given a non-zero algebra homomorphism $\lambda : \mathbb{T} \longrightarrow \mathbb{C}$, there is exactly one eigenform f up to scaling, which satisfies $Tf = \lambda(T)f$, for all $T \in \mathbb{T}$.*

Sketch of proof: The proof of the existence of f is an exercise in commutative algebra (localize $S_2(\Gamma)$ at the kernel of λ), and the uniqueness is clear from the formula above. □

We call an eigenform satisfying $a_1 = 1$ a *normalized* eigenform.

Atkin-Lehner theory: It is natural to ask whether $S_2(\Gamma)$ can be decomposed into a basis consisting of distinct normalized eigenforms. Unfortunately, this is not always possible, as the following exercise illustrates.

Exercise 1.18 Suppose that p^3 divides N exactly. Let $\mathbb{T}'$ be the algebra of Hecke operators (generated by the operators T_q with $q \nmid N/p^3$, and U_q with $q \mid N/p^3$) acting on $S_2(N/p^3)$. Let $f = \sum_{n=1}^{\infty} a_n q^n$ be a $\mathbb{T}'$-eigenform of level N/p^3 in $S_2(N/p^3)$. Show that the space S_f spanned by the forms $f(\tau)$, $f(p\tau)$, $f(p^2\tau)$, and $f(p^3\tau)$ is contained in $S_2(N)$, and is stable for the action of the Hecke operators T_q, $q \nmid N$, and U_q, $q \mid N$. Show that S_f has no basis of simultaneous eigenforms for the Hecke algebra $\mathbb{T}$ of level N, so that the action of $\mathbb{T}$ on S_f is not semi-simple.

Let $\mathbb{T}^0$ denote the subalgebra of $\mathbb{T}$ generated only by the "good" Hecke operators T_q with $q \nmid N$, and $\langle d \rangle$.

Proposition 1.19 *If q does not divide N, the adjoint of the Hecke operator T_q with respect to the Petersson scalar product is the operator $\langle q \rangle^{-1} T_q$, and the adjoint of $\langle q \rangle$ is $\langle q \rangle^{-1}$. In particular, the Hecke operators commute with their adjoints.*

Proof: See [Kn], th. 9.18 and 8.22, or [Ogg]. □

Proposition 1.19 implies, by the spectral theorem for commuting operators that commute with their adjoints:

Proposition 1.20 *The algebra $\mathbb{T}^0$ is semi-simple (i.e, it is isomorphic to a product $\mathbb{C} \times \cdots \times \mathbb{C}$ of a certain number of copies of $\mathbb{C}$), and there is a basis of $S_2(\Gamma)$ consisting of simultaneous eigenvectors for the operators T_q.*

Thus, $\mathbb{T}^0$ has the merit of being semi-simple, while $\mathbb{T}$ is not in general. The cost of replacing $\mathbb{T}$ by $\mathbb{T}^0$, however, is that one loses "multiplicity one", i.e., the eigenspaces for $\mathbb{T}^0$ need not be one-dimensional. For example, the space S_f defined in the previous exercise is contained in a single eigenspace for $\mathbb{T}^0$.

The theory of Atkin-Lehner [AL] gives essentially a complete understanding of the structure of the algebra $\mathbb{T}$, and the structure of the space of eigenforms. To motivate the main result, observe that the problem in the exercise above seems to be caused by forms of level N that are coming from forms of lower level N/p^3 by a straightforward operation, and are therefore not "genuinely" of level N. They are the analogues, in the context of modular forms, of non-primitive Dirichlet characters.

Definition 1.21 We define the *old* subspace of $S_2(\Gamma)$ to be the space spanned by those functions which are of the form $g(az)$, where g is in $S_2(\Gamma_1(M))$ for some $M < N$ and aM divides N. We define the *new* subspace of $S_2(\Gamma)$ to be the orthogonal complement of the old subspace with respect to the Petersson scalar product. A normalized eigenform in the new subspace is called a *newform of level N*.

The following result is the main consequence of the theory of Atkin-Lehner. It gives a complete answer to the question of what is the structure of the algebra $\mathbb{T}$ acting on $S_2(\Gamma)$.

Theorem 1.22 *If f is in the new subspace of $S_2(\Gamma)$ and is an eigenvector for all the operators in $\mathbb{T}^0$, then it is also an eigenform for $\mathbb{T}$, and hence is unique up to scaling. More generally, if f is a newform of level $N_f | N$, then the space S_f defined by*

$$S_f = \{g \in S_2(\Gamma) \text{ such that } Tg = \lambda_f(T)g, \text{ for all } T \in \mathbb{T}^0\}$$

is stable under the action of all the Hecke operators in $\mathbb{T}$. *It is spanned by the linearly independent forms* $f(az)$ *where* a *ranges over the divisors of* N/N_f. *Furthermore, we have*

$$S_2(\Gamma) = \oplus_f S_f,$$

where the sum is taken over all newforms f *of some level* N_f *dividing* N.

See [AL] for the proof in the case $\Gamma = \Gamma_0(N)$, and [La2], ch. VIII for the general case. (See also [DI], sec. 6 for an overview.)

Exercise 1.23 Consider the case where $\Gamma = \Gamma_0(22)$. Show that $X_0(22)$ is of genus 2, and hence that $S_2(22)$ has dimension 2. Show that $S_2(22)$ is equal to S_f, where $f = (\eta(\tau)\eta(11\tau))^2$ is a newform of level 11, so that in particular there are no newforms of level 22 on Γ. Show that $\mathbb{T}^0$ is isomorphic to $\mathbb{C}$ in this case, and that $\mathbb{T}$ is isomorphic to the semisimple algebra $\mathbb{C} \times \mathbb{C}$.

Action on homology and Jacobians: Note that the Hecke operators act on $V = S_2(\Gamma)^\vee$ by duality. One checks (cf. [Kn], props. 11.23, 11.24) that the sublattice Λ of V is stable under the action of all the Hecke operators T_n, and of the diamond operators $\langle d\rangle$. Therefore the operators T_n and $\langle d\rangle$ give rise to endomorphisms of the torus V/Λ, and hence the Jacobian variety J_Γ, in a natural way. The involution $\tau \mapsto -\bar{\tau}$ gives rise to an involution on $X_\Gamma(\mathbb{C})$ (which is complex conjugation on the model of X_Γ over $\mathbb{R}$ deduced from the $\mathbb{Q}$- model defined in section 1.5). Since complex conjugation is continuous it also acts on the integral homology $\Lambda = H_1(X_\Gamma(\mathbb{C}), \mathbb{Z})$. Let Λ^+ and Λ^- be the sublattices of Λ on which complex conjugation acts by $+1$ and -1. These are sublattices of Λ of rank g which are stable under the Hecke operators, since complex conjugation commutes with the Hecke action.

A more algebraic description of the action of T_p on J_Γ is given via the notion of an *algebraic correspondence*. A correspondence on a curve X is a divisor C on $X \times X$ which is taken modulo divisors of the form $\{P\} \times X$ and $X \times \{Q\}$. Let π_1 and π_2 denote the projections of $X \times X$ onto each factor. Then the correspondence C induces a map on divisors of X, by setting

$$C(D) = \pi_2(\pi_1^{-1}(D) \cdot C).$$

(For the definition of the intersection $D_1 \cdot D_2$ of two divisors, see [Wel].) The map C preserves divisors of degree 0, and sends principal divisors on X to principal divisors. It gives a well defined algebraic endomorphism of the Jacobian variety Jac (X). Given a correspondence C, its transpose $C^\vee$ is defined to be the divisor of $X \times X$ obtained by interchanging the two factors of $X \times X$. One can define a natural notion of composition for correspondences, and the set of correspondences forms a ring. The general theory of correspondences and the proofs of the above facts are given in [Wel], particularly the second chapter.

The Hecke correspondence T_n is defined as the closure in $X_\Gamma \times X_\Gamma$ of the locus of points (A, B) in $Y_\Gamma \times Y_\Gamma$ such that there is a degree n cyclic isogeny of elliptic curves with Γ-structure from A to B. For example, if p is a prime not dividing N, then T_p is an algebraic curve in $X_1(N) \times X_1(N)$ which is birational to $X_{\Gamma_1(N)\cap\Gamma_0(p)}$. The induced map on divisors in this case satisfies

$$T_p((E,P)) = \sum(E/C, P \bmod C)$$

where the sum runs over the subgroups C of E having order p. Note also that if (A,B) belongs to T_p, then the isogeny dual to $A \longrightarrow B$ gives a p-isogeny from B to $\langle p\rangle A$, so that

$$T_p^\vee = \langle p\rangle^{-1}T_p.$$

1.4 The L-function associated to a cusp form

For this section, let f in $S_2(\Gamma_1(N))$ be a cusp form with Fourier expansion at $i\infty$ given by $\sum_n a_n q^n$. One has the following estimate for the size of the Fourier coefficients a_n:

Theorem 1.24 *The coefficients $a_n \in \mathbb{C}$ satisfy the inequality*

$$|a_n| \le c(f)\sigma_0(n)\sqrt{n},$$

where $c(f)$ is a constant depending only on f, and $\sigma_0(n)$ denotes the number of positive divisors of n.

Sketch of proof: This follows from proposition 1.51 of section 1.7 which relates the p-th Fourier coefficients of eigenforms, for p a prime not dividing the level of Γ, to the number of points on certain abelian varieties over the finite field $\mathbb{F}_p$. The estimates of Hasse and Weil for the number of points on abelian varieties over finite fields (stated in theorem 1.4 of section 1.1 for the special case of elliptic curves; see [Wel], §IV for the general case) thus translate into asymptotic bounds for the Fourier coefficients of these eigenforms. We note that the cruder estimate $|a_n| = O(n)$, which is enough for the purposes of this section, can be derived by a more elementary, purely analytic argument; cf. [Ogg], ch. IV, prop. 16. □

The L-function associated to f is defined by the formula:

$$L(f,s) = \sum a_n n^{-s}.$$

As in exercise 1.8, one can show that the infinite sum defining $L(f,s)$ converges absolutely in the right half-plane $\mathrm{Re}\,(s) > \frac{3}{2}$. A much better insight is gained into the function $L(f,s)$ by noting that it is essentially the Mellin transform of the modular form f. More precisely, if we set $\Lambda(f,s) = N^{s/2}(2\pi)^{-s}\Gamma(s)L(f,s)$, then we have

$$\Lambda(f,s) = N^{s/2}\int_0^\infty f(iy)y^s dy/y \tag{1.4.1}$$

Exercise 1.25 Check the formula above.

This integral representation for $L(f,s)$ gives the analytic continuation of $L(f,s)$ to the entire complex plane. The modular invariance of f translates into a functional equation for $L(f,s)$: more precisely, let w_N be the Atkin-Lehner involution defined by $w_N(\tau) = -1/N\tau$. The reader may check that w_N induces an involution of X_Γ, and hence of $\Omega^1(X_\Gamma) = S_2(\Gamma)$. One finds that $L(f,s)$ satisfies the functional equation:

$$\Lambda(f,s) = -\Lambda(w_N(f), 2-s).$$

For a proof of this, see [Ogg], ch. V, lemma 1. Eigenforms for $\mathbb{T}$ in $S_2(N,\chi)$ have a great importance in the theory because their associated L-functions have an Euler product expansion, in addition to an analytic continuation and functional equation:

Theorem 1.26 *If $f = \sum a_n q^n$ is a normalized eigenform in $S_2(N,\chi)$ for all the Hecke operators, then the L-function $L(f,s) = \sum a_n n^{-s}$ has the Euler product expansion*

$$\prod_{p \nmid N} (1 - a_p p^{-s} + \chi(p) p^{1-2s})^{-1} \prod_{p|N} (1 - a_p p^{-s})^{-1}.$$

Proof: This follows directly from equation (1.3.1) of section 1.3. □

If f is a newform of level N, then it is also an eigenform for w_N, so that the functional equation may be viewed as relating $L(f,s)$ and $L(f,2-s)$. We can also state the following more precise version of theorem 1.24 (see lemma 3.2 of [Hi2] for example for parts (b), (c) and (d)).

Theorem 1.27 *Suppose that f is a newform of level N_f and let N_χ denote the conductor of its character χ.*

(a) *If p does not divide N_f then $|a_p| \leq 2\sqrt{p}$.*

(b) *If $p || N_f$ and p does not divide N_χ then $a_p^2 = \chi_0(p)$ where χ_0 is the primitive character associated to χ.*

(c) *If p divides N_f and p does not divide N_f/N_χ then $|a_p| = \sqrt{p}$.*

(d) *If p^2 divides N_f and p divides N_f/N_χ then $a_p = 0$.*

1.5 Modular curves and modular forms over $\mathbb{Q}$

Modular curves: For Γ between $\Gamma_0(N)$ and $\Gamma_1(N)$, the modular curve X_Γ has a model over $\mathbb{Q}$. We describe such a model in the case of $\Gamma = \Gamma_0(N)$; the construction for general Γ follows from similar considerations.

The key remark here is that, as was noted in section 1.2, the complex points on the curve $Y_0(N)$ have a natural interpretation as moduli of elliptic curves together with a cyclic subgroup of order N, given by sending the point $\tau \in \mathcal{H}/\Gamma_0(N)$ to the pair $(\mathbb{C}/\langle 1,\tau\rangle, \langle 1/N\rangle)$.

Consider the "universal elliptic curve"

$$E_j : y^2 + xy = x^3 - \frac{36}{j-1728}x - \frac{1}{j-1728}.$$

It is an elliptic curve over the function field $\mathbb{Q}(j)$, with j-invariant j. Let d be the order of $\mathbb{P}^1(\mathbb{Z}/N\mathbb{Z})$, and let $C_1, \ldots, C_d$ denote the set of all cyclic subgroups of E_j of order N, defined over $\overline{\mathbb{Q}(j)}$, an algebraic closure of $\mathbb{Q}(j)$. Fix one of these subgroups, C. The Galois group $\mathrm{Gal}\,(\overline{\mathbb{Q}(j)}/\mathbb{Q}(j))$ permutes the C_i in a natural way. Let F_N be the smallest extension of $\mathbb{Q}(j)$ (viewed as embedded in $\overline{\mathbb{Q}(j)}$) with the property that $\sigma(C) = C$, for all $\sigma \in \mathrm{Gal}\,(\overline{\mathbb{Q}(j)}/F_N)$. It can be seen (cf. [Shi2], thm. 6.6) that the Galois action on the C_i is transitive so that $F_N/\mathbb{Q}(j)$ is of degree d, and that it is a *regular* extension, i.e., $F_N \cap \bar{\mathbb{Q}} = \mathbb{Q}$. Geometrically, F_N can be viewed as the function field of a curve $X_{/\mathbb{Q}}$ over $\mathbb{Q}$, with the inclusion of $\mathbb{Q}(j)$ corresponding to a map from $X_{/\mathbb{Q}}$ to the projective "j-line" over $\mathbb{Q}$. The pair (E_j, C) is an elliptic curve over $\mathbb{Q}(j)$ with a subgroup of order N defined over F_N. Using (E_j, C), each complex point of X gives an elliptic curve over $\mathbb{C}$ with a subgroup of order N, provided the point does not lie over $j = 0$, 1728 or ∞. The resulting map to $X_0(N)$ extends to an isomorphism from X to $X_0(N)$. The curve X thus constructed, together with this identification, is the desired model of $X_0(N)$ over $\mathbb{Q}$.

More concretely, the functions $j = j(\tau)$ and $j_N = j(N\tau)$ are related by a polynomial equation $\Phi_N(j, j_N) = 0$ with coefficients in $\mathbb{Q}$, of bidegree d. The field F_N is the function field of the affine curve $\Phi_N(X, Y) = 0$, and the mapping $\tau \mapsto (j(\tau), j(N\tau))$ gives a birational equivalence between $\mathcal{H}/\Gamma_0(N)$ and the complex curve defined by the equation Φ_N. In practice it is not feasible to write down the polynomial Φ_N, except for certain very small values of N. To study the models over $\mathbb{Q}$ of $X_0(N)$, more indirect methods are needed, which rely crucially on the moduli interpretation of $X_0(N)$. Similar remarks hold for $X_1(N)$.

Models over $\mathbb{Z}$: The work of Igusa [Ig], Deligne-Rapoport [DR], Drinfeld [Dr], and Katz-Mazur [KM] uses the moduli-theoretic interpretation to describe a canonical proper model for X_Γ over Spec $\mathbb{Z}$. These models allow us to talk about the reduction of X_Γ over finite fields $\mathbb{F}_p$, for p prime. The curve has good reduction at primes p not dividing N, with the "non-cuspidal" points of $X_{\Gamma/\mathbb{F}_p}$ corresponding to elliptic curves over $\bar{\mathbb{F}}_p$ with Γ-structure. The singular fibers at primes p dividing N can also be described precisely; an important special case (see [DR]) is that of $\Gamma_0(N)$ with p exactly dividing N. For further discussion and references, see [DI], sec. 8, 9.

From now on, when we write X_Γ, $X_0(N)$, or $X_1(N)$, we will mean the curve over $\mathbb{Q}$ which are the models described above for the complex curves defined in section 1.2.

Remark 1.28 When considering q-expansions, it is more convenient to use a different set of models over $\mathbb{Z}$ for these complex curves. We define $X_1^\mu(N)$ in the case of $\Gamma_1(N)$ as a model over $\mathbb{Z}$ which parametrizes pairs (E, i) where i is an embedding of μ_N in the (generalized) elliptic curve E. (So $X_1^\mu(N)$ is the model denoted $X_\mu(N)$ in [DI], sec. 9.3, assuming $N > 4$.) For Γ between $\Gamma_0(N)$ and $\Gamma_1(N)$, we define X_Γ^μ as the corresponding quotient of $X_1^\mu(N)$. This model has good reduction at primes p not dividing N, but unlike the models mentioned above, its fibers at primes p dividing N are smooth and irreducible, but not proper. In the case of $\Gamma = \Gamma_0(N)$, the curve $X_{\Gamma,\mathbb{Q}}^\mu$ can be identified with $X_0(N)$. However, this is not the case in general: the cusp ∞ is a rational point of $X_{\Gamma,\mathbb{Q}}^\mu$ but not necessarily of X_Γ.

Jacobians: Weil's theory [Wel] of the Jacobian shows that the Jacobians J_Γ defined in section 1.2 as complex tori also admit models over $\mathbb{Q}$. When we speak of J_Γ, $J_0(N)$ and $J_1(N)$ from now on, we will refer to these as abelian varieties defined over $\mathbb{Q}$. Thus, the points in $J_\Gamma(K)$, for any $\mathbb{Q}$-algebra K, are identified with the divisor classes on X_Γ of degree 0, defined over K.

We let $J_{\Gamma/\mathbb{Z}}$, denote the Néron model of the Jacobian J_Γ over Spec $(\mathbb{Z})$. Using this model we define $J_{\Gamma/A}$ for arbitrary rings A. In particular we can consider $J_{\Gamma/\mathbb{F}_p}$, the reduction of the Jacobian in characteristic p, which is closely related to the reduction of the integral model of the curve X_Γ mentioned above. In particular, if p does not divide the level of Γ, then $J_{\Gamma/\mathbb{F}_p}$ can be identified with the Jacobian of $X_{\Gamma/\mathbb{F}_p}$. For a treatment of the case $\Gamma = \Gamma_0(N)$ with p exactly dividing N, see the appendix of [Maz1]; for more general discussion and references, see [DI], sec. 10, especially sec. 10.3.

Hecke operators: The Hecke operators have a natural moduli interpretation, which was already touched upon in section 1.3. In particular, one finds that the operator T_n arises from a correspondence which is defined over $\mathbb{Q}$, and gives rise to an endomorphism of the Jacobian J_Γ which is defined over $\mathbb{Q}$. This in turn gives rise to an endomorphism of the Néron model $J_{\Gamma/\mathbb{Z}}$, and we can then consider the endomorphism T_n on the reduction of the Jacobian in characteristic p. Recall that if p is a prime not dividing N, we may identify this reduction with the Jacobian of $X_{\Gamma/\mathbb{F}_p}$. (Cf. [MW], ch. 2, sec. 1, prop. 2.) Furthermore, one can show that the moduli-theoretic interpretation of the Hecke operator remains valid in characteristic p; i.e., the endomorphism T_n of $J_{\Gamma/\mathbb{F}_p}$ is induced by a map on divisors satisfying, for all ordinary elliptic curves A with Γ-structure:

$$T_n(A) = \sum_i i(A),$$

where the sum is taken over all cyclic isogenies of degree n. (See [DI], sec. 10.2 for further discussion and references.)

This description allows one to analyse, for example, the Hecke operator T_p over $\mathbb{F}_p$, when $(p, N) = 1$. Let us work with $\Gamma = \Gamma_1(N)$, to illustrate the idea.

For a variety X over $\bar{\mathbb{F}}_p$, let ϕ_X be the Frobenius morphism on X defined by raising coordinates to the pth power. Thus if (E, P) corresponds to a point of $X_1(N)_{/\mathbb{F}_p}$, then ϕ_E is an isogeny of degree p from (E, P) to the pair $(E_\infty, P_\infty) = \phi_{X_1(N)}(E, P)$. The graph of $\phi_{X_1(N)}$ in $(X_1(N) \times X_1(N))_{/\mathbb{F}_p}$ is a correspondence of degree p, which we call F. Let F' be the transpose of this correspondence. The endomorphism F of J_Γ induced by F is the Frobenius endomorphism ϕ_{J_Γ}, and the endomorphism F' is the dual endomorphism (in the sense of duality of abelian varieties). Now consider the divisor

$$F'((E, P)) = (E_1, P_1) + \cdots + (E_p, P_p),$$

where the (E_i, P_i) are elliptic curves with Γ-structure in characteristic p. Since ϕ_{E_i} is an isogeny of degree p from (E_i, P_i) to (E, P), we also have the dual isogeny from (E, P) to (E_i, pP_i). If E is ordinary at p, then the points (E_∞, P_∞), (E_1, pP_1), ..., (E_p, pP_p) are a complete list of the distinct curves with Γ-structure which are p-isogenous to (E, P). Hence one has the equality of divisors on $X_1(N)_{/\mathbb{F}_p}$:

$$T_p((E, P)) = (E_\infty, P_\infty) + (E_1, pP_1) + \cdots + (E_p, pP_p) = (F + \langle p \rangle F')((E, P)).$$

Since the ordinary points are dense on $X_1(N)_{/\mathbb{F}_p}$, we deduce that $T_p = (F + \langle p \rangle F')$ as endomorphisms of $J_1(N)_{/\mathbb{F}_p}$. This equation, known as Eichler-Shimura congruence relation, plays a central role in the theory. (For more details, see [DI], sec. 10.2, 10.3.)

Theorem 1.29 *If $p \nmid N$ then the endomorphism T_p of $J_{\Gamma/\mathbb{F}_p}$ satisfies*

$$T_p = F + \langle p \rangle F'.$$

Remark 1.30 This was proved by Eichler [Ei] to hold for all but finitely many p in the case of $\Gamma_0(N)$, and by Shimura ([Shi1], see also [Shi2], ch. 7) in the case of $\Gamma_1(N)$. The fact that it holds for all p not dividing N follows from work of Igusa [Ig].

Modular forms: In the same way that modular curves have models over $\mathbb{Q}$ and over $\mathbb{Z}$, the Fourier coefficients of modular forms also have natural rationality and integrality properties. We start by sketching the proof of:

Theorem 1.31 *The space $S_2(\Gamma)$ has a basis consisting of modular forms with integer Fourier coefficients.*

Proof: The Hecke operators act on the integral homology Λ^+ in a way that is compatible with the action on $S_2(\Gamma)$ and respects the natural (Poincaré) duality between these two spaces. Hence, if $\{\lambda_n\}_{n \in \mathbb{N}}$ is a system of eigenvalues for the T_n, then the λ_n are algebraic integers in some finite extension K of $\mathbb{Q}$, and the system $\{\lambda_n^\sigma\}_{n \in \mathbb{N}}$ is a system of eigenvalues for the T_n for any Galois automorphism σ of $\bar{\mathbb{Q}}/\mathbb{Q}$. Hence, we have shown:

Proposition 1.32 *If $f \in S_2(M,\chi)$ is a newform of some level M dividing N, then its Fourier coefficients lie in a finite extension K of $\mathbb{Q}$. Moreover, if $\sigma \in \mathrm{Gal}(\bar{\mathbb{Q}}/\mathbb{Q})$ is any Galois automorphism, then the Fourier series f^σ obtained by applying σ to the Fourier coefficients is a newform in $S_2(M,\chi^\sigma)$.*

The explicit description of $S_2(\Gamma)$ given in section 1.3 implies that $S_2(\Gamma)$ is spanned by forms having Fourier coefficients which are algebraic integers in some finite (Galois) extension K of $\mathbb{Q}$, and that the space of forms with Fourier coefficients in K is stable under the natural action of $\mathrm{Gal}(K/\mathbb{Q})$ on Fourier expansions. An application of Hilbert's theorem 90 shows that $S_2(\Gamma)$ has a basis consisting of forms with rational Fourier expansions, and the integrality of the Fourier coefficients of eigenforms yields the integrality statement of theorem 1.31. □

We define $S_2(\Gamma,\mathbb{Z})$ to be the space of modular forms with integral Fourier coefficients in $S_2(\Gamma)$. Theorem 1.31 states that $S_2(\Gamma,\mathbb{Z})\otimes\mathbb{C} = S_2(\Gamma)$. Given any ring A, we define

$$S_2(\Gamma,A) = S_2(\Gamma,\mathbb{Z})\otimes A,$$

and define $S_2(N,A)$ and $S_2(N,\chi,A)$ (where χ now is a character with values in $A^\times$) in the obvious way. If A is contained in $\mathbb{C}$, the q-expansion principle below allows us to identify $S_2(\Gamma,A)$ with the set of modular forms in $S_2(\Gamma)$ with Fourier coefficients in A.

The q-expansion principle: Because the modular curve $X_0(N)$ has a model over $\mathbb{Q}$, the space of modular forms $S_2(N) = \Omega^1(X_0(N))$ has a natural rational structure, given by considering the differential forms on $X_0(N)$ defined over $\mathbb{Q}$. The fundamental q-expansion principle (see [DR], ch. 7, or [Kat], sec. 1.6) says that these algebraic structures are the same as those obtained analytically by considering q-expansions at ∞. More generally, using the model X_Γ^μ, we obtain the q-expansion principle over $\mathbb{Z}$ for cusp forms on Γ (cf. [DI], sec. 12.3).

Theorem 1.33 *The map $S_2(\Gamma) \longrightarrow \Omega^1(X_\Gamma)$ defined by $f \mapsto \omega_f$ induces an isomorphism from $S_2(\Gamma,\mathbb{Z})$ to $\Omega^1(X_\Gamma^\mu)$. Furthermore, if A is flat over $\mathbb{Z}$ or N is invertible in A, then the induced map $S_2(\Gamma,A) \to \Omega^1(X_{\Gamma,A}^\mu)$ is an isomorphism. Furthermore, if A is a subring of $\mathbb{C}$, then this isomorphism identifies $S_2(\Gamma,A)$ with set of modular forms in $S_2(\Gamma)$ having coefficients in A.*

1.6 The Hecke algebra

It follows directly from the formulas for the Hecke operators acting on q-expansions that the T_n leave $S_2(\Gamma_0(N),\mathbb{Z})$ stable, as well as the subspace of $S_2(N,\chi)$ with coefficients in $\mathbb{Z}[\chi]$. Using the q-expansion principle (theorem 1.33), one can also show ([DI], sec. 12.4) that the diamond operators preserve the spaces of cusp forms on Γ with integral Fourier expansions, and hence that the space $S_2(\Gamma, \mathbb{Z})$ is preserved by all the Hecke operators.

We define $\mathbb{T}_{\mathbb{Z}}$ to be the ring generated over $\mathbb{Z}$ by the Hecke operators T_n and $\langle d \rangle$ acting on the space $S_2(\Gamma, \mathbb{Z})$. More generally, if A is any ring, we define $\mathbb{T}_A$ to be the A-algebra $\mathbb{T}_{\mathbb{Z}} \otimes A$. This Hecke ring acts on the space $S_2(\Gamma, A)$ in a natural way. Before studying the structure of the Hecke rings $\mathbb{T}_A$ as we vary the rings A, we note the following general result (Cf. [Shi2], ch. 3.):

Lemma 1.34 *The space* $S_2(\Gamma, A)^\vee = \mathrm{Hom}_A(S_2(\Gamma, A), A)$ *is a free* $\mathbb{T}_A$*-module of rank one.*

Sketch of proof: One checks that the pairing $\mathbb{T}_{\mathbb{Z}} \times S_2(\Gamma, \mathbb{Z}) \longrightarrow \mathbb{Z}$ defined by $(T, f) \mapsto a_1(Tf)$ sets up a perfect, $\mathbb{T}_{\mathbb{Z}}$-equivariant duality between $\mathbb{T}_{\mathbb{Z}}$ and $S_2(\Gamma, \mathbb{Z})$. The result for arbitrary A follows. □

Hecke rings over $\mathbb{C}$: If $A = \mathbb{C}$, then the structure of the ring $\mathbb{T} = \mathbb{T}_{\mathbb{C}}$ is completely described by theorem 1.22. More precisely, if $\mathbb{T}_{\mathbb{C},f}$ denotes the image of the Hecke algebra acting on the space S_f defined in section 1.3, then

$$\mathbb{T}_{\mathbb{C}} = \oplus_f \mathbb{T}_{\mathbb{C},f},$$

where the direct sum ranges over all distinct newforms f of some level N_f dividing N. Furthermore, the algebra $\mathbb{T}_{\mathbb{C},f}$ can be described explicitly. In particular, if f is a newform of level N then $\mathbb{T}_{\mathbb{C},f}$ is isomorphic to $\mathbb{C}$, but if N_f is not equal to N then the ring $\mathbb{T}_{\mathbb{C},f}$ need not be a semi-simple algebra over $\mathbb{C}$.

Lemma 1.34 in the case $A = \mathbb{C}$ says that $V = S_2(\Gamma)^\vee$ is a free $\mathbb{T}_{\mathbb{C}}$-module of rank one, but we also have:

Lemma 1.35 *The module* $S_2(\Gamma)$ *is a free* $\mathbb{T}_{\mathbb{C}}$*-module of rank one.*

Proof: Let $g_1, \ldots, g_t$ be a complete system of newforms of levels $N_1, \ldots, N_t$ dividing N. One can check that the form

$$g = g_1(N/N_1\tau) + \cdots + g_t(N/N_t\tau)$$

generates $S_2(\Gamma)$ as a $\mathbb{T}_{\mathbb{C}}$-module. The map $\mathbb{T}_{\mathbb{C}} \longrightarrow S_2(\Gamma)$ defined by $T \mapsto Tg$ gives an isomorphism from $\mathbb{T}_{\mathbb{C}}$ to $S_2(\Gamma)$, as $\mathbb{T}_{\mathbb{C}}$-modules. □

Remark 1.36 Lemmas 1.34 and 1.35 imply that $\mathbb{T}_{\mathbb{C}}$ is a *Gorenstein* $\mathbb{C}$-algebra (of finite rank), i.e., $\mathrm{Hom}_{\mathbb{C}}(\mathbb{T}_{\mathbb{C}}, \mathbb{C})$ is isomorphic to $\mathbb{T}_{\mathbb{C}}$ as a $\mathbb{T}_{\mathbb{C}}$-module.

Hecke rings over $\mathbb{Q}$: Let $[f]$ be the Galois orbit (under the action of $G_{\mathbb{Q}} = \mathrm{Gal}(\bar{\mathbb{Q}}/\mathbb{Q})$) of a normalized newform f of some level N_f dividing N, and let K_f be the field extension of $\mathbb{Q}$ generated by the Fourier coefficients of f. The space $\oplus_{g \in [f]} S_g$ is a vector space of dimension $[K_f : \mathbb{Q}]\sigma_0(N/N_f)$, which is spanned by modular forms with rational Fourier coefficients. Let $S_{[f]}$ be the $\mathbb{Q}$-subspace of forms in $\oplus_{g\in[f]}S_g$ with rational Fourier coefficients. The space $S_{[f]}$ is stable under the action of $\mathbb{T}_{\mathbf{Q}}$, and letting $\mathbb{T}_{\mathbf{Q},[f]}$ be the image of $\mathbb{T}_{\mathbf{Q}}$ acting on $S_{[f]}$, we obtain the direct sum decomposition

$$\mathbb{T}_{\mathbf{Q}} = \oplus_{[f]}\mathbb{T}_{\mathbf{Q},[f]},$$

where the sum is taken over the distinct $G_{\mathbf{Q}}$-orbits of normalized newforms f of some level N_f dividing N. If N_f is equal to N, then the algebra $\mathbb{T}_{\mathbf{Q},[f]}$ is isomorphic to the field K_f. If N_f is a proper divisor of N, then, as in the complex case, the algebra $\mathbb{T}_{\mathbf{Q},[f]}$ is a (not necessarily semi-simple) algebra over $\mathbb{Q}$ of rank $\sigma_0(N/N_f)[K_f : \mathbb{Q}]$. The nature of the fields K_f, and in general the structure of $\mathbb{T}_{\mathbf{Q}}$, is very poorly understood at this stage; for example, one does not know how to characterize the number fields that occur as a K_f for some f (but they are all known to be totally real or CM fields).

The ring $\mathbb{T}_{\mathbf{Q}}$ acts naturally on the rational homology $H_1(X_\Gamma, \mathbb{Q}) = \Lambda\otimes\mathbb{Q}$, and we have

Lemma 1.37 *The module $\Lambda \otimes \mathbb{Q}$ is free of rank two over $\mathbb{T}_{\mathbf{Q}}$.*

Sketch of proof: The modules $\Lambda^+ \otimes \mathbb{C} \simeq V$ and $\Lambda^- \otimes \mathbb{C} \simeq V$ are free of rank one over $\mathbb{T}_{\mathbf{C}}$, by lemma 1.34. This implies that $\Lambda^+ \otimes \mathbb{Q}$ and $\Lambda^- \otimes \mathbb{Q}$ are both free of rank one over $\mathbb{T}_{\mathbf{Q}}$. □

Hecke rings over $\mathbb{Z}$: The ring $\mathbb{T}_{\mathbf{Z}}$ is a certain (not necessarily maximal) order in $\mathbb{T}_{\mathbf{Q}}$. One still has an injection

$$\mathbb{T}_{\mathbf{Z}} \hookrightarrow \oplus_{[f]}\mathbb{T}_{\mathbf{Z},[f]},$$

where now $\mathbb{T}_{\mathbf{Z},[f]}$ denotes the ring generated over $\mathbb{Z}$ by the Hecke operators acting on $S_{[f]}$. Of course the structure of $\mathbb{T}_{\mathbf{Z}}$ is even more mysterious than that of $\mathbb{T}_{\mathbf{Q}}$! The ring $\mathbb{T}_{\mathbf{Z}}$ acts naturally on Λ, but it is not the case in general that Λ is free of rank two over $\mathbb{T}_{\mathbf{Z}}$, i.e., that the integral analogue of lemma 1.37 is true. (See remark 1.42 below.)

Hecke rings over $\mathbb{Q}_\ell$: The study of the algebras $\mathbb{T}_{\mathbf{Z}_\ell}$ and $\mathbb{T}_{\mathbf{Q}_\ell}$ arises naturally because of the Hecke action on the Tate module $\mathcal{T}_\ell(J_\Gamma)$,

$$\mathcal{T}_\ell(J_\Gamma) := \lim_{\leftarrow}(J_\Gamma)[\ell^n],$$

where the inverse limit is taken with respect to the multiplication by ℓ maps. The action of $\mathbb{T}_{\mathbf{Z}_\ell}$ on $\mathcal{T}_\ell(J_\Gamma)$ is compatible with that of $G_{\mathbf{Q}}$, and it is this pair of actions on the Tate module which is used to associate two-dimensional Galois representations to modular forms.

It will sometimes be more convenient to consider the ring $\mathbb{T}_{\mathbf{Q}_\ell}$ and its action on

$$\mathcal{V} = \mathcal{T}_\ell(J_\Gamma) \otimes_{\mathbf{Z}_\ell} \mathbb{Q}_\ell.$$

We first record a useful duality property enjoyed by the Tate modules. The Weil pairings on the groups $J_\Gamma[\ell^n]$ for $n \geq 1$ induce a perfect pairing

$$\langle\ ,\ \rangle : \mathcal{T}_\ell(J_\Gamma) \times \mathcal{T}_\ell(J_\Gamma) \to \mathbb{Z}_\ell.$$

Since each Hecke operator T is adjoint to wTw where $w = w_N$ is the Atkin-Lehner involution, we have the following lemma.

Lemma 1.38 *The map $x \mapsto \phi_x$ where $\phi_x(y) = \langle x, wy\rangle$ defines an isomorphism of $\mathbb{T}_{\mathbb{Z}_\ell}$-modules,*

$$\mathcal{T}_\ell(J_\Gamma) \cong \mathcal{T}_\ell(J_\Gamma)^\vee = \mathrm{Hom}_{\mathbb{Z}_\ell}(\mathcal{T}_\ell(J_\Gamma), \mathbb{Z}_\ell),$$

and hence an isomorphism of $\mathbb{T}_{\mathbb{Q}_\ell}$-modules

$$\mathcal{V} \cong \mathcal{V}^\vee = \mathrm{Hom}_{\mathbb{Q}_\ell}(\mathcal{V}, \mathbb{Q}_\ell).$$

The following lemma allows us to regard $\mathbb{T}_{\mathbb{Q}_\ell}$ as a coefficient ring for a two-dimensional Galois representation

$$\mathrm{Gal}\,(\bar{\mathbb{Q}}/\mathbb{Q}) \to \mathrm{Aut}_{\,\mathbb{T}_{\mathbb{Q}_\ell}}(\mathcal{V}) \cong GL_2(\mathbb{T}_{\mathbb{Q}_\ell}).$$

Lemma 1.39 *The module $\mathcal{V}$ is free of rank 2 over $\mathbb{T}_{\mathbb{Q}_\ell}$.*

Proof: Lemma 1.37 implies directly that the module $\mathcal{V} = \Lambda \otimes \mathbb{Q}_\ell$ is free of rank two over $\mathbb{T}_{\mathbb{Q}_\ell}$. □

Corollary 1.40 *The ring $\mathbb{T}_{\mathbb{Q}_\ell}$ is a Gorenstein $\mathbb{Q}_\ell$-algebra; i.e.,*

$$\mathbb{T}_{\mathbb{Q}_\ell}^\vee = \mathrm{Hom}_{\mathbb{Q}_\ell}(\mathbb{T}_{\mathbb{Q}_\ell}, \mathbb{Q}_\ell)$$

is free of rank one over $\mathbb{T}_{\mathbb{Q}_\ell}$.

Proof: Choosing a basis for $\mathcal{V}$ over $\mathbb{T}_{\mathbb{Q}_\ell}$, we obtain an isomorphism

$$\mathbb{T}_{\mathbb{Q}_\ell} \oplus \mathbb{T}_{\mathbb{Q}_\ell} \cong \mathbb{T}_{\mathbb{Q}_\ell}^\vee \oplus \mathbb{T}_{\mathbb{Q}_\ell}^\vee.$$

Decomposing $\mathbb{T}_{\mathbb{Q}_\ell}$ as $\prod_i R_i$ where each factor R_i is a finite-dimensional local $\mathbb{Q}_\ell$-algebra, we obtain an isomorphism

$$R_i \oplus R_i \cong R_i^\vee \oplus R_i^\vee$$

for each i. At least one of the four maps $R_i^\vee \longrightarrow R_i$ deduced from this isomorphism must be surjective, and by counting dimensions, we see that it must be injective as well. It follows that $\mathbb{T}_{\mathbb{Q}_\ell}$ is isomorphic to $\mathbb{T}_{\mathbb{Q}_\ell}^\vee$. □

Recall that for primes p not dividing N, the Jacobian J_Γ has good reduction mod p, and the Eichler-Shimura relation, theorem 1.29, states that on $J_{\Gamma/\mathbb{F}_p}$, we have

$$T_p = F + \langle p\rangle F'.$$

For primes p not dividing $N\ell$, we may identify $\mathcal{T}_\ell(J_\Gamma)$ with the ℓ-adic Tate module of the reduction (see [ST]) and consider the Frobenius endomorphism F on the free rank two $\mathbb{T}_{\mathbb{Q}_\ell}$-module $\mathcal{V}$. As a consequence of the Eichler-Shimura relation, we find:

Theorem 1.41 *For p not dividing $N\ell$, the characteristic polynomial of F on the $\mathbb{T}_{\mathbf{Q}_\ell}$-module $\mathcal{V}$ is*

$$X^2 - T_p X + \langle p \rangle p.$$

Proof: (We are grateful to Brian Conrad for showing us this argument.) Since $FF' = p$, it follows from the Eichler-Shimura relation that

$$F^2 - T_p F + \langle p \rangle p = 0.$$

To conclude that this is in fact the characteristic polynomial, it suffices to compute the trace of F. To do so, we use the $\mathbb{T}_{\mathbf{Q}_\ell}$ isomorphism

$$\mathcal{V} \to \mathcal{V}^\vee$$

defined by the modified pairing $\langle \cdot, w \cdot \rangle$. Under this modified pairing, F is adjoint to $wF'w = \langle p \rangle F'$, so the trace of F on $\mathcal{V}$ is the same as that of $\phi \mapsto \phi \circ (\langle p \rangle F')$ on $\mathcal{V}^\vee$. Choosing bases for $\mathcal{V}$ and $\mathbb{T}^\vee_{\mathbf{Q}_\ell}$, one sees that this is the same as the trace of $\langle p \rangle F'$. Hence

$$2\,\mathrm{tr}\,F = \mathrm{tr}\,F + \mathrm{tr}\,(\langle p \rangle F') = \mathrm{tr}\,(T_p) = 2\,T_p.$$

$\square$

Hecke rings over $\mathbb{Z}_\ell$: The ring $\mathbb{T}_{\mathbf{Z}_\ell}$ is free of finite rank over $\mathbb{Z}_\ell$. It therefore decomposes as

$$\mathbb{T}_{\mathbf{Z}_\ell} = \prod \mathbb{T}_{\mathfrak{m}},$$

where the product runs over the maximal ideals $\mathfrak{m}$ of $\mathbb{T}_{\mathbf{Z}_\ell}$, and $\mathbb{T}_{\mathfrak{m}}$ is the localization of $\mathbb{T}_{\mathbf{Z}_\ell}$ at $\mathfrak{m}$. For each $\mathfrak{m}$, $\mathbb{T}_{\mathfrak{m}}$ is a complete local $\mathbb{Z}_\ell$-algebra, free of finite rank as a $\mathbb{Z}_\ell$-module.

Remark 1.42 While the analogue of lemma 1.39 does not always hold for $\mathbb{T}_{\mathbf{Z}_\ell}$ (see [MRi], sec. 13) we shall see that it holds for certain localizations $\mathbb{T}_{\mathfrak{m}}$. Results of this type are much deeper than lemma 1.39 and were first obtained by Mazur [Maz1], sec. 14, 15. We shall return in chapter 4 to explain Mazur's result and its generalizations, which play a role in the arguments of [W3] and [TW].

Example 1.43 The curve $X_0(19)$ has genus 1, and $X_0(57)$ has genus 5. By consulting the tables in the Antwerp volume [Ant4] or Cremona's book [Cr], one finds that there is exactly one newform of level 19, and that there are three newforms of level 57, which all have rational Fourier coefficients. Their Fourier coefficients a_p, for the first few primes p, are listed in the following table:

	2	3	5	7	11	13	17	19	23	29	31
$19A$	0	−2	3	−1	3	−4	−3	1	0	6	−4
$57A$	−2	−1	−3	−5	1	2	−1	−1	−4	−2	−6
$57B$	1	1	−2	0	0	6	−6	−1	4	2	8
$57C$	−2	1	1	3	−3	−6	3	−1	4	−10	2

Setting $f = 19A$, we find that a basis of simultaneous eigenforms in $S_2(57, \mathbb{C})$ for the Hecke operators T_p, $l \neq 3, 19$, and U_3, U_{19} is:

$$f(\tau) + (1 + \sqrt{-2})f(3\tau), \quad f(\tau) + (1 - \sqrt{-2})f(3\tau), \quad 57A, \quad 57B, \quad 57C.$$

It appears from the table that the Fourier coefficients corresponding to the forms $57B$ and $57C$ are congruent modulo 3. This is in fact the case. One finds that the Hecke ring $\mathbb{T}_{\mathbb{Z}_3}$ generated over $\mathbb{Z}_3$ by the Hecke operators acting on $S_2(57, \mathbb{Z}_3)$ is isomorphic to the subalgebra of $\mathbb{Z}_3^5$:

$$\{(x, y, z, t, w) \text{ such that } t \equiv w \pmod 3\}.$$

The isomorphism sends T_p (for $p \neq 3, 19$) to the element

$$(a_p(19A), a_p(19A), a_p(57A), a_p(57B), a_p(57C)).$$

It sends U_3 to $(-1 + \sqrt{-2}, -1 - \sqrt{-2}, -1, 1, 1)$ and U_{19} to $(1, 1, -1, -1, -1)$. Thus there are exactly 4 distinct maximal ideals of $\mathbb{T}_{\mathbb{Z}_3}$, and the localizations at these maximal ideals are the rings $\mathbb{Z}_3$, $\mathbb{Z}_3$, $\mathbb{Z}_3$, and

$$\mathbb{T}_{\mathfrak{m}} = \{(t, w) \text{ such that } t \equiv w \pmod 3\},$$

where $\mathfrak{m}$ is the ideal generated by 3 and $T_n - a_n(57B)$ for all n.

1.7 The Shimura construction

Let $f = \sum a_n q^n$ be an eigenform on Γ with (not necessarily rational) Fourier coefficients, corresponding to a surjective algebra homomorphism

$$\lambda_f : \mathbb{T}_{\mathbf{Q}} \longrightarrow K_f,$$

where K_f is the field generated over $\mathbb{Q}$ by the Fourier coefficients of f. We briefly review in this section a construction of Shimura ([Shi2], ch. 7) which associates to f (or rather, to the orbit $[f]$ of f under $G_{\mathbf{Q}}$) an abelian variety A_f defined over $\mathbb{Q}$ and of dimension $[K_f : \mathbb{Q}]$.

Let $I_f \subset \mathbb{T}_{\mathbf{Z}}$ be the ideal $\ker(\lambda_f) \cap \mathbb{T}_{\mathbf{Z}}$. The image $I_f(J_\Gamma)$ is a (connected) subabelian variety of J_Γ which is stable under $\mathbb{T}_{\mathbf{Z}}$ and is defined over $\mathbb{Q}$.

Definition 1.44 The abelian variety A_f associated to f is the quotient

$$A_f = J_\Gamma / I_f J_\Gamma$$

From this definition one sees that A_f is defined over $\mathbb{Q}$ and depends only on $[f]$, and that its endomorphism ring contains $\mathbb{T}_{\mathbf{Z}}/I_f$ which is isomorphic to an order in K_f.

Remark 1.45 Using theorem 1.22, one can show that $J_0(N)$ is isogenous to $\prod_{[f]} A_{[f]}^{\sigma_0(N/N_f)}$.

We now describe the abelian variety A_f as a complex torus. Let V_f be the subspace of $V = S_2(\Gamma)^\vee$ on which $\mathbb{T}$ acts by λ_f. Theorem 1.22 and lemma 1.34 show that V_f is a one-dimensional complex vector space. Let π_f be the orthogonal projection of V to V_f, relative to the Petersson scalar product. The projector π_f belongs naturally to $\mathbb{T}_{K_f}$.

Let $[f]$ be the set of all eigenforms whose Fourier coefficients are Galois conjugate to those of f. The number of forms in $[f]$ is equal to the degree d of K_f over $\mathbb{Q}$. Now, we set

$$V_{[f]} = \oplus_{g\in[f]} V_g, \quad \pi_{[f]} = \sum_{g\in[f]} \pi_g.$$

Note that $\pi_{[f]}$ is simply the orthogonal projection of V to $V_{[f]}$. Note also that $\pi_{[f]}$ belongs to the Hecke algebra $\mathbb{T}_{\mathbb{Q}}$.

Lemma 1.46 *The abelian variety A_f is isomorphic over $\mathbb{C}$ to the complex torus $V_{[f]}/\pi_{[f]}(\Lambda)$, with the map $\pi_{[f]} : V/\Lambda \longrightarrow V_{[f]}/\pi_{[f]}(\Lambda)$ corresponding to the natural projection from J_Γ to A_f.*

In particular, one sees that A_f is an abelian variety of dimension $d = [K_f : \mathbb{Q}]$. Hence if f has rational Fourier coefficients, then the abelian variety A_f is an elliptic curve. This elliptic curve is called the *strong modular elliptic curve* associated to f if also f is a newform of level N and $\Gamma = \Gamma_0(N)$.

Example 1.47 If $\Gamma = \Gamma_0(26)$, one checks that the genus of X_Γ is two, and that there are two distinct normalized eigenforms f_1 and f_2 in $S_2(26)$. From the tables in [Ant4] or [Cr], one sees that f_1 and f_2 have integral Fourier coefficients, whose values for the primes ≤ 31 are:

	2	3	5	7	11	13	17	19	23	29	31
f_1	−1	1	−3	−1	6	1	−3	2	0	6	−4
f_2	1	−3	−1	1	−2	−1	−3	6	−4	2	4

Hence the abelian varieties A_{f_1} and A_{f_2} are elliptic curves. The above table suggests (and this can be checked directly by looking at the equations for these curves given in [Ant4] or [Cr], or by using the discussion in [DO], lemma 2.1) that the Fourier coefficients of f_1 and f_2 are congruent modulo 2. The natural projection $J_0(26) \longrightarrow A_{f_1} \oplus A_{f_2}$ is an isogeny whose kernel is isomorphic to $\mathbb{Z}/2\mathbb{Z} \times \mathbb{Z}/2\mathbb{Z}$, and $J_0(26)$ is *not* isomorphic to $A_{f_1} \oplus A_{f_2}$. More generally, one knows that a Jacobian can never decompose as a non-trivial direct product of two principally polarized abelian varieties, cf. [Maz1], prop. 10.6. (There are no non-zero homomorphisms from A_{f_1} to A_{f_2}, and so a non-trivial product decomposition of J_Γ would have to induce a decomposition as a product of principally polarized abelian varieties.)

Let $\mathcal{T}_\ell(A_f)$ be the Tate module of the abelian variety A_f,

$$\mathcal{T}_\ell(A_f) := \varprojlim (A_f)[\ell^n],$$

where the inverse limit is taken with respect to the multiplication by ℓ maps. This module is naturally a module for $\mathbb{T}_f \otimes \mathbb{Z}_\ell$, and $\mathcal{T}_\ell(A_f) \otimes \mathbb{Q}_\ell$ is a module for $K_f \otimes \mathbb{Q}_\ell$ as well.

Lemma 1.48 *The module $\mathcal{T}_\ell(A_f) \otimes \mathbb{Q}_\ell$ is a free module of rank 2 over $K_f \otimes \mathbb{Q}_\ell$.*

Proof: This follows directly from lemma 1.39. □

Proposition 1.49 *The algebra $\mathrm{End}_{\mathbb{Q}}(A_f) \otimes \mathbb{Q}$ is isomorphic to K_f. In particular A_f is simple over $\mathbb{Q}$.*

The proof is given in [R2], cor. 4.2. The main ingredient is the *irreducibility* of the Galois representation attached to A_f as in section 3.1. (Cf. theorem 3.1.)

Properties of A_f: good reduction:

Theorem 1.50 *If f is an eigenform of level N, and p is a prime that does not divide N, then the abelian variety A_f has good reduction at p.*

Proof: This follows from the fact that J_Γ has good reduction at such primes, which in turn is a consequence of the good reduction of X_Γ. □

If p is a prime not dividing N, it then becomes natural to study the number $N_{f,p}$ of points on the abelian variety A_f over the finite field $\mathbb{F}_p$. It is given by the following formula:

Proposition 1.51 *The number of points $N_{f,p}$ is given by the formula*

$$N_{f,p} = \mathrm{Norm}_{K_f/\mathbb{Q}}(\lambda_f(1 - a_p(f) + \langle p \rangle p)),$$

where $a_f(p) \in K_f$ is the p-th Fourier coefficient of f.

Proof: By Weil's theory [Wel], this number is given by the determinant

$$N_{f,p} = \det(1 - F),$$

where F is the Frobenius endomorphism acting on the ℓ-adic Tate module $\mathcal{T}_\ell(A_f)$. So theorem 1.41 implies

$$\det(1 - F) = \mathrm{Norm}_{K_f/\mathbb{Q}}(\lambda_f(1 - T_p + \langle p \rangle p)),$$

and proposition 1.51 follows. □

Defining the local Hasse-Weil L-function of A_f over $\mathbb{F}_p$ by the formula

$$L(A_f/\mathbb{F}_p, s) = \det(1 - Fp^{-s})^{-1},$$

the proof above gives the formula:

$$L(A_f/\mathbb{F}_p, s) = \prod_\sigma L_p(f^\sigma, s),$$

where the product is taken over all complex embeddings $\sigma : K_f \hookrightarrow \mathbb{C}$, and $L_p(f^\sigma, s)$ is the Euler factor at p in the L-function that was associated to f^σ in section 1.4.

In particular, if $A_f = E$ is an elliptic curve, i.e., f has rational Fourier coefficients and f is on $\Gamma_0(N)$, then the number of points on E over $\mathbb{F}_p$ is given by the formula

$$\#E(\mathbb{F}_p) = p + 1 - a_p(f). \tag{1.7.1}$$

Properties of A_f: bad reduction: A fundamental quantity associated to any abelian variety A over $\mathbb{Q}$ is its *arithmetic conductor*, which measures the amount of bad reduction of A. If we factor this conductor as a product of prime powers, $\prod_p p^{m_p}$, then the exponents m_p are equal to 0 precisely when A has good reduction at p. The definition of $m_p(A)$ was already given in section 1.1 when A is an abelian variety of dimension 1, i.e., an elliptic curve. In general, the exponent m_p coincides with $m_p(\rho_{A,\ell})$ (see section 2.1 below), where $\rho_{A,\ell}$ is the Galois representation on the ℓ-adic Tate module of A.

Regarding the bad reduction of the abelian variety A_f, one has the following consequence of the results of Langlands, Deligne and Carayol [Ca3] discussed below in section 3.1.

Theorem 1.52 *If f is a newform of level N, then the conductor of the abelian variety A_f is equal to N^g.*

1.8 The Shimura-Taniyama conjecture

Let E be any elliptic curve defined over $\mathbb{Q}$, and let N denote its arithmetic conductor, defined as in section 1.1. Then we have:

Proposition 1.53 *The following are equivalent:*

(a) *The curve E is isogenous over $\mathbb{Q}$ to A_f, for some newform f on some congruence group Γ.*

(b) *The curve E is isogenous over $\mathbb{Q}$ to A_f, for a newform f on $\Gamma_0(N)$.*

(c) *There is a non-constant morphism defined over $\mathbb{Q}$, from $X_0(N)$ to E.*

Sketch of proof: The implication $(b) \Rightarrow (a)$ is immediate, and $(a) \Rightarrow (b)$ follows from the work of Carayol [Ca3] (cf. proposition 3.20 below). If (b) holds, then there is a surjective map $J_0(N) \longrightarrow E$. Composing with the Abel-Jacobi map $X_0(N) \longrightarrow J_0(N)$ gives a map to E which is not constant, by the definition of the Jacobian, and hence (c) holds. Conversely, a non-constant map of curves $\pi : X_0(N) \longrightarrow E$ induces, by Albanese (covariant) functoriality, a surjective map of Jacobians $\pi_* : J_0(N) \longrightarrow E$, where we have identified E with its own Jacobian in the natural way. Since $J_0(N)$

is isogenous to a product (possibly with multiplicities) of abelian varieties A_f where f run over newforms of level N_f dividing N (cf. remark 1.45), it follows that there is a surjective map $A_f \longrightarrow E$ for some A_f. Since A_f is simple (proposition 1.49), this map is an isogeny, and part (*b*) follows. □

We call an elliptic curve over $\mathbb{Q}$ satisfying the equivalent properties above a *modular elliptic curve.* A startling conjecture that was first proposed by Taniyama in the 1950's and made more precise by Shimura predicts that the Shimura construction explained in the previous section is surjective:

Conjecture 1.54 *All elliptic curves defined over* $\mathbb{Q}$ *are modular.*

This conjecture is now known to be true for a very wide class of elliptic curves. The results of Wiles [W3] and Taylor-Wiles [TW] imply it is true for all semi-stable elliptic curves, and a strengthening of the method, [Di2], implies the conjecture for all elliptic curves which have semi-stable reduction at 3 and 5.

Remark 1.55 See [DI], sec. 13 for a more thorough list of equivalent forms of the conjecture.

Relation with L-functions: Define numbers $a_p = a_p(E)$ as in section 1.1. Recall the (global) Hasse-Weil L-function defined in section 1.1 by equations (1.1.2), (1.1.4), (1.1.5), and (1.1.6). The following proposition reveals some of the importance of the Shimura-Taniyama conjecture:

Proposition 1.56 *If E is modular, i.e., is associated to a newform f by the Shimura construction, then*

$$L(E/\mathbb{Q}, s) = L(f, s).$$

In particular, $L(E/\mathbb{Q}, s)$ has an analytic continuation to the entire complex plane, and a functional equation.

Sketch of proof: If E is isogenous to an elliptic curve A_f associated to a newform f on $\Gamma_0(N)$ by the Shimura construction, then the two L-functions $L(A_f/\mathbb{Q}, s)$ and $L(E/\mathbb{Q}, s)$ are equal. On the other hand, formula (1.7.1) directly implies that $L(A_f/\mathbb{Q}, s)$ is equal to $L(f, s)$, at least up to finitely many Euler factors (corresponding precisely, by the work of Igusa, to the primes dividing N). The equality of all the Euler factors follows from the work of Deligne, Langlands, and Carayol (cf. [Ca1]). □

Knowing the analytic continuation and functional equation of the Hasse-Weil L-function of an elliptic curve is of great importance in the theory. For example, the conjecture of Birch and Swinnerton-Dyer (which was stated in a weak form in section 1.1, conjecture 1.9) relates arithmetic invariants of E such as the rank of its Mordell-Weil group and the order of its Shafarevich-Tate group to the behaviour at $s = 1$ of $L(E/\mathbb{Q}, s)$. (Note that $s = 1$ is outside the domain of absolute convergence of the infinite product used to define $L(E/\mathbb{Q}, s)$.)

One consequence of this conjecture is that $E(\mathbb{Q})$ is finite if $L(E/\mathbb{Q}, 1)$ is non-zero. This was proved by Coates and Wiles in [CW] for elliptic curves with complex multiplication, and by Gross-Zagier and Kolyvagin [Kol] for modular elliptic curves. (Recently, a different proof for modular elliptic curves has been announced by K. Kato.) Thanks to the breakthroughs of [W3], [TW], and [Di2], the results of by Gross-Zagier, Kolyvagin and Kato are now unconditional for a very large class of elliptic curves.

The Shimura-Taniyama conjecture for abelian varieties: We may view conjecture 1.54 as asserting that the map

$$\left\{\begin{array}{c}\text{Newforms of weight 2 on } X_0(N)\\ \text{with rational Fourier coefficients}\end{array}\right\} \longrightarrow \left\{\begin{array}{c}\text{Isogeny classes of elliptic curves}\\ \text{over } \mathbb{Q} \text{ of conductor N}\end{array}\right\}$$

provided by the Shimura construction is a bijection. It is natural to extend this result to all eigenforms, not just those having rational Fourier expansions. We say that an abelian variety defined over $\mathbb{Q}$ is *modular* if it is isogenous to an abelian variety of the form A_f for some newform f on $\Gamma_1(N)$. It would not be reasonable to expect that all abelian varieties over $\mathbb{Q}$ are modular: the abelian varieties arising from the Shimura construction are very special, in many respects. Most importantly, they have a very large ring of endomorphisms. Following Ribet [R7], we make the definition:

Definition 1.57 An abelian variety over $\mathbb{Q}$ of dimension g is said to be of GL_2-type if its endomorphism ring over $\mathbb{Q}$ contains an order in a field of degree g over $\mathbb{Q}$.

Then we have the following generalized Shimura-Taniyama conjecture:

Conjecture 1.58 *Every simple abelian variety A over $\mathbb{Q}$ which is of GL_2-type is modular.*

This generalization of the Shimura-Taniyama conjecture is far from being proved, and tackling it still seems to require some major new ideas, even after the ideas introduced by Wiles.

2 Galois theory

2.1 Galois representations

If F is a perfect field, we will let $\bar{F}$ denote an algebraic closure of F and $G_F = \mathrm{Gal}(\bar{F}/F)$ its absolute Galois group. Recall that G_F is a profinite group: more precisely, $G_F = \varprojlim \mathrm{Gal}(L/F)$ as L runs over finite Galois extensions of F contained in $\bar{F}$. Hence G_F carries a natural topology. If ℓ is a prime different from the characteristic of F we will let $\epsilon_\ell : G_F \to \mathbb{Z}_\ell^\times$

denote the ℓ-adic cyclotomic character, i.e. if ζ is an ℓ-power root of unity in $\bar{F}$ then $\sigma(\zeta) = \zeta^{\epsilon_\ell(\sigma)}$ for all $\sigma \in G_F$. We will simply write ϵ if the choice of ℓ is clear from the context.

The finite fields $\mathbb{F}_p$: The group $G_{\mathbb{F}_p}$ is isomorphic to $\hat{\mathbb{Z}}$, the profinite completion of $\mathbb{Z}$. A natural topological generator is provided by the Frobenius element Frob_p, the automorphism of $\bar{\mathbb{F}}_p$ which raises elements of $\bar{\mathbb{F}}_p$ to their p^{th} power. If ℓ is different from p, then $\epsilon_\ell(\mathrm{Frob}_p) = p$.

The p-adic fields $\mathbb{Q}_p$: The p-adic valuation $v_p : \mathbb{Q}_p^\times \twoheadrightarrow \mathbb{Z}$ extends uniquely to a valuation $v_p : \bar{\mathbb{Q}}_p^\times \twoheadrightarrow \mathbb{Q}$. Let $\mathcal{O}_{\bar{\mathbb{Q}}_p}$ denote the ring of integers of $\bar{\mathbb{Q}}_p$ and $\mathfrak{m}_{\bar{\mathbb{Q}}_p}$ its maximal ideal. The residue field $\mathcal{O}_{\bar{\mathbb{Q}}_p}/\mathfrak{m}_{\bar{\mathbb{Q}}_p}$ is an algebraic closure of $\mathbb{F}_p$, which we will identify with $\bar{\mathbb{F}}_p$. The valuation v_p is compatible with the action of $G_{\mathbb{Q}_p}$, and hence $\mathcal{O}_{\bar{\mathbb{Q}}_p}$ and $\mathfrak{m}_{\bar{\mathbb{Q}}_p}$ are stable under $G_{\mathbb{Q}_p}$. In particular $G_{\mathbb{Q}_p}$ acts on $\mathcal{O}_{\bar{\mathbb{Q}}_p}/\mathfrak{m}_{\bar{\mathbb{Q}}_p}$ and we obtain a map $\varrho : G_{\mathbb{Q}_p} \to G_{\mathbb{F}_p}$, which is in fact a surjection. We call the kernel the *inertia group* and denote it I_p. It has a unique maximal pro-p subgroup: the *wild inertia group*, denoted P_p. There is a canonical isomorphism

$$I_p/P_p \cong \prod_{\ell \neq p} \mathbb{Z}_\ell(1),$$

where the (1) indicates that if $f \in G_{\mathbb{Q}_p}/P_p$ is a lifting of Frobenius and if $\sigma \in I_p/P_p$ then $f\sigma f^{-1} = \sigma^p$.

The inertia group I_p is filtered by a series of subgroups called the higher ramification groups. More precisely, there are closed normal subgroups I_p^u in $G_{\mathbb{Q}_p}$ for any $u \in [-1, \infty]$, with the following properties:

- If $u \leq v$ then $I_p^u \supset I_p^v$, i.e., the I_p^u form a *decreasing filtration.*
- For $u \leq 0$, $I_p^u = I_p$ while $I_p^\infty = \{1\}$.
- $P_p = \bigcup_{u>0} I_p^u$.
- $I_p^u = \bigcap_{v<u} I_p^v$.

These groups are defined in section 3 of chapter IV of [Se2], where they are denoted $G_{\mathbb{Q}_p}^u$ (except that there, $G_{\mathbb{Q}_p}^{-1} = G_{\mathbb{Q}_p}$ whereas we have $I_p^{-1} = I_p$).

The local Kronecker-Weber theorem ([Se2], ch. 14, sec. 7, for example) asserts that the map

$$\varrho \times \epsilon_p : G_{\mathbb{Q}_p}^{\mathrm{ab}} \longrightarrow G_{\mathbb{F}_p} \times \mathbb{Z}_p^\times.$$

is an isomorphism. (We will use G^{ab} to denote the abelianisation of a profinite group G, i.e. the unique maximal abelian continuous image.) Under it I_p goes to $\mathbb{Z}_p^\times$ and for $u > 0$, I_p^u goes to $(1 + p^{\lceil u \rceil}\mathbb{Z}_p) \subset \mathbb{Z}_p^\times$, where $\lceil u \rceil$ denotes the least integer greater than or equal to u. If $\ell \neq p$ then ϵ_ℓ is trivial on I_p and takes Frob_p to $p \in \mathbb{Z}_\ell^\times$.

The field $\mathbb{Q}$: If p is a rational prime, the usual p-adic valuation $v_p : \mathbb{Q}^\times \to \mathbb{Z}$ extends to a valuation $v : \bar{\mathbb{Q}}^\times \to \mathbb{Q}$. This extension is not unique, but all extensions are permuted transitively by $G_\mathbb{Q}$. Fix one such extension v_p and let G_p denote its stabiliser in $G_\mathbb{Q}$. Then G_p acts on the completion $(\bar{\mathbb{Q}})_{v_p}$ and preserves the subfield of elements algebraic over $\mathbb{Q}_p$. This field is an algebraic closure of $\mathbb{Q}_p$, which we shall identify with $\bar{\mathbb{Q}}_p$. One can check that the resulting map $G_p \to G_{\mathbb{Q}_p}$ is an isomorphism. Thus we obtain subgroups $G_p \supset I_p \supset I_p^u$ for $u \in [-1,\infty]$ and a distinguished element $\mathrm{Frob}_p \in G_p/I_p$. These objects depend on the choice of v_p and vary by conjugation in $G_\mathbb{Q}$ as this choice is varied. We call an algebraic extension $F/\mathbb{Q}$ *unramified at p* if all conjugates of I_p lie in $G_F \subset G_\mathbb{Q}$; otherwise we say that $F/\mathbb{Q}$ is *ramified.* If $F/\mathbb{Q}$ is Galois and unramified at p, then there is a well-defined conjugacy class $[\mathrm{Frob}_p] \subset \mathrm{Gal}(F/\mathbb{Q})$.

By replacing the p-adic completion by an Archimedean completion, one has a well-defined conjugacy class $[c]$ in $G_\mathbb{Q}$ consisting of those elements that arise as complex conjugation for some embedding $\bar{\mathbb{Q}} \hookrightarrow \mathbb{C}$. We will denote by G_∞ the subgroup $\{1, c\}$ for one such element c.

We have the following fundamental results concerning the structure of $G_\mathbb{Q}$ (see for instance [La1]).

Theorem 2.1 *If $F/\mathbb{Q}$ is a finite extension then F is only ramified at finitely many primes (those dividing the discriminant of $F/\mathbb{Q}$).*

Theorem 2.2 (Hermite-Minkowski) *If S is a finite set of primes and if $d \in \mathbb{Z}_{>0}$ then there are only finitely many extensions $F/\mathbb{Q}$ of degree d which are unramified outside S (contained in a fixed algebraic closure $\bar{\mathbb{Q}}$).*

Theorem 2.3 (Chebotarev) *If $F/\mathbb{Q}$ is a Galois extension unramified outside a finite set S of primes then $\bigcup_{p \notin S} [\mathrm{Frob}_p]$ is dense in $\mathrm{Gal}(F/\mathbb{Q})$.*

Theorem 2.4 (Kronecker-Weber) *The product of the p-adic cyclotomic characters gives an isomorphism*

$$\prod_p \epsilon_p : G_\mathbb{Q}^{\mathrm{ab}} \xrightarrow{\sim} \prod_p \mathbb{Z}_p^\times.$$

We remark that the Chebotarev density theorem (theorem 2.3) applied to the extension of $\mathbb{Q}$ generated by the mth roots of unity implies Dirichlet's theorem that if n and m are coprime integers then there are infinitely many primes p with $p \equiv n \bmod m$. We also remark that the Kronecker-Weber theorem is equivalent to the fact that the maximal abelian extension of $\mathbb{Q}$ is obtained by adjoining all roots of unity to $\mathbb{Q}$.

Representations: A d-dimensional representation of $G_\mathbb{Q}$ is a homomorphism

$$G_\mathbb{Q} \longrightarrow GL_d(K),$$

where K is any field. (In our later discussion, we will also consider representations with coefficients in a ring.) Often K (and hence $GL_d(K)$) comes equipped with a natural topology, and it is then customary to restrict one's attention to continuous homomorphisms $G_{\mathbb{Q}} \longrightarrow GL_d(K)$.

Since any one-dimensional representation has abelian image, theorem 2.4 allows one to give a complete description of the one-dimensional representations of $G_{\mathbb{Q}}$ together with the behaviour of these representations on the decomposition groups at all primes. The aim of this article is to discuss attempts to give a similar theory for the two-dimensional representations of $G_{\mathbb{Q}}$.

We will call a representation ρ of $G_{\mathbb{Q}}$ unramified at p if it is trivial on the inertia group I_p. Otherwise we say that it is ramified at p. If ρ is unramified at p, then $\rho(\mathrm{Frob}_p)$ is well-defined (and its conjugacy class is independent of the choice of v_p).

If ρ is a representation of a group G and i is a non-negative integer, then we let $\wedge^i\rho$ denote the representation of G on the ith exterior power of the underlying module of ρ. If H is a subgroup of G, we will let ρ^H (resp. ρ_H) denote the H-invariants (resp. H-coinvariants) of the underlying module of ρ. If H is normal in G, then we shall also use ρ^H and ρ_H to denote the corresponding representation of G/H. In particular if p is a prime and ρ is a representation of $G_{\mathbb{Q}}$ or $G_{\mathbb{Q}_p}$, then $\rho^{I_p}(\mathrm{Frob}_p)$ and $\rho_{I_p}(\mathrm{Frob}_p)$ are well-defined.

We will be primarily interested in three types of representation of $G_{\mathbb{Q}}$.

- **Artin representations**, i.e. continuous representations $G_{\mathbb{Q}} \to GL_d(\mathbb{C})$. Since all compact totally disconnected subgroups of $GL_d(\mathbb{C})$ are finite, Artin representations have finite image. Hence they are semi-simple and are unramified at all but finitely many primes, by theorem 2.1.

- **Mod ℓ representations**, i.e. continuous representations $G_{\mathbb{Q}} \to GL_d(k)$, where k is a finite field of characteristic ℓ. These always have finite image and hence, like Artin representations, are unramified at all but finitely many primes.

- **ℓ-adic representations**, i.e. continuous representations $G_{\mathbb{Q}} \to GL_d(K)$, where K is a finite extension of $\mathbb{Q}_\ell$. We require that an ℓ-adic representation be unramified at all but finitely many primes.

Remark 2.5 (a) Continuous representations $G_{\mathbb{Q}} \to GL_d(K)$, unlike those to $GL_d(\mathbb{C})$ or $GL_d(k)$, may be ramified at infinitely many primes. We shall not need to consider such representations. (One rarely if ever does.) However, the term "ℓ-adic representation" is often used in the more general sense elsewhere in the literature.

(b) Note that the image of an ℓ-adic representation can very well be infinite. For instance this is the case if $d = 1$ and the representation is the cyclotomic character.

(c) Note that in the case of mod ℓ and ℓ-adic representations, the representations need not be semi-simple.

Proposition 2.6 *Let S be any finite set of primes.*

(a) *An Artin representation $\rho : G_{\mathbf{Q}} \to GL_d(\mathbf{C})$ is determined by the values of $\operatorname{tr}\rho(\operatorname{Frob}_p)$ on the primes $p \notin S$ at which ρ is unramified.*

(b) *A semi-simple mod ℓ representation $\rho : G_{\mathbf{Q}} \to GL_d(k)$ is determined by the values of $\operatorname{tr} \wedge^i \rho(\operatorname{Frob}_p)$ $(i = 1, ..., d)$ on the primes $p \notin S$ at which ρ is unramified. If $\ell > d$ then it is determined by $\operatorname{tr}\rho(\operatorname{Frob}_p)$ on the primes $p \notin S$ at which ρ is unramified.*

(c) *A semi-simple ℓ-adic representation $\rho : G_{\mathbf{Q}} \to GL_d(K)$ is determined by the values of $\operatorname{tr}\rho(\operatorname{Frob}_p)$ on the primes $p \notin S$ at which ρ is unramified.*

Proof: Combining the Chebotarev density theorem (theorem 2.3) with the continuity of ρ we see that we must show that $\operatorname{tr}\rho$ (resp. $\operatorname{tr} \wedge^i \rho$) determines ρ up to conjugacy in the various settings. For the characteristic zero representations, see [Bour], ch. 8, sec. 12.1, prop. 3; for mod ℓ representations, this is the Brauer-Nesbitt theorem ([CR], (30.16)). □

Let K denote a finite extension of $\mathbf{Q}_\ell$, let $\mathcal{O}$ be its ring of integers, λ the maximal ideal of $\mathcal{O}$ and k the residue field. If $\rho : G_{\mathbf{Q}} \to GL_d(K)$ is an ℓ-adic representation, then the image of ρ is compact, and hence ρ can be conjugated to a homomorphism $G_{\mathbf{Q}} \to GL_d(\mathcal{O})$. Reducing modulo the maximal ideal λ gives a *residual representation* $\bar{\rho} : G_{\mathbf{Q}} \to GL_d(k)$. This representation may depend on the particular $GL_d(K)$-conjugate of ρ chosen, but its semi-simplification $\bar{\rho}^{ss}$ (i.e. the unique semi-simple representation with the same Jordan-Hölder factors) is uniquely determined by ρ, by proposition 2.6 (b).

Note that the kernel of the reduction map $GL_d(\mathcal{O}) \to GL_d(k)$ is a pro-ℓ-group. In particular we see that if $p \neq \ell$ then $\rho(P_p)$ is finite and $\rho(P_p)$ is isomorphic to $\bar{\rho}(P_p)$.

Suppose ρ is a representation of $G_{\mathbf{Q}}$ of one of the three sorts above. If $p \neq \ell$ then we define the conductor, $m_p(\rho)$, of ρ at p by

$$m_p(\rho) = \int_{-1}^{\infty} \operatorname{codim}\rho^{I_p^u}\, du = \operatorname{codim}\rho^{I_p} + \int_0^{\infty} \operatorname{codim}\rho^{I_p^u}\, du.$$

This is well-defined as $\rho(P_p)$ is a finite group. If ρ is an Artin representation this makes sense also for $p = \ell$. (It would even make sense for a mod ℓ representation if $p = \ell$, but in this case it does not seem to be a useful notion.) It is known that $m_p(\rho)$ is an integer (see chapter VI of [Se2]). Moreover it is easily seen that $m_p(\rho) = 0$ if and only if ρ is unramified at p. We define the conductor $N(\rho)$ of ρ to be

$$\prod_p p^{m_p(\rho)},$$

where the product is over all primes $p \neq \ell$ in the case of an ℓ-adic or mod ℓ representation and over all primes in the case of an Artin representation. This makes sense because ρ is unramified almost everywhere.

The following lemma is an exercise.

Lemma 2.7 *Suppose that $\rho : G_{\mathbb{Q}} \to GL_d(K)$ is an ℓ-adic representation and that $\bar{\rho} : G_{\mathbb{Q}} \to GL_d(k)$ is a reduction of ρ. Then for each prime $p \neq \ell$ we have $\int_0^\infty \operatorname{codim}\rho^{I_p^u}\, du = \int_0^\infty \operatorname{codim}\bar{\rho}^{I_p^u}\, du$. Thus*

$$N(\rho) = N(\bar{\rho}) \prod_{p \neq \ell} p^{(\dim \bar{\rho}^{I_p} - \dim \rho^{I_p})},$$

and in particular $N(\bar{\rho})$ divides $N(\rho)$.

We take this opportunity to introduce some important notation. Let G denote a group and R a ring. If $\rho : G \to GL_d(R)$ is a representation we shall let M_ρ denote the underlying $R[G]$-module (so that $M_\rho \cong R^d$ as an R-module). If M is an $R[G]$-module we shall let $\operatorname{End}(M)$ denote the module of R-linear endomorphisms of M. It is also an $R[G]$-module with the G-action being defined by

$$(g(\phi))(m) = g(\phi(g^{-1}m)).$$

If M is a finitely generated free R-module we will let $\operatorname{End}^0(M)$ denote the sub-$R[G]$-module of $\operatorname{End}(M)$ consisting of endomorphisms of trace zero. We will use $\operatorname{ad}\rho$ to denote $\operatorname{End}(M_\rho)$ and $\operatorname{ad}^0\rho$ to denote $\operatorname{End}^0(M_\rho)$. Note that if d is invertible in R then $\operatorname{ad}\rho \cong \operatorname{ad}^0\rho \oplus R$ as $R[G]$-modules, where R has the trivial action of G. Note also then that the kernel of $\operatorname{ad}^0\rho$ is the same as the kernel of the composite map $G \xrightarrow{\rho} GL_d(R) \to PGL_d(R)$.

We also remark that if R is an algebraically closed field of characteristic other than 2 and $\rho : G \to GL_2(R)$ is irreducible, then either $\operatorname{ad}^0\rho$ is irreducible or there is a subgroup $H \subset G$ of index 2 and a character $\chi : H \to R^\times$ such that $\rho \cong \operatorname{Ind}_H^G \chi$. In the latter case $\operatorname{ad}^0\rho \cong \delta \oplus \operatorname{Ind}_H^G(\chi/\chi')$, where δ is the nontrivial character of G/H and χ' is the composite of χ and conjugation by an element of $G - H$. Moreover, either $\operatorname{Ind}_H^G(\chi/\chi')$ is irreducible or $\operatorname{ad}^0\rho \cong \delta_1 \oplus \delta_2 \oplus \delta_3$ where the δ_i are distinct characters of G of order 2. In the latter case, ρ is induced from a character from each of the subgroups $H_i = \ker \delta_i$ of index 2 in G. We leave the verification of these facts as an exercise to the reader.

2.2 Representations associated to elliptic curves

Perhaps the simplest examples of non-abelian ℓ-adic representations arise from elliptic curves defined over $\mathbb{Q}$.

We will let $E[n](\bar{\mathbb{Q}})$ denote the group of n-torsion points on $E(\bar{\mathbb{Q}})$. By proposition 1.1 there is a non-canonical isomorphism $E[n](\bar{\mathbb{Q}}) \cong (\mathbb{Z}/n\mathbb{Z})^2$.

Furthermore, $E[n](\bar{\mathbb{Q}})$ carries a natural action of $G_{\mathbb{Q}}$ and so we get a representation (defined up to conjugation)

$$\bar{\rho}_{E,n} : G_{\mathbb{Q}} \to GL_2(\mathbb{Z}/n\mathbb{Z}).$$

If ℓ is a prime then we set $\mathcal{T}_\ell E = \varprojlim E[\ell^n](\bar{\mathbb{Q}})$, which is non-canonically isomorphic to $\mathbb{Z}_\ell^2$. Again it has a natural continuous action of $G_{\mathbb{Q}}$, and so we get a representation (defined up to conjugation)

$$\rho_{E,\ell} : G_{\mathbb{Q}} \to GL_2(\mathbb{Z}_\ell).$$

Note that $\bar{\rho}_{E,\ell^n} \cong (\rho_{E,\ell} \bmod \ell^n)$.

Global properties: We have the following basic properties of the representations $\bar{\rho}_{E,\ell^n}$ and $\rho_{E,\ell}$:

Proposition 2.8 (a) *The determinant of $\rho_{E,\ell}$ is ϵ_ℓ.*

(b) *The representation $\rho_{E,\ell}$ is absolutely irreducible for all ℓ. For fixed E, $\bar{\rho}_{E,\ell}$ is absolutely irreducible for all but finitely many ℓ.*

(c) *If E does not have complex multiplication then $\rho_{E,\ell}$ (and hence $\bar{\rho}_{E,\ell}$) is surjective for all but finitely many ℓ.*

Proof: Part (a) follows from the existence of the non-degenerate alternating Galois-equivariant Weil pairing

$$\mathcal{T}_\ell E \times \mathcal{T}_\ell E \longrightarrow \mathbb{Z}_\ell(1) := \varprojlim \mu_{\ell^n}.$$

Part (b) is proved in [Se5], ch. IV. Part (c) is the main result of [Se6]. □

The following deep result, which is stronger than the second assertion of part (b) of proposition 2.8, is due to Mazur [Maz2], thms. 1 and 2. (Parts (b) and (c) can actually be deduced directly from theorem 1.7, using the fact that a semi-stable curve with a rational subgroup of order ℓ is necessarily isogenous to a curve with a rational point of order ℓ.)

Theorem 2.9 *Suppose that $E/\mathbb{Q}$ is an elliptic curve.*

(a) *If $\ell > 163$ is a prime then $\bar{\rho}_{E,\ell}$ is irreducible.*

(b) *If E is semistable then $\bar{\rho}_{E,\ell}$ is irreducible for $\ell > 7$.*

(c) *If E is semistable and $\bar{\rho}_{E,2}$ is trivial then $\bar{\rho}_{E,\ell}$ is irreducible for $\ell > 3$.*

Remark 2.10 Combined with the results of Serre [Se6], Mazur's results imply that if E is semistable everywhere then $\bar{\rho}_{E,\ell}$ is surjective for $\ell > 7$ ([Maz2], thm. 4).

Local behaviour of $\rho_{E,\ell}$ and $\bar{\rho}_{E,\ell}$:

Proposition 2.11 *Suppose E has good reduction at p.*

(a) *If $\ell \neq p$, then $\rho_{E,\ell}$ is unramified at p, and we have the formula*

$$\operatorname{tr} \rho_{E,\ell}(\operatorname{Frob}_p) = p + 1 - \#\bar{E}_p(\mathbb{F}_p).$$

In particular $\operatorname{tr} \rho_{E,\ell}(\operatorname{Frob}_p)$ *belongs to $\mathbb{Z}$ and is independent of $\ell \neq p$.*

(b) *For all $n \geq 1$ there is a finite flat group scheme $\mathcal{F}_n/\mathbb{Z}_p$ such that*

$$E[p^n](\bar{\mathbb{Q}}_p) \cong \mathcal{F}_n(\bar{\mathbb{Q}}_p)$$

as G_p-modules.

(c)
- *If E has good ordinary reduction at p (i.e., a_p is not divisible by p) then $\bar{E}_p[p](\bar{\mathbb{F}}_p)$ has order p and*

$$\rho_{E,p}|_{I_p} \sim \begin{pmatrix} \epsilon_p & * \\ 0 & 1 \end{pmatrix};$$

- *If E has supersingular reduction at p, then $\bar{E}_p[p](\bar{\mathbb{F}}_p)$ is trivial and $\bar{\rho}_{E,p}|_{G_p}$ is irreducible.*

Proofs: The proof of part (a) can be found in chapters V and VII of [Sil]. For part (b) one considers the finite flat group scheme $\mathcal{F}_n = \mathcal{E}[p^n]$, where $\mathcal{E}$ is the model for E over $\mathbb{Z}_p$ defined by ($W^{\min}$). (See [Sha] for an introduction to theory of finite flat group schemes.) Part (c) follows from the results in chapters V and VII of [Sil] together with basic results on finite flat group schemes.

Proposition 2.12 *Suppose that E has multiplicative reduction at p. Let $q = q_{E,p} \in p\mathbb{Z}_p$ be the multiplicative Tate period attached to E, defined as in section 1.1. Let $\delta : G_p/I_p \to \{\pm 1\}$ be the unique non-trivial unramified quadratic character of G_p if E has non-split multiplicative reduction, and let δ be the trivial character if E has split multiplicative reduction. Then*

(a)

$$\rho_{E,\ell}|_{G_p} \sim \begin{pmatrix} \epsilon_\ell & * \\ 0 & 1 \end{pmatrix} \otimes \delta,$$

and if $\ell \neq p$ then $m_p(\rho_{E,\ell}) = 1$.

(b)

$$\bar{\rho}_{E,\ell}|_{G_p} \sim \begin{pmatrix} \epsilon_\ell & \Psi \\ 0 & 1 \end{pmatrix} \otimes \delta,$$

and $\Psi \in H^1(G_{\mathbb{Q}_p}, \mathbb{Z}/\ell\mathbb{Z}(1)) \cong \mathbb{Q}_p^\times/(\mathbb{Q}_p^\times)^\ell$ corresponds to the Tate period $q_{E,p} \in \mathbb{Q}_p^\times/(\mathbb{Q}_p^\times)^\ell$.

(c) *If $\ell \neq p$ then $m_p(\bar{\rho}_{E,\ell}) = 0$ (i.e., $\bar{\rho}_{E,\ell}$ is unramified at p) if and only if $\ell | v_p(\Delta_E^{min}) = -v_p(j_E)$.*

(d) *There is a finite flat group scheme $\mathcal{F}/\mathbb{Z}_p$ such that $E[p](\bar{\mathbb{Q}}_p) \cong \mathcal{F}(\bar{\mathbb{Q}}_p)$ if and only if $p | v_p(\Delta_E^{min}) = -v_p(j_E)$.*

Sketch of proof: By Tate's proposition 1.5 of section 1.1, we have

$$E(\bar{\mathbb{Q}}_p) \cong (\bar{\mathbb{Q}}_p^\times / q^{\mathbb{Z}})(\delta)$$

as G_p-modules. (The (δ) indicates that G_p acts on $(\bar{\mathbb{Q}}_p^\times / q^{\mathbb{Z}})(\delta)$ by $\sigma : x \mapsto \sigma(x)^{\delta(\sigma)}$.) The proposition follows directly from this, together with [Edi], prop. 8.2. for the last part. (For further details, see ch. VII, prop. 5.1 of [Si1], ch. IV, thm. 10.2. and ch. V of [Si2], and sec. 2.9 of [Se7].) □

Proposition 2.13 *Suppose that E has additive reduction at p. Then for any $\ell \neq p$ the conductor of $\rho_{E,\ell}|_{G_p}$ is at least 2, and if $p > 3$ then it is equal to 2.*

Remark 2.14 For an elliptic curve E with any type of reduction at p, the conductor of $\rho_{E,\ell}|_{G_p}$ (for $\ell \neq p$) coincides with the local conductor of E at p given by a formula of Ogg (proved by Saito [Sa] in the case $p = 2$). In particular the conductor of $\rho_{E,\ell}|_{G_p}$ is independent of $\ell \neq p$. For further discussion and references, see ch. IV, secs. 10 and 11 of [Si2].

The Frey curve: We now consider the elliptic curve

$$E_{A,B} : Y^2 Z = X(X - AZ)(X + BZ),$$

where A and B are non-zero coprime integers with $A+B \neq 0$. Set $C = A+B$. This equation has discriminant $\Delta = 16(ABC)^2$. If p is an odd prime then this curve has good reduction at p if $p \nmid (ABC)$ and has multiplicative reduction at p if $p | ABC$. Thus $\Delta_{E_{A,B}}^{\min} = 2^{4-12n}(ABC)^2$ for some $n \in \mathbb{Z}_{\geq 0}$. If for instance $A \equiv -1 \bmod 4$ and $B \equiv 0 \bmod 16$ then $E_{A,B}$ has a minimal Weierstrass equation

$$Y^2 Z + XYZ = X^3 + \frac{B - A - 1}{4} X^2 Z - (AB/16) X Z^2.$$

In this special case $E_{A,B}$ has semi-stable reduction everywhere and $\Delta_{E_{A,B}}^{\min} = 2^{-8}(ABC)^2$.

The connection with Fermat's Last Theorem is provided by the elliptic curve

$$E : Y^2 Z = X(X - a^\ell Z)(x + b^\ell Z)$$

considered by Hellegouarch [He] and Frey [Fr], where $a^\ell + b^\ell = c^\ell$ is a hypothetical non-trivial solution to Fermat's Last theorem, and where we have supposed without loss of generality that a, b and c are pairwise coprime, that b is even and that $a \equiv -1 \bmod 4$. Frey suggested that properties of this hypothetical elliptic curve E could lead to a contradiction if the curve were also known to be modular. This suggestion was made precise by Serre [Se7], who observed that the following list of properties would yield the desired contradiction, when combined with his conjectures on Galois representations associated to modular forms (to be discussed in section 3.2).

Theorem 2.15 *Suppose that $\ell > 3$ is prime. The representation $\bar{\rho}_{E,\ell}$ has the following properties.*

(a) *$\bar{\rho}_{E,\ell}$ is irreducible.*

(b) *$\bar{\rho}_{E,\ell}$ is unramified outside ℓ and 2.*

(c) *There is a finite flat group scheme $\mathcal{F}/\mathbb{Z}_\ell$ such that $\mathcal{F}(\bar{\mathbb{Q}}_\ell) \cong E[\ell](\bar{\mathbb{Q}}_\ell)$ as G_ℓ-modules.*

(d) *$\#\bar{\rho}_{E,\ell}(I_2) = \ell$.*

Proof: Part (a) follows from Mazur's theorem (theorem 2.9, (c)). To prove (b), (c), and (d), we note the key fact that $\Delta_E^{\min} = 2^{-8}(abc)^{2\ell}$ is a perfect ℓ-th power, up to powers of 2. The fact that $\bar{\rho}_{E,\ell}$ is unramified at $p \neq \ell, 2$ (part (b)) follows from this, from proposition 2.11, (a) (for primes $p \neq \ell$ of good reduction for E) and from proposition 2.12 (c) (for primes $p \neq \ell, 2$ of multiplicative reduction for E). Part (c) is a consequence of proposition 2.12 (d), and part (d) follows from proposition 2.12 (c). □

Remark 2.16 Part (a) implies that the image of $\bar{\rho}_{E,\ell}$ is large, while parts (b) and (c) state that the image has very limited ramification. We remark that the conclusion of the theorem ensures that $m_p(\bar{\rho}_{E,\ell}) = 0$ for $p \neq 2, \ell$ and that $m_2(\bar{\rho}_{E,\ell}) = 1$.

2.3 Galois cohomology

In this section M will denote a continuous discrete $G_\mathbb{Q}$-module of finite cardinality. If M and N are two such modules we will endow the space $\mathrm{Hom}(M, N)$ of homomorphisms of abelian groups with an action of $G_\mathbb{Q}$ via the formula

$$(g(\phi))(m) = g(\phi(g^{-1}m)).$$

We will use M^* to denote $\mathrm{Hom}(M, \mu_n(\bar{\mathbb{Q}}))$, where n is an integer such that $nM = (0)$ and where $\mu_n(\bar{\mathbb{Q}})$ denotes the group of nth roots of unity in $\bar{\mathbb{Q}}$. We will use $\vee$ to denote Pontryagin duals. All cohomology groups of profinite groups will mean continuous cohomology. The general references for this section are Milne [Mi] and Serre [Se1].

We start by recalling Tate's local duality theorem (see chapter 1, section 2 of [Mi]).

Theorem 2.17 *Let v be a place of $\mathbb{Q}$ (i.e. either a prime number or ∞).*

(a) *$H^i(G_v, M)$ is finite for all i.*

(b) *For all integers n there are compatible embeddings*

$$H^2(G_v, \mu_n(\bar{\mathbb{Q}})) \hookrightarrow \mathbb{Q}/\mathbb{Z}$$

(i.e. compatible with the maps coming from $\mu_n(\bar{\mathbb{Q}}) \hookrightarrow \mu_m(\bar{\mathbb{Q}})$ if $n|m$).

(c) *For $i = 0, 1, 2$ the cup product and the above embeddings give rise to a perfect pairing*

$$H^i(G_v, M) \times H^{2-i}(G_v, M^*) \to H^2(G_v, \mu_n(\bar{\mathbb{Q}})) \hookrightarrow \mathbb{Q}/\mathbb{Z}.$$

(d) *If $v \neq \infty$ is a prime, then $H^i(G_v, M) = (0)$ for $i > 2$ and*

$$\#H^1(G_v, M) = \#H^0(G_v, M)\#H^2(G_v, M)\#(M \otimes \mathbb{Z}_v).$$

(e) *If $v \neq \infty$ and $M \otimes \mathbb{Z}_v = 0$ then $H^1(G_v/I_v, M^{I_v})$ and $H^1(G_v/I_v, M^{*I_v})$ are the annihilators of each other under the pairing*

$$H^1(G_v, M) \times H^1(G_v, M^*) \to \mathbb{Q}/\mathbb{Z}.$$

By a collection of local conditions for M we shall mean a collection $\mathcal{L} = \{L_v\}$ of subgroups $L_v \subset H^1(G_v, M)$ as v runs over all places of $\mathbb{Q}$, such that $L_v = H^1(G_v/I_v, M^{I_v})$ for all but finitely many v. Note that if $\mathcal{L}$ is a collection of local conditions for M then $\mathcal{L}^* = \{L_v^\perp\}$ is a collection of local conditions for M^*. If $\mathcal{L}$ is a collection of local conditions for M we define the corresponding Selmer group, $H^1_{\mathcal{L}}(\mathbb{Q}, M)$, to be the subgroup of $x \in H^1(G_{\mathbb{Q}}, M)$ such that for all places v of $\mathbb{Q}$ we have

$$\mathrm{res}_v x \in L_v \subset H^1(G_v, M).$$

Suppose that v is a finite place of $\mathbb{Q}$. We have an exact sequence

$$(0) \to H^0(G_v, M) \to M^{I_v} \overset{\mathrm{Frob}_v - 1}{\longrightarrow} M^{I_v} \to H^1(G_v/I_v, M^{I_v}) \to (0),$$

and hence we see that $\#H^1(G_v/I_v, M^{I_v}) = \#H^0(G_v, M)$.

We have the following important observation of Wiles [W3], inspired by a formula of Greenberg [Gre]. An important theme in number theory has been the calculation of Selmer groups. This result, although it does not allow the absolute calculation of Selmer groups, allows the comparison of two (dual) Selmer groups. In the applications in this paper we shall apply the theorem where the various data have been chosen to make one of the Selmer groups trivial. In such a situation it allows the exact calculation of the order of the non-trivial Selmer group.

Theorem 2.18 *If $\mathcal{L}$ is a collection of local conditions for M then the Selmer group $H^1_{\mathcal{L}}(\mathbb{Q}, M)$ is finite. Moreover we have the formula*

$$\frac{\#H^1_{\mathcal{L}}(\mathbb{Q}, M)}{\#H^1_{\mathcal{L}^*}(\mathbb{Q}, M^*)} = \frac{\#H^0(G_{\mathbb{Q}}, M)}{\#H^0(G_{\mathbb{Q}}, M^*)} \prod_v \frac{\#L_v}{\#H^0(G_v, M)}.$$

We note that all but finitely many terms in the product are 1 so that it makes sense.

Proof of theorem 2.18: Choose a finite set, S, of places of $\mathbb{Q}$, which contains ∞, all the places whose residue characteristic divides the order of M, all places at which M is ramified and all places at which $L_v \neq H^1(G_v/I_v, M^{I_v})$. Let $\mathbb{Q}_S$ denote the maximal extension of $\mathbb{Q}$ unramified outside S and let $G_S = \mathrm{Gal}\,(\mathbb{Q}_S/\mathbb{Q})$. Then we have an exact sequence

$$(0) \to H^1_{\mathcal{L}}(\mathbb{Q}, M) \to H^1(G_S, M) \to \bigoplus_{v\in S} H^1(G_v, M)/L_v.$$

Because $H^1(G_S, M)$ is finite (see cor. 4.15 of ch. 1 of [Mi]), we see that $H^1_{\mathcal{L}}(\mathbb{Q}, M)$ is finite.

Dualising this sequence for M^* and using theorem 2.17, we see that we have an exact sequence

$$\bigoplus_{v\in S} L_v \to H^1(G_S, M^*)^\vee \to H^1_{\mathcal{L}^*}(\mathbb{Q}, M^*)^\vee \to (0).$$

On the other hand we have the following part of the Poitou-Tate nine term sequence (see thm. 4.10 of ch. 1 of [Mi]):

$$\begin{array}{ccccc} (0) \to & H^0(G_S, M) & \to \left(\bigoplus_{v\in S} H^0(G_v, M)\right)/(1+c)M & \to & H^2(G_S, M^*)^\vee \\ & & & & \downarrow \\ & H^1(G_S, M^*)^\vee \leftarrow & \bigoplus_{v\in S} H^1(G_v, M) & \leftarrow & H^1(G_S, M), \end{array}$$

where we regard $(1+c)M$ as contained in $H^0(G_\infty, M)$. Replacing $H^1(G_v, M)$ by L_v for $v \in S$ and combining this with the previous exact sequence we get

$$\begin{array}{ccccc} (0) & \to & H^0(G_S, M) & \to & \left(\bigoplus_{v\in S} H^0(G_v, M)\right)/(1+c)M \\ & & & & \downarrow \\ \bigoplus_{v\in S} L_v & \leftarrow & H^1_{\mathcal{L}}(\mathbb{Q}, M) & \leftarrow & H^2(G_S, M^*)^\vee \\ \downarrow & & & & \\ H^1(G_S, M^*)^\vee & \to & H^1_{\mathcal{L}^*}(\mathbb{Q}, M^*)^\vee & \to & (0). \end{array}$$

Theorem 2.18 follows from this and the global Euler characteristic formula (see thm. 5.1 of ch. 1 of [Mi]):

$$\frac{\#H^1(G_S, M^*)}{\#H^0(G_S, M^*)\#H^2(G_S, M^*)} = \frac{(\#M^*)}{\#H^0(G_\infty, M^*)} = \#(1+c)M.$$

□

We remark, but will not use elsewhere in this article, that the above argument works over any number field:

Theorem 2.19 *Let F be a number field and M a discrete G_F module of finite cardinality. For each place v of F let G_v denote a decomposition group at v and if v is finite let $I_v \subset G_v$ denote the inertia group. Fix subgroups $L_v \subset H^1(G_v, M)$, and such that for all but finitely many v,*

$$L_v = H^1(G_v/I_v, M^{I_v}).$$

Let $H^1_{\mathcal{L}}(F, M)$ (respectively, $H^1_{\mathcal{L}^*}(F, M^*)$) denote the inverse image of $\prod_v L_v$ (respectively, $\prod_v L_v^\perp$) under the map $H^1(G_F, M) \to \prod_v H^1(G_v, M)$ (respectively, $H^1(G_F, M^*) \to \prod_v H^1(G_v, M^*)$). Then $H^1_{\mathcal{L}}(G, M)$ and $H^1_{\mathcal{L}^*}(G, M^*)$ are finite and we have the formula

$$\frac{\# H^1_{\mathcal{L}}(F, M)}{\# H^1_{\mathcal{L}^*}(F, M^*)} = \frac{\# H^0(G_F, M)}{\# H^0(G_F, M^*)} \prod_v \frac{\# L_v}{\# H^0(G_v, M)}.$$

2.4 Representations of $G_{\mathbb{Q}_\ell}$

In this section we assume that ℓ is an **odd** prime. If G is any topological group then by a finite $\mathcal{O}[G]$-module we shall mean a discrete $\mathcal{O}$-module of finite cardinality with a continuous action of G. By a profinite $\mathcal{O}[G]$-module we shall mean an inverse limit of finite $\mathcal{O}[G]$-modules.

If M is a profinite $\mathcal{O}[G_\ell]$-module then we will call M

- **good** if for every discrete quotient M' of M there is a finite flat group scheme $\mathcal{F}/\mathbb{Z}_\ell$ such that $M' \cong F(\bar{\mathbb{Q}}_\ell)$ as $\mathbb{Z}_\ell[G_\ell]$-modules;

- **ordinary** if there is an exact sequence

$$(0) \to M^{(-1)} \to M \to M^{(0)} \to (0)$$

  of profinite $\mathcal{O}[G_\ell]$-modules such that I_ℓ acts trivially on $M^{(0)}$ and by ϵ on $M^{(-1)}$ (equivalently, if for all $\sigma, \tau \in I_\ell$ we have $(\sigma - \epsilon(\sigma))(\tau - 1) = 0$ on M);

- **semi-stable** if M is either good or ordinary.

Suppose that R is a complete Noetherian local $\mathcal{O}$-algebra with residue field k. We will call a continuous representation $\rho : G_\ell \to GL_2(R)$ good, ordinary or semistable, if

$$\det \rho|_{I_\ell} = \epsilon \tag{2.4.1}$$

and if the underlying profinite $\mathcal{O}[G_\ell]$-module, M_ρ is good, ordinary or semi-stable. We write $\bar{\rho}$ for $\rho \bmod \mathfrak{m}_R$. We record the following consequence of Nakayama's lemma.

Lemma 2.20 *If M_ρ and $\bar{\rho}$ are ordinary, then $M_\rho^{(-1)}$ and $M_\rho^{(0)}$ are each free of rank one over R and ρ is ordinary.*

Remark 2.21 (a) These definitions are somewhat ad hoc, but at the moment that is all that seems to be available (though the work of Fontaine and Laffaille [FL] and its generalisations may well provide a more systematic setting).

(b) For part of the motivation for our definitions, see proposition 2.23 and remark 2.24. For further motivation, we shall see in the next chapter that representations arising from certain modular forms are semistable. Moreover the Fontaine-Mazur conjecture predicts that, conversely, any representation of $G_{\mathbb{Q}}$ with semistable restriction to G_ℓ arises from such a modular form. We shall state a weak form of the Fontaine-Mazur conjecture below (conjecture 3.17).

(c) The terminology in [W3] to describe representations of G_ℓ is slightly different. In particular, we impose the condition (2.4.1) in our definitions of good and ordinary, as this is all we shall need here. Assuming ρ satisfies this condition, the notion of ordinary in [W3] coincides with the one here, and ρ is flat in the sense of [W3] if and only if it is good but not ordinary. (Note that a representation may be both good and ordinary, for instance $\rho_{E,\ell}$ for an elliptic curve with good, ordinary reduction; see proposition 2.23.)

The following assertions, in the good case, are consequences of results of Raynaud [Ray] (in particular, see sec. 2.1, prop. 2.3.1 and cor. 3.3.6 of [Ray]). In the ordinary case they are elementary.

Lemma 2.22 (a) *Good profinite $\mathcal{O}[G_\ell]$-modules are closed under taking sub-objects, quotients and direct products. The same is true for ordinary profinite $\mathcal{O}[G_\ell]$-modules.*

(b) *Suppose that M is a profinite $\mathcal{O}[G_\ell]$-module and that $\{M_i\}$ is a family of sub-objects with trivial intersection, such that each M/M_i is good (resp. ordinary). Then M is good (resp. ordinary).*

(c) *If M is a finite $\mathcal{O}[G_\ell]$-module then M is good if and only if there is a finite flat group scheme $\mathcal{F}/\mathbb{Z}_\ell$ such that $M \cong \mathcal{F}(\bar{\mathbb{Q}}_\ell)$ as $\mathbb{Z}_\ell[G_\ell]$-modules.*

(d) *If M and M' are profinite $\mathcal{O}[G_\ell]$-modules with $M' \cong M$ as $\mathbb{Z}_\ell[I_\ell]$-modules then M is good (resp. ordinary) if and only if M' is good (resp. ordinary).*

(e) *Suppose that M is a profinite $\mathcal{O}[G_\ell]$-module which is finite and free over $\mathcal{O}$. Then M is good if and only if there exists an ℓ-divisible group $\mathcal{F}/\mathbb{Z}_\ell$ such that M is isomorphic to the Tate module of $\mathcal{F}$ as a $\mathbb{Z}_\ell[G_\ell]$-module.*

Together with results stated in section 2.2, we have the following.

Proposition 2.23 *Suppose that E is an elliptic curve over $\mathbb{Q}$ and $\mathcal{O} = \mathbb{Z}_\ell$.*

- *if E has good (resp. semistable) reduction at ℓ, then $\rho_{E,\ell}$ and $\bar{\rho}_{E,\ell}$ are good (resp. semistable);*

- *if E has semistable reduction at ℓ, then $\rho_{E,\ell}$ is ordinary if and only if $\bar{\rho}_{E,\ell}$ is ordinary if and only if either E has multiplicative reduction or E has good ordinary reduction.*

Remark 2.24 It is also true that if $\rho_{E,\ell}$ is good (resp. semistable), then E has good (resp. semistable) reduction at ℓ, but the result is more difficult and we shall not need it.

We will need a few more definitions. We will let $\psi : I_\ell \to \bar{\mathbb{F}}_\ell^\times$ denote the character $\sigma \mapsto (\sigma\varpi)/\varpi \bmod \varpi$ where $\varpi = \sqrt[\ell^2-1]{\ell}$. If F is a field of characteristic other than ℓ and if M is a profinite $\mathbb{Z}_\ell[G_\ell]$-module then we set $M(1) = \varprojlim M \otimes_{\mathbb{Z}_\ell} \mu_{\ell^n}(\bar{F})$. If $\rho : G_\ell \to GL_2(R)$ is ordinary the extension

$$(0) \to R(1) \to M_\rho \to R \to (0)$$

of $R[I_\ell]$-modules gives rise to a class $c_\rho \in H^1(I_\ell, R(1))$. Kummer theory and the valuation on $(\bar{\mathbb{Q}}_\ell^{I_\ell})^\times$ give rise to a map $v : H^1(I_\ell, R(1)) \to H \otimes_{\mathbb{Z}_\ell} R \to R$, where H denotes the ℓ-adic completion of $(\bar{\mathbb{Q}}_\ell^{I_\ell})^\times$. Then we also have the following lemma.

Lemma 2.25 (a) *If $\bar{\rho} : G_\ell \to GL_2(k)$ is good then either $\bar{\rho}$ is ordinary or $\bar{\rho}|_{I_\ell} \otimes \bar{k} = \psi \oplus \psi^\ell$.*

(b) *If $\rho : G_\ell \to GL_2(R)$ is such that M_ρ is good and $\bar{\rho}$ is ordinary, then ρ is good and ordinary.*

(c) *If $\rho : G_\ell \to GL_2(R)$ is ordinary then ρ is good if and only if $v(c_\rho) = 0$.*

Note that we need only consider the case that R has finite cardinality to prove the lemma. Parts (a) and (b) again follow from Raynaud's results [Ray] (for part (b) consider the connected-étale sequences for M_ρ and $M_{\bar{\rho}}$).

We sketch the proof of part (c) using an argument suggested to us by Edixhoven. As in [Edi], prop. 8.2 it suffices to determine which extensions

$$(0) \longrightarrow R(1) \longrightarrow M \longrightarrow R \longrightarrow (0) \qquad (2.4.2)$$

of $R[I_\ell]$-modules arise from finite flat group schemes over $\mathbb{Z}_\ell^{\mathrm{nr}}$, the ring of integers of the maximal unramified extension of $\mathbb{Z}_\ell$. By prop. 17.4 of [Oo], the extension (2.4.2) arises from a finite flat group scheme if and only if it corresponds to an extension of sheaves of R-modules in the fpqc topology over $\mathbb{Z}_\ell^{\mathrm{nr}}$. Therefore we must compute the image of

$$\mathrm{Ext}^1_{R\text{-mod/fpqc}}(R, R(1)) \to \mathrm{Ext}^1_{R[I_\ell]}(R, R(1)).$$

Since the sheaf Ext^1 of R by $R(1)$ vanishes, this is equivalent to computing the image of

$$H^1_{\mathrm{fpqc}}(\mathbb{Z}_\ell^{\mathrm{nr}}, R(1)) \to H^1(I_\ell, R(1)).$$

Part (c) follows from the fact this is precisely the kernel of v.

Remark 2.26 The authors expect that the theory of Fontaine and Laffaille [FL] discussed in the next section could be used to prove that if

$$\rho : G_\ell \to GL_2(R)$$

is such that M_ρ is good and $\bar{\rho}$ is good, then $\det \rho|_{I_\ell}$ is cyclotomic, i.e., ρ is good.

For semistable representations $\rho : G_\ell \to GL_2(\mathcal{O}/\lambda^n)$ we shall define $\mathcal{O}$-submodules

$$H^1_{\mathrm{f}}(G_\ell, \mathrm{ad}\rho) \subset H^1_{\mathrm{ss}}(G_\ell, \mathrm{ad}\rho) \subset H^1(G_\ell, \mathrm{ad}\rho).$$

Before doing so, let us consider more generally a continuous representation ρ of a profinite group G with values in $GL_d(\mathcal{O}/\lambda^n)$. For each continuous cocycle $\xi : G \to \mathrm{ad}\rho$, we define a representation

$$\begin{aligned} \rho_\xi : G \to \quad & GL_2(R_n) \\ g \mapsto & (1 + \varepsilon\xi(g))\rho(g) \end{aligned}$$

where $R_n = (\mathcal{O}/\lambda^n)[\varepsilon]/(\varepsilon^2)$. We find that the map $\xi \to \rho_\xi$ induces a bijection between $H^1(G, \mathrm{ad}\rho)$ and equivalence classes of representations $\rho' : G \to GL_2(R_n)$ such that $\rho = \rho' \bmod \varepsilon R_n$, where ρ'_1 and ρ'_2 are deemed equivalent if they are conjugate by an element of $1 + \varepsilon M_2(R_n)$. Now let M_ρ denote the $(\mathcal{O}/\lambda^n)[G]$-module corresponding to ρ, and for each continuous cocycle ξ, let E_ξ denote the $R_n[G]$-module corresponding to ρ_ξ. Note that multiplication by ε defines an isomorphism from $M_\rho = E_\xi/\varepsilon E_\xi$ to εE_ξ of $(\mathcal{O}/\lambda^n)[G]$-modules. We thus obtain an extension

$$0 \to M_\rho \to E_\xi \to M_\rho \to 0$$

and hence a class in $\mathrm{Ext}^1(M_\rho, M_\rho)$ in the category of profinite $(\mathcal{O}/\lambda^n)[G]$-modules. Moreover if ρ_{ξ_1} and ρ_{ξ_2} are equivalent, then E_{ξ_1} and E_{ξ_2} define the same extension class. We thus obtain a map

$$H^1(G, \mathrm{ad}\rho) \to \mathrm{Ext}^1(M_\rho, M_\rho),$$

which the reader can check is an $(\mathcal{O}/\lambda^n)$-linear isomorphism. Classes in $H^1(G, \mathrm{ad}^0\rho)$ correspond to the equivalence classes of representations ρ' satisfying $\det \rho' = \det \rho$, and ϕ in $\mathrm{Hom}(G, \mathcal{O}/\lambda^n)$ corresponds to the twist of ρ by the character $g \mapsto 1 + \varepsilon\phi(g)$. In the case $\ell \nmid d$ where we have the decomposition

$$H^1(G, \mathrm{ad}\rho) = H^1(G, \mathrm{ad}^0\rho) \oplus \mathrm{Hom}(G, \mathcal{O}/\lambda^n),$$

we see that if ρ' corresponds to a class in $H^1(G, \mathrm{ad}\rho)$, then the projection to $\mathrm{Hom}(G, \mathcal{O}/\lambda^n)$ is the homomorphism ϕ such that $\det\rho' = (1+\varepsilon d\phi)\det\rho$.

We now return to the case of a semistable representation

$$\rho : G_\ell \to GL_2(\mathcal{O}/\lambda^n)$$

and define the cohomology groups $H^1_{\mathrm{f}}(G_\ell, \mathrm{ad}\rho)$ and $H^1_{\mathrm{ss}}(G_\ell, \mathrm{ad}\rho)$ as follows. Let $H^1_{\mathrm{ss}}(G_\ell, \mathrm{ad}\rho)$ denote the natural image in $H^1(G_\ell, \mathrm{ad}\rho)$ of $\mathrm{Ext}^1(M_\rho, M_\rho)$ taken in the category of semi-stable profinite $\mathcal{O}/\lambda^n[G_\ell]$-modules. If ρ is not good then $H^1_{\mathrm{f}}(G_\ell, \mathrm{ad}\rho)$ is defined to be $H^1_{\mathrm{ss}}(G_\ell, \mathrm{ad}\rho)$. If, however, ρ is good then $H^1_{\mathrm{f}}(G_\ell, \mathrm{ad}\rho)$ will denote the natural image in $H^1(G_\ell, \mathrm{ad}\rho)$ of the group $\mathrm{Ext}^1(M_\rho, M_\rho)$ taken in the category of good profinite $\mathcal{O}/\lambda^n[G_\ell]$-modules. We define $H^1_{\mathrm{f}}(G_\ell, \mathrm{ad}^0\rho)$ (resp. $H^1_{\mathrm{ss}}(G_\ell, \mathrm{ad}^0\rho)$) as the intersection of $H^1_{\mathrm{f}}(G_\ell, \mathrm{ad}\rho)$ (resp. $H^1_{\mathrm{ss}}(G_\ell, \mathrm{ad}\rho)$) and $H^1(G_\ell, \mathrm{ad}^0\rho)$. Note that if ρ is good (resp. semistable) and $\xi : G_\ell \to \mathrm{ad}\rho$ is a cocycle, then E_ξ is good (resp. semistable) if and only if ρ_ξ is.

Suppose that $\rho : G_\ell \to GL_2(\mathcal{O}/\lambda^n)$ is ordinary. Consider the exact sequence

$$0 \to M_\rho^{(-1)} \to M_\rho \to M_\rho^{(0)} \to 0,$$

where $M_\rho^{(-1)}$ denote the maximal submodule of M_ρ where I_ℓ acts by ϵ. Let $\mathrm{ad}^{(-1)}\rho$ denote the submodule $\mathrm{Hom}(M_\rho^{(0)}, M_\rho^{(-1)})$ of $\mathrm{ad}^0\rho$. Then

$$H^1_{\mathrm{ss}}(G_\ell, \mathrm{ad}\rho) = \ker\left(H^1(G_\ell, \mathrm{ad}\rho) \to H^1(I_\ell, \mathrm{ad}\rho/\mathrm{ad}^{(-1)}\rho)\right).$$

The same is true with ad^0 replacing ad.

If $\rho : G_\ell \to GL_2(\mathcal{O})$ is semi-stable then we set

$$H^1_{\mathrm{f}}(G_\ell, \mathrm{ad}\rho \otimes K/\mathcal{O}) = \varinjlim H^1_{\mathrm{f}}(G_\ell, \mathrm{ad}\rho \otimes (\lambda^{-n}/\mathcal{O})) \subset H^1(G_\ell, \mathrm{ad}\rho \otimes K/\mathcal{O}).$$

We define $H^1_{\mathrm{f}}(G_\ell, \mathrm{ad}^0\rho \otimes K/\mathcal{O})$ as

$$H^1_{\mathrm{f}}(G_\ell, \mathrm{ad}\rho \otimes K/\mathcal{O}) \cap H^1(G_\ell, \mathrm{ad}^0\rho \otimes K/\mathcal{O}).$$

We make similar definitions for $H^1_{\mathrm{ss}}(G_\ell, \mathrm{ad}\rho \otimes K/\mathcal{O})$ and $H^1_{\mathrm{ss}}(G_\ell, \mathrm{ad}^0\rho \otimes K/\mathcal{O})$. We will need the following calculations.

Proposition 2.27 (a) *Suppose that $\bar\rho : G_\ell \to GL_2(k)$ is semi-stable. Then*

$$\#H^1_{\mathrm{f}}(G_\ell, \mathrm{ad}^0\bar\rho) \le \#H^0(G_\ell, \mathrm{ad}^0\bar\rho)\#k$$

and equality holds if $\bar\rho$ is ordinary[3].

(b) *Suppose that $\rho : G_\ell \to GL_2(\mathcal{O})$ is both good and ordinary. Let χ_1 and χ_2 be the unramified characters such that $\rho \sim \begin{pmatrix} \chi_1\epsilon & * \\ 0 & \chi_2 \end{pmatrix}$. Let*

$$c_\ell = (\chi_1/\chi_2)(\mathrm{Frob}\,\ell) - 1.$$

[3]The authors expect equality to hold without this hypothesis; cf. remark 2.26.

If $c_\ell \neq 0$, then

$$H^1_{\mathrm{ss}}(G_\ell, \mathrm{ad}^0\rho \otimes K/\mathcal{O})/H^1_{\mathrm{f}}(G_\ell, \mathrm{ad}^0\rho \otimes K/\mathcal{O})$$

is finite of order $\#(\mathcal{O}/c_\ell\mathcal{O})$.

(c) *If $\rho \bmod \lambda$ is either not good or not ordinary, then*

$$H^1_{\mathrm{ss}}(G_\ell, \mathrm{ad}^0\rho \otimes K/\mathcal{O}) = H^1_{\mathrm{f}}(G_\ell, \mathrm{ad}^0\rho \otimes K/\mathcal{O}).$$

Proof: Part (c) is clear, and for parts (a) and (b) it suffices to prove the following two results.

Proposition 2.28 *Suppose that $\rho : G_\ell \to GL_2(\mathcal{O}/\lambda^n)$ is good. Then*

$$\#H^1_{\mathrm{f}}(G_\ell, \mathrm{ad}\rho) = \#H^0(G_\ell, \mathrm{ad}^0\rho)(\#\mathcal{O}/\lambda^n)^2.$$

Lemma 2.29 *Suppose that $\rho : G_\ell \to GL_2(\mathcal{O}/\lambda^n)$ is ordinary, so that we have $\rho \sim \begin{pmatrix} \chi_1\epsilon & * \\ 0 & \chi_2 \end{pmatrix}$, for some unramified characters χ_1 and χ_2.*

(a) *We have*

$$\#H^1_{\mathrm{ss}}(G_\ell, \mathrm{ad}^0\rho) \leq \#H^0(G_\ell, \mathrm{ad}^0\rho)\#(\mathcal{O}/\lambda^n)\#(\mathcal{O}/(\lambda^n, c_\ell))$$

where $c_\ell = (\chi_1/\chi_2)(\mathrm{Frob}_\ell) - 1$. Moreover if ρ is good, then equality holds.

(b) *If $n = 1$ and ρ is not good then $\#H^1_{\mathrm{ss}}(G_\ell, \mathrm{ad}^0\rho) = \#k$.*

End of proof of proposition 2.27: To deduce proposition 2.27 (in the good case) from proposition 2.28 and lemma 2.29, note that if $\rho : G_\ell \to GL_2(\mathcal{O}/\lambda^n)$ is good, then

$$H^1(G_\ell/I_\ell, \mathcal{O}/\lambda^n) \subset H^1_{\mathrm{f}}(G_\ell, \mathrm{ad}\rho) \cap H^1(G_\ell, \mathcal{O}/\lambda^n)$$

by lemma 2.22 (d), and this gives the inequality in proposition 2.27(a). Furthermore if ρ is also ordinary, then the above group is contained in

$$\mathrm{Im}\,(H^1_{\mathrm{f}}(G_\ell, \mathrm{ad}\rho) \to H^1(G_\ell, \mathcal{O}/\lambda^n)) \subset H^1(G_\ell/I_\ell, \mathcal{O}/\lambda^n)$$

where the last inclusion comes from lemma 2.25 (b). Therefore

$$\#H^1_{\mathrm{f}}(G_\ell, \mathrm{ad}^0\rho) = \#H^0(G_\ell, \mathrm{ad}^0\rho)\#(\mathcal{O}/\lambda^n).$$

□

We will prove proposition 2.28 in the next section using the theory of Fontaine and Laffaille. The proof of lemma 2.29 is a somewhat technical exercise in the Galois cohomology of local fields, for which we refer the reader to ch. 1 of [W3], prop. 1.9, parts (iii) and (iv). We remark that for our purposes, inequalities would suffice in part (b) of proposition 2.27 and part (a) of lemma 2.29 (and this is all that is proved in [W3]). We have included the precise formulas for the sake of completeness, since they are not much more difficult to obtain. The only additional observation required is that if ρ is good, then the composite of the natural maps

$$H^1(G_\ell/I_\ell, (\mathrm{ad}^0\rho/\mathrm{ad}^{(-1)}\rho)^{I_\ell}) \to H^1(G_\ell, \mathrm{ad}^0\rho/\mathrm{ad}^{(-1)}\rho) \to H^2(G_\ell, \mathrm{ad}^{(-1)}\rho)$$

is trivial. To prove this, rewrite the composite as

$$H^1(G_\ell/I_\ell, \mathcal{O}/\lambda^n) \to H^1(G_\ell, \mathcal{O}/\lambda^n) \to H^2(G_\ell, \mathrm{ad}^{(-1)}\rho)$$

with the second map given by $\cup c_\rho$ where c_ρ is the class in $H^1(G_\ell, \mathrm{ad}^{(-1)}\rho)$ defining the extension M_ρ and apply lemma 2.25 (c). (In fact, one only needs the easier half of lemma 2.25 (c): if ρ is good then $v(c_\rho) = 0$.)

2.5 The theory of Fontaine and Laffaille

In this section we again assume that ℓ is an odd prime. We mentioned in the last section the importance of understanding good representations of G_ℓ. However the definition of good is somewhat indirect, and this makes computations difficult. The key result we use to address the problem is an equivalence between the category $\mathcal{FF}_\mathcal{O}$ of good finite $\mathcal{O}[G_\ell]$-modules and a category $\mathcal{MF}_\mathcal{O}$ which we define below following Fontaine and Laffaille [FL]. The beauty and utility of the result stems from the elementary algebraic nature of the definition of $\mathcal{MF}_\mathcal{O}$; we can convert questions about good representations into questions in linear algebra.

Remark 2.30 Let $\mathcal{GF}_\mathcal{O}$ denote the category of finite flat commutative group schemes over $\mathbb{Z}_\ell$ with an action of $\mathcal{O}$. By results of Raynaud [Ray], taking $\bar{\mathbb{Q}}_\ell$ points defines an equivalence between the categories $\mathcal{GF}_\mathcal{O}$ and $\mathcal{FF}_\mathcal{O}$. The equivalence between $\mathcal{GF}_\mathcal{O}$ and a category closely related to $\mathcal{MF}_\mathcal{O}$ was first established by Fontaine [Fo2] and [Fo1]. While Fontaine's results would suffice for our purposes here, our formulation will be closer to that in [FL], where an equivalence between $\mathcal{MF}_\mathcal{O}$ and $\mathcal{FF}_\mathcal{O}$ is defined as part of a more general construction of representations of G_ℓ from linear-algebraic data. We caution however that our formulation is not exactly the same as that of [FL] since we wish to work with covariant functors.

We now turn to the definition of $\mathcal{MF}_\mathcal{O}$. The objects are $\mathcal{O}$-modules D of finite cardinality together with a distinguished submodule D^0 and $\mathcal{O}$-linear maps $\phi_{-1} : D \to D$ and $\phi_0 : D^0 \to D$ which satisfy:

- $\phi_{-1}|_{D^0} = \ell\phi_0$,
- $\mathrm{Im}\,\phi_{-1} + \mathrm{Im}\,\phi_0 = D$.

The morphisms are $\mathcal{O}$-linear maps compatible with the additional data of the distinguished submodules and maps ϕ.

It is useful to note that if D is an object of $\mathcal{MF}_{\mathcal{O}}$, then there is a surjection

$$\phi_{-1} \oplus \phi_0 : D/D^0 \oplus D^0/\ell D^0 \twoheadrightarrow D/\ell\phi_0(D^0),$$

and on counting orders we see that this is in fact an isomorphism. Thus there is an isomorphism

$$D/(D^0 + \lambda D) \oplus D^0/\lambda D^0 \xrightarrow{\sim} D/\lambda D.$$

It follows that $D^0/\lambda D^0 \to D/\lambda D$ is injective, and hence that D^0 is (non-canonically) a direct summand of D as an $\mathcal{O}$-module. Note also that ϕ_0 is injective, and if $D = D^0 \oplus D'$ as $\mathcal{O}$-modules, then also $D = \phi_0(D^0) \oplus \phi_{-1}(D')$ as $\mathcal{O}$-modules.

It is then straightforward to check that there is a contravariant functor * from $\mathcal{MF}_{\mathcal{O}}$ to itself defined by:

- $D^* = \mathrm{Hom}(D, \mathbb{Q}_\ell/\mathbb{Z}_\ell)$;
- $(D^*)^0 = \mathrm{Hom}(D/D^0, \mathbb{Q}_\ell/\mathbb{Z}_\ell)$;
- $\phi^*_{-1}(f)(z) = f(\ell x + y)$, where $z = \phi_{-1}(x) + \phi_0(y)$;
- $\phi^*_0(f)(z) = f(x \bmod D^0)$, where $z \equiv \phi_{-1}(x) \bmod (\phi_0 D^0)$.

Moreover the canonical isomorphism $D \cong (D^*)^*$ of $\mathcal{O}$-modules defines a natural isomorphism in $\mathcal{MF}_{\mathcal{O}}$.

We leave it as an exercise for the reader to use the above observations to define cokernels and then kernels of morphisms in $\mathcal{MF}_{\mathcal{O}}$ and verify that it is an abelian category (or see [FL], sec. 1).

Theorem 2.31 *There is a covariant functor $\mathbb{D} : \mathcal{FF}_{\mathcal{O}} \to \mathcal{MF}_{\mathcal{O}}$ which defines an $\mathcal{O}$-additive equivalence of categories. Moreover if M is an object of $\mathcal{FF}_{\mathcal{O}}$, then we have*

(a) *M and $\mathbb{D}(M)$ have the same cardinality;*

(b) *$\mathbb{D}(M) = \mathbb{D}(M)^0$ if and only if M is unramified.*

Remark 2.32 (a) It follows on applying part (a) to $M/\lambda^i M$ for each i that M and $\mathbb{D}(M)$ are in fact (non-canonically) isomorphic as $\mathcal{O}$-modules.

(b) As mentioned above, our formulation differs from that of Fontaine and Laffaille in that we are using covariant functors. To deduce theorem 2.31 from the results in [FL], sec. 9, we define $\mathbb{D}$ as a quasi-inverse of the functor $\underline{U}_S(\cdot^*[1])$, where $\underline{U}_S$ is defined in [FL] and [1] indicates a shift by 1 in filtration degrees. In particular, if $\mathcal{F}$ is a finite flat group scheme over $\mathbb{Z}_\ell$ with an action of $\mathcal{O}$, then the underlying $\mathcal{O}$-module of $\mathbb{D}(\mathcal{F}(\bar{\mathbb{Q}}_\ell))$ can be identified with the covariant Dieudonné module of $\mathcal{F}_{/\mathbf{F}_\ell}$, and ϕ_{-1} with F.

Suppose now that $\rho : G_\ell \to GL_2(\mathcal{O}/\lambda^n)$ is a good continuous representation. Let $D_\rho = \mathbb{D}(M_\rho)$ be the corresponding object of $\mathcal{MF}_\mathcal{O}$. Then $D_\rho \cong (\mathcal{O}/\lambda^n)^2$ as an $\mathcal{O}$-module while $D_\rho^0 \cong \mathcal{O}/\lambda^n$.

Lemma 2.33 *The following are isomorphic.*

(a) *The group of extensions of M_ρ by itself in the category of good finite $\mathcal{O}/\lambda^n[G_\ell]$-modules.*

(b) *The group of extensions of D_ρ by itself in the full subcategory of $\mathcal{MF}_\mathcal{O}$ consisting of objects which are $\mathcal{O}/\lambda^n$-modules.*

(c) *Pairs (α_{-1}, α_0) where $\alpha_{-1} \in \mathrm{Hom}_\mathcal{O}(D_\rho, D_\rho)$, $\alpha_0 \in \mathrm{Hom}_\mathcal{O}(D_\rho^0, D_\rho)$ and $\ell\alpha_0 = \alpha_{-1}|_{D_\rho^0}$, modulo pairs of the form*

$$(a\phi_{\rho,-1} - \phi_{\rho,-1}a, a\phi_{\rho,0} - \phi_{\rho,0}a|_{D^0})$$

where $a \in \mathrm{Hom}_\mathcal{O}(D_\rho, D_\rho)$ and $aD_\rho^0 \subset D_\rho^0$.

Proof: The first two groups of extensions are isomorphic by the Fontaine-Laffaille theorem. Following Ramakrishna [Ram] we explain how to calculate the second group of extensions. We will write D for D_ρ. If

$$(0) \to D \to E \to D \to (0)$$

is an extension then we have as $\mathcal{O}$-modules that $E \cong D^2$ by an isomorphism such that $E^0 \xrightarrow{\sim} (D^0)^2$. Then E is determined by giving elements $\alpha_{E,-1} \in \mathrm{Hom}(D, D)$ and $\alpha_{E,0} \in \mathrm{Hom}(D^0, D)$ with $\ell\alpha_{E,0} = \alpha_{E,-1}|_{D^0}$. Explicitly

$$\phi_{E,-1} = \begin{pmatrix} \phi_{-1} & \alpha_{E,-1} \\ 0 & \phi_{-1} \end{pmatrix} \quad \text{and} \quad \phi_{E,0} = \begin{pmatrix} \phi_0 & \alpha_{E,0} \\ 0 & \phi_0 \end{pmatrix}.$$

Two such extensions E and E' are isomorphic if there is an element $a \in \mathrm{End}\,(D)$ such that $aD^0 \subset D^0$,

$$\begin{pmatrix} 1 & a \\ 0 & 1 \end{pmatrix}\begin{pmatrix} \phi_{-1} & \alpha_{E,-1} \\ 0 & \phi_{-1} \end{pmatrix} = \begin{pmatrix} \phi_{-1} & \alpha_{E',-1} \\ 0 & \phi_{-1} \end{pmatrix}\begin{pmatrix} 1 & a \\ 0 & 1 \end{pmatrix},$$

and

$$\begin{pmatrix} 1 & a \\ 0 & 1 \end{pmatrix}\begin{pmatrix} \phi_0 & \alpha_{E,0} \\ 0 & \phi_0 \end{pmatrix} = \begin{pmatrix} \phi_0 & \alpha_{E',0} \\ 0 & \phi_0 \end{pmatrix}\begin{pmatrix} 1 & a|_{D^0} \\ 0 & 1 \end{pmatrix};$$

or equivalently if $aD^0 \subset D^0$, $\alpha_{E,-1} - \alpha_{E',-1} = \phi_{-1}a - a\phi_{-1}$ and $\alpha_{E,0} - \alpha_{E',0} = \phi_0 a|_{D^0} - a\phi_0$. The lemma follows. □

Corollary 2.34 *The group, $H^1_\mathrm{f}(G_\ell, \mathrm{ad}\rho)$, of extensions of M_ρ by itself in the category of good, finite $\mathcal{O}/\lambda^n[G_\ell]$-modules is (non-canonically) isomorphic to $(\mathcal{O}/\lambda^n)^2 \oplus H^0(I_\ell, M_\rho)$.*

Proof: Choose generators e_0, e_{-1} of D, such that e_0 is a generator of D^0. With respect to this basis let ϕ_0 have matrix $\binom{x}{z}$ and ϕ_{-1} have matrix $\begin{pmatrix} \ell x & y \\ \ell z & w \end{pmatrix}$. If a has matrix $\begin{pmatrix} a_1 & a_2 \\ 0 & a_3 \end{pmatrix}$ then a direct calculation shows that

$$(a\phi_{-1} - \phi_{-1}a, a\phi_0 - \phi_0 a|_{D^0}) = \left(\begin{pmatrix} \ell a_2 z & (a_1 - a_3)y + a_2(w - \ell x) \\ (a_3 - a_1)\ell z & -\ell z a_2 \end{pmatrix}, \begin{pmatrix} a_2 z \\ (a_3 - a_1)z \end{pmatrix} \right).$$

Thus the Ext group we want is the quotient of the set of pairs of matrices (α_{-1}, α_0) as above by the submodule generated by

$$\left(\begin{pmatrix} \ell z & w - \ell x \\ 0 & -\ell z \end{pmatrix}, \begin{pmatrix} z \\ 0 \end{pmatrix} \right) \quad \text{and} \quad \left(\begin{pmatrix} 0 & -y \\ \ell z & 0 \end{pmatrix}, \begin{pmatrix} 0 \\ z \end{pmatrix} \right).$$

Note that either z or w is a unit in $\mathcal{O}/\lambda$ from which it follows that the Ext group we want is isomorphic to $(\mathcal{O}/\lambda^n)^2 \oplus \mathcal{O}/(z, \lambda^n)$. On the other hand $H^0(I_\ell, M_\rho)$ corresponds to the largest submodule $C \subset D^0$ such that $\phi_0 C = C$, i.e. $C = \{d \in D^0 : zd = 0\} \cong \mathcal{O}/(z, \lambda^n)$. □

Corollary 2.35 *Suppose that $\rho : G_\ell \to GL_2(\mathcal{O}/\lambda^n)$ is a continuous good representation. Then $H^1_{\mathrm{f}}(G_\ell, \mathrm{ad}\rho)$ is isomorphic to $(\mathcal{O}/\lambda^n)^2 \oplus H^0(G_\ell, \mathrm{ad}^0\rho)$.*

Proof: If ρ is not ordinary then $H^0(G_\ell, \mathrm{ad}^0\rho)$ and $H^0(I_\ell, M_\rho)$ are both trivial, so suppose that ρ is ordinary. In this case, let ρ' denote the ordinary representation defined by

$$M_{\rho'} = \mathrm{Hom}(M_\rho^{(0)}, M_\rho) \subset \mathrm{ad}^0(\rho),$$

and let $M_{\rho'}^{(-1)} = \mathrm{ad}^0(\rho)^{(-1)}$. Then

$$H^0(G_\ell, \mathrm{ad}^0\rho) = H^0(G_\ell, M_{\rho'}) \subset H^0(I_\ell, M_{\rho'}) \cong H^0(I_\ell, M_\rho).$$

Since $H^0(I_\ell, M_{\rho'}^{(-1)})$ is trivial, the restriction homomorphism

$$H^1(G_\ell, M_{\rho'}^{(-1)}) \to H^1(I_\ell, M_{\rho'}^{(-1)})$$

is injective. We deduce from the long exact sequences associated to

$$(0) \to M_{\rho'}^{(-1)} \to M_{\rho'} \to \mathcal{O}/\lambda^n \to 0$$

that $H^0(G_\ell, M_{\rho'}) = H^0(I_\ell, M_{\rho'})$. □

2.6 Deformations of representations

In this section we shall review Mazur's theory of deformations of representations of profinite groups (see [Maz3]).

Let $\mathcal{C}_{\mathcal{O}}$ denote the category whose objects are complete noetherian local $\mathcal{O}$-algebras with residue field k and whose morphisms are $\mathcal{O}$-algebra homomorphisms which are local (i.e. take maximal ideals into maximal ideals). (The structure maps from $\mathcal{O}$ to every object of $\mathcal{C}_{\mathcal{O}}$ are also assumed to be local.) Let G denote a topologically finitely generated profinite group and let $\bar{\rho}$ denote an absolutely irreducible representation of G into $GL_d(k)$. (In fact all we shall use in the sequel is that k is the centraliser in $M_d(k)$ of the image of $\bar{\rho}$.) Let $\mathcal{D}_0$ denote the category of profinite $\mathcal{O}[G]$-modules with continuous morphisms. We will let $\mathcal{D}$ denote a full subcategory of $\mathcal{D}_0$ which is closed under taking sub-objects, quotients and direct products and which contains $M_{\bar{\rho}}$. Note that if M is an object of $\mathcal{D}_0$ and M_i is a collection of subobjects which have trivial intersection and such that each M/M_i is an object of $\mathcal{D}$, then M is an object of $\mathcal{D}$, since $M \subset \prod_i M/M_i$.

Let $\chi : G \to \mathcal{O}^\times$ be a continuous character such that $\det \bar{\rho} = \chi \bmod \lambda$. By a lifting of $\bar{\rho}$ of type $D = (\mathcal{O}, \chi, \mathcal{D})$ we shall mean an object R of $\mathcal{C}_{\mathcal{O}}$ and a continuous representation $\rho : G \to GL_d(R)$ such that:

(a) $\rho \bmod \mathfrak{m}_R = \bar{\rho}$,

(b) $\det \rho = \chi$,

(c) M_ρ is an object of $\mathcal{D}$.

Note that if $\phi : R \to R'$ and $\rho : G \to GL_d(R)$ is of type D so is $\phi \circ \rho$.

Theorem 2.36 *There is a lifting $\rho_D^{\mathrm{univ}} : G \to GL_d(R_D)$ of $\bar{\rho}$ of type D such that if $\rho : G \to GL_d(R)$ is any lifting of $\bar{\rho}$ of type D then there is a unique homomorphism of $\mathcal{O}$-algebras $\phi : R_D \to R$ such that ρ is conjugate to $\phi \circ \rho_D^{\mathrm{univ}}$.*

The representation ρ_D^{univ} is referred to as the *universal deformation* of type D. Mazur [Maz3] proved this theorem for $\mathcal{D}_0$ and certain other categories $\mathcal{D}$. Ramakrishna [Ram] observed that the arguments work with any category $\mathcal{D}$ satisfying the above hypotheses. We will sketch a proof which was suggested by Faltings. (We remark that another explicit construction of the deformation ring has been given by de Smit and Lenstra in [dSL].)

Proof of theorem 2.36: Choose a sequence $g_1, ..., g_r$ of topological generators of G and liftings $A_1, ..., A_r$ of $\bar{\rho}(g_1), ..., \bar{\rho}(g_r)$ to $M_d(\mathcal{O})$. Define a mapping $\iota : M_d(\mathcal{O}) \to M_d(\mathcal{O})^r$ which sends x to $(xA_1 - A_1x, ..., xA_r - A_rx)$. Since ι has torsion-free cokernel, so we can decompose

$$M_d(\mathcal{O})^r = \iota(M_d(\mathcal{O})) \oplus V,$$

for some submodule $V \subset M_d(\mathcal{O})^r$. If $\rho : G \to GL_d(R)$ is a lifting of $\bar{\rho}$ of type D set $v_\rho = (\rho(g_1) - A_1, ..., \rho(g_r) - A_r) \in M_d(R)^r$. Note that $v_\rho \equiv 0 \bmod \mathfrak{m}_R$ and that v_ρ completely determines ρ. We call the lifting ρ well-placed if v_ρ belongs to $V \otimes_{\mathcal{O}} R \subset M_d(R)^r$. The crucial observation is the following result.

Lemma 2.37 *If $\rho : G \to GL_d(R)$ is a lifting of $\bar{\rho}$ of type D then there is a unique conjugate ρ' of ρ which is well-placed.*

The lemma is first proved for algebras R such that $\mathfrak{m}_R^n = (0)$ by induction on n, and then one deduces the general case.

In virtue of the lemma it suffices to find a universal well-placed lifting of type D. Let $e_1, ..., e_s$ be a basis of V as an $\mathcal{O}$-module. If ρ is a well-placed lifting of $\bar{\rho}$ of type D then we can write $v_\rho = \sum_{i=1}^s v_{\rho,i} e_i$ for unique elements $v_{\rho,i} \in \mathfrak{m}_R$ and we can define a homomorphism

$$\theta_\rho : \mathcal{O}[[T_1, ..., T_s]] \to R$$

sending T_i to $v_{\rho,i}$. Note that ρ is completely determined by θ_ρ ($\rho(g_i) = A_i + \sum_{j=1}^s \theta_\rho(T_j) e_{ji}$, where $e_j = (e_{j1}, ..., e_{jr})$). Let I denote the intersection of all ideals J of $\mathcal{O}[[T_1, ..., T_s]]$ such that there is a representation $\rho_J : G \to GL_d(\mathcal{O}[[T_1, ..., T_s]]/J)$ of type D with $\rho_J(g_i) = A_i + \sum_{j=1}^s T_j e_{ji}$ for all i. Let R_D denote the quotient of $\mathcal{O}[[T_1, ..., T_s]]$ by I. Then one can check that there is a representation $\rho^{\text{univ}} : G \to GL_d(R_D)$ with $\rho^{\text{univ}}(g_i) = A_i + \sum_{j=1}^s T_j e_{ji}$ for all i, and that this is the desired universal representation. □

We will need a few elementary properties of these universal deformations. More precisely we will need to know how these universal rings change when we change the base field, we will need to know how to calculate the equi-characteristic tangent space of these rings and more generally how to calculate $\wp/\wp^2$ for certain prime ideals $\wp$. The first of these lemmas is a remark of Faltings, the second is due to Mazur [Maz3] and the third to Wiles [W3].

Lemma 2.38 *Let K'/K be a finite extension with ring of integers $\mathcal{O}'$ and residue field k'. Let $\mathcal{D}'$ denote the full sub-category of the category of profinite $\mathcal{O}'[G]$-modules such that the underlying object of $\mathcal{D}_0$ is actually an object of $\mathcal{D}$. Let $D' = (\mathcal{O}', \chi, \mathcal{D}')$. Then $R_{D'} = R_D \otimes_{\mathcal{O}} \mathcal{O}'$ and $\rho_{D'}^{\text{univ}} = \rho_D^{\text{univ}} \otimes 1$.*

Proof: Let $\tilde{R}_{D'}$ denote the subring of $R_{D'}$ consisting of elements which reduce modulo the maximal ideal to an element of $k \subset k'$. Similarly let $\tilde{R}_D$ denote the subring of $R_D \otimes_{\mathcal{O}} \mathcal{O}'$ consisting of elements which reduce modulo the maximal ideal to an element of $k \subset k'$. Then $\rho_{D'}^{\text{univ}}$ is in fact valued in $GL_d(\tilde{R}_{D'})$ and $\rho_D^{\text{univ}} \otimes 1$ is in fact valued in $GL_d(\tilde{R}_D)$. The universal properties give natural maps $\alpha : R_D \to \tilde{R}_{D'}$ and $\beta : R_{D'} \to R_D \otimes_{\mathcal{O}} \mathcal{O}'$. Moreover they show that the composite $(\alpha \otimes 1) \circ \beta : R_{D'} \to R_{D'}$ is the identity and that $\beta \circ \alpha : R_D \to \tilde{R}_D$ is the natural embedding. Thus β is an isomorphism. □

We will let $\mathcal{D}^{(n)}$ denote the full subcategory of $\mathcal{D}$ whose objects are killed by λ^n. Suppose that M is an object of $\mathcal{D}_0$ which is finite and free over $\mathcal{O}/\lambda^n$. Recall from section 2.4 that $H^1(G, \text{End}\,(M))$ may be identified with $\text{Ext}^1_{\mathcal{D}_0^{(n)}}(M, M)$. If M is an object of $\mathcal{D}^{(n)}$ which is finite and free over $\mathcal{O}/\lambda^n$, then we have a natural inclusion

$$\text{Ext}^1_{\mathcal{D}^{(n)}}(M, M) \subset \text{Ext}^1_{\mathcal{D}_0^{(n)}}(M, M) \cong H^1(G, \text{End}\,(M)).$$

We define $H^1_{\mathcal{D}}(G, \mathrm{End}\,(M))$ to be the image of $\mathrm{Ext}^1_{\mathcal{D}^{(n)}}(M, M)$ in the group $H^1(G, \mathrm{End}\,(M))$, and $H^1_{\mathcal{D}}(G, \mathrm{End}^0(M))$ to be the intersection

$$H^1(G, \mathrm{End}^0(M)) \cap H^1_{\mathcal{D}}(G, \mathrm{End}\,(M)).$$

Lemma 2.39 *There is a canonical isomorphism of k-vector spaces*

$$\mathrm{Hom}_k(\mathfrak{m}_{R_D}/(\lambda, \mathfrak{m}^2_{R_D}), k) \cong H^1_{\mathcal{D}}(G, \mathrm{ad}^0 \bar{\rho}).$$

Proof: There is a natural bijection between $\mathrm{Hom}_k(\mathfrak{m}_{R_D}/(\lambda, \mathfrak{m}^2_{R_D}), k)$ and the set of $\mathcal{O}$-algebra homomorphisms from R_D to the algebra $k[\varepsilon]$ where $\varepsilon^2 = 0$ (the correspondence associates $\phi : R_D \to k[\varepsilon]$ to $\phi|_{\mathfrak{m}_{R_D}}$). Hence there is a bijection to the set of liftings $\rho : G \to GL_d(k[\varepsilon])$ of $\bar{\rho}$ of type D, modulo conjugation by elements of $1 + \varepsilon M_d(k)$. On the other hand, recall from section 2.4 that there is a natural bijection between $\mathrm{Ext}^1_{\mathcal{D}^{(1)}_0}(M_{\bar{\rho}}, M_{\bar{\rho}})$ and the set of of all continuous liftings $\rho : G \to GL_d(k[\varepsilon])$ of $\bar{\rho}$ modulo conjugation by elements of $1 + \varepsilon M_d(k)$. Moreover a lifting ρ is type D if and only if $\det \rho = \det \bar{\rho}$ and the corresponding extension M_ρ is an object of $\mathcal{D}$. The lemma follows on checking linearity. □

Now suppose that $\theta : R_D \to \mathcal{O}$ is an $\mathcal{O}$-algebra homomorphism. Let $\wp$ denote the kernel of θ and let $\rho = \theta \circ \rho^{\mathrm{univ}}_D$. We set

$$H^1_{\mathcal{D}}(G, \mathrm{ad}^0\rho \otimes K/\mathcal{O}) = \varinjlim H^1_{\mathcal{D}}(G, \mathrm{ad}^0\rho \otimes \lambda^{-n}/\mathcal{O}) \subset H^1(G, \mathrm{ad}^0\rho \otimes K/\mathcal{O}).$$

Lemma 2.40 *There is a canonical $\mathcal{O}$-linear isomorphism*

$$\mathrm{Hom}_{\mathcal{O}}(\wp/\wp^2, K/\mathcal{O}) \cong H^1_{\mathcal{D}}(G, \mathrm{ad}^0\rho \otimes K/\mathcal{O}).$$

Proof: One shows in a very similar manner to the proof of lemma 2.39 that for all n there is a natural isomorphism.

$$\mathrm{Hom}_{\mathcal{O}}(\wp/\wp^2, \mathcal{O}/\lambda^n) \cong H^1_{\mathcal{D}}(G, \mathrm{ad}^0\rho \otimes \mathcal{O}/\lambda^n).$$

□

2.7 Deformations of Galois representations

Again in this section we assume that ℓ is an odd prime. Let $\bar{\rho} : G_{\mathbb{Q}} \to GL_2(k)$ denote a continuous absolutely irreducible representation. Suppose moreover that $\det \bar{\rho} = \epsilon$ and that $\bar{\rho}$ is semi-stable in the sense that

- $\bar{\rho}|_{G_\ell}$ is semi-stable,
- and if $p \neq \ell$ then $\#\bar{\rho}(I_p)|\ell$.

Note that if $E/\mathbb{Q}$ is a semistable elliptic curve then $\bar{\rho}_{E,\ell} : G_{\mathbb{Q}} \to GL_2(\mathbb{F}_\ell)$, satisfies these conditions if it is irreducible. By a theorem of Mazur (theorem 2.9) this will be the case if $\ell > 7$.

Let Σ denote a finite set of prime numbers. If R is an object of $\mathcal{C}_{\mathcal{O}}$ then we say that a continuous lifting $\rho : G_{\mathbb{Q}} \to GL_2(R)$ of $\bar{\rho}$ is of type Σ if the following hold.

- $\det \rho = \epsilon$.
- $\rho|_{G_\ell}$ is semi-stable.
- If $\ell \notin \Sigma$ and $\bar{\rho}|_{G_\ell}$ is good then $\rho|_{G_\ell}$ is good.
- If $p \notin \Sigma \cup \{\ell\}$ and $\bar{\rho}$ is unramified at p then ρ is unramified at p.
- If $p \notin \Sigma \cup \{\ell\}$ then $\rho|_{I_p} \sim \begin{pmatrix} 1 & * \\ 0 & 1 \end{pmatrix}$.

Roughly speaking we require that at primes $p \notin \Sigma$, ρ is as unramified as could be hoped for and we require that $\rho|_{G_\ell}$ is semi-stable. Note that if $\Sigma \subset \Sigma'$ and ρ is a lifting of type Σ then it is also a lifting of type Σ'. Note also that if $E/\mathbb{Q}$ is an elliptic curve which is semi-stable at ℓ and for which $\bar{\rho}_{E,\ell}$ is irreducible and semi-stable, then $\rho_{E,\ell} : G_\mathbb{Q} \to GL_2(\mathbb{Z}_\ell)$ is a lifting of type Σ if Σ contains all the primes for which E has bad reduction.

Now suppose that $\rho : G_\mathbb{Q} \to GL_2(\mathcal{O}/\lambda^n)$ is a lifting of $\bar{\rho}$ of type Σ. We will write $H^1_\Sigma(\mathbb{Q}, \mathrm{ad}^0\rho)$ for $H^1_{\mathcal{L}_\Sigma}(\mathbb{Q}, \mathrm{ad}^0\rho)$, where

- $L_{\Sigma,p} = H^1(G_p/I_p, (\mathrm{ad}^0\rho)^{I_p})$ if $p \notin \Sigma \cup \{\ell\}$;
- $L_{\Sigma,p} = H^1(G_p, (\mathrm{ad}^0\rho))$ if $p \in \Sigma$ and $p \neq \ell$;
- $L_{\Sigma,\ell} = H^1_{\mathrm{f}}(G_\ell, (\mathrm{ad}^0\rho))$ if $\ell \notin \Sigma$;
- $L_{\Sigma,\ell} = H^1_{\mathrm{ss}}(G_\ell, (\mathrm{ad}^0\rho))$ if $\ell \in \Sigma$.

Note that the pairing $\mathrm{ad}^0\rho \times \mathrm{ad}^0\rho \to \mathcal{O}/\lambda^n$ given by $(a, b) \mapsto \mathrm{tr}\,(ab)$ is perfect and respects the action of $G_\mathbb{Q}$ (i.e. $\mathrm{tr}\,((\mathrm{ad}^0\rho)(g)a, (\mathrm{ad}^0\rho)(g)b) = \mathrm{tr}\,(ab)$ for all $g \in G_\mathbb{Q}$). Choosing a generator for the $\mathcal{O}$-module $\mathrm{Hom}(\mathcal{O}/\lambda^n, \mathbb{Q}_\ell/\mathbb{Z}_\ell)$, we obtain an isomorphism of $\mathcal{O}[G_\mathbb{Q}]$-modules

$$\mathrm{ad}^0\rho(1) \cong \mathrm{Hom}_\mathcal{O}(\mathrm{ad}^0\rho, \mathcal{O}/\lambda^n)(1) \xrightarrow{\sim} \mathrm{Hom}(\mathrm{ad}^0\rho, \mathbb{Q}_\ell/\mathbb{Z}_\ell)(1) = (\mathrm{ad}^0\rho)^*.$$

The $\mathcal{O}$-submodule of $H^1(\mathbb{Q}, \mathrm{ad}^0\rho(1))$ corresponding to $H^1_{\mathcal{L}^*_\Sigma}(\mathbb{Q}, (\mathrm{ad}^0\rho)^*)$ is independent of the choice of generator and we denote it $H^1_{\Sigma^*}(\mathbb{Q}, \mathrm{ad}^0\rho(1))$. Thus $H^1_{\Sigma^*}(\mathbb{Q}, \mathrm{ad}^0\rho(1))$ is defined by the local conditions $\{L^\perp_{\Sigma,v}\}$ where the orthogonality is with respect to the pairing

$$H^1(G_v, \mathrm{ad}^0\rho) \times H^1(G_v, \mathrm{ad}^0\rho(1)) \to \mathbb{Q}_\ell/\mathbb{Z}_\ell$$

arising from the above isomorphism. We may equivalently regard the orthogonality as being with respect to the natural perfect $\mathcal{O}$-bilinear pairing defined by the composition

$$\begin{aligned} H^1(G_v, \mathrm{ad}^0\rho) \times H^1(G_v, \mathrm{ad}^0\rho(1)) &\to H^2(G_v, \mathrm{ad}^0\rho \otimes_\mathcal{O} \mathrm{ad}^0\rho(1)) \\ &\to H^2(G_v, \mathcal{O}/\lambda^n(1)) \cong \mathcal{O}/\lambda^n. \end{aligned}$$

Note that if $p \neq \ell$ we have that

- $L^{\perp}_{\Sigma,p} = H^1(G_p/I_p, (\mathrm{ad}^0\rho)(1)^{I_p})$ if $p \notin \Sigma$;
- $L^{\perp}_{\Sigma,p} = (0)$ if $p \in \Sigma$.

If $\rho : G_{\mathbb{Q}} \to GL_2(\mathcal{O})$ is a lifting of $\bar{\rho}$ of type Σ, we will write $H^1_{\Sigma}(\mathbb{Q}, \mathrm{ad}^0\rho \otimes K/\mathcal{O})$ for the direct limit $\varinjlim H^1_{\Sigma}(\mathbb{Q}, \mathrm{ad}^0\rho \otimes \lambda^{-n}/\mathcal{O})$, and $H^1_{\Sigma}(\mathbb{Q}, \mathrm{ad}^0\rho(1) \otimes K/\mathcal{O})$ for the direct limit $\varinjlim H^1_{\Sigma}(\mathbb{Q}, \mathrm{ad}^0\rho(1) \otimes \lambda^{-n}/\mathcal{O})$.

The main result we need about deformations of $\bar{\rho}$ is the following.

Theorem 2.41 *There is a universal lifting $\rho^{\mathrm{univ}}_{\Sigma} : G_{\mathbb{Q}} \to GL_2(R_{\Sigma})$ of $\bar{\rho}$ of type Σ, i.e. $\rho^{\mathrm{univ}}_{\Sigma}$ is a lifting of type Σ and if $\rho : G_{\mathbb{Q}} \to GL_2(R)$ is any lifting of type Σ then there is a unique $\mathcal{O}$-algebra homomorphism $\phi : R_{\Sigma} \to R$ such that $\rho \sim \phi \circ \rho^{\mathrm{univ}}_{\Sigma}$. Moreover we have the following.*

(a) *If K'/K is a finite extension and R'_{Σ} is the corresponding deformation ring then $R'_{\Sigma} = R_{\Sigma} \otimes_{\mathcal{O}} \mathcal{O}'$ and $(\rho^{\mathrm{univ}}_{\Sigma})' = \rho^{\mathrm{univ}}_{\Sigma} \otimes 1$.*

(b) *R_{Σ} can be topologically generated as an $\mathcal{O}$-algebra by $\dim_k H^1_{\Sigma}(\mathbb{Q}, \mathrm{ad}^0\bar{\rho})$ elements.*

(c) *If $\phi : R_{\Sigma} \twoheadrightarrow \mathcal{O}$ is a $\mathcal{O}$-algebra homomorphism, if $\rho = \phi \circ \rho_{\Sigma}$ and if $\wp = \ker \phi$ then $\mathrm{Hom}(\wp/\wp^2, K/\mathcal{O}) \cong H^1_{\Sigma}(\mathbb{Q}, \mathrm{ad}^0\rho \otimes K/\mathcal{O})$.*

Proof: Let L_0 denote the fixed field of $\bar{\rho}$. Let L_n (for $n \in \mathbb{Z}_{>0}$) denote the maximal elementary abelian ℓ-extension of L_{n-1} unramified outside Σ, $\{\ell\}$ and the primes where $\bar{\rho}$ ramifies. Let $L_{\infty} = \bigcup_n L_n$ and let $G = \mathrm{Gal}(L_{\infty}/\mathbb{Q})$. Note that any lifting of $\bar{\rho}$ of type Σ factors through G. $\mathrm{Gal}(L_{\infty}/L_0)$ is a pro-ℓ-group and its maximal elementary abelian quotient, $\mathrm{Gal}(L_1/L_0)$, is finite by theorem 2.2. We deduce from the following lemma that $\mathrm{Gal}(L_{\infty}/L_0)$ and hence G are topologically finitely generated. (See for instance [Koc] Satz 4.10 for a proof of this lemma.)

Lemma 2.42 *Let H be a pro-ℓ-group and $\bar{H}$ its maximal elementary abelian quotient. Suppose $h_1, ..., h_r \in H$ map to a set of generators of $\bar{H}$, then $h_1, ..., h_r$ topologically generate H.*

Let $\mathcal{D}$ denote the category of profinite $\mathcal{O}[G]$-modules M for which

- M is semi-stable as an $\mathcal{O}[G_{\ell}]$-module,
- if $\ell \notin \Sigma$ and if $\bar{\rho}$ is good then M is good as an $\mathcal{O}[G_{\ell}]$-module,
- if $p \notin \Sigma \cup \{\ell\}$ and if $\bar{\rho}$ is ramified at p then there is an exact sequence of $\mathcal{O}[I_p]$-modules

$$(0) \to M^{(-1)} \to M \to M^{(0)} \to (0),$$

such that I_p acts trivially on $M^{(-1)}$ and $M^{(0)}$.

Then we see that a lifting $\rho : G_{\mathbb{Q}} \to GL_2(R)$ of $\bar{\rho}$ is of type Σ if and only if

- ρ factors through G,
- $\det \rho = \epsilon$,
- M_ρ is an object of $\mathcal{D}$.

The existence part of theorem 2.41 now follows from theorem 2.36 and part (a) follows from lemma 2.38. Parts (b) and (c) follow from lemmas 2.39 and 2.40, and the following observation:
If $\rho : G_{\mathbb{Q}} \to GL_2(\mathcal{O}/\lambda^n)$ is a lifting of type Σ and if $p \notin \Sigma \cup \{l\}$ is a prime where $\bar{\rho}$ ramifies then

$$\ker\left(H^1(G_p, \mathrm{ad}^0\rho) \to H^1(I_p, \mathrm{ad}^0\rho/(\mathrm{ad}^0\rho)^{I_p})\right) = H^1(G_p/I_p, (\mathrm{ad}^0\rho)^{I_p}). \tag{2.7.1}$$

Equation (2.7.1) follows from the fact that the natural map

$$H^1(I_p, \mathrm{ad}^0\rho) \to H^1(I_p, \mathrm{ad}^0\rho/(\mathrm{ad}^0\rho)^{I_p})$$

is an injection (in fact an isomorphism). It is nothing other than the map

$$(\mathrm{ad}^0\rho)_{I_p} \to (\mathrm{ad}^0\rho/(\mathrm{ad}^0\rho)^{I_p})_{I_p}.$$

This completes the proof of theorem 2.41. □

Corollary 2.43 *Suppose that if $\ell = 3$ then $\bar{\rho}|_{G_{\mathbb{Q}(\sqrt{-3})}}$ is absolutely irreducible. Then R_Σ can be topologically generated as an $\mathcal{O}$-algebra by*

$$\dim_k H^1_\Sigma(\mathbb{Q}, \mathrm{ad}^0\bar{\rho}(1)) + d_\ell + \sum_{p \in \Sigma - \{\ell\}} \dim_k H^0(\mathbb{Q}_p, \mathrm{ad}^0\bar{\rho}(1))$$

elements, where $d_\ell = \dim_k H^1_{\mathrm{ss}}(\mathbb{Q}_\ell, \mathrm{ad}^0\bar{\rho}) - \dim_k H^1_{\mathrm{f}}(\mathbb{Q}_\ell, \mathrm{ad}^0\bar{\rho})$ if $\ell \in \Sigma$, while $d_\ell = 0$ if $\ell \notin \Sigma$.

Proof: Note that $H^0(\mathbb{Q}, \mathrm{ad}^0\bar{\rho}(1)) = 0$ unless $\ell = 3$ and $\bar{\rho}|_{G_{\mathbb{Q}(\sqrt{-3})}}$ is not absolutely irreducible. Thus according to theorem 2.18, $\dim_k H^1_\Sigma(\mathbb{Q}, \mathrm{ad}^0\bar{\rho})$ is the sum of the following terms.

- $\dim_k H^1_\Sigma(\mathbb{Q}, \mathrm{ad}^0\bar{\rho}(1))$.
- $\dim_k H^1_{\mathrm{f}}(\mathbb{Q}_\ell, \mathrm{ad}^0\bar{\rho}) - \dim_k H^0(\mathbb{Q}_\ell, \mathrm{ad}^0\bar{\rho}) - \dim_k H^0(\mathbb{R}, \mathrm{ad}^0\bar{\rho}) \leq 0$ by proposition 2.27.
- d_ℓ.
- For each $p \in \Sigma - \{l\}$,

  $$\dim_k H^1(\mathbb{Q}_p, \mathrm{ad}^0\bar{\rho}) - \dim_k H^0(\mathbb{Q}_p, \mathrm{ad}^0\bar{\rho}) = \dim_k H^0(\mathbb{Q}_p, \mathrm{ad}^0\bar{\rho}(1))$$

  by the local Euler characteristic formula (see theorem 2.17).

□

2.8 Special cases

In this section we will restrict attention to the deformation problems associated with sets Q of certain special primes q. These sets Q are chosen so that the associated deformation ring R_Q enjoys a number of special properties which will be crucial in the sequel. In particular we will only consider sets Q of primes q which satisfy the following two properties

- $q \equiv 1 \bmod \ell$,
- $\bar{\rho}$ is unramified at q and $\bar{\rho}(\mathrm{Frob}_q)$ has distinct k-rational eigenvalues.

For such primes q we will let Δ_q denote the maximal quotient of $(\mathbb{Z}/q\mathbb{Z})^\times$ of ℓ-power order. Then Δ_q is naturally a quotient of both G_q and $G_\mathbb{Q}$ via the map

$$\chi_q : G_\mathbb{Q} \twoheadrightarrow \mathrm{Gal}\,(\mathbb{Q}(\zeta_q)/\mathbb{Q}) \cong (\mathbb{Z}/q\mathbb{Z})^\times \twoheadrightarrow \Delta_q.$$

We will let $\Delta_Q = \prod_{q\in Q} \Delta_q$, $\chi_Q = \prod_{q \in Q} \chi_q : G_\mathbb{Q} \twoheadrightarrow \Delta_Q$ and $\mathfrak{a}_Q$ denote the augmentation ideal of $\mathcal{O}[\Delta_Q]$. Note that there is an isomorphism

$$\mathcal{O}[\Delta_Q] \cong \mathcal{O}[[S_q : q \in Q]]/((1+S_q)^{\#\Delta_q} - 1 : q \in Q)$$

under which $\mathfrak{a}_Q$ corresponds to $(S_q : q \in Q)$.

For each $q \in Q$, we choose an eigenvalue α_q of $\bar{\rho}(\mathrm{Frob}_q)$ and denote the other β_q.

Lemma 2.44 *If $q \in Q$ then $\rho_Q^{\mathrm{univ}}|_{G_q}$ is conjugate to $\begin{pmatrix} \xi & 0 \\ 0 & \epsilon\xi^{-1}\end{pmatrix}$ for some character ξ such that $\bar{\xi}(\mathrm{Frob}_q) = \alpha_q$.*

Proof: Choose a lifting $f \in G_q$ of Frob_q. As $\bar{\rho}(f)$ has distinct k-rational eigenvalues it is a simple application of Hensel's lemma to see that we can choose a basis such that $\rho_Q^{\mathrm{univ}}(f) = \begin{pmatrix} \tilde{\alpha}_q & 0 \\ 0 & \tilde{\beta}_q \end{pmatrix}$ where $\tilde{\alpha}_q$ and $\tilde{\beta}_q$ reduce to α_q and β_q in k. It suffices to show that for any $\sigma \in I_q$, $\rho_Q^{\mathrm{univ}}(\sigma)$ is diagonal in this basis. Suppose $\rho_Q^{\mathrm{univ}}(\sigma) = 1_2 + \begin{pmatrix} a & b \\ c & d \end{pmatrix}$ with $a, b, c, d \in \mathfrak{m}_{R_Q}$. Because ρ_Q^{univ} is tamely ramified at q we see that $\rho_Q^{\mathrm{univ}}(f)\rho_Q^{\mathrm{univ}}(\sigma)\rho_Q^{\mathrm{univ}}(f)^{-1} = \rho_Q^{\mathrm{univ}}(\sigma)^q$. Thus $(\tilde{\alpha}_q/\tilde{\beta}_q - q)b$ and $(\tilde{\beta}_q/\tilde{\alpha}_q - q)c$ lie in $\mathfrak{m}_{R_Q}(b,c)$ so that $(b,c) = \mathfrak{m}_{R_Q}(b,c)$ and (by Nakayama's lemma) $b = c = 0$. □

We let $\xi_{q,Q}$ denote the character $\xi : G_q \to R_Q^\times$ in the conclusion of the lemma. The restriction of $\xi_{q,Q}$ to I_q factors through χ_q. We let ξ_Q denote the character $G_\mathbb{Q} \to R_Q^\times$ which is unramified outside the primes of Q and whose restriction to I_q is $\xi_{q,Q}$ for each q in Q. Thus ξ_Q factors through χ_Q. We wish to regard R_Q as an $\mathcal{O}[\Delta_Q]$-algebra, and it will be most convenient to do so via the map which gives rise to ξ_Q^{-2}. Note the following consequence of the lemma 2.44.

Corollary 2.45 *The natural map $R_Q \to R_\emptyset$ gives rise to an isomorphism $R_Q/\mathfrak{a}_Q R_Q \tilde{\to} R_\emptyset$.*

Lemma 2.46 (a) *If $q \in Q$ then $H^0(\mathbb{F}_q, \mathrm{ad}^0\bar{\rho}) = H^0(\mathbb{F}_q, \mathrm{ad}^0\bar{\rho}(1)) = k$ and $H^1(\mathbb{F}_q, \mathrm{ad}^0\bar{\rho}) = H^1(\mathbb{F}_q, \mathrm{ad}^0\bar{\rho}(1)) = k$.*

(b) *R_Q can be topologically generated as an $\mathcal{O}$-algebra by*

$$\#Q + \dim_k H^1_Q(\mathbb{Q}, \mathrm{ad}^0\bar{\rho}(1))$$

elements.

(c) *If*

$$H^1_\emptyset(\mathbb{Q}, \mathrm{ad}^0\bar{\rho}(1)) \tilde{\to} \bigoplus_{q\in Q} H^1(\mathbb{F}_q, \mathrm{ad}^0\bar{\rho}(1))$$

then $\#Q = \dim_k H^1_\emptyset(\mathbb{Q}, \mathrm{ad}^0\bar{\rho}(1))$ and R_Q can be topologically generated as an $\mathcal{O}$-algebra by $\#Q$ elements.

Proof: The first part is a direct calculation using the fact that Frob_q acts semi-simply on $\mathrm{ad}^0\bar{\rho}$ with eigenvalues $x, 1, x^{-1}$ for some $x \in \bar{k}\backslash\{0,1\}$. The same is true for $\mathrm{ad}^0\bar{\rho}(1)$ as $q \equiv 1 \bmod \ell$. The second and third parts follow from this and corollary 2.43. □

For the proof of the last theorem of the chapter, we shall need two results on finite groups.

Theorem 2.47 (a) *If H is a finite subgroup of $PGL_2(\mathbb{C})$ then H is isomorphic to one of the following groups: the cyclic group C_n of order n ($n \in \mathbb{Z}_{>0}$), the dihedral group D_{2n} of order $2n$ ($n \in \mathbb{Z}_{>1}$), A_4, S_4 or A_5.*

(b) *If H is a finite subgroup of $PGL_2(\bar{\mathbb{F}}_\ell)$ then one of the following holds:*

- *H is conjugate to a subgroup of the upper triangular matrices;*
- *H is conjugate to $PSL_2(\mathbb{F}_{\ell^r})$ or $PGL_2(\mathbb{F}_{\ell^r})$ for some $r \in \mathbb{Z}_{>0}$;*
- *H is isomorphic to A_4, S_4, A_5 or the dihedral group D_{2r} of order $2r$ for some $r \in \mathbb{Z}_{>1}$ not divisible by ℓ.*

In fact we shall only need part (b) which is due to Dickson [Dic2], secs. 255, 260 (see also [Hu] II.8.27), but we have included part (a) for later reference.

Lemma 2.48 *Let $\mathbb{F}$ be a finite field of odd characteristic ℓ. If $\#\mathbb{F} \neq 5$, then*

$$H^1(SL_2(\mathbb{F}), \mathrm{End}^0(\mathbb{F}^2)) = 0.$$

Proof. This is a special case of results of [CPS], table 4.5 (assuming $\#\mathbb{F} \neq 3$). In fact we shall only need it in the case $\ell = 3$, but the proof in the general case is no more difficult, and we sketch it here for the reader's convenience.

We let B (resp. U) denote the group of upper-triangular (resp. unipotent) matrices in $G = SL_2(\mathbb{F})$. Since ℓ does not divide the index of B in G, the restriction homomorphism

$$H^1(G, \mathrm{End}^0(\mathbb{F}^2)) \to H^1(B, \mathrm{End}^0(\mathbb{F}^2))$$

is injective, so it suffices to prove the latter group vanishes. Since ℓ does not divide the index of U in B, we have

$$H^i(B, M) \cong H^i(U, M)^{B/U}$$

all integers $i \geq 0$ and $\mathbb{F}[B]$-modules M. If $\#\mathbb{F} = 3$, then one checks directly that for $M = \mathrm{End}^0(\mathbb{F}_3^2)$,

$$H^1(U, M) \cong \ker N/(\sigma - 1)M = 0$$

where σ generates U and $N = 1 + \sigma + \sigma^2$ on M. If $\#\mathbb{F} > 5$, then one proceeds by writing

$$(0) = M_0 \subset M_1 \subset M_2 \subset M_3 = \mathrm{End}^0(\mathbb{F}^2)$$

as $\mathbb{F}[B]$-modules where the successive quotients M_i/M_{i-1}, (for $i = 1, 2, 3$) are one-dimensional over $\mathbb{F}$. The calculation is then straightforward using long exact sequences of cohomology, except in the case $\#\mathbb{F} = 9$ where one must also check that the one-dimensional space of classes in $H^1(U, M_3/M_2)$ fixed by B/U maps injectively to $H^2(U, M_2/M_1)$ via the connecting homomorphism. □

Theorem 2.49 *Keep the assumptions of the last section and suppose moreover that if $L = \mathbb{Q}(\sqrt{(-1)^{(\ell-1)/2}\ell})$ then $\bar{\rho}|_{G_L}$ is absolutely irreducible. Then there exists a non-negative integer r such that for any $n \in \mathbb{Z}_{>0}$ we can find a finite set of primes Q_n with the following properties.*

(a) *If $q \in Q_n$ then $q \equiv 1 \bmod \ell^n$.*

(b) *If $q \in Q_n$ then $\bar{\rho}$ is unramified at q and $\bar{\rho}(\mathrm{Frob}_q)$ has distinct eigenvalues.*

(c) *$\#Q_n = r$.*

(d) *R_{Q_n} can be topologically generated by r elements as a $\mathcal{O}$-algebra.*

Proof. Take $r = \dim_k H^1_\emptyset(\mathbb{Q}, \mathrm{ad}^0\bar{\rho}(1))$. It suffices to find a set Q_n with the following properties.

(a) If $q \in Q_n$ then $q \equiv 1 \bmod \ell^n$.

(b) If $q \in Q_n$ then $\bar{\rho}$ is unramified at q and $\bar{\rho}(\mathrm{Frob}_q)$ has distinct eigenvalues.

(c) $H^1_\emptyset(\mathbb{Q}, \mathrm{ad}^0\bar{\rho}(1)) \tilde{\to} \bigoplus_{q\in Q_n} H^1(\mathbb{F}_q, \mathrm{ad}^0\bar{\rho}(1))$.

As each $H^1(\mathbb{F}_q, \mathrm{ad}^0\bar{\rho}(1))$ is one dimensional we may replace the last condition with

$$H^1_\emptyset(\mathbb{Q}, \mathrm{ad}^0\bar{\rho}(1)) \hookrightarrow \bigoplus_{q\in Q_n} H^1(\mathbb{F}_q, \mathrm{ad}^0\bar{\rho}(1)).$$

Thus what we need show is that for each non-zero class $[\psi] \in H^1_\emptyset(\mathbb{Q}, \mathrm{ad}^0\bar{\rho}(1))$ there is a prime q (depending on $[\psi]$) such that

(a) $q \equiv 1 \bmod \ell^n$,

(b) $\bar{\rho}$ is unramified at q and $\bar{\rho}(\mathrm{Frob}_q)$ has distinct eigenvalues,

(c) $\mathrm{res}_q[\psi] \in H^1(\mathbb{F}_q, \mathrm{ad}^0\bar{\rho}(1))$ is nontrivial.

Using the Chebotarev density theorem we see that it will do to find $\sigma \in G_{\mathbf{Q}}$ such that

(a) $\sigma|_{\mathbf{Q}(\zeta_{\ell^n})} = 1$,

(b) $\mathrm{ad}^0\bar{\rho}(\sigma)$ has an eigenvalue other than 1,

(c) $\psi(\sigma) \notin (\sigma - 1)\mathrm{ad}^0\bar{\rho}(1)$.

For $m \geq 0$, let F_m denote the extension of $\mathbb{Q}(\zeta_{\ell^m})$ cut out by $\mathrm{ad}^0\bar{\rho}$; i.e., the field fixed by the kernel of the representation $\mathrm{ad}^0\bar{\rho}$ restricted to $G_{\mathbf{Q}(\zeta_{\ell^m})}$. We will show that $\psi(G_{F_n})$ is non-trivial. For this it suffices to prove that $H^1(\mathrm{Gal}\,(F_n/\mathbb{Q}), \mathrm{ad}^0\bar{\rho}(1)) = (0)$. Consider the inflation-restriction exact sequence

$$\begin{aligned}(0) \to H^1(\mathrm{Gal}\,(F_0/\mathbb{Q}), (\mathrm{ad}^0\bar{\rho}(1))^{G_{F_0}}) &\to H^1(\mathrm{Gal}\,(F_n/\mathbb{Q}), \mathrm{ad}^0\bar{\rho}(1)) \\ &\to H^1(\mathrm{Gal}\,(F_n/F_0), \mathrm{ad}^0\bar{\rho}(1))^{G_{\mathbf{Q}}}.\end{aligned}$$

Since F_1/F_0 is an extension of degree prime to ℓ, and since $G_{\mathbf{Q}}$ acts trivially on $\mathrm{Gal}\,(F_n/F_1)$, we have:

$$H^1(\mathrm{Gal}\,(F_n/F_0), \mathrm{ad}^0\bar{\rho}(1))^{G_{\mathbf{Q}}} \simeq \mathrm{Hom}(\mathrm{Gal}\,(F_n/F_1), \mathrm{ad}^0\bar{\rho}(1)^{G_{\mathbf{Q}}}). \tag{2.8.1}$$

Since $\bar{\rho}|_{G_L}$ is absolutely irreducible, the cohomology group in equation (2.8.1) vanishes. On the other hand, G_{F_0} acts trivially on $\mathrm{ad}^0\bar{\rho}$ so the first term vanishes as well unless $\mathrm{Gal}\,(F_0/\mathbb{Q})$ has order divisible by ℓ and has $\mathrm{Gal}\,(\mathbb{Q}(\zeta_\ell)/\mathbb{Q})$ as a quotient. Recall that $\mathrm{Gal}\,(F_0/\mathbb{Q})$ is isomorphic to the projective image of $\bar{\rho}$, so by theorem 2.47 we are reduced to the case $\ell = 3$ and the map

$$\mathrm{Gal}\,(F_0/\mathbb{Q}(\zeta_3)) \to PSL_2(\bar{k})$$

has image conjugate to $PSL_2(\mathbb{F}_{3^r})$ for some $r \geq 1$. It suffices to prove that in this case

$$H^1(\mathrm{Gal}\,(F_0/\mathbb{Q}(\zeta_3)), \mathrm{ad}^0\bar{\rho} \otimes_k \bar{k}) = (0),$$

but this follows directly from lemma 2.48.

Now note that since $\bar{\rho}|_{G_L}$ is absolutely irreducible, so is $\bar{\rho}|_{G_{\mathbb{Q}(\zeta_{\ell^n})}}$. Regarding $\psi(G_{F_n})$ as a $\mathrm{Gal}\,(F_n/\mathbb{Q}(\zeta_{\ell^n}))$-submodule of $\mathrm{ad}^0(\bar{\rho})$, we find that some $g \in \mathrm{Gal}\,(F_n/\mathbb{Q}(\zeta_{\ell^n}))$ of order not dividing ℓ fixes a non-zero element of $\psi(G_{F_n})$. Let σ_0 be a lifting of g to $G_{\mathbb{Q}(\zeta_{\ell^n})}$. We will look for $\sigma = \tau\sigma_0$ with $\tau \in G_{F_n}$. We only need choose τ so that $\psi(\tau\sigma_0) = \psi(\tau) + \psi(\sigma_0) \notin (\sigma_0 - 1)\mathrm{ad}^0\bar{\rho}(1)$. This is possible because $\psi(G_F) \not\subset (g-1)\mathrm{ad}^0\bar{\rho}(1)$. □

3 Modular forms and Galois representations

3.1 From modular forms to Galois representations

We suppose in this section that $f = \sum a_n(f)q^n$ is a newform of weight two and level N_f (see definition 1.21). Let K_f denote the number field in $\mathbb{C}$ generated by the Fourier coefficients $a_n(f)$. Let ψ_f denote the character of f, i.e., the homomorphism $(\mathbb{Z}/N_f\mathbb{Z})^\times \to K_f^\times$ defined by mapping d to the eigenvalue of $\langle d\rangle$ on f.

Recall that a construction of Shimura (section 1.7) associates to f an abelian variety A_f of dimension $[K_f : \mathbb{Q}]$. This abelian variety is a certain quotient of $J_1(N_f)$, and the action of the Hecke algebra on $J_1(N_f)$ provides an embedding

$$K_f \hookrightarrow \mathrm{End}_{\mathbb{Q}}(A_f) \otimes \mathbb{Q}.$$

We saw also that for each prime ℓ the Tate module $T_\ell(A_f) \otimes_{\mathbb{Z}_\ell} \mathbb{Q}_\ell$ becomes a free module of rank two over $K_f \otimes \mathbb{Q}_\ell$ (lemma 1.48). The action of the Galois group $G_{\mathbb{Q}}$ on the Tate module commutes with that of K_f, so that a choice of basis for the Tate module provides a representation

$$G_{\mathbb{Q}} \to GL_2(K_f \otimes \mathbb{Q}_\ell). \tag{3.1.1}$$

As $K_f \otimes \mathbb{Q}_\ell$ can be identified with the product of the completions of K_f at its primes over ℓ, we obtain from f certain 2-dimensional ℓ-adic representations of $G_{\mathbb{Q}}$.

ℓ-adic representations: In this discussion, we fix a prime ℓ and a finite extension K of $\mathbb{Q}_\ell$. We let $\mathcal{O}$ denote the ring of integers of K, λ the maximal ideal and k the residue field. We shall consider ℓ-adic representations with coefficients in finite extensions of our fixed field K. We regard K as a subfield of $\bar{\mathbb{Q}}_\ell$ and fix embeddings $\bar{\mathbb{Q}} \hookrightarrow \bar{\mathbb{Q}}_\ell$ and $\bar{\mathbb{Q}} \hookrightarrow \bar{\mathbb{C}}$. If K' is a finite extension of K with ring of integers $\mathcal{O}'$, then we say that an ℓ-adic representation $G_\ell \to GL_2(K')$ is good (respectively, ordinary, semistable) if it is conjugate over K' to a representation $G_\ell \to GL_2(\mathcal{O}')$ which is good (respectively, ordinary, semistable) in the sense of section 2.4.

Let K'_f denote the K-algebra in $\bar{\mathbb{Q}}_\ell$ generated by the Fourier coefficients of f. Thus K'_f is a finite extension of K, and it contains the completion of K_f at the prime over ℓ determined by our choice of embeddings. We let $\mathcal{O}'_f$ denote the ring of integers of K'_f and write k_f for its residue field. We define

$$\rho_f : G_\mathbb{Q} \to GL_2(K'_f)$$

as the pushforward of (3.1.1) by the natural map $K_f \otimes \mathbb{Q}_\ell \to K'_f$. We assume the basis is chosen so that ρ_f factors through $GL_2(\mathcal{O}'_f)$. We also let ψ'_f denote the finite order ℓ-adic character

$$G_\mathbb{Q} \twoheadrightarrow \mathrm{Gal}\,(\mathbb{Q}(\zeta_{N_f})/\mathbb{Q}) \to (K'_f)^\times$$

obtained from ψ_f.

The following theorem lists several fundamental properties of the ℓ-adic representations ρ_f obtained from Shimura's construction. The result is a combination of the work of many mathematicians. We discuss some of the proofs and provide references below. In the statement we fix f as above and write simply N, a_n, ρ, ψ, ψ' and K' for N_f, $a_n(f)$, ρ_f, ψ_f, ψ'_f and K'_f respectively.

Theorem 3.1 *The ℓ-adic representation*

$$\rho : G_\mathbb{Q} \to GL_2(K')$$

has the following properties.

(a) *If $p \not| N\ell$ then ρ is unramified at p and $\rho(\mathrm{Frob}_p)$ has characteristic polynomial*
$$X^2 - a_p X + p\psi(p).$$

(b) *$\det(\rho)$ is the product of ψ' with the ℓ-adic cyclotomic character ϵ, and $\rho(c)$ is conjugate to $\begin{pmatrix} 1 & 0 \\ 0 & -1 \end{pmatrix}$.*

(c) *ρ is absolutely irreducible.*

(d) *The conductor $N(\rho)$ is the prime-to-ℓ-part of N.*

(e) *Suppose that $p \neq \ell$ and $p||N$. Let χ denote the unramified character $G_p \to (K')^\times$ satisfying $\chi(\mathrm{Frob}_p) = a_p$. If p does not divide the conductor of ψ, then $\rho|_{G_p}$ is of the form*

$$\begin{pmatrix} \chi\epsilon & * \\ 0 & \chi \end{pmatrix}.$$

If p divides the conductor of ψ, then $\rho|_{G_p}$ is of the form

$$\chi^{-1}\epsilon\psi'|_{G_p} \oplus \chi.$$

(f) *If $\ell \not| 2N$, then $\rho|_{G_\ell}$ is good. Moreover $\rho|_{G_\ell}$ is ordinary if and only if a_ℓ is a unit in the ring of integers of K', in which case $\rho_{I_\ell}(\mathrm{Frob}_\ell)$ is the unit root of the polynomial $X^2 - a_\ell X + \ell\psi(\ell)$.*

(g) *If ℓ is odd and $\ell||N$, but the conductor of ψ is not divisible by ℓ, then $\rho|_{G_\ell}$ is ordinary and $\rho_{I_\ell}(\mathrm{Frob}_\ell) = a_\ell$.*

Proof: Part (a) was established by Shimura ([Shi2], [Shi3]). The key ingredient is the Eichler-Shimura congruence relation, theorem 1.29. Recall that $J_1(N)$ has good reduction at primes p not dividing N. So the action of G_p on $\mathcal{T}_\ell(A_f) \otimes_{\mathbf{Z}_\ell} \mathbb{Q}_\ell$ is unramified and is in fact described by the action of $\mathrm{Frob}_p \in G_{\mathbf{F}_p}$ on the Tate module of the reduction. But this is given by the Frobenius endomorphism F whose characteristic polynomial is computed in corollary 1.41.

The first assertion of (b) follows from (a) on applying the Chebotarev density theorem. The second assertion then follows on noting that $\psi(-1) = 1$.

Part (c) was proved by Ribet (see section 2 of [R3]). Assuming reducibility of the representation, he applies algebraicity results of Lang and Serre to obtain a contradiction to the estimate on the Fourier coefficients stated in theorem 1.24.

Parts (d) and (e) follow from a deep result of Carayol [Ca1], Thm. (A), building on the work of Langlands [Ll1], Deligne and others. In fact, this result and the local Langlands correspondence characterize $\rho|_{G_p}$ in terms of $\psi|_{G_p}$ and the L- and ϵ-factors at p of twists of f. The descriptions in the case of $p||N$ are based on the analysis of Deligne-Rapoport of the reduction mod p of $J_1(N)$ (see [DR], [Ll1]).

The first assertion of (f) follows from the fact that A_f has good reduction at ℓ if ℓ does not divide N. The second assertion of (f) (respectively, all of (g)) follows from the Eichler-Shimura congruence relation (respectively, the work of Deligne-Rapoport), and general results on ℓ-divisible groups and the reduction of abelian varieties; see thm. 2.2 of [W2], lemma 2.1.5 of [W1], §12 of [Gro] and thms. 2.5 and 2.6 of [Edi]. The restriction to odd ℓ in (f) and (g) is made primarily out of lack of suitable definitions. □

Mod ℓ representations: We maintain the notation used in the discussion of ℓ-adic representations. Define

$$\bar{\rho}_f : G_{\mathbb{Q}} \to GL_2(k_f)$$

to be the *semi-simplification* of the reduction of ρ_f. (See the discussion following proposition 2.6.) Assertions analogous to those in theorem 3.1 hold for $\bar{\rho} = \bar{\rho}_f$, except that

- The representation need not be absolutely irreducible (as in (c)). However if ℓ is odd, one checks using (b) that $\bar{\rho}$ is irreducible if and only if it is absolutely irreducible.

- In (d) one only has divisibility of the prime-to-ℓ part of N_f by $N(\bar{\rho})$ in general.

The various possibilities for $m_p(\bar{\rho})$ to be strictly less than the exponent of p in N (where $p \neq \ell$) were classified independently by Carayol [Ca2] and Livné [Liv]. We record the following consequence of their results (cf. the introduction of [DT1]):

Proposition 3.2 *Suppose that p is a prime such that $p|N_f$, $p \not\equiv 1 \bmod \ell$ and $\bar{\rho}_f$ is unramified at p. Then* $\operatorname{tr}(\bar{\rho}_f(\operatorname{Frob}_p))^2 = (p+1)^2$ *in k_f.*

Artin representations: The theory of Hecke operators and newforms (see section 1.3) extends to modular forms on $\Gamma_1(N)$ of arbitrary weight. The construction of ℓ-adic representations associated to newforms was generalized to weight greater than 1 by Deligne [De] using étale cohomology. There are also Galois representations associated to newforms of weight 1 by Deligne and Serre [DS], but an essential difference is that these are Artin representations.

Theorem 3.3 *If $g = \sum a_n(g)q^n$ is a newform of weight one, level N_g and character ψ_g, then there is an irreducible Artin representation*

$$\rho_g : G_{\mathbb{Q}} \longrightarrow GL_2(\mathbb{C})$$

of conductor N_g with the following property: If $p \nmid N_g$, then the characteristic polynomial of $\rho_g(\operatorname{Frob}_p)$ is

$$X^2 - a_p(g)X + \psi_g(p).$$

Remark 3.4 Note that $\det(\rho_g)$ is the character of $G_{\mathbb{Q}}$ corresponding to ψ and $\rho_g(c)$ is conjugate to $\begin{pmatrix} 1 & 0 \\ 0 & -1 \end{pmatrix}$ (see theorem 3.1).

Remark 3.5 A basis can be chosen so that the representation ρ_g takes values in $GL_2(K_g)$ (where K_g is the number field generated by the $a_n(g)$). Moreover suppose that K is a finite extension of $\mathbb{Q}_\ell$ in $\bar{\mathbb{Q}}_\ell$ and we have fixed embeddings of $\bar{\mathbb{Q}}$ in $\mathbb{C}$ and $\bar{\mathbb{Q}}_\ell$). If K_g is contained in K, then we can view ρ_g as giving rise to an ℓ-adic representation $G_{\mathbb{Q}} \to GL_2(K)$ and hence a mod ℓ representation $G_{\mathbb{Q}} \to GL_2(k)$.

Remark 3.6 A key idea in the construction of ρ_g is to first construct the mod ℓ representations using those already associated to newforms of higher weight. More precisely, suppose that $K_g \hookrightarrow K$ as in remark 3.5. One can show that for some newform f of weight 2 and level N_f dividing $N\ell$ we have

$$a_p(g) \equiv a_p(f), \qquad \psi_g(p)) \equiv p\psi_f(p)$$

for all $p \nmid N\ell$, the congruence being modulo the maximal ideal of the ring of integers of K'_f. Thus $\bar{\rho}_f$ is the semisimplification of the desired mod ℓ representation (with scalars extended to k_f).

3.2 From Galois representations to modular forms

It is conjectured that certain types of two-dimensional representations of $G_{\mathbb{Q}}$ always arise from the constructions described in section 3.1. We now state some of the conjectures and the results known prior to [W3] and [TW].

Artin representations:

Conjecture 3.7 *Let $\rho : G_{\mathbb{Q}} \to GL_2(\mathbb{C})$ be a continuous irreducible representation with* $\det(\rho(c)) = -1$. *Then ρ is equivalent to ρ_g for some newform g of weight one.*

Recall that ρ_g is the Artin representation associated to g by the Deligne-Serre construction (theorem 3.3).

Remark 3.8 Conjecture 3.7 is equivalent to the statement that the Artin L-functions attached to ρ and to all its twists by one-dimensional characters are entire. (The Artin conjecture predicts that the Artin L-function $L(s, \rho)$ is entire, for an arbitrary irreducible, non-trivial Artin representation $\rho : G_{\mathbb{Q}} \longrightarrow GL_d(\mathbb{C})$.)

A large part of conjecture 3.7 was proved by Langlands in [Ll2], and the results were extended by Tunnell [Tu].

Theorem 3.9 *Let $\rho : G_{\mathbb{Q}} \to GL_2(\mathbb{C})$ be a continuous irreducible representation such that $\rho(G_{\mathbb{Q}})$ is solvable and* $\det(\rho(c)) = -1$. *Then ρ is equivalent to ρ_g for some newform g of weight one.*

Remark 3.10 Note that by theorem 2.47, part (a), the solvability hypothesis excludes only the case where the projective image of ρ is isomorphic to A_5.

Remark 3.11 If the projective image of ρ is dihedral, then ρ is induced from a character of a quadratic extension of $\mathbb{Q}$. In this case the result can already be deduced from work of Hecke.

Mod ℓ representations: We fix notation as in the discussion of ℓ-adic and mod ℓ representations in section 3.1.

Definition 3.12 We say that a representation $\bar{\rho} : G_{\mathbb{Q}} \to GL_2(k)$ is *modular (of level N)* if for some newform f of weight 2 (and level N), $\bar{\rho}$ is equivalent over k_f to $\bar{\rho}_f$.

By proposition 1.32 the notion is independent of the choices in section 3.1 of embeddings $K \hookrightarrow \bar{\mathbb{Q}}_\ell$, $\bar{\mathbb{Q}} \hookrightarrow \bar{\mathbb{Q}}_\ell$ and $\bar{\mathbb{Q}} \hookrightarrow \mathbb{C}$. Moreover if K' is a finite extension of K with residue field k', then $\bar{\rho}$ is modular if and only if $\bar{\rho} \otimes_k k'$ is modular.

The following was conjectured by Serre [Se7], (3.2.3). (See also [Dal] for further discussion and references.)

Conjecture 3.13 *Let $\bar{\rho} : G_{\mathbb{Q}} \to GL_2(k)$ be a continuous absolutely irreducible representation with* $\det(\bar{\rho}(c)) = -1$. *Then $\bar{\rho}$ is modular.*

Some cases of Serre's conjecture can be deduced from theorem 3.9.

Theorem 3.14 *Let $\bar{\rho} : G_{\mathbb{Q}} \to GL_2(k)$ be a continuous absolutely irreducible representation with* $\det(\bar{\rho}(c)) = -1$. *Suppose that one of the following holds:*

(a) $k = \mathbb{F}_3$;

(b) *the projective image of $\bar{\rho}$ is dihedral.*

Then $\bar{\rho}$ is modular.

Sketch of proof: For case (a), we consider the surjection

$$GL_2(\mathbb{Z}[\sqrt{-2}]) \to GL_2(\mathbb{F}_3)$$

defined by reduction mod $(1 + \sqrt{-2})$. One checks that there is a section $s : GL_2(\mathbb{F}_3) \to GL_2(\mathbb{Z}[\sqrt{-2}])$ and applies theorem 3.9 to $s \circ \bar{\rho}$. The resulting representation arises from a weight one newform, and hence its reduction $\bar{\rho}$ is equivalent to $\bar{\rho}_f$ for some f (see remark 3.6).

In case (b), $\bar{\rho}$ is equivalent to a representation of the form $\mathrm{Ind}_{G_F}^{G_{\mathbb{Q}}} \bar{\xi}$ where F is a quadratic extension of $\mathbb{Q}$ and $\bar{\xi}$ is a character $G_F \to k^\times$. (We have here enlarged K if necessary.) Let n be the order of $\bar{\xi}$, choose an embedding $\mathbb{Q}(e^{2\pi i/n}) \hookrightarrow K$ and lift $\bar{\xi}$ to a character $\xi : G_F \to \mathbb{Z}[e^{2\pi i/n}]^\times$. We may always choose ξ so that the Artin representation $\rho = \mathrm{Ind}_{G_F}^{G_{\mathbb{Q}}} \xi$ is odd, i.e., $\det(\rho(c)) = -1$. (In the case $\ell = 2$ and F real quadratic, we may have to multiply ξ by a suitable quadratic character of G_F.) We then apply theorem 3.9 to ρ and deduce as in case (a) that $\bar{\rho}$ is modular. □

Serre also proposed a refinement ([Se7], (3.2.4)) of the conjecture which predicts that $\bar{\rho}$ is associated to a newform of specified weight, level and character. Through work of Mazur, Ribet [R5], Carayol [Ca2], Gross [Gro] and others, this refinement is now known to be equivalent to conjecture 3.13 if ℓ is odd. (One also needs to impose a mild restriction in the case $\ell = 3$.) See [R6] and [Dil] for statements of the results and further references; here we give a variant which applies to newforms of weight two. Before doing so, we assume ℓ is odd and define an integer $\delta(\bar{\rho})$ as follows:

- $\delta(\bar{\rho}) = 0$ if $\bar{\rho}|_{G_\ell}$ is good;
- $\delta(\bar{\rho}) = 1$ if $\bar{\rho}|_{G_\ell}$ is not good and $\bar{\rho}|_{I_\ell} \otimes_k \bar{k}$ is of the form

$$\begin{pmatrix} \epsilon^a & * \\ 0 & 1 \end{pmatrix}, \quad \begin{pmatrix} \epsilon & * \\ 0 & \epsilon^a \end{pmatrix} \quad \text{or} \quad \begin{pmatrix} \psi^a & 0 \\ 0 & \psi^{\ell a} \end{pmatrix}$$

for some positive integer $a < \ell$. (Recall that ϵ is the cyclotomic character and ψ is the character of I_ℓ defined after lemma 2.22.)

- $\delta(\bar{\rho}) = 2$ otherwise.

Theorem 3.15 *Suppose that ℓ is odd and $\bar{\rho}$ is absolutely irreducible and modular. If $\ell = 3$, then suppose also that $\bar{\rho}|_{G_{\mathbf{Q}(\sqrt{-3})}}$ is absolutely irreducible. Then there exists a newform f of weight two such that*

- *$\bar{\rho}$ is equivalent over k_f to $\bar{\rho}_f$;*
- *$N_f = N(\bar{\rho})\ell^{\delta(\bar{\rho})}$;*
- *the order of ψ_f is not divisible by ℓ.*

Proof: The existence of such an f follows from [Di1] thm. 1.1, thm. 5.1 and lemma 2.1, but with N_f dividing $N(\bar{\rho})\ell^{\delta(\bar{\rho})}$. By lemma 2.7 above, we see that N_f is divisible by $N(\bar{\rho})$. The divisibility of N_f by $\delta(\bar{\rho})$ follows from results in sec. 8 of [Gro] and sec. 2.4 of [Edi] (cf. sec. 4 of [Kh]). □

ℓ-adic representations: We again use the notation of section 3.1. Let $\rho : G_{\mathbf{Q}} \to GL_2(K)$ be an ℓ-adic representation.

Definition 3.16 We say that ρ is *modular* if for some weight 2 newform f, ρ is equivalent over K'_f to ρ_f.

The notion is independent of the choices of embeddings and well-behaved under extension of scalars by proposition 1.32 (cf. definition 3.12).

The following is a special case of a conjecture of Fontaine and Mazur [FM].

Conjecture 3.17 *If $\rho : G_{\mathbf{Q}} \to GL_2(K)$ is an absolutely irreducible ℓ-adic representation and $\rho|_{G_\ell}$ is semistable (in the sense of section 2.4), then ρ is modular.*

(Recall that for us ℓ-adic representations are defined to be unramified at all but finitely many primes. Recall also that if $\rho|_{G_\ell}$ is semistable, then by definition $\det \rho|_{I_\ell}$ is the cyclotomic character ϵ.)

Remark 3.18 Relatively little was known about this conjecture before Wiles' work [W3]. Wiles proves that under suitable hypotheses, the modularity of $\bar{\rho}$ implies that of ρ.

Remark 3.19 The conjecture stated in [FM] is stronger than the one here; in particular, the semistability hypothesis could be replaced with a suitable notion of potential semistability. On the other hand, one expects that if $\rho|_{G_\ell}$ is semistable, then it is equivalent to ρ_f (over K'_f) for some f on $\Gamma_1(N(\rho)) \cap \Gamma_0(\ell)$ (and on $\Gamma_1(N(\rho))$ if $\rho|_{G_\ell}$ is good).

The Shimura-Taniyama conjecture: Conjecture 1.54 can be viewed in the framework of the problem of associating modular forms to Galois representations. Let E be an elliptic curve defined over $\mathbb{Q}$. For each prime ℓ, we let $\rho_{E,\ell}$ denote the ℓ-adic representation $G_{\mathbf{Q}} \to GL_2(\mathbb{Q}_\ell)$ defined by the action of $G_{\mathbf{Q}}$ on the Tate module of E (see section 2.2).

Proposition 3.20 *The following are equivalent:*

(a) *E is modular.*

(b) *$\rho_{E,\ell}$ is modular for all primes ℓ.*

(c) *$\rho_{E,\ell}$ is modular for some prime ℓ.*

Proof: If E is modular, then E is isogenous to A_f for some weight two newform f with $K_f = \mathbb{Q}$ (see section 1.8). It follows that for each prime ℓ, $\rho_{E,\ell}$ is equivalent to the ℓ-adic representation ρ_f. Hence (a) $\Rightarrow$ (b) $\Rightarrow$ (c).

To show (c) $\Rightarrow$ (b), suppose that for some ℓ and some f, the representations $\rho_{E,\ell}$ and ρ_f are equivalent. First observe that for all but finitely primes p, we have

$$\operatorname{tr}(\rho_f(\operatorname{Frob}_p)) = \operatorname{tr}(\rho_{E,\ell}(\operatorname{Frob}_p)).$$

We deduce from proposition 2.11 and theorem 3.1, part (a) that for all but finitely many primes p

$$a_p(f) = p + 1 - \#\tilde{E}_p(\mathbb{F}_p) \in \mathbb{Z}. \tag{3.2.1}$$

Applying proposition 2.6, we find that for each prime ℓ, $\rho_{E,\ell}$ is equivalent to ρ_f and is therefore modular.

We finally show that (b) $\Rightarrow$ (a). The equality (3.2.1) holds for all primes p not dividing N_f, which by theorem 3.1, part (d), is the conductor of E. Since $\det(\rho_f) = \det(\rho_{E,\ell}) = \epsilon$, we see by 3.1, Part (b) that ψ_f is trivial. By theorem 1.27 parts (b) and (d) (or [AL] thm. 3), a_p is in $\{0, \pm 1\}$ for primes p dividing N_f. Thus $K_f = \mathbb{Q}$ and A_f is an elliptic curve. Faltings' isogeny theorem (see [CS], sec. II.5) now tells us that E and A_f are isogenous and we conclude that E is modular. □

Remark 3.21 Note that the equivalence (b) $\Leftrightarrow$ (c) does not require Faltings' isogeny theorem.

Proposition 3.22 *If the Fontaine-Mazur conjecture (conjecture 3.17) holds for some prime ℓ, then the Shimura-Taniyama conjecture (conjecture 1.54) holds. If Serre's conjecture (conjecture 3.13) holds for infinitely many ℓ, then conjecture 1.54 holds.*

Proof: The first assertion is immediate from proposition 3.20 and the irreducibility of $\rho_{E,\ell}$. See [Se7], sec. 4.6 for a proof of the second. (We have implicitly chosen the field K to be $\mathbb{Q}_\ell$ in the statements of conjectures 3.17 and 3.13, but it may be replaced by a finite extension.) □

Remark 3.23 Note that to prove a given elliptic curve E is modular, it suffices to prove that conjecture 3.17 holds for a single ℓ at which E has semistable reduction. Wiles' approach is to show that certain cases of conjecture 3.13 imply cases of conjecture 3.17 and hence cases of conjecture 1.54.

3.3 Hecke algebras

In this section fix the following notation. Let ℓ be an odd prime, let K be a finite extension of $\mathbb{Q}_\ell$, let $\mathcal{O}$ denote the ring of integers of K, let λ denote its maximal ideal and k its residue field. Fix embeddings $K \hookrightarrow \bar{\mathbb{Q}}_\ell$, $\bar{\mathbb{Q}} \hookrightarrow \bar{\mathbb{Q}}_\ell$ and $\bar{\mathbb{Q}} \hookrightarrow \mathbb{C}$. Let $\bar{\rho} : G_\mathbb{Q} \to GL_2(k)$ denote a continuous representation with the following properties

(a) $\bar{\rho}$ is irreducible,

(b) $\bar{\rho}$ is modular,

(c) $\det \bar{\rho} = \epsilon$,

(d) $\bar{\rho}|_{G_\ell}$ is semi-stable,

(e) and if $p \neq \ell$ then $\#\bar{\rho}(I_p)|\ell$.

Let us first record the following lemma.

Lemma 3.24 *The representation* $\bar{\rho}|_{G_L}$ *is absolutely irreducible where* $L = \mathbb{Q}(\sqrt{(-1)^{(\ell-1)/2}\ell})$.

Proof: If it were not then we see that $\ell \not| \#\bar{\rho}(G_\mathbb{Q})$ and so $\bar{\rho}$ is unramified at all $p \neq \ell$. Moreover we can check that $\bar{\rho}|_{I_\ell} \sim \begin{pmatrix} \epsilon & 0 \\ 0 & 1 \end{pmatrix}$. If $\ell > 3$ we can use theorem 3.15 to deduce that $\bar{\rho}$ is modular of weight 2 and level 1 and hence obtain a contradiction. If $\ell = 3$ we see that the splitting field of $\bar{\rho}$ is everywhere unramified over $\mathbb{Q}(\sqrt{-3})$ and hence must equal $\mathbb{Q}(\sqrt{-3})$, a contradiction. □

Let Σ denote a finite set of primes. For the application to modularity of semistable elliptic curves, it suffices to consider sets Σ contained in $\Sigma_{\bar{\rho}}$ where $\Sigma_{\bar{\rho}}$ is defined as follows

Definition 3.25 For a representation $\bar{\rho}$ as above, we let $\Sigma_{\bar{\rho}}$ denote the set of primes p satisfying

- $p = \ell$ and $\bar{\rho}|_{G_\ell}$ is good and ordinary; or
- $p \neq \ell$ and $\bar{\rho}$ is unramified at p.

We shall sometimes assume that $\Sigma \subset \Sigma_{\bar{\rho}}$ in order to simplify statements and proofs. Let $\mathcal{N}_\Sigma$ denote the set of newforms f such that

$$\rho_f : G_\mathbb{Q} \to GL_2(\mathcal{O}'_f)$$

is equivalent to a lifting of $\bar{\rho}\otimes_k k_f$ of type Σ and $\ell^2 \not| N_f$ (the last condition is presumably not necessary, cf. remark 3.19). From theorem 3.1 and lemma 2.7 one deduces the following description of $\mathcal{N}_\Sigma$.

Lemma 3.26 *The set $\mathcal{N}_\Sigma$ consists of newforms f such that*

- $\bar{\rho}_f \cong \bar{\rho} \otimes_k k_f$,
- ψ_f *is trivial,*
- N_f *divides* $\ell^\delta N(\bar{\rho}) \prod_{p\in\Sigma-\{\ell\}} p^{\dim \bar{\rho}^{I_p}}$, *where* $\delta = 0$ *if* $\bar{\rho}$ *is good and* $\ell \notin \Sigma$ *and* $\delta = 1$ *otherwise.*

As $\bar{\rho}$ is modular it follows from theorem 3.15 that for all Σ, $\mathcal{N}_\Sigma \neq \emptyset$. Set $\tilde{\mathbb{T}}_\Sigma = \prod_{f\in\mathcal{N}_\Sigma} \mathcal{O}'_f$. If p is a prime not in Σ and not dividing $\ell N(\bar{\rho})$, we let T_p denote the element $(a_p(f))_f$ in $\tilde{\mathbb{T}}_\Sigma$. Then define $\mathbb{T}_\Sigma$ to be the $\mathcal{O}$-subalgebra of $\tilde{\mathbb{T}}_\Sigma$ generated by the elements T_p for such primes p. Then $\mathbb{T}_\Sigma$ is a complete noetherian local $\mathcal{O}$-algebra with residue field k. Moreover it is reduced and it is a finitely generated free $\mathcal{O}$-module.

Lemma 3.27 *There is a continuous representation*

$$\rho_\Sigma^{\mathrm{mod}} : G_\mathbb{Q} \to GL_2(\mathbb{T}_\Sigma)$$

such that if $p \not| \ell N(\bar{\rho})$ and $p \notin \Sigma$ then $\rho_\Sigma^{\mathrm{mod}}$ is unramified at p and we have $\operatorname{tr} \rho_\Sigma^{\mathrm{mod}}(\mathrm{Frob}_p) = T_p$. *Moreover we have the following.*

(a) $\rho_\Sigma^{\mathrm{mod}}$ *is a lift of $\bar{\rho}$ of type Σ and there is a unique surjection* $\phi_\Sigma : R_\Sigma \twoheadrightarrow \mathbb{T}_\Sigma$ *such that* $\rho_\Sigma^{\mathrm{mod}} \sim \phi_\Sigma \circ \rho_\Sigma^{\mathrm{univ}}$.

(b) *If $\Sigma' \supset \Sigma$ then there is a unique surjection $\mathbb{T}_{\Sigma'} \twoheadrightarrow \mathbb{T}_\Sigma$ such that $\rho_{\Sigma'}^{\mathrm{mod}}$ pushes forward to $\rho_\Sigma^{\mathrm{mod}}$ and T_p maps to T_p for $p \not| \ell N(\bar{\rho})$ and $p \notin \Sigma'$.*

(c) *If K' is a finite extension of K and $\mathbb{T}'_\Sigma$ is constructed in the same way as $\mathbb{T}_\Sigma$ but with K' replacing K then $\mathbb{T}'_\Sigma \cong \mathbb{T}_\Sigma \otimes_\mathcal{O} \mathcal{O}_{K'}$.*

Proof: Consider $\tilde{\rho}_\Sigma^{\mathrm{mod}} = \prod \rho_f : G_\mathbb{Q} \to GL_2(\tilde{\mathbb{T}}_\Sigma)$. Choose some complex conjugation, c, and conjugate $\bar{\rho}$ so that $\bar{\rho}(c) = \begin{pmatrix} 1 & 0 \\ 0 & -1 \end{pmatrix}$. By conjugating $\tilde{\rho}_\Sigma^{\mathrm{mod}}$ we may assume that modulo every maximal ideal it reduces to $\bar{\rho}$ and that $\tilde{\rho}_\Sigma^{\mathrm{mod}}(c) = \begin{pmatrix} 1 & 0 \\ 0 & -1 \end{pmatrix}$. Call the corresponding basis $\{e_+, e_-\}$. Note that by continuity and the Chebotarev density theorem $\operatorname{tr} \tilde{\rho}_\Sigma^{\mathrm{mod}}$ is valued in $\mathbb{T}_\Sigma$. Thus for any $g \in G_\mathbb{Q}$ the diagonal entries of $\tilde{\rho}_\Sigma^{\mathrm{mod}}(g)$ lie in $\mathbb{T}_\Sigma$, because $\operatorname{tr} \tilde{\rho}_\Sigma^{\mathrm{mod}}(g)$ and $\operatorname{tr} \tilde{\rho}_\Sigma^{\mathrm{mod}}(cg)$ both do.

By irreducibility of $\bar{\rho}$ we can find some $\sigma \in G_\mathbb{Q}$ such that $\bar{\rho}(\sigma) = \begin{pmatrix} * & b \\ * & * \end{pmatrix}$ where $b \neq 0$. Rescaling e_+ we may assume that $\tilde{\rho}_\Sigma^{\mathrm{mod}}(\sigma) = \begin{pmatrix} a & 1 \\ c & d \end{pmatrix}$. Then for all $g \in G_{\mathbb{Q}}$ the lower left entry of $\tilde{\rho}_{\Sigma}^{\mathrm{mod}}(g)$ lies in $\mathbb{T}_{\Sigma}$ (look at the upper left entry of $\tilde{\rho}_{\Sigma}^{\mathrm{mod}}(\sigma g)$). Again using the irreducibility of $\bar{\rho}$ we can find $\tau \in G_{\mathbb{Q}}$ such that $\tilde{\rho}_{\Sigma}^{\mathrm{mod}}(\tau) = \begin{pmatrix} * & * \\ e & * \end{pmatrix}$ where $e \in \mathbb{T}_{\Sigma}^{\times}$. Looking at the lower right entry of $\tilde{\rho}_{\Sigma}^{\mathrm{mod}}(\tau g)$ we see that for any $g \in G_{\mathbb{Q}}$ the upper right entry of $\tilde{\rho}_{\Sigma}^{\mathrm{mod}}(g)$ lies in $\mathbb{T}_{\Sigma}$. Thus $\tilde{\rho}_{\Sigma}^{\mathrm{mod}}$ is now in fact valued in $GL_2(\mathbb{T}_{\Sigma})$ and will be our candidate for $\rho_{\Sigma}^{\mathrm{mod}}$. We leave the verification of the other properties of $\rho_{\Sigma}^{\mathrm{mod}}$ as an exercise. □

Example 3.28 Let $\bar{\rho} = \bar{\rho}_{f_{57B},3}$ where f_{57B} is the newform of level 57 discussed in the example of section 1.6. As f_{57B} is not congruent modulo 3 to any form of level 19 or 3 we see by theorem 3.15 that $\bar{\rho}$ is ramified at 19 and $\bar{\rho}|_{G_3}$ is not good. On the other hand $\bar{\rho}$ is semi-stable. The facts that $\bar{\rho}(\mathrm{Frob}_2)$ has order 8 (see the table below) and $3|\#\bar{\rho}(I_{19})$ (from the discussion above) imply that $\bar{\rho} : G_{\mathbb{Q}} \twoheadrightarrow GL_2(\mathbb{F}_3)$. The table below also shows that $\mathbb{T}_{\emptyset} \cong \{(x,y) \in \mathbb{Z}_3^2 : x \equiv y \bmod 3\}$. There is a unique continuous representation $\rho_{\emptyset}^{\mathrm{mod}} : G_{\mathbb{Q}} \to GL_2(\mathbb{T}_{\emptyset})$ such that if $p \nmid 57$ then $\rho_{\emptyset}^{\mathrm{mod}}$ is unramified at p and $\operatorname{tr} \rho_{\emptyset}^{\mathrm{mod}}(\mathrm{Frob}_p) = T_p$. We have an isomorphism $\mathbb{T}_{\emptyset}/3\mathbb{T}_{\emptyset} \xrightarrow{\sim} \mathbb{F}_3[\varepsilon]$ (where $\varepsilon^2 = 0$) given by $(x,y) \mapsto x + \frac{y-x}{3}\varepsilon$. Thus we get a representation $\rho : G_{\mathbb{Q}} \to GL_2(\mathbb{F}_3[\varepsilon])$. The following table lists the traces of Frobenius elements for the first few unramified primes under these representations:

p	2	5	7	11	13	17	23	29
$\bar{\rho}$	1	1	0	0	0	0	1	-1
$\rho_{\emptyset}^{\mathrm{mod}}$	$(1,-2)$	$(-2,1)$	$(0,3)$	$(0,-3)$	$(6,-6)$	$(-6,3)$	$(4,4)$	$(2,-10)$
ρ	$1-\varepsilon$	$1+\varepsilon$	ε	$-\varepsilon$	$-\varepsilon$	0	1	$-1-\varepsilon$

Exercise 3.29 Show that there are three algebra homomorphisms $\mathbb{T}_{\emptyset} \to \mathbb{Z}/9\mathbb{Z}$ and hence show that there are at least three liftings of $\bar{\rho}$ of type $\emptyset$ to $\mathbb{Z}/9\mathbb{Z}$.

Exercise 3.30 What is the image of ρ?

We will need two deeper properties of the Hecke algebras $\mathbb{T}_{\Sigma}$ (theorem 3.31 and theorem 3.36 below). These results will be proved in the next chapter (sections 4.3 and 4.4).

Theorem 3.31 *Let Q be a finite set of primes as in section 2.8. Then $\mathbb{T}_Q$ is a free $\mathcal{O}[\Delta_Q]$-module, where $\mathbb{T}_Q$ inherits the structure of an $\mathcal{O}[\Delta_Q]$-module from R_Q via the homomorphism $\phi_Q : R_Q \twoheadrightarrow \mathbb{T}_Q$.*

Corollary 3.32 $\mathbb{T}_{\emptyset} = \mathbb{T}_Q/\mathfrak{a}_Q$.

Proof: From the definitions and corollary 2.45 we have that

$$\mathbb{T}_{\emptyset} \otimes_{\mathcal{O}} K = (\mathbb{T}_Q \otimes_{\mathcal{O}} K)/\mathfrak{a}_Q,$$

and from the theorem we have that $\mathbb{T}_Q/\mathfrak{a}_Q$ is torsion free. The corollary follows. □

For the second of these results we will need some additional notation and we restrict our attention to sets of primes contained in $\Sigma_{\bar{\rho}}$. Suppose that $\Sigma \subset \Sigma_{\bar{\rho}}$ and that f is an element of $\mathcal{N}_\Sigma$ whose Fourier coefficients are in $\mathcal{O}$. Then $\mathcal{O}'_f = \mathcal{O}$ and projection to the component corresponding to f gives rise to an $\mathcal{O}$-algebra homomorphism

$$\pi = \pi_f : \mathbb{T}_\Sigma \to \mathcal{O}.$$

The pushforward of $\rho_\Sigma^{\mathrm{mod}}$ by π is equivalent to

$$\rho_f : G_{\mathbb{Q}} \to GL_2(\mathcal{O}).$$

Remark 3.33 Most of the objects considered in the rest of the section will depend on the choice of newform f. We also remark that to give an $\mathcal{O}$-algebra homomorphism $\mathbb{T}_\Sigma \to \mathcal{O}$ is equivalent to giving a newform f in $\mathcal{N}_\Sigma$ with coefficients in $\mathcal{O}$. Indeed given such a homomorphism, there exists a newform f in $\mathcal{N}_\Sigma$ such that the homomorphism is defined by $T_p \mapsto a_p(f)$ for all $p \notin \Sigma$ with p not dividing $\ell N(\bar{\rho})$. The fact that $a_p(f) \in \mathcal{O}$ for such p implies (using for example parts (b) and (d) of theorem 1.27 and lemma 4.1 below) that all the Fourier coefficients of f are in $\mathcal{O}$. The uniqueness of f is a consequence of the theory of newforms, theorem 1.22.

For Σ' satisfying

$$\Sigma \subset \Sigma' \subset \Sigma_{\bar{\rho}},$$

let $\pi_{\Sigma'}$ denote the composite $\mathbb{T}_{\Sigma'} \to \mathbb{T}_\Sigma \to \mathcal{O}$. Let

$$\eta_{\Sigma'} = \pi_{\Sigma'}(\operatorname{Ann}_{\mathbb{T}_{\Sigma'}}(\ker \pi_{\Sigma'})). \tag{3.3.1}$$

Note that because $\mathbb{T}_{\Sigma'}$ is reduced, $\eta_{\Sigma'} \neq (0)$. Also let $\wp_{\Sigma'}$ denote the kernel of $\phi_{\Sigma'} \circ \pi_{\Sigma'}$. Recall that

$$\#\wp_{\Sigma'}/\wp_{\Sigma'}^2 = \#H^1_{\Sigma'}(\mathbb{Q}, \mathrm{ad}^0\rho_f \otimes_{\mathcal{O}} K/\mathcal{O}).$$

Remark 3.34 We have not yet shown that these groups are finite, but if either is finite, then so is the other and their cardinality is the same.

Note that if ℓ is in $\Sigma_{\bar{\rho}} - \Sigma$ then $T_\ell := (a_\ell(g))_g \in \tilde{\mathbb{T}}_\Sigma$ is actually in $\mathbb{T}_\Sigma$. This follows from theorem 3.1(f), which shows that $T_\ell = \alpha_\ell + \ell\alpha_\ell^{-1}$ where α_ℓ is the eigenvalue of Frob_ℓ on the free rank one $\mathbb{T}_\Sigma$-module $M^{(0)}$ where M is the module underlying $\rho_\Sigma^{\mathrm{mod}}$ (see lemma 2.20). Now that we have shown that $T_\ell \in \mathbb{T}_\Sigma$, we may characterize α_ℓ as the unit root in $\mathbb{T}_\Sigma$ of

$$X^2 - T_\ell X + \ell = 0. \tag{3.3.2}$$

For primes p in $\Sigma_{\bar{\rho}} - \Sigma$ we define an element $c_p \in \mathbb{T}_\Sigma$ by

$$c_p = (p-1)(T_p^2 - (p+1)^2).$$

Note that $\pi(c_p) = (p-1)(a_p(f)^2 - (p+1)^2)$, which is non-zero by theorem 1.27(a). A calculation using theorem 2.17 and theorem 3.1(a) shows that if p is in $\Sigma_{\bar\rho} - (\Sigma \cup \ell)$ then

$$\begin{aligned}\#H^1(G_p, \mathrm{ad}^0\rho_f \otimes_{\mathcal{O}} K/\mathcal{O})/H^1(G_p/I_p, \mathrm{ad}^0\rho_f \otimes_{\mathcal{O}} K/\mathcal{O})\\ = \#H^0(G_p, \mathrm{ad}^0\rho_f \otimes_{\mathcal{O}} K/\mathcal{O}(1)) = \#\mathcal{O}/\pi(c_p).\end{aligned}$$

If ℓ is in $\Sigma_{\bar\rho} - \Sigma$ then we see from proposition 2.27 and theorem 3.1(f) that

$$\#H^1_{\mathrm{ss}}(G_\ell, \mathrm{ad}^0\rho_f \otimes_{\mathcal{O}} K/\mathcal{O})/H^1_{\mathrm{f}}(G_\ell, \mathrm{ad}^0\rho_f \otimes_{\mathcal{O}} K/\mathcal{O}) = \#\mathcal{O}/\pi(c_\ell).$$

For the next proposition define groups H_p for $p \in \Sigma_{\bar\rho} - \Sigma$ by

- $H_p = H^1(G_p, \mathrm{ad}^0\rho_f \otimes_{\mathcal{O}} K/\mathcal{O})/H^1(G_p/I_p, \mathrm{ad}^0\rho_f \otimes_{\mathcal{O}} K/\mathcal{O})$ if $p \neq \ell$
- and $H_\ell = H^1_{\mathrm{ss}}(G_\ell, \mathrm{ad}^0\rho_f \otimes_{\mathcal{O}} K/\mathcal{O})/H^1_{\mathrm{f}}(G_\ell, \mathrm{ad}^0\rho_f \otimes_{\mathcal{O}} K/\mathcal{O})$.

Then we have the following result.

Proposition 3.35 *If $\Sigma \subset \Sigma' \subset \Sigma_{\bar\rho}$, then*

$$\#H^1_{\Sigma'}(\mathbb{Q}, \mathrm{ad}^0\rho_f \otimes_{\mathcal{O}} K/\mathcal{O})/H^1_{\Sigma}(\mathbb{Q}, \mathrm{ad}^0\rho_f \otimes_{\mathcal{O}} K/\mathcal{O}) \le \#(\mathcal{O}/\pi(\prod_{p\in\Sigma'-\Sigma} c_p)).$$

Moreover if we have equality then the sequence

$$(0) \to H^1_{\Sigma}(\mathbb{Q}, \mathrm{ad}^0\rho_f \otimes_{\mathcal{O}} K/\mathcal{O}) \to H^1_{\Sigma'}(\mathbb{Q}, \mathrm{ad}^0\rho_f \otimes_{\mathcal{O}} K/\mathcal{O}) \to \bigoplus_{p\in\Sigma'-\Sigma} H_p \to (0)$$

is exact.

Finally we state the second theorem on Hecke algebras which shall be proved in section 4.4.

Theorem 3.36 *If $\Sigma \subset \Sigma' \subset \Sigma_{\bar\rho}$ and f is a newform in $\mathcal{N}_\Sigma$ with coefficients in $\mathcal{O}$, then we have*

$$\eta_{\Sigma'} \subset \pi(\prod_{p\in\Sigma'-\Sigma} c_p)\eta_\Sigma.$$

Corollary 3.37 *With the above notation*

$$\#H^1_{\Sigma'}(\mathbb{Q}, \mathrm{ad}^0\rho_f \otimes_{\mathcal{O}} K/\mathcal{O})/H^1_{\Sigma}(\mathbb{Q}, \mathrm{ad}^0\rho_f \otimes_{\mathcal{O}} K/\mathcal{O}) \le \#(\eta_\Sigma/\eta'_\Sigma).$$

3.4 Isomorphism criteria

The main thrust of Wiles' approach is to prove that in many circumstances the map $\phi_\Sigma : R_\Sigma \to \mathbb{T}_\Sigma$ is an isomorphism. For this we will need two criteria from commutative algebra. The first was found by Wiles [W3] (but is presented here in a slightly stronger form due to Lenstra [Len]); the second was developed by Faltings, from the original arguments of [TW]. In both criteria the notion of complete intersection plays a vital part.

Proofs of all the results in this section, together with some background, references and examples, is given in chapter 5.

Definition 3.38 Suppose that A is an object of $\mathcal{C}_\mathcal{O}$ which is finitely generated and free as an $\mathcal{O}$-module. Then we call A a *complete intersection* if and only if for some $r \in \mathbb{Z}_{\geq 0}$ and some $f_1, ..., f_r \in \mathcal{O}[[X_1, ..., X_r]]$ we have

$$A \cong \mathcal{O}[[X_1, ..., X_r]]/(f_1, ..., f_r)$$

(i.e. there are the same number of generators as relations).

We first record a lemma about complete intersections.

Lemma 3.39 *Suppose that K'/K is a finite extension with ring of integers $\mathcal{O}'$ and that A is an object of $\mathcal{C}_\mathcal{O}$ which is finitely generated and free as an $\mathcal{O}$-module. Then A is a complete intersection if and only if $A \otimes_\mathcal{O} \mathcal{O}'$ is.*

For the proof, see chapter 5, lemma 5.30.

Now fix objects R and T of $\mathcal{C}_\mathcal{O}$ and a surjection of $\mathcal{O}$-algebras $\phi : R \twoheadrightarrow T$. Also assume that T *is a finitely generated, free $\mathcal{O}$-module.* The first criterion is as follows.

Theorem 3.40 *Suppose that $\pi : T \twoheadrightarrow \mathcal{O}$. Let $\wp = \ker(\pi \circ \phi) \lhd R$ and let $\eta = \pi(\mathrm{Ann}_T(\ker \pi)) \lhd \mathcal{O}$. Suppose also that $\eta \neq (0)$. Then the following are equivalent.*

(a) *The inequality $\#\wp/\wp^2 \leq \#\mathcal{O}/\eta$ is satisfied.*

(b) *The equality $\#\wp/\wp^2 = \#\mathcal{O}/\eta$ is satisfied.*

(c) *The rings R and T are complete intersections, and the map $\phi : R \longrightarrow T$ is an isomorphism.*

The proof is explained in chapter 5, sections 5.1 to 5.8. (See theorem 5.3.)

For the second criterion let us also fix a non-negative integer r. If $J \lhd \mathcal{O}[[S_1, ..., S_r]]$ is an ideal contained in $(S_1, ..., S_r)$, then by a J-structure we mean a commutative diagram in $\mathcal{C}_\mathcal{O}$

$$\begin{array}{ccccc} & & \mathcal{O}[[S_1, ..., S_r]] & & \\ & & \downarrow & & \\ \mathcal{O}[[X_1, ..., X_r]] & \twoheadrightarrow & R' & \twoheadrightarrow & T' \\ & & \downarrow & & \downarrow \\ & & R & \twoheadrightarrow & T, \end{array}$$

such that

(a) $T'/(S_1, ..., S_r) \tilde{\rightarrow} T$ and $R'/(S_1, ..., S_r) \twoheadrightarrow R$,

(b) for each ideal $I \supset J$, $I = \ker(\mathcal{O}[[S_1, ..., S_r]] \to T'/I)$.

Theorem 3.41 *Suppose there exist a sequence of ideals $J_n \lhd \mathcal{O}[[S_1, ..., S_r]]$ such that $J_0 = (S_1, ..., S_r)$, $J_n \supset J_{n+1}$, $\bigcap_n J_n = (0)$ and for each n there exists a J_n-structure. Then the map $R \longrightarrow T$ is an isomorphism and these rings are complete intersections.*

The proof of this statement is explained in sections 5.9 and 5.10.

3.5 The main theorem

We are now in a position to deduce the main theorems. We will keep the notation and assumptions from the start of section 3.3.

Theorem 3.42 *Keep the notation and assumptions of section 3.3. Then, for all finite sets $\Sigma \subset \Sigma_{\bar\rho}$, $\phi_\Sigma : R_\Sigma \longrightarrow \mathbb{T}_\Sigma$ is an isomorphism and these rings are complete intersections.*

Remark 3.43 There seems to be a deep link between the fact that ϕ_Σ is an isomorphism and the fact that $\mathbb{T}_\Sigma$ is a complete intersection. The proof of the theorem divides into two parts. One first proves it in the minimal case where $\Sigma = \emptyset$. One then deduces the full theorem from this special case by a different argument. In both these steps the facts that ϕ_Σ is an isomorphism and that $\mathbb{T}_\Sigma$ is a complete intersection are proved simultaneously.

Proof of theorem 3.42: Note that to prove the theorem we may extend scalars if necessary (by lemma 3.39) and hence assume that both of the following hold:

- The eigenvalues of all elements of the image of ρ are rational over k.
- There is a newform f in $\mathcal{N}_\emptyset$ with coefficients in $\mathcal{O}$, hence an $\mathcal{O}$-algebra homomorphism $\mathbb{T}_\emptyset \to \mathcal{O}$.

We first prove that $\phi_\emptyset : R_\emptyset \longrightarrow \mathbb{T}_\emptyset$ is an isomorphism and that the rings $R_\emptyset$ and $\mathbb{T}_\emptyset$ are complete intersections.

Note that according to theorem 2.49 and lemma 3.24 we can find an integer $r \geq 0$ and for each $n \in \mathbb{Z}_{>0}$ we can find a set Q_n of r primes such that

- if $q \in Q_n$ then $q \equiv 1 \bmod \ell^n$;
- if $q \in Q_n$ then $\bar\rho$ is unramified at q and $\bar\rho(\mathrm{Frob}_q)$ has distinct eigenvalues;

- R_{Q_n} can be topologically generated by r elements as a $\mathcal{O}$-algebra.

Let $J'_n = ((S_i+1)^{\ell^{n_i}} - 1 : i = 1, \dots, r) \lhd \mathcal{O}[[S_1, \dots, S_r]]$, where the n_i are chosen such that $\mathcal{O}[[S_1, \dots, S_r]]/J_n \cong \mathcal{O}[\Delta_{Q_n}]$. Then for each n we have a diagram

$$\begin{array}{ccccc} & & \mathcal{O}[[S_1, \dots, S_r]] & & \\ & & \downarrow & & \\ \mathcal{O}[[X_1, \dots, X_r]] & \twoheadrightarrow & R_{Q_n} & \twoheadrightarrow & \mathbb{T}_{Q_n} \\ & & \downarrow & & \downarrow \\ & & R_\emptyset & \twoheadrightarrow & \mathbb{T}_\emptyset, \end{array}$$

where

(a) $R_{Q_n}/(S_1, \dots, S_r) \xrightarrow{\sim} R_\emptyset$ (see corollary 2.45),

(b) $\mathbb{T}_{Q_n}$ is finite and free over $\mathcal{O}[[S_1, \dots, S_r]]/J'_n$ (see theorem 3.31),

(c) $\mathbb{T}_{Q_n}/(S_1, \dots, S_r) \xrightarrow{\sim} \mathbb{T}_\emptyset$ (see corollary 3.32).

Let $J_n = ((S_i+1)^{\ell^n} - 1 : i = 1, \dots, r)$. Replacing $\mathbb{T}_{Q_n}$ and R_{Q_n} by $\mathbb{T}_{Q_n}/J_n$ and R_{Q_n}/J_n we see that we have a J_n-structure for every n. Theorem 3.42 for $\Sigma = \emptyset$ now follows from the criterion of theorem 3.41.

We now turn to the proof of theorem 3.42 in the general case. By theorem 2.41 and theorem 3.40 we see that

$$\#H^1_\emptyset(\mathbb{Q}, \mathrm{ad}^0 \rho_f \otimes_\mathcal{O} K/\mathcal{O}) = \#\mathcal{O}/\eta_\emptyset, \tag{3.5.1}$$

and so applying corollary 3.37 we see that for any $\Sigma \subset \Sigma_{\bar{\rho}}$

$$\#H^1_\Sigma(\mathbb{Q}, \mathrm{ad}^0 \rho_f \otimes_\mathcal{O} K/\mathcal{O}) \le \#\mathcal{O}/\eta_\Sigma. \tag{3.5.2}$$

A second application of theorems 2.41 and 3.40 allows us to deduce theorem 3.42. □

Remark 3.44 In certain cases where $\eta_\emptyset = (1)$, the bound (3.5.1) on the order of the Selmer group $H^1_\emptyset(\mathbb{Q}, \mathrm{ad}^0 \rho_f \otimes_\mathcal{O} K/\mathcal{O})$ also follows from the previous work of Flach, by a different method. See [Fl1] for details.

Corollary 3.45 *Keep the notation of theorem 3.42 and suppose that f is a newform in $\mathcal{N}_\Sigma$ with coefficients in $\mathcal{O}$.*

(a) *We have*

$$\#H^1_\Sigma(\mathbb{Q}, \mathrm{ad}^0 \rho_f \otimes_\mathcal{O} K/\mathcal{O}) = \#\mathcal{O}/\eta_\Sigma < \infty,$$

where η_Σ was defined in equation (3.3.1) after remark 3.33.

(b) *If $\Sigma \subset \Sigma' \subset \Sigma_{\bar{\rho}}$ then*

$$\begin{aligned}(0) \to H^1_\Sigma(\mathbb{Q}, \mathrm{ad}^0 \rho_f \otimes_\mathcal{O} K/\mathcal{O}) &\to H^1_{\Sigma'}(\mathbb{Q}, \mathrm{ad}^0 \rho_f \otimes_\mathcal{O} K/\mathcal{O}) \\ &\to \textstyle\bigoplus_{p \in \Sigma' - \Sigma} H_p \to (0)\end{aligned}$$

is exact, where the groups H_p and H_ℓ were defined before proposition 3.35.

Proof: The first part now follows as a direct consequence of theorem 3.42 and another application of theorems 2.41 and 3.40. The second part follows from the first, together with proposition 3.35 and theorem 3.36. □

Corollary 3.46 *Suppose* $\rho : G_{\mathbb{Q}} \to GL_2(K)$ *is a continuous representation and let* $\bar{\rho}$ *denote its reduction. Suppose also that*

(a) $\bar{\rho}$ *is irreducible and modular,*

(b) *if* $p \neq \ell$ *then* $\rho|_{I_p} \sim \begin{pmatrix} 1 & * \\ 0 & 1 \end{pmatrix}$,

(c) $\rho|_{G_\ell}$ *is semi-stable,*

(d) $\det \rho = \epsilon$.

Then ρ *is modular.*

Proof: We let Σ denote the set of primes in $\Sigma_{\bar{\rho}}$ at which ρ is ramified. Then $\rho : G_{\mathbb{Q}} \to GL_2(\mathcal{O})$ is a deformation of $\bar{\rho}$ of type Σ, so there is an $\mathcal{O}$-algebra homomorphism $R_\Sigma \to \mathcal{O}$ such that $\rho = \rho_\Sigma^{\text{univ}} \otimes_{R_\Sigma} \mathcal{O}$. Since $\phi_\Sigma : R_\Sigma \to \mathbb{T}_\Sigma$ is an isomorphism by theorem 3.42, it follows that there is a homomorphism $\mathbb{T}_\Sigma \to \mathcal{O}$ sending T_p to $\text{tr}\,(\rho(\text{Frob}_p))$. Since such a homomorphism is necessarily of the form $T_p \mapsto a_p(f)$ for some newform f, it follows that ρ is equivalent to ρ_f and hence is modular. □

Corollary 3.47 *Suppose that* $E/\mathbb{Q}$ *is a semi-stable elliptic curve such that* $\bar{\rho}_{E,3}$ *is irreducible. Then* E *is modular.*

Proof: One need only apply the last corollary with $\ell = 3$ and theorem 3.14. □

3.6 Applications

The Shimura-Taniyama conjecture for semi-stable elliptic curves:

Theorem 3.48 *If* $E/\mathbb{Q}$ *is a semistable elliptic curve, then* E *is modular.*

Proof: By corollary 3.47, it is enough to show that E is modular when its associated mod 3 representation $\bar{\rho}_{E,3}$ is *reducible*, i.e., when E has a subgroup of order 3 defined over $\mathbb{Q}$. Consider the group $E[5]$ of 5-division points of E. The mod 5 Galois representation $\bar{\rho}_{E,5}$ associated to $E[5]$ is irreducible: for otherwise, E would have a subgroup of order 15 defined over $\mathbb{Q}$, and would give rise to a (non-cuspidal) rational point on the modular curve $X_0(15)$. This curve is of genus one, and is known to have only 4 non-cuspidal rational points, which do not correspond to semi-stable elliptic curves (and, at any rate, are known to correspond to modular elliptic curves). Hence we know that $\bar{\rho}_{E,5}$ satisfies all the assumptions of corollary 3.46, except the (crucial!) modularity property. To show that $\bar{\rho}_{E,5}$ is modular, one starts with

Lemma 3.49 *There is an auxiliary (semi-stable) elliptic curve $A/\mathbb{Q}$ which satisfies*

(a) $A[5] \simeq E[5]$ *as $G_{\mathbb{Q}}$-modules.*

(b) $A[3]$ *is an irreducible $G_{\mathbb{Q}}$-module.*

Proof: Let $Y'(5)$ be the curve over $\mathbb{Q}$ which classifies elliptic curves A together with an isomorphism $E[5] \simeq A[5]$ compatible with Weil pairings. Elliptic curves over $\mathbb{Q}$ satisfying (a) correspond to rational points in $Y'(5)(\mathbb{Q})$. Adjoining a finite set of points to $Y'(5)$ yields its compactification $X'(5)$ which is a twist of the modular curve $X(5)$ with full level 5 structure. (I.e., it becomes isomorphic to this curve, over $\bar{\mathbb{Q}}$.) As was shown by Klein, the modular curve $X(5)$ over $\mathbb{C}$ has genus 0. Since $X'(5)$ has a point x_0 defined over $\mathbb{Q}$ corresponding to E, it is isomorphic over $\mathbb{Q}$ to $\mathbb{P}^1$. The rational points of $Y'(5)$ therefore give a plentiful supply of elliptic curves satisfying condition (a). Now consider the curve $Y'(5,3)$ classifying elliptic curves A with an isomorphism $E[5] \simeq A[5]$ (respecting Weil pairings) and a subgroup of A of order 3. One checks that the compactification of $Y'(5,3)$ has genus greater than 1, hence has only finitely many rational points by Faltings' theorem (the Mordell conjecture). It follows that only finitely many points $Y'(5)(\mathbb{Q})$ are in the image of $Y'(5,3)(\mathbb{Q})$ under the natural map $Y'(5,3) \to Y'(5)$. Hence for all but finitely many points x in $Y'(5)(\mathbb{Q})$, the corresponding elliptic curve A satisfies (b) since it has no rational subgroup of order 3. Choosing x arbitrarily close in the 5-adic topology to x_0, we find that the elliptic curve A associated to x is semistable and satisfies the two conditions in the lemma. □

We can now finish the proof of theorem 3.48. Applying corollary 3.47 to the curve A, we find that A is modular. Hence so is the mod 5 representation $\bar{\rho}_{A,5} \simeq \bar{\rho}_{E,5}$. Now applying corollary 3.46 with $\ell = 5$ and ρ the representation of $G_{\mathbb{Q}}$ acting on the 5-adic Tate module of E, we find that $\rho_{E,5}$ is modular, and hence, so is E, as was to be shown.

Remark 3.50 Wiles' original argument uses Hilbert's irreducibility theorem where we have used Faltings' theorem. The alternative presented here is based on a remark of Karl Rubin.

Remark 3.51 The results of [W3] and [TW] actually apply to a larger class of elliptic curves than those which are semistable. In [Di2], their methods are further strengthened to prove that all elliptic curves which have semi-stable reduction at 3 and 5 are modular.

Remark 3.52 Rubin and Silverberg observed that an elliptic curve of the form $y^2 = x(x-a)(x+b)$ has a twist with semi-stable reduction at all odd primes, hence is modular by [Di2]. In fact it is shown in [DK] that their observation together with the general results of [W3] and [TW] already imply modularity.

Fermat's Last Theorem: As was already mentioned in the introduction, the Shimura-Taniyama conjecture for semi-stable elliptic curves (and, more precisely, for the elliptic curves that arise in Frey's construction explained in section 2.2) implies Fermat's Last Theorem.

More precisely, suppose that there is a non-trivial solution to the Fermat equation $x^\ell + y^\ell = z^\ell$, with $\ell > 3$. By theorem 2.15 the Frey curve constructed from this solution (cf. section 2.2) is a semistable elliptic curve $E/\mathbb{Q}$ whose associated mod ℓ representation $\bar{\rho}_{E,\ell}$ is irreducible, unramified outside 2ℓ and is good at ℓ. Serre's conjecture predicts that $\bar{\rho}_{E,\ell}$ arises from a newform of weight 2 and level 2; the "lowering the level" result of Ribet [R5] (cf. theorem 3.15) actually proves this, once we know that E is modular, i.e., $\bar{\rho}_{E,\ell}$ arises from a modular form of weight 2 and *some* level. But this is a contradiction, since there are no modular forms of weight two and level two: such forms would correspond to holomorphic differentials on the modular curve $X_0(2)$ which is of genus 0. This contradiction completes the proof.

Values of L-functions: Also mentioned in the introduction was the relationship between the calculation of the Selmer group (3.5.1) and certain cases of a conjecture of Bloch-Kato [BK], called the Tamagawa number conjecture. It was in this context that partial results were obtained by Flach in [Fl1] (cf. remark 3.44).

If f is a newform of weight 2, then one can associate to f a certain "symmetric square" L-function $L(\mathrm{Symm}^2 f, s)$. We shall recall the definition in section 4.4 and explain how a method of Hida establishes a relationship between $L(\mathrm{Symm}^2 f, 2)$ and $\mathcal{O}/\eta_\Sigma$ in the setting of corollary 3.45. We may therefore regard part (a) of that corollary as a relationship between $L(\mathrm{Symm}^2 f, 2)$ and the size of a Selmer group. While the result is in the spirit of the Tamagawa number conjecture of [BK], we have not verified that the relevant cases of the conjecture can be deduced from it. We shall however state a partial result in the context of semistable elliptic curves. The reader can consult [Fl1] and [Fl2] for a discussion of the relation to the Tamagawa number conjecture.

Suppose that E is a semistable elliptic curve over $\mathbb{Q}$ of conductor N_E and (minimal) discriminant

$$\Delta_E = \prod_{p|N_E} p^{d_p}.$$

The symmetric square L-function associated to E is defined by

$$L(\mathrm{Symm}^2 E, s) = \prod_p L_p(\mathrm{Symm}^2 E, s)$$

where the Euler factors $L_p(\mathrm{Symm}^2 E, s)$ are defined as follows:

- If $p \nmid N_E$, then

$$L_p(\mathrm{Symm}^2 E, s) = ((1 - \alpha_p^2 p^{-s})(1 - p^{1-s})(1 - \beta_p^2 p^{-s}))^{-1}$$

where α_p and β_p are the roots roots of $X^2 - a_pX + p$ with $a_p = p+1-N_p$ as in section 1.1).

- If $p|N_E$, then $L_p(\mathrm{Symm}^2 E, s) = (1 - p^{-s})^{-1}$.

Let

$$\Omega_E = \int_{E(\mathbb{C})} \omega_E^{\mathrm{Neron}} \wedge \bar{\omega}_E^{\mathrm{Neron}}$$

where $\omega_E^{\mathrm{Neron}}$ is the Néron differential defined in section 1.1. Since E is modular by theorem 3.48, a method of Shimura (see [Shi4] and the introduction of [St]) establishes the analytic continuation of $L(\mathrm{Symm}^2 E, s)$ to an entire function and shows that $L(\mathrm{Symm}^2 E, 2)$ is a non-zero rational multiple of $i\pi\Omega_E$. We now explain how to deduce the following theorem from Wiles' results and a formula of Hida, corollary 4.21.

Theorem 3.53 *Suppose that E is a semistable elliptic curve and ℓ is a prime such that*

- *$\bar{\rho}_{E,\ell}$ is irreducible, and*
- *ℓ does not divide $2 \prod_{p|N_E} d_p$.*

Then the ℓ-part of

$$\frac{N_E L(\mathrm{Symm}^2 E, 2)}{\pi i \Omega_E}$$

is the order of

$$H^1_\emptyset(\mathbb{Q}, \mathrm{ad}^0 \rho_{E,\ell} \otimes_{\mathbb{Z}_\ell} \mathbb{Q}_\ell/\mathbb{Z}_\ell).$$

Sketch of proof: Since E is semistable, it is modular by theorem 3.48. Letting f denote the associated newform, we have $N_f = N_E$ and $L(\mathrm{Symm}^2 E, s) = L(\mathrm{Symm}^2 f, s)$. Furthermore the representations $\rho_{E,\ell}$ and

$$\rho_f : G_{\mathbb{Q}} \to GL_2(\mathbb{Z}_\ell)$$

are equivalent, since they are equivalent over $\mathbb{Q}_\ell$ and $\bar{\rho} = \bar{\rho}_{E,\ell}$ is irreducible. The conditions on E and ℓ ensure that $\bar{\rho}$ and f satisfy the hypotheses of corollary 3.45 with $\Sigma = \emptyset$, so that

$$\# H^1_\emptyset(\mathbb{Q}, \mathrm{ad}^0 \rho_{E,\ell} \otimes_{\mathbb{Z}_\ell} \mathbb{Q}_\ell/\mathbb{Z}_\ell) = \#(\mathbb{Z}_\ell/\eta_\emptyset).$$

To apply Hida's formula, corollary 4.21, it remains to relate $\det A$ and Ω_E where the matrix $A \in GL_2(\mathbb{C})$ is defined in section 4.4. Using that E is semistable, ℓ is odd and $\bar{\rho}_{E,\ell}$ is irreducible, one obtains a modular parametrization $\pi : X_0(N_E) \to E$ such that

- $\pi^* : H^1(E, \mathbb{Z}_\ell) \to H^1(X_0(N_E), \mathbb{Z}_\ell)$ has torsion-free cokernel;
- the Manin constant for π is not divisible by ℓ ([Maz2], sec. 4(a)).

One can then verify that $\Omega_E^{-1} \det A$ is an ℓ-adic unit and theorem 3.53 follows from corollary 4.21. □

4 Hecke algebras

4.1 Full Hecke algebras

Suppose that K is a finite extension of $\mathbb{Q}_\ell$ for some prime ℓ. Let $\mathcal{O}$ denote its ring of integers and let $k = \mathcal{O}/\lambda$ where λ is the maximal ideal of $\mathcal{O}$. Fix embeddings $K \hookrightarrow \bar{\mathbb{Q}}_\ell$, $\bar{\mathbb{Q}} \hookrightarrow \bar{\mathbb{Q}}_\ell$ and $\bar{\mathbb{Q}} \hookrightarrow \mathbb{C}$. Recall that we defined "Hecke algebras" over $\mathcal{O}$ in two different contexts:

- In section 1.6 as an algebra $\mathbb{T}_\mathcal{O}$ generated by the full set of Hecke operators acting on a space of modular forms;

- In section 3.3 as a certain subring $\mathbb{T}_\Sigma$ of a product of fields of Fourier coefficients of newforms giving rise to the same mod ℓ representation.

The first of these provides a concrete geometric description useful for establishing properties of the fine structure of the algebra; the second yields a reduced ring which is more easily interpreted as the coefficient ring of a Galois representation. In the next section we shall relate the two notions by identifying the "reduced Hecke algebras" of the form $\mathbb{T}_\Sigma$ as localizations of the "full Hecke algebras" of the form $\mathbb{T}_\mathcal{O}$. Before doing so we need to recall some fundamental properties of the algebras $\mathbb{T}_K$, $\mathbb{T}_\mathcal{O}$ and $\mathbb{T}_k$.

Let $\Gamma = \Gamma_H(N)$ for some positive integer N and subgroup H of $(\mathbb{Z}/N\mathbb{Z})^\times$ (see section 1.2). We let $\mathbb{T}_\mathbf{Z}$ denote the subring of End$\,(S_2(\Gamma))$ generated by the operators T_n for all positive integers n and $\langle d\rangle$ for all $d \in (\mathbb{Z}/N\mathbb{Z})^\times$. If R is a ring, then $\mathbb{T}_R$ denotes the R-algebra $\mathbb{T}_\mathbf{Z}\otimes R$. Recall that $\mathbb{T}_R$ acts faithfully on $S_2(\Gamma, R)$ and is finitely generated and free as an R-module. (This holds for $R = \mathbb{Z}$, hence for arbitrary R.)

We first record the following lemma:

Lemma 4.1 (a) *$\mathbb{T}_R$ is generated as an R-algebra by either of the following sets of elements:*

- *T_n for all positive integers n.*
- *T_p for all primes p and $\langle d\rangle$ for all d in $(\mathbb{Z}/N\mathbb{Z})^\times$.*

(b) *Suppose that D is a positive integer relatively prime to N. If either D is odd or 2 is invertible in R, then $\mathbb{T}_R$ is generated as an R-algebra by either of the following sets of elements:*

- *T_n for all positive integers n relatively prime to D.*
- *T_p for all primes p not dividing D, and $\langle d\rangle$ for all d in $(\mathbb{Z}/N\mathbb{Z})^\times$.*

For a proof of (a), see [DI], prop. 3.5.1; for (b), see p. 491 of [W3].

The spectrum of $\mathbb{T}_\mathcal{O}$: First note that $\mathbb{T}_K$ and $\mathbb{T}_k$ are Artinian, hence have only a finite number of prime ideals, all of which are maximal. Since $\mathbb{T}_\mathcal{O}$ is finitely generated and free as an $\mathcal{O}$-module, its maximal (resp. minimal) prime ideals are those lying over the prime λ (resp. (0)) of $\mathcal{O}$. (This follows from the going-up and going-down theorems, [Mat] thms. 9.4 and 9.5, for example.) It follows that the natural maps

$$\mathbb{T}_{\mathcal{O}} \hookrightarrow \mathbb{T}_{\mathcal{O}} \otimes_{\mathcal{O}} K \cong \mathbb{T}_K; \quad \text{and} \quad \mathbb{T}_{\mathcal{O}} \twoheadrightarrow \mathbb{T}_{\mathcal{O}} \otimes_{\mathcal{O}} k \cong \mathbb{T}_k$$

induce bijections

$$\begin{array}{lcll} \{\text{maximal ideals of } \mathbb{T}_K\} & \leftrightarrow & \{\text{minimal primes of } \mathbb{T}_{\mathcal{O}}\} & \text{and} \\ \{\text{maximal ideals of } \mathbb{T}_k\} & \leftrightarrow & \{\text{maximal primes of } \mathbb{T}_{\mathcal{O}}\}. & \end{array}$$

Moreover, since $\mathcal{O}$ is complete we have (by [Mat] thms. 8.7 and 8.15, for example) that the natural map

$$\mathbb{T}_{\mathcal{O}} \to \prod_{\mathfrak{m}} \mathbb{T}_{\mathfrak{m}}$$

is an isomorphism, where the product is over the finite set of maximal ideals $\mathfrak{m}$ of $\mathbb{T}_{\mathcal{O}}$ and $\mathbb{T}_{\mathfrak{m}}$ denotes the localization of $\mathbb{T}_{\mathcal{O}}$ at $\mathfrak{m}$. Furthermore each $\mathbb{T}_{\mathfrak{m}}$ is a complete local $\mathcal{O}$-algebra which is finitely generated and free as an $\mathcal{O}$-module, and each minimal prime $\mathcal{P}$ of $\mathbb{T}_{\mathcal{O}}$ is contained in a unique $\mathfrak{m}$.

Now suppose that $f = \sum a_n q^n$ is a normalized eigenform in $S_2(\Gamma, \bar{K})$ for the operators T_n for all $n \geq 1$. Then $T_n \mapsto a_n$ defines a map $\mathbb{T}_{\mathbf{Z}} \to \bar{K}$ and induces a K-algebra homomorphism $\theta_f : \mathbb{T}_K \to \bar{K}$. The image is the finite extension of K generated by the a_n, and the kernel is a maximal ideal of $\mathbb{T}_K$ which depends only on the G_K-conjugacy class of f. Similarly a G_k-conjugacy class of normalized eigenforms in $S_2(\Gamma, \bar{k})$ gives rise to a maximal ideal of $\mathbb{T}_k$.

Recall also that a normalized eigenform f in $S_2(\Gamma, \bar{K})$ has coefficients in $\mathcal{O}_{\bar{K}}$, hence gives rise by reduction to a normalized eigenform $\bar{f}$ in $S_2(\Gamma, \bar{k})$. Furthermore if f and g are G_K-conjugate, then $\bar{f}$ and $\bar{g}$ are G_k-conjugate.

We have thus constructed a diagram of maps of finite sets whose commutativity is easily verified.

$$\begin{array}{ccc} \left\{ \begin{array}{c} \text{normalized eigenforms in} \\ S_2(\Gamma, \bar{K}) \text{ modulo } G_K\text{-conjugacy} \end{array} \right\} & \to & \left\{ \begin{array}{c} \text{normalized eigenforms in} \\ S_2(\Gamma, \bar{k}) \text{ modulo } G_k\text{-conjugacy} \end{array} \right\} \\ \downarrow & & \downarrow \\ \{\text{maximal ideals of } \mathbb{T}_K\} & & \{\text{maximal ideals of } \mathbb{T}_k\} \\ \updownarrow & & \updownarrow \\ \{\text{minimal primes of } \mathbb{T}_{\mathcal{O}}\} & \twoheadrightarrow & \{\text{maximal primes of } \mathbb{T}_{\mathcal{O}}\}. \end{array} \tag{4.1.1}$$

Proposition 4.2 *The vertical maps are bijective, and the horizontal maps are surjective.*

Proof: For the injectivity of the upper-left vertical arrow, note that if $\mathfrak{p}$ is a maximal ideal of $\mathbb{T}_K$, then all K-algebra homomorphisms $\mathbb{T}_K/\mathfrak{p} \hookrightarrow \bar{K}$ are obtained from a single one by composing with an element of G_K. For the surjectivity, let $K' = \mathbb{T}_K/\mathfrak{p}$ and $\mathfrak{p}'$ denote the kernel of the natural K'-algebra homomorphism $\mathbb{T}_{K'} \twoheadrightarrow K'$. Since $\mathbb{T}_{K'}$ acts faithfully on $S_2(\Gamma, K')$, the localization $S_2(\Gamma, K')_{\mathfrak{p}}$ is non-zero, hence so is $S_2(\Gamma, K')[\mathfrak{p}']$. (For an R-module M and an ideal I of R, we write $M[I]$ for the intersection over the elements r in I of the kernels of $r : M \to M$.) It follows that there is a normalized eigenform f in $S_2(\Gamma, K')$ so that $\mathfrak{p}'$ is the kernel of the θ'_f : $\mathbb{T}_{K'} \to K'$, and therefore $\mathfrak{p}$ is the kernel of θ_f. To prove that the upper-right vertical arrow is bijective, note that the above arguments carry over with K replaced by k. □

Remark 4.3 The surjectivity of the top arrow is called the Deligne-Serre lifting lemma ([DS], lemma 6.11).

Suppose that $\mathfrak{m}$ is a maximal ideal of $\mathbb{T}_{\mathcal{O}}$. Note that the maximal ideals of $\mathbb{T}_K$ mapping to $\mathfrak{m}$ are precisely those $\mathfrak{p}$ for which $\mathfrak{p} \cap \mathbb{T}_{\mathcal{O}}$ is contained in $\mathfrak{m}$. Note also that the natural map

$$\alpha_{\mathfrak{m}} : \mathbb{T}_{\mathfrak{m}} \otimes_{\mathcal{O}} K \to \prod_{\mathfrak{p}} \mathbb{T}_{\mathfrak{p}}$$

is an isomorphism, where the product is over such $\mathfrak{p}$.

It is straightforward to check that the above constructions are well-behaved with respect to replacing the field K by an extension K'. More precisely, for each set $\mathcal{S}$ in the above diagram, there is a natural surjective map from $\varpi : \mathcal{S}' \to \mathcal{S}$ where $\mathcal{S}'$ is defined by replacing K with K', and these maps are compatible with the maps in the diagram. Furthermore the maps

$$\mathbb{T}_{\mathfrak{p}} \otimes_K K' \to \prod_{\mathfrak{p}' \in \varpi^{-1}(\mathfrak{p})} \mathbb{T}_{\mathfrak{p}'} \quad \text{and} \quad \mathbb{T}_{\mathfrak{m}} \otimes_{\mathcal{O}} \mathcal{O}' \to \prod_{\mathfrak{m}' \in \varpi^{-1}(\mathfrak{m})} \mathbb{T}_{\mathfrak{m}'}$$

are isomorphisms by which $\alpha_{\mathfrak{m}} \otimes_K K'$ can be identified with $\prod_{\mathfrak{m}' \in \varpi^{-1}(\mathfrak{m})} \alpha_{\mathfrak{m}'}$.

Associated Galois representations: Suppose that $\mathfrak{p}$ is a maximal ideal of $\mathbb{T}_K$ and $\mathfrak{m}$ is the associated maximal ideal of $\mathbb{T}_{\mathcal{O}}$. By lemma 1.39,

$$\mathcal{T}_\ell(J_H(N)) \otimes_{\mathbf{Z}_\ell} K$$

is free of rank two over $\mathbb{T}_K$, so reduction mod $\mathfrak{p}$ yields a two-dimensional vector space over the field $\mathbb{T}_K/\mathfrak{p}$ endowed with an action of $G_{\mathbf{Q}}$. The resulting Galois representation

$$\rho_{\mathfrak{p}} : G_{\mathbf{Q}} \to GL_2(\mathbb{T}_K/\mathfrak{p})$$

is unramified at all primes p not dividing $N\ell$, and for such p the characteristic polynomial of $\rho_{\mathfrak{p}}(\mathrm{Frob}_p)$ is

$$X^2 - T_p X + p\langle p\rangle \mod \mathfrak{p}.$$

If ℓ is odd, then $\bar{\rho}_{\mathfrak{p}}$ is defined over $\mathbb{T}_{\mathcal{O}}/\mathfrak{m}$ and we write $\rho_{\mathfrak{m}}$ for its semisimplification

$$\rho_{\mathfrak{m}} : G_{\mathbf{Q}} \to GL_2(\mathbb{T}_{\mathcal{O}}/\mathfrak{m}).$$

Thus $\rho_\mathfrak{m}$ is unramified at primes p not dividing $N\ell$ and the characteristic polynomial of Frob_p is

$$X^2 - T_pX + p\langle p\rangle \mod \mathfrak{m}.$$

Suppose that g is in the G_K-conjugacy class of eigenforms in $S_2(\Gamma)$ corresponding to $\mathfrak{p}$. If g is a newform then $\mathbb{T}_K/\mathfrak{p}$ is isomorphic to the field denoted K'_g in section 3.1 and $\rho_\mathfrak{p}$ can be identified with the representation

$$\rho_g : G_\mathbf{Q} \to GL_2(K'_g)$$

considered in theorem 3.1. If also ℓ is odd then $\bar\rho_g$ is obtained from $\rho_\mathfrak{m}$ by extending scalars. More generally suppose that g is not necessarily a newform and consider the associated newform f. Let D denote the product of the primes which divide N but not N_f. Let $\mathbb{T}_K^{(D)}$ denote the K-subalgebra of $\mathbb{T}_K$ generated by the operators T_n for n relatively prime to D and $\langle d\rangle$ for d in $(\mathbb{Z}/N\mathbb{Z})^\times$. Let $\Gamma' = \Gamma_{H'}(N_f)$ where H' is the image of H in $(\mathbb{Z}/N_f\mathbb{Z})^\times$, and let $\mathbb{T}'_K = \mathbb{T}'_\mathbf{Z} \otimes K$ where $\mathbb{T}'_\mathbf{Z}$ is the Hecke algebra acting on $S_2(\Gamma')$. Restriction of operators defines a natural homomorphism $\mathbb{T}_K^{(D)} \to \mathbb{T}'_K$ which is surjective by lemma 4.1. The composite with $\mathbb{T}'_K \twoheadrightarrow K'_f$ factors through the field $\mathbb{T}_K^{(D)}/(\mathfrak{p}\cap\mathbb{T}_K^{(D)})$, so we may identify K'_f with a subfield of $\mathbb{T}_K/\mathfrak{p}$ and $\rho_\mathfrak{p}$ is then equivalent to the extension of scalars of ρ_f. If ℓ is odd, then $\rho_\mathfrak{m}$ and $\bar\rho_f$ are defined and equivalent over a common subfield of $\mathbb{T}_\mathcal{O}/\mathfrak{m}$ and k_f.

The structure of $\mathbb{T}_K$: We now give an explicit description of $\mathbb{T}_K$ in the case that K contains the coefficients of all eigenforms of level dividing N. Let $\mathcal{N}_\Gamma$ denote the set of newforms in $S_2(\Gamma)$, i.e., the set of newforms f of level N_f dividing N such that H is contained in the kernel of the character

$$(\mathbb{Z}/N\mathbb{Z})^\times \to (\mathbb{Z}/N_f\mathbb{Z})^\times \overset{\psi_f}{\to} K^\times.$$

By theorem 1.22

$$S_2(\Gamma, K) = \bigoplus_{f\in\mathcal{N}_\Gamma} S_{K,f},$$

where $S_{K,f}$ is spanned by the linear independent elements

$$\{\, f(a\tau) \mid a \text{ divides } N/N_f\,\}.$$

For each $f = \sum a_n(f)q^n$ in $\mathcal{N}_\Gamma$, let

- $\mathbb{T}_{K,f}$ denote the image of $\mathbb{T}_K$ in $\mathrm{End}_K S_{K,f}$;
- $A_{K,f}$ denote the polynomial ring over K in the variables $u_{f,p}$ indexed by the prime divisors of N/N_f;
- $I_{K,f}$ denote the ideal in $A_{K,f}$ generated by the polynomials

$$P_{f,p}(u_{f,p}) = u_{f,p}^{v_p(N/N_f)-1}(u_{f,p}^2 - a_p(f)u_{f,p} + \psi_f(p)p),$$

for primes p dividing N/N_f (setting $\psi_f(p) = 0$ if p divides N_f).

Consider the K-algebra homomorphism $A_{K,f} \twoheadrightarrow \mathbb{T}_{K,f}$ defined by mapping $u_{f,p}$ to the operator T_p. Since $P_{f,p}$ is the characteristic polynomial of T_p on the span of $\{\, f(ap^i\tau) \,|\, i = 1, \dots, v_p(N/N_f)\,\}$ for each a dividing $N/(N_f p^{v_p(N/N_f)})$, we see that $I_{K,f}$ is contained in the kernel of $A_{K,f} \twoheadrightarrow \mathbb{T}_{K,f}$. Taking the product over f in $\mathcal{N}_\Gamma$, we have a surjective K-algebra homomorphism

$$\prod_f A_{K,f}/I_{K,f} \twoheadrightarrow \prod_f \mathbb{T}_{K,f}.$$

Since the natural map $\mathbb{T}_K \to \prod_f \mathbb{T}_{K,f}$ is injective and

$$\dim_K \mathbb{T}_K = \dim_{\mathbb{C}} S_2(\Gamma, \mathbb{C}) = \sum_f \sigma_0(N/N_f) = \sum_f \dim_K A_{K,f}/I_{K,f},$$

we conclude

Lemma 4.4 *There is an isomorphism of K-algebras:*

$$\alpha : \mathbb{T}_K \to \prod_{f \in \mathcal{N}_\Gamma} A_{K,f}/I_{K,f}$$

defined by

- $\alpha(T_p)_f = a_p(f)$ *if p is a prime not dividing N/N_f;*
- $\alpha(T_p)_f = u_{f,p} \bmod I_{K,f}$ *if p is a prime dividing N/N_f;*
- $\alpha(\langle d \rangle)_f = \psi_f(d)$ *if d is relatively prime to N.*

Remark 4.5 It follows that the algebra $\mathbb{T}_K$ is a "complete intersection" over K in the sense that it is a finite-dimensional K-algebra of the form

$$K[X_1, \dots, X_r]/(P_1, \dots, P_r)$$

for some r.

4.2 Reduced Hecke algebras

As in section 4.1, we suppose K is a finite extension of $\mathbb{Q}_\ell$ with ring of integers $\mathcal{O}$ and residue field k, and fix embeddings $K \hookrightarrow \bar{\mathbb{Q}}_\ell$, $\bar{\mathbb{Q}} \hookrightarrow \bar{\mathbb{Q}}_\ell$ and $\bar{\mathbb{Q}} \hookrightarrow \mathbb{C}$. We assume in this section that ℓ is odd and we fix a representation

$$\bar{\rho} : G_{\mathbb{Q}} \to GL_2(k)$$

which is modular (definition 3.12). Thus $\bar{\rho}$ is equivalent to $\rho_{\mathfrak{m}}$ (over $\mathbb{T}_{\mathcal{O}}/\mathfrak{m}$) for some $\Gamma = \Gamma_H(N)$ and maximal ideal $\mathfrak{m}$ of $\mathbb{T}_{\mathcal{O}}$.

We suppose also that $\bar{\rho}$ has the properties listed at the beginning of section 3.3 and that Σ is a finite set of primes contained in $\Sigma_{\bar{\rho}}$ (definition 3.25).

We shall show that Γ and $\mathfrak{m}$ can be chosen so that the reduced Hecke algebra $\mathbb{T}_\Sigma$ can be identified with a localization $\mathbb{T}_\mathfrak{m}$ of the full Hecke algebra $\mathbb{T}_\mathcal{O}$. The main result is due to Wiles ([W3], prop. 2.15), but we also explain an important variant ([TW], lemma 1) which arises when considering special sets of primes Q as in section 2.8 above.

$\mathbb{T}_\Sigma$ and the full Hecke algebra: Let $\Gamma = \Gamma_0(N_\Sigma)$ where

$$N_\Sigma = \ell^\delta N(\bar\rho) \prod_{p\in\Sigma-\{\ell\}} p^{\dim \bar\rho^{I_p}} = \ell^\delta \prod_{p|N(\bar\rho)} p \prod_{p\in\Sigma-\{\ell\}} p^2 \tag{4.2.1}$$

with $\delta = 0$ if $\bar\rho$ is good and $\ell \notin \Sigma$ and $\delta = 1$ otherwise.

Suppose that f is a newform in $\mathcal{N}_\Sigma$. Recall that $\mathcal{N}_\Sigma$, defined in section 3.3, is the set of newforms f in $S_2(\Gamma)$ such that $\bar\rho$ is equivalent to $\bar\rho_f$ over k_f (lemma 3.26). Note that these representations are equivalent if and only if

$$a_p(f) \bmod \lambda' = \operatorname{tr}(\bar\rho(\mathrm{Frob}_p)) \qquad \text{for all } p \nmid N_\Sigma \ell,$$

where λ' is maximal ideal of the ring of integers of K'_f. There is then a normalized eigenform g in $S_2(\Gamma, K'_f)$ characterized by

- $a_p(g) = a_p(f)$ if p does not divide N_Σ/N_f;
- $a_p(g) = 0$ if $p \neq \ell$ and p divides N_Σ/N_f;
- $a_\ell(g)$ is the unit root of $X^2 - a_\ell(f)X + \ell$ if ℓ divides N_Σ/N_f.

Note that if ℓ divides N_Σ/N_f, then $\delta = 1$ and $\bar\rho$ is flat, hence $\bar\rho$ is ordinary and $a_\ell(f)$ is a unit by theorem 3.1 (f).

Lemma 4.6 *For f and g as above, the reduction $\bar g$ is the normalized eigenform in $S_2(\Gamma, k')$ characterized by:*

- $a_p(\bar g) = \operatorname{tr} \bar\rho_{I_p}(\mathrm{Frob}_p)$ *if $p = \ell$ or p is not in Σ (recall that $\bar\rho_{I_p}$ denotes the representation on I_p-coinvariants);*
- $a_p(\bar g) = 0$ *otherwise.*

Furthermore, if K is as in lemma 4.4, then g is the unique normalized eigenform in $S_{K,f}$ with this reduction.

Proof: If p is not in $\Sigma \cup \{\ell\}$ then the formula for $a_p(\bar g) = a_p(\bar f)$ follows from theorem 3.1, parts (a) and (e). If $p = \ell$, then we use parts (f) and (g) of theorem 3.1. Otherwise, we have $a_p(\bar g) = 0$ by theorem 1.27(d) (if $p^2|N_f$ and f has trivial character, then $a_p(f) = 0$). The uniqueness follows from the fact that if $p||N_f$, then $a_p(f) = \pm 1$ (by theorem 1.27(b)). □

Note in particular that $\bar g$ has coefficients in k and is independent of the choice of f in $\mathcal{N}_\Sigma$. We let $\mathfrak{m}$ denote the corresponding maximal ideal of $\mathbb{T}_\mathcal{O}$.

Proposition 4.7 *There is an isomorphism* $\mathbb{T}_\Sigma \xrightarrow{\sim} \mathbb{T}_\mathfrak{m}$ *of* $\mathcal{O}$*-algebras such that* $T_p \mapsto T_p$ *for all primes* p *not dividing* $N_\Sigma \ell$.

Proof: Replacing K by a larger field K', we have $\mathbb{T}_\Sigma \otimes_\mathcal{O} \mathcal{O}' \xrightarrow{\sim} \mathbb{T}'_\Sigma$ (lemma 3.27) and $\mathbb{T}_\mathfrak{m} \otimes_\mathcal{O} \mathcal{O}' \xrightarrow{\sim} \mathbb{T}'_{\mathfrak{m}'}$ (since there is a unique maximal ideal $\mathfrak{m}'$ of $\mathbb{T}_{\mathcal{O}'}$ over $\mathfrak{m}$). We are therefore reduced to the case where K is as in lemma 4.4. Note that $\mathcal{N}_\Sigma$ is the set of newforms f in $\mathcal{N}_\Gamma$ such that

$$a_p(\bar{f}) = \mathrm{tr}\,(\bar{\rho}(\mathrm{Frob}_p)) \qquad \text{for all } p \nmid N_\Sigma \ell.$$

(We have $K = K'_f$ and write $\bar{f}$ for f mod λ.)

We now define an isomorphism of K-algebras

$$\kappa : \mathbb{T}_\mathfrak{m} \otimes K \xrightarrow{\sim} \prod_{f \in \mathcal{N}_\Sigma} K,$$

such that for each $f \in \mathcal{N}_\Sigma$:

- $\kappa(T_p)_f = a_p(f)$ if p is not in Σ;
- $\kappa(T_p)_f = 0$ if p is in $\Sigma - \{\ell\}$;
- $\kappa(T_\ell)_f$ is the unit root of $X^2 - a_\ell(f)X + \psi_f(\ell)\ell$ if ℓ is in Σ.

Recall that $\mathbb{T}_\mathfrak{m} \otimes K \cong \prod_\mathfrak{p} \mathbb{T}_\mathfrak{p}$ where $\mathfrak{p}$ runs over the primes of $\mathbb{T}_K$ whose preimage in $\mathbb{T}_\mathcal{O}$ is contained in $\mathfrak{m}$. Thus according to lemma 4.4 we have

$$\mathbb{T}_\mathfrak{m} \otimes K \xrightarrow{\sim} \prod_{f \in \mathcal{N}_\Gamma} \prod_{\mathfrak{p} \in \mathcal{M}_f} (A_{K,f}/I_{K,f})_\mathfrak{p},$$

where $\mathcal{M}_f$ is the set of prime ideals in $A_{K,f}/I_{K,f}$ whose preimage in $\mathbb{T}_\mathcal{O}$ is contained in $\mathfrak{m}$. If f is not in $\mathcal{N}_\Sigma$, then $\mathcal{M}_f$ is empty. If f is in $\mathcal{N}_\Sigma$, then $\mathcal{M}_f$ consists only of the kernel $\mathfrak{p}_f$ of the map defined by $u_{p,f} \mapsto a_p(g)$ where g is the eigenform of lemma 4.6. Furthermore the maps $u_{p,f} \mapsto a_p(g)$ induce isomorphisms $(A_{K,f}/I_{K,f})_{\mathfrak{p}_f} \xrightarrow{\sim} K$, and taking the product over $f \in \mathcal{N}_\Sigma$, we obtain the desired isomorphism κ.

Identifying $\mathbb{T}_\Sigma$ with the $\mathcal{O}$-subalgebra of $\prod_{f \in \mathcal{N}_\Sigma} K$ generated by the elements $T_p = (a_p(f))_f$ for all p not dividing $N_\Sigma \ell$, it suffices to prove that $\mathbb{T}_\Sigma$ contains $\kappa(T_p)$ for all p dividing $N(\bar{\rho})\ell$. Now observe that for each f in $\mathcal{N}_\Sigma$ the representation ρ_f is isomorphic to

$$G_\mathbb{Q} \to GL_2(\mathbb{T}_\Sigma) \to GL_2(K)$$

obtained by composing $\rho_\Sigma^{\mathrm{mod}}$ with the projection $\mathbb{T}_\Sigma \to K$. It follows that for p dividing $N(\bar{\rho})\ell^\delta$, the image of $(\rho_\Sigma^{\mathrm{mod}})_{I_p}(\mathrm{Frob}_p)$ in K is $(\rho_f)_{I_p}(\mathrm{Frob}_p) = a_p(f)_f$ and therefore

$$(\rho_\Sigma^{\mathrm{mod}})_{I_p}(\mathrm{Frob}_p) = \kappa(T_p).$$

Finally if $\delta = 0$, then we conclude from lemma 4.1 that $\mathbb{T}_\Sigma$ contains $\kappa(T_\ell)$. $\square$

A twisted variant: Suppose now that Σ is a finite set of primes contained in $\Sigma_{\bar\rho} - \{\ell\}$ such that if p is in Σ then $\bar\rho(\mathrm{Frob}_p)$ has eigenvalues α_p and β_p in k satisfying

$$\alpha_p/\beta_p \neq p^{\pm 1}. \tag{4.2.2}$$

Note that (4.2.2) is equivalent to

$$\mathrm{tr}\,(\bar\rho(\mathrm{Frob}_p))^2 \neq (p+1)^2. \tag{4.2.3}$$

Let Q denote the set of primes in Σ such that $q \equiv 1 \bmod \ell$. The set of primes Q is therefore as in section 2.8. Choosing an eigenvalue α_q of $\bar\rho(\mathrm{Frob}_q)$ for each $q \in Q$ as in section 2.8, we regard R_Q and hence $\mathbb{T}_Q$ as an $\mathcal{O}[\Delta_Q]$-algebra. (Recall that Δ_Q is the the maximal quotient of $(\mathbb{Z}/(\prod_{q\in Q} q)\mathbb{Z})^\times$ of ℓ-power order.)

Instead of working with $\Gamma_0(N_\Sigma)$ as in proposition 4.7, we shall now work with the group

$$\Gamma = \Gamma_0(N_\emptyset) \cap \Gamma_1(M), \tag{4.2.4}$$

where

$$M = \prod_{p\in\Sigma-Q} p^2 \prod_{q\in Q} q.$$

Remark 4.8 We are about to relate $\mathbb{T}_Q$ to a localization of $\mathbb{T}_\mathcal{O}$, where $\mathbb{T}_\mathcal{O}$ is now defined using the Hecke operators on $S_2(\Gamma)$, with Γ defined by (4.2.4). Recall that $\mathbb{T}_Q$ is defined using modular forms with trivial character, but note that the modular forms involved in the definition of $\mathbb{T}_\mathcal{O}$ may have non-trivial character. The purpose of establishing this relationship is to give a concrete realization of the image of Δ_Q in $\mathbb{T}_Q$ for the purpose of proving theorem 3.31.

Suppose that f is a newform pair in $\mathcal{N}_\emptyset$. For each q in Q, let $\tilde\alpha_q(f)$ be the root of $X^2 - a_q(f)X + q = 0$ (in K'_f) whose image in k_f is α_q. We let g denote the unique normalized eigenform in $S_2(\Gamma, K'_f)$ of trivial character such that

- $a_p(g) = a_p(f)$ if p is not in Σ;
- $a_p(g) = \tilde\alpha_p(f)$ if p is in Q;
- $a_p(g) = 0$ otherwise.

The reduction $\bar g$ is the unique normalized eigenform in $S_2(\Gamma, k')$ of trivial character such that

- $a_p(\bar g) = \mathrm{tr}\,\bar\rho_{I_p}(\mathrm{Frob}_p)$ if p is not in Σ;

- $a_p(\bar{g}) = \alpha_p$ if p is in Q;
- $a_p(\bar{g}) = 0$ otherwise.

Thus $\bar{g}$ has coefficients in k and is independent of the choice of f in $\mathcal{N}_\emptyset$. We let $\mathfrak{m}$ denote the corresponding maximal ideal of $\mathbb{T}_\mathcal{O}$.

Suppose for the moment that K contains the coefficients of all eigenforms of level dividing N_Σ. Let $\mathcal{N}'$ denote the set of newforms g in $\mathcal{N}_\Gamma$ such that

- $\bar{\rho}_g \sim \bar{\rho}$;
- $a_p(\bar{g}) = \alpha_p$ for all p dividing $N_g/N_\emptyset$.

Suppose we are given a newform $g \in \mathcal{N}'$. Let ψ_g denote its character and write Q_g for the conductor of ψ_g. Note that ψ_g has trivial reduction and hence ℓ-power order, and that Q_g divides Q. By proposition 3.2 we see that if $p \in \Sigma - Q$, then ρ_g is unramified at p and hence N_g is not divisible by p. Furthermore, by theorem 3.1 (e) and (4.2.2), $N_g = N_\emptyset Q_g$. Let ξ_g denote the character of $(\mathbb{Z}/Q_g\mathbb{Z})^\times$ of ℓ-power order such that ξ_g^{-2} is the primitive character associated to ψ_g. Then

$$g \otimes \xi_g = \sum \xi_g(n) a_n(g) q^n$$

is in $\mathcal{N}_{Q_g} \subset \mathcal{N}_Q$.

Lemma 4.9 *The map $g \mapsto g \otimes \xi_g$ defines a bijection between $\mathcal{N}'$ and $\mathcal{N}_Q$.*

Proof: Suppose we are given a newform f in $\mathcal{N}_Q$; i.e., f is in $\mathcal{N}_{\Gamma_0(Q)}$ and $\bar{\rho}_f \sim \bar{\rho}$. For $q \in Q$, we have (by lemma 2.44 for example) that

$$\rho_f|_{G_q} \sim \begin{pmatrix} \xi_{f,q} & 0 \\ 0 & \epsilon\xi_{f,q}^{-1} \end{pmatrix},$$

where $\xi_{f,q} : G_q \to K^\times$ is a character whose reduction is the unramified character sending Frob_q to α_q. Note that the characters $\xi_{f,q}|_{I_q}$ have ℓ-power order. In particular $N_f/N_\emptyset$ is Q_f^2 where Q_f is the product of the primes $q \in Q$ such that $\xi_{f,q}$ is ramified (theorem 3.1 (d)). Note also that there is a unique character

$$\xi_f : (\mathbb{Z}/Q_f\mathbb{Z})^\times \to K^\times$$

such that

$$I_q \to G_\mathbb{Q} \twoheadrightarrow \mathrm{Gal}\,(\mathbb{Q}(\zeta_{Q_f})/\mathbb{Q}) \xrightarrow{\xi_f} K^\times$$

coincides with the restriction of $\xi_{f,q}$ to I_q for each $q|Q_f$. (We have written ξ_f for the character of $G_\mathbb{Q}$ as well as the corresponding Dirichlet character. We shall also write ξ_f for the corresponding character of Δ_Q.) Let g denote the newform associated to the eigenform

$$\sum \xi_f^{-1}(n) a_n(f) q^n \in S_2(\Gamma_0(N_\emptyset) \cap \Gamma_1(Q_f^2), K).$$

By proposition 2.6 and theorem 3.1 we have

- $\rho_g \sim \rho_f \otimes \xi_f^{-1}$,
- $\psi_g = \xi_f^{-2}$,
- $N_g = N_\emptyset Q_f$ and
- $a_q(g) = (\xi_f|_{G_q}^{-1}\xi_{f,q})(\mathrm{Frob}_q)$ has reduction $\bar{\xi}_{f,q}(\mathrm{Frob}_q) = \alpha_q$ for all q dividing Q_f.

Therefore g is in $\mathcal{N}'$ and $f = g \otimes \xi_g$. In particular $g \mapsto g \otimes \xi_g$ is surjective and $\xi_g = \xi_{g\otimes\xi_g}$. Injectivity follows as well on noting that if $g \otimes \xi_g = g' \otimes \xi_{g'}$ then $\xi_g = \xi_{g'}$, hence $g = g'$. □

Recall that $\mathbb{T}_\mathcal{O}$ contains the operators $\langle d\rangle$ for d in $(\mathbb{Z}/M\mathbb{Z})^\times$, and note that $\langle d\rangle - 1 \in \mathfrak{m}$ for all d. Let H' denote the kernel of

$$(\mathbb{Z}/M\mathbb{Z})^\times \twoheadrightarrow (\mathbb{Z}/(\prod_{q\in Q} q)\mathbb{Z})^\times \twoheadrightarrow \Delta_Q.$$

Since the order of H' is not divisible by ℓ, we find that $\sum_{d\in H'}\langle d\rangle \notin \mathfrak{m}$. It follows that if $d \in H'$ then $\langle d\rangle = 1$ in $\mathbb{T}_\mathfrak{m}$. We may therefore regard $\mathbb{T}_\mathfrak{m}$ as an $\mathcal{O}[\Delta_Q]$-algebra via the map $d \mapsto \langle d\rangle$. Recall that $\mathbb{T}_Q$ is considered an $\mathcal{O}[\Delta_Q]$-algebra via the map

$$\mathcal{O}[\Delta_Q] \to R_Q \to \mathbb{T}_Q.$$

If p is a prime not in Q, let x_p denote the unique element of Δ_Q such that $x_p^{-2} = \bar{p}$.

Proposition 4.10 *There is an isomorphism $\mathbb{T}_Q \xrightarrow{\sim} \mathbb{T}_\mathfrak{m}$ of $\mathcal{O}[\Delta_Q]$-algebras such that $T_p \mapsto x_pT_p$ for all primes $p \notin \Sigma$ with $p \nmid \ell N(\bar{\rho})$.*

Proof: We may enlarge K so that we are in the setting of lemma 4.9. We then define an isomorphism of K-algebras

$$\kappa : \mathbb{T}_\mathfrak{m} \otimes K \xrightarrow{\sim} \prod_{g\in\mathcal{N}'} K$$

such that for each $g \in \mathcal{N}'$:

- $\kappa(T_p)_g = a_p(g)$ if p is not in Σ or if p divides Q_g,
- $\kappa(T_p)_g = 0$ if p is in $\Sigma - Q$,
- $\kappa(T_p)_g$ is the root of $X^2 - a_p(g) + \psi_g(p)p$ with reduction α_p if p is in Q but does not divide Q_g, and
- $\kappa(d)_g = \psi_g(d)$ for all $d \in \Delta_Q$.

The existence of such an isomorphism follows from lemma 4.4 which gives

$$\mathbb{T}_{\mathfrak{m}} \otimes K \xrightarrow{\sim} \prod_{g \in \mathcal{N}_\Gamma} \prod_{\mathfrak{p} \in \mathcal{M}_g} (A_{K,g}/I_{K,g})_{\mathfrak{p}},$$

where $\mathcal{M}_g$ is the set of prime ideals in $A_{K,g}/I_{K,g}$ whose preimage in $\mathbb{T}_{\mathcal{O}}$ is contained in $\mathfrak{m}$. If g is not in $\mathcal{N}'$, then $\mathcal{M}_g$ is empty. If g is in $\mathcal{N}'$, then $\mathcal{M}_g$ consists of the prime ideal corresponding to the eigenform whose eigenvalues are prescribed as above, and one checks that $(A_{K,g}/I_{K,g})_{\mathfrak{p}} = K$.

Viewing $\mathbb{T}_Q$ as a subalgebra of $\prod_{f \in \mathcal{N}_Q} K$ and matching indices via the bijection $f = g \otimes \xi_g \leftrightarrow g$, we obtain an injective homomorphism of $\mathcal{O}[\Delta_Q]$-algebras

$$\kappa' : \mathbb{T}_Q \longrightarrow \prod_{g \in \mathcal{N}'} K$$

such that

$$\kappa'(T_p) = (\xi_g(p) a_p(g))_g = \kappa(x_p T_p)$$

for primes $p \notin \Sigma$ such that $p \nmid \ell N_\emptyset$. Since $\mathbb{T}_Q$ is generated over $\mathcal{O}$ by the set of such T_p, we see that $\kappa'(\mathbb{T}_Q)$ is contained in $\kappa(\mathbb{T}_{\mathfrak{m}})$. On the other hand, the image of κ' contains the image of Δ_Q, hence contains $\kappa(\langle d \rangle)$ for d in $(\mathbb{Z}/M\mathbb{Z})^\times$ as well as $\kappa(T_p)$ for $p \notin \Sigma$ with $p \nmid N_\emptyset \ell$.

It remains to prove that $\kappa(T_p)$ is in the image of $\kappa'(\mathbb{T}_Q)$ for p dividing $N_\emptyset \ell \prod_{q \in Q} q$. Consider the composite

$$R_Q \xrightarrow{\phi_Q} \mathbb{T}_Q \xrightarrow{\kappa'} \prod_{g \in \mathcal{N}'} K.$$

The pushforward of ρ_Q^{univ} is equivalent to $\prod \rho_g \otimes \xi_g$ and therefore $\prod \rho_g$ is equivalent to the pushforward of

$$\rho_Q^{\mathrm{univ}} \otimes \xi_Q^{-1}$$

(where ξ_Q was defined in section 2.8). For primes p dividing $N_\emptyset$, we recover $\kappa(T_p)$ as the image of Frob_p on the I_p-coinvariants, hence $\kappa(T_p)$ is in $\kappa'(\mathbb{T}_Q)$. For $q \in Q$ and $g \in \mathcal{N}'$, the pushforward of $\xi_Q|_{G_q}^{-1} \cdot \xi_{q,Q}$ to the g-component is an unramified summand of $\rho_g|_{G_q}$ whose reduction sends Frob_q to α_q. It follows that this character maps Frob_q to $\kappa(T_q)$ and we conclude that $\kappa(T_q) \in \kappa'(\mathbb{T}_Q)$. Finally in the case that ℓ does not divide $N_\emptyset$, we appeal to lemma 4.1 to conclude that $\kappa(T_\ell) \in \kappa'(\mathbb{T}_Q)$. □

Auxiliary primes: For Σ as in proposition 4.10, i.e. a set of primes satisfying (4.2.2), a somewhat simpler argument provides a similar a description of $\mathbb{T}_\emptyset$ with Γ replaced by

$$\Gamma' = \Gamma_0(N_\emptyset) \cap \Gamma_H(M) \qquad (4.2.5)$$

where H is the ℓ-Sylow subgroup of $(\mathbb{Z}/M\mathbb{Z})^\times$. While we shall make no direct use of this, the group Γ' will play a role in the proof of theorem 3.31 and we shall need to choose Σ so that Γ' has no elliptic elements; i.e., non-trivial elements of finite order. This is the case for example if Σ contains a prime $p \not\equiv 1 \bmod \ell$ with $p > 3$. The group Γ' is then contained in $\Gamma_1(N)$ for some integer $N > 3$, and therefore has no elliptic elements. Indeed if $\begin{pmatrix} a & b \\ c & d \end{pmatrix} \in \Gamma$ has finite order then the roots of $X^2 - (a+d)X + 1$ are roots of unity and we deduce that this matrix is the identity.

In order to show that Σ can be so chosen, we appeal to the following lemma (cf. [DT2], lemma 3).

Lemma 4.11 *Suppose that G is a finite group, $\chi : G \to \bar{k}^\times$ is a character of order d and $\bar{\rho} : G \to GL_2(\bar{k})$ is a representation. Suppose that for all $g \notin \ker \chi$ we have*

$$(\operatorname{tr} \bar{\rho}(g))^2 / \det \bar{\rho}(g) = (1 + \chi(g))^2 / \chi(g). \tag{4.2.6}$$

- *If $d > 3$, then $\bar{\rho}$ is reducible.*
- *If $d = 3$, then $\bar{\rho}$ is reducible or has projective image isomorphic to A_4.*
- *If $d = 2$, then $\bar{\rho}|_{\ker \chi}$ is reducible.*

Proof: Note that if $\bar{\rho}(g)$ is a scalar then $\chi(G) = 1$. Hence χ induces a surjective homomorphism $\chi' : G' \to C_d$ where G' is the projective image of $\bar{\rho}$ and C_d is cyclic of order d. Furthermore if $d = 2$, then every element of $G' - \ker \chi'$ has order 2. The lemma then follows from theorem 2.47(b). □

4.3 Proof of theorem 3.31

We shall give a proof of theorem 3.31 which is based on the q-expansion principle rather than the method of de Shalit [dS] employed in [TW]. It will be more convenient to consider the action of the Hecke operators on the full space of modular forms $M_2(\Gamma)$. The Riemann-Roch theorem shows that the dimension of $M_2(\Gamma)$ is $g + s - 1$ where g is the genus of the modular curve X_Γ associated to Γ and s is the number of cusps on X_Γ (see for example [Shi2] thm. 2.23 or [DI] (12.1.5)).

Eisenstein maximal ideals: One can give an explicit description of a space of Eisenstein series $G_2(\Gamma)$ so that

$$M_2(\Gamma) = S_2(\Gamma) \oplus G_2(\Gamma).$$

(See for example [Hi3], lemma 5.2.)

There is a natural action on $M_2(\Gamma)$ by the Hecke operators $\langle d \rangle$ for all $d \in (\mathbb{Z}/N\mathbb{Z})^\times$ and T_p for all primes p. We shall only need to use the operators $\langle d \rangle$ and T_p for p not dividing N, and we let $\tilde{\mathbb{T}}_{\mathbb{Z}}$ denote the subring of End $(M_2(\Gamma))$

these generate. The ring $\tilde{\mathbb{T}}_{\mathbf{Z}}$ is commutative and is finitely generated and free as a $\mathbb{Z}$-module. That $\tilde{\mathbb{T}}_{\mathbf{Z}}$ is finitely generated is proved for example by showing that $M_2(\Gamma, \mathbb{Z})$, the set of forms with integer Fourier coefficients at ∞, is stable under the Hecke operators and contains a basis for $M_2(\Gamma)$ (see [DI], cor. 12.3.12 and prop. 12.4.1). Alternatively, one can show that $\tilde{\mathbb{T}}_{\mathbf{Z}}$ acts faithfully on the cohomology of the non-compact modular curve Y_Γ (cf. section 1.3).

For any ring A, we write $M_2(\Gamma, A)$ for $M_2(\Gamma, \mathbb{Z}) \otimes A$ and regard this as a module for $\tilde{\mathbb{T}}_A := \tilde{\mathbb{T}}_{\mathbf{Z}} \otimes A$. If $\mathfrak{n}$ is a maximal ideal of $\tilde{\mathbb{T}}_{\mathcal{O}}$ we write $\tilde{\mathbb{T}}_{\mathfrak{n}}$ for the localization. If $\mathfrak{m}$ is a maximal ideal of $\mathbb{T}_{\mathcal{O}}$ then we let $\tilde{\mathfrak{m}}$ denote its preimage in $\tilde{\mathbb{T}}_{\mathcal{O}}$.

We say that a maximal ideal $\mathfrak{n}$ of $\tilde{\mathbb{T}}_{\mathcal{O}}$ is *Eisenstein* if

$$T_p \equiv p + 1 \bmod \mathfrak{n} \quad \text{for all } p \equiv 1 \bmod N.$$

One sees from the explicit description of the Eisenstein series that $T_p = p+1$ on $G_2(\Gamma)$ if $p \equiv 1 \bmod N$. We shall be interested in the *non-Eisenstein* maximal ideals because of the following lemma (see [R5], thm. 5.2(c)).

Lemma 4.12 *Suppose that ℓ is odd. The representation*

$$\rho_{\mathfrak{m}} : G_{\mathbf{Q}} \to GL_2(\mathbb{T}_{\mathcal{O}}/\mathfrak{m})$$

is absolutely irreducible if and only if $\tilde{\mathfrak{m}}$ is not Eisenstein.

Proof: If $\rho_{\mathfrak{m}}$ is not absolutely irreducible then $\rho_{\mathfrak{m}} \sim \chi_1 \oplus \chi_2$ with N divisible by the product of the conductors of χ_1 and χ_2. Furthermore if ℓ does not divide N, then one of the characters is unramified at ℓ while the other coincides with the cyclotomic character ϵ on I_ℓ. Therefore $\tilde{\mathfrak{m}}$ is Eisenstein. Conversely if $\tilde{\mathfrak{m}}$ is Eisenstein then proposition 2.6 implies that $\rho_{\mathfrak{m}}$ restricted to $G_{\mathbf{Q}(\zeta_{N\ell})}$ has trivial semisimplification from which it follows that $\rho_{\mathfrak{m}}$ is also reducible. □

Differentials: Suppose now that Σ is a finite set of primes satisfying (4.2.2) Moreover we assume that Σ contains a prime $p > 3$ with $p \not\equiv 1 \bmod \ell$. Let Γ and Γ' be defined as in (4.2.4) and (4.2.5); i.e.

$$\begin{aligned} \Gamma &= \Gamma_0(N_\emptyset) \cap \Gamma_1(M) \\ \Gamma' &= \Gamma_0(N_\emptyset) \cap \Gamma_H(M) \end{aligned}$$

where H is the ℓ-Sylow subgroup of $(\mathbb{Z}/M\mathbb{Z})^\times$. Let X and X' denote the modular curves associated to Γ and Γ'.

We first consider the case where $\ell \nmid N_\emptyset$. The curve X has a smooth proper model $\mathcal{X}$ over $\mathbb{Z}[1/N_\emptyset M]$ such that the complement of the cusps parametrizes cyclic isogenies $(E_1, i_1) \to (E_2, i_2)$ of degree $N_\emptyset$ where E_j is an elliptic curve and i_j is an embedding $\mu_M \hookrightarrow E_j$. (See [DR] or [Kat] for example.) The action of H on X extends to $\mathcal{X}$ and the quotient $\mathcal{X}' = \mathcal{X}/H$ is a smooth model over $\mathbb{Z}[1/N_\emptyset M]$ for X'. The natural projection $\mathcal{X} \to \mathcal{X}'$ is étale. (The fact that it is étale on the complement of the cusps follows from the natural moduli-theoretic description of $\mathcal{X}'$ using the fact that $\Gamma_H(M) \subset \Gamma_1(p)$ for some $p > 3$. One then need only verify that $X \to X'$ is unramified at the cusps.)

For a $\mathbb{Z}[1/N_\emptyset M]$-algebra A, we define

- $\Omega_A = H^0(\mathcal{X}_A, \Omega^1_{\mathcal{X}_A/A})$,
- $\Omega'_A = H^0(\mathcal{X}'_A, \Omega^1_{\mathcal{X}'_A/A})$,
- $\tilde{\Omega}_A = H^0(\mathcal{X}_A, \Omega^1_{\mathcal{X}_A/A}(\mathcal{D}))$,
- $\tilde{\Omega}'_A = H^0(\mathcal{X}'_A, \Omega^1_{\mathcal{X}'_A/A}(\mathcal{D}'))$,

where $\mathcal{D}$ (resp. $\mathcal{D}'$) denotes the reduced divisor defined by the cusps of $\mathcal{X}$ (resp. $\mathcal{X}'$). The q-expansion principle and standard base-change arguments allow us to identify these with $S_2(\Gamma, A)$, $S_2(\Gamma', A)$, $M_2(\Gamma, A)$ and $M_2(\Gamma', A)$ (see [Kat], sec. 1.6, 1.7). We may therefore regard the first two of these as modules for $\mathbb{T}_A$ and all of them as modules for $\tilde{\mathbb{T}}_A$.

Lemma 4.13 *Suppose that $A = k$ or K.*

(a) $\tilde{\Omega}_{\mathcal{O}} \otimes_{\mathcal{O}} A \cong \tilde{\Omega}_A$ *and* $\tilde{\Omega}'_{\mathcal{O}} \otimes_{\mathcal{O}} A \cong \tilde{\Omega}'_A$.

(b) *The natural map $\mathcal{X} \to \mathcal{X}'$ induces an isomorphism*

$$\tilde{\Omega}'_A \xrightarrow{\sim} \tilde{\Omega}^H_A.$$

(c) *If $\mathfrak{n}$ is a non-Eisenstein maximal ideal of $\tilde{\mathbb{T}}_{\mathcal{O}}$, then the localization at $\mathfrak{n}$ of $\Omega_{\mathcal{O}} \to \tilde{\Omega}_{\mathcal{O}}$ is an isomorphism.*

(d) *If $\mathfrak{m}$ is a maximal ideal of $\mathbb{T}_A$ then $\Omega_A[\mathfrak{m}]$ is one-dimensional over $\mathbb{T}_A/\mathfrak{m}$.*

Sketch of proof.

(a) In the case $A = K$ this follows from the fact that K is flat over $\mathcal{O}$, so suppose $A = k$. Identify $\tilde{\Omega}_{\mathcal{O}}/\lambda\tilde{\Omega}_{\mathcal{O}}$ with the direct image of $\tilde{\Omega}_k$ under $\mathcal{X}_k \to \mathcal{X}_{\mathcal{O}}$ and use that $H^1(\mathcal{X}_k, \Omega^1_{\mathcal{X}_k/k}(\mathcal{D}))$ vanishes by Serre duality. The case of $\tilde{\Omega}'$ is similar. (See [Maz1] sec. II.3.)

(b) Using the fact that $\mathcal{X} \to \mathcal{X}'$ is étale one identifies the pull-back of $\Omega^1_{\mathcal{X}'_A/A}$ with $\Omega^1_{\mathcal{X}_A/A}$ and that of $\mathcal{D}'$ with $\mathcal{D}$.

(c) The map is injective and one proves that its cokernel is free and annihilated by the operators $T_p - (p+1)$ for all $p \equiv 1 \bmod N_\emptyset M$. To prove the latter assertion, first observe that it holds with $\mathcal{O}$ replaced by $\mathbb{C}$, then by $\mathbb{Z}[1/N_\emptyset M]$.

(d) This follows from

$$\Omega_A[\mathfrak{m}] \cong S_2(\Gamma, A)[\mathfrak{m}] \cong \mathrm{Hom}_A(\mathbb{T}_A, A)[\mathfrak{m}] \cong \mathrm{Hom}_A(\mathbb{T}_A/\mathfrak{m}, A),$$

where the middle isomorphism is that of proposition 1.34, but let us reformulate the argument in a way more easily generalized to the case of ℓ dividing $N_\emptyset$ discussed below. Note that $\Omega_A[\mathfrak{m}]$ is non-zero since $\mathbb{T}_\mathfrak{m}$ acts faithfully on $\Omega_\mathfrak{m}$. To prove that the dimension is at most one we can enlarge the field A and assume $\mathbb{T}_A/\mathfrak{m} \cong A$. One then shows that an eigenform f in Ω_A (for all the Hecke operators T_p and $\langle d\rangle$) is determined by its eigenvalues and the first coefficient of its q-expansion. This follows from the fact that f is determined by its q-expansion and for all n, $a_n(f) = a_1(T_n f)$ and T_n can be expressed in terms of the T_p and $\langle d\rangle$. (See [Maz1], sec. II.9.) □

We now explain how the situation changes if ℓ divides $N_\emptyset$. In that case X has a regular model $\mathcal{X}$ over $\mathbb{Z}[\ell/N_\emptyset M]$ with the same moduli-theoretic description as above. This model is smooth over $\mathbb{Z}[1/N_\emptyset M]$, but $\mathcal{X}_{\mathbb{F}_\ell}$ has two smooth irreducible components crossing at ordinary double points as in [DR] sec. V.1. The quotient $\mathcal{X}' = \mathcal{X}/H$ is a regular model for X' over $\mathbb{Z}[\ell/N_\emptyset M]$ and $\mathcal{X} \to \mathcal{X}'$ is étale. For a $\mathbb{Z}[\ell/N_\emptyset M]$-algebra A we define Ω_A, Ω'_A, $\tilde{\Omega}_A$ and $\tilde{\Omega}'_A$ as before, except that Ω^1 is replaced by the sheaf of regular differentials (see [DR] sec. I.2, [Maz1] sec. II.6 or [MRi] sec. 7). Formation of these modules again commutes with change of the base A, but we can no longer identify them with $S_2(\Gamma, A)$, $S_2(\Gamma', A)$, $M_2(\Gamma, A)$ and $M_2(\Gamma', A)$ if ℓ is not invertible in A. For every A, there is a natural action of $\mathbb{T}_A$ (resp. $\tilde{\mathbb{T}}_A$) on Ω_A and Ω'_A (resp. $\tilde{\Omega}_A$ and $\tilde{\Omega}'_A$). In the case of $\mathbb{T}_A$ this is proved by identifying Ω_A (resp. Ω'_A) with the cotangent space at the origin for the Néron model over A for the Jacobian of X (resp. X'). In the case of $\tilde{\mathbb{T}}_A$ the action is defined using Grothendieck-Serre duality (as in [Maz1] sec. II.6 or [MRi] sec. 7).

Lemma 4.14 *If ℓ divides $N_\emptyset$ then lemma 4.13 carries over with the above notation and the additional hypothesis $T_\ell \notin \mathfrak{m}$ for part* (d).

The proof is essentially the same except that part (d) is more delicate in the case $A = k$. For the proof in that case we refer the reader to [W3], lemma 2.2 (from which the result stated here is immediate). We remark however that we shall only use lemma 4.14 if $\rho_\mathfrak{m}$ is not good. In that case one can also deduce the statement here from the argument used in the proof of [MRi] prop. 20, which shows instead that $\dim_k \Omega_k[\mathfrak{m}'] = 1$ where $\mathfrak{m}'$ is a certain maximal ideal corresponding to $\mathfrak{m}$, but contained in an algebra defined using the operators $\langle d\rangle$, T_p for $p \neq \ell$ and w_ℓ.

Remark 4.15 The approach in [TW] to proving theorem 3.31 is based on the result that under certain hypotheses $(\mathcal{T}_\ell(J) \otimes_{\mathbb{Z}_\ell} \mathcal{O})_\mathfrak{m}$ is free of rank two over $\mathbb{T}_\mathfrak{m}$ (see [W3], thm. 2.1 and its corollaries). This result generalizes work of Mazur Ribet, and Edixhoven (see secs. 14 and 15 of ch. II of [Maz1], thm. 5.2 of [R5], [MRi] and sec. 9 of [Edi]), and the key to its proof is lemmas 4.13(d) and 4.14. We shall instead give a proof of theorem 3.31 which is based directly on lemma 4.13, (d).

The fact that $(\mathcal{T}_\ell(J)\otimes_{\mathbf{Z}_\ell}\mathcal{O})_\mathfrak{m}$ is free of rank two over $\mathbb{T}_\mathfrak{m}$ actually underlies Wiles' approach to many intermediate results along the way to proving the Shimura-Taniyama conjecture for semistable elliptic curves. We shall appeal to a special case in the course of proving theorem 3.36 below.

Proof of the theorem: Suppose we are given a representation $\bar\rho$ as in section 3.3 and a set of primes Q as in the statement of theorem 3.31. We apply lemma 4.11 with $G = G_\mathbf{Q}$ and $\chi = \epsilon$ to choose an auxiliary prime p not dividing $6N_\emptyset\ell\prod_{q\in Q} q$ such that

- $p \not\equiv 1 \bmod \ell$
- $\mathrm{tr}\,(\bar\rho(\mathrm{Frob}_p))^2 \neq (p+1)^2$.

To prove theorem 3.31 we may replace K by a larger field and assume that k contains the eigenvalues of $\bar\rho(\mathrm{Frob}_p)$.

Let $\Sigma = Q \cup \{p\}$. Thus Σ is a set of primes as in proposition 4.10 and Σ contains a prime $p > 3$ such that $p \not\equiv 1 \bmod \ell$. We let Γ be as in (4.2.4); thus

$$\Gamma = \Gamma_0(N_\emptyset) \cap \Gamma_1(M)$$

where $M = p^2\prod_{q\in Q} q$, and we choose the maximal ideal $\mathfrak{m}$ of $\mathbb{T}_\mathcal{O}$ as in proposition 4.10. Let H denote the ℓ-Sylow subgroup of $(\mathbb{Z}/M\mathbb{Z})^\times$ and regard $\mathbb{T}_\mathcal{O}$ and hence $\mathbb{T}_\mathfrak{m}$ as an $\mathcal{O}[H]$-algebra via $d \mapsto \langle d\rangle$. In view of proposition 4.10, theorem 3.31 is equivalent to the following:

Theorem 4.16 *$\mathbb{T}_\mathfrak{m}$ is free over $\mathcal{O}[H]$.*

Proof: Consider the $\tilde{\mathbb{T}}_\mathcal{O}$-module

$$\tilde L = \mathrm{Hom}_\mathcal{O}(\tilde\Omega_\mathcal{O}, \mathcal{O})$$

and the $\mathbb{T}_\mathcal{O}$-module

$$L = \mathrm{Hom}_\mathcal{O}(\Omega_\mathcal{O}, \mathcal{O}).$$

Since $X \to X'$ is unramified, the Riemann-Hurwitz formula implies that

$$\dim M_2(\Gamma) = g + s - 1 = \#H(g' + s' - 1) = \#H \dim M_2(\Gamma'),$$

where g (resp. g') is the genus and s (resp. s') is the number of cusps of X (resp. X'). It follows from lemma 4.13 (a) (and lemma 4.14) that

$$\dim_K(\tilde L \otimes_\mathcal{O} K) = \#H \dim_k \tilde\Omega'_k \qquad (4.3.1)$$

On the other hand, by (b) we have that

$$\tilde{L}/(\mathfrak{a},\lambda)\tilde{L} \cong \mathrm{Hom}_k(\tilde{\Omega}_k^H, k) \cong \mathrm{Hom}_k(\tilde{\Omega}'_k, k) \tag{4.3.2}$$

where $\mathfrak{a} = \ker(\mathcal{O}[H] \to \mathcal{O})$ is the augmentation ideal. By Nakayama's lemma and (4.3.2), $\tilde{L}$ is generated as an $\mathcal{O}[H]$-module by d elements where $d = \dim_k \tilde{\Omega}'_k$. By (4.3.1) any surjective homomorphism

$$\mathcal{O}[H]^d \to \tilde{L}$$

is in fact an isomorphism. Hence $\tilde{L}$ is free over $\mathcal{O}[H]$, as is

$$L_{\mathfrak{n}} = \mathrm{Hom}_{\mathcal{O}}(\Omega_{\mathfrak{n}}, \mathcal{O}) \cong \mathrm{Hom}_{\mathcal{O}}(\tilde{\Omega}_{\mathfrak{n}}, \mathcal{O})$$

for each non-Eisenstein maximal ideal $\mathfrak{n}$ of $\tilde{\mathbb{T}}_{\mathcal{O}}$ (by part (c) of lemmas 4.13 and 4.14).

Since $\rho_{\mathfrak{m}} \sim \bar{\rho}$ is absolutely irreducible, $\tilde{\mathfrak{m}}$ is not Eisenstein (by lemma 4.12), and it follows that $L_{\mathfrak{m}}$ is free over $\mathcal{O}[H]$.

Since $L \otimes_{\mathcal{O}} K$ is free of rank one over $\mathbb{T}_K$ (lemma 1.34), we have that $L_{\mathfrak{m}} \otimes_{\mathcal{O}} K$ is free of rank one over $\mathbb{T}_{\mathfrak{m}} \otimes_{\mathcal{O}} K$. If $\ell \nmid N_\emptyset$ then lemma 4.13 implies that

$$L/\mathfrak{m}L \cong \mathrm{Hom}_k(\Omega_k[\mathfrak{m}], k)$$

is one-dimensional over k. The same assertion holds if $\ell | N_\emptyset$ by lemma 4.14 (from the definition of $\mathfrak{m}$ in this case, we see that it does not contain T_ℓ). It follows from Nakayama's lemma that $L_{\mathfrak{m}}$ is free of rank one over $\mathbb{T}_{\mathfrak{m}}$. Therefore $\mathbb{T}_{\mathfrak{m}}$ is free over $\mathcal{O}[H]$. □

4.4 Proof of theorem 3.36

Our proof of theorem 3.36 is based on Wiles' arguments in ch. 2 of [W3], which in turn are based on a method of Ribet [R4]. We shall reformulate the proof somewhat to underscore the relationship observed by Doi and Hida [Hi1] between the size of $\mathcal{O}/\eta$ and the value of an L-function, but after doing so we shall also sketch the more direct argument used by Wiles. Two important ingredients appear in both versions of the argument, and we shall discuss their proofs in section 4.5. These ingredients are

- the generalization of a result of Ihara [Ih] used in Ribet's argument;
- the generalization of a result of Mazur [Maz1] on the structure of the Tate module $\mathcal{T}_\ell(J_0(N))$ as a module for $\mathbb{T}_{\mathbf{Z}_\ell}$.

Assume now that we are in the setting of theorem 3.36. In particular, $\bar{\rho}$ is a representation as in section 3.3, Σ is a finite set of primes contained in $\Sigma_{\bar{\rho}}$ and π is an $\mathcal{O}$-algebra homomorphism $\mathbb{T}_\Sigma \to \mathcal{O}$ arising from a newform f in $\mathcal{N}_\Sigma$ with coefficients in $\mathcal{O}$.

The symmetric square L-function: For each prime p let $\alpha_p(f)$ and $\beta_p(f)$ denote the roots of the polynomial $X^2 - a_p(f)X + \delta_p p = 0$, where $\delta_p = 0$ or 1 according to whether p divides N_f. Thus we have

$$L_p(f,s) = (1 - \alpha_p(f)p^{-s})^{-1}(1 - \beta_p(f)p^{-s})^{-1}.$$

If $p = \ell$ divides N/N_f, then we require that $\alpha_\ell(f)$ be the root which is a unit in $\mathcal{O}$, i.e. $a_\ell(g)$. We define the symmetric square L-function associated to f as

$$L(\mathrm{Symm}^2 f, s) = \prod_p L_p(\mathrm{Symm}^2 f, s),$$

where

$$L_p(\mathrm{Symm}^2 f,s) = (1-\alpha_p^2(f)p^{-s})^{-1}(1-\alpha_p(f)\beta_p(f)p^{-s})^{-1}(1-\beta_p^2(f)p^{-s})^{-1}.$$

(We caution the reader we have defined the Euler factors rather naively at primes p such that p^2 divides N_f.) The product converges absolutely for real part of $s > 2$ and can be analytically continued to an entire function.

The calculation of η_Σ: We write $\wp$ for the kernel of π and I for the annihilator of $\wp$ in $\mathbb{T}_\Sigma$. Since $\mathbb{T}_\Sigma$ is reduced, we have $\wp \cap I = 0$ and $\wp \oplus I$ has finite index in $\mathbb{T}_\Sigma$. Note that

$$\mathcal{O}/\eta_\Sigma \cong \mathbb{T}_\Sigma/(\wp \oplus I).$$

Suppose that we are given a $\mathbb{T}_\Sigma$-module L which is finitely generated and free over $\mathcal{O}$, and such that $L \otimes K$ is free of rank d over $\mathbb{T}_\Sigma \otimes_{\mathcal{O}} K$. Suppose also that L is endowed with a perfect $\mathcal{O}$-bilinear pairing

$$\begin{array}{rcl} L \times L & \to & \mathcal{O} \\ (x,y) & \mapsto & \langle x,y \rangle \end{array}$$

such that $\langle Tx, y\rangle = \langle x, Ty\rangle$ for all $x, y \in L$ and $T \in \mathbb{T}_\Sigma$. Note that to give such a pairing is equivalent to giving an isomorphism

$$\begin{array}{rcl} L & \xrightarrow{\sim} & \mathrm{Hom}_{\mathcal{O}}(L,\mathcal{O}) \\ x & \mapsto & \langle x, \cdot \rangle \end{array}$$

of $\mathbb{T}_\Sigma$-modules. We shall refer to the module L together with the pairing $\langle\ ,\ \rangle$ as a *self-dual* $\mathbb{T}_\Sigma$-module of *rank d*. We then can give a lower bound for the size of $\mathcal{O}/\eta_\Sigma$ in terms of a basis $\{x_1, x_2, \ldots, x_d\}$ for the free $\mathcal{O}$-module $L[\wp]$:

Lemma 4.17 *We have*

$$\eta_\Sigma^d \subset \mathcal{O}\det(\langle x_i, x_j\rangle)_{i,j},$$

and equality holds if L is free over $\mathbb{T}_\Sigma$.

Proof: The modules $L[\wp]$ and $L/L[I]$ are free of rank d over $\mathcal{O}$ and the pairing $\langle\ ,\ \rangle$ induces an isomorphism

$$L/L[I] \xrightarrow{\sim} \mathrm{Hom}_{\mathcal{O}}(L[\wp], \mathcal{O}). \tag{4.4.1}$$

The $\mathcal{O}$-module $M = L/(L[\wp] \oplus L[I])$ is annihilated by η_Σ and is generated by d elements. Furthermore M is isomorphic to $(\mathcal{O}/\eta_\Sigma)^d$ if L is free over $\mathbb{T}_\Sigma$. It follows that

$$\#(\mathcal{O}/\eta_\Sigma)^d \geq \#M,$$

with equality if L is free. The cardinality of M is that of the cokernel of the map

$$L[\wp] \hookrightarrow \mathrm{Hom}_{\mathcal{O}}(L[\wp], \mathcal{O})$$

arising from the pairing $\langle\ ,\ \rangle$, and this is precisely

$$\mathcal{O}/\det(\langle x_i, x_j\rangle)_{i,j}.$$

□

Recall that proposition 4.7 establishes an isomorphism between $\mathbb{T}_\Sigma$ and $\mathbb{T}_{\mathfrak{m}}$, the localization at a certain maximal ideal $\mathfrak{m}$ of $\mathbb{T}_{\mathcal{O}} = \mathbb{T}_{\mathbf{Z}} \otimes \mathcal{O}$, where

$$\mathbb{T}_{\mathbf{Z}} \subset \mathrm{End}\,(S_2(\Gamma_0(N))),$$

and where $N = N_\Sigma$ is defined in (4.2.1). Write X for $X_0(N)$ and J for $J_0(N)$. Recall that $H_1(X, \mathbb{Z})$ is endowed with the structure of a $\mathbb{T}_{\mathbf{Z}}$-module and

$$\mathcal{T}_\ell(J) \otimes_{\mathbf{Z}_\ell} \mathcal{O} \cong H_1(X, \mathbb{Z}) \otimes \mathcal{O}$$

with that of a $\mathbb{T}_{\mathcal{O}}$-module. We also regard

$$H^1(X, \mathcal{O}) \cong \mathrm{Hom}(H_1(X, \mathbb{Z}), \mathcal{O})$$

as a $\mathbb{T}_{\mathcal{O}}$-module; as such it is naturally isomorphic to

$$\mathrm{Hom}_{\mathcal{O}}(\mathcal{T}_\ell(J) \otimes_{\mathbf{Z}_\ell} \mathcal{O}, \mathcal{O}).$$

The Weil pairing defines a perfect pairing

$$\mathcal{T}_\ell(J) \times \mathcal{T}_\ell(J) \to \mathbb{Z}_\ell$$

(with $(e^{2\pi i/\ell^n})_n$ as the chosen generator for $\varprojlim \mu_{\ell^n}(\mathbb{C})$). The composite

$$\mathcal{T}_\ell(J) \overset{w_*}{\to} \mathcal{T}_\ell(J) \to \mathrm{Hom}_{\mathbf{Z}_\ell}(\mathcal{T}_\ell(J), \mathbb{Z}_\ell) \tag{4.4.2}$$

is an isomorphism of $\mathbb{T}_{\mathbf{Z}_\ell}$-modules where w_* is induced by the involution $w = w_N$ of X (see section 1.4). Tensoring with $\mathcal{O}$ and localizing at $\mathfrak{m}$, we regard

$$L_{\mathcal{T}} = (\mathcal{T}_\ell(J) \otimes_{\mathbf{Z}_\ell} \mathcal{O})_{\mathfrak{m}}$$

as a self-dual $\mathbb{T}_\Sigma$-module of rank 2 via the isomorphism $\mathbb{T}_\Sigma \cong \mathbb{T}_\mathfrak{m}$ of proposition 4.7. We shall write $\langle,\rangle_\mathcal{T}$ for the pairing obtained from (4.4.2). Similarly, using w^*, the cup product and Poincaré duality, we regard

$$L_H = H^1(X, \mathcal{O})_\mathfrak{m}$$

as a self-dual $\mathbb{T}_\Sigma$-module of rank 2, with $\langle x, y\rangle_H$ defined by the image of $x \cup w^* y$ under the canonical isomorphism of $H^2(X, \mathcal{O})$ with $\mathcal{O}$. Note that L_H is canonically isomorphic as a $\mathbb{T}_\Sigma$-module to $\mathrm{Hom}_\mathcal{O}(L_\mathcal{T}, \mathcal{O})$. In the next section we shall discuss the following generalization of a result of Mazur [Maz1].

Theorem 4.18 *The $\mathbb{T}_\Sigma$-modules $L_\mathcal{T}$ and L_H are free.*

Corollary 4.19 *If $\{x, y\}$ is a basis for $L_\mathcal{T}[\wp]$ (resp. $L_H[\wp]$), then η_Σ is generated by $\langle x, y\rangle_\mathcal{T}$ (resp. $\langle x, y\rangle_H$).*

This follows from lemma 4.17, theorem 4.18 and skew-symmetry of the pairings.

Hida's formula: We now explain how the value of $\langle x, y\rangle_H$ in corollary 4.19 is related to $L(\mathrm{Symm}^2 f, 2)$ by a formula of Hida (see [Hi1] sec. 5, [W3] sec. 4.1).

Recall that the homomorphism

$$\mathbb{T}_\mathcal{O} \to \mathbb{T}_\mathfrak{m} \cong \mathbb{T}_\Sigma \xrightarrow{\pi} \mathcal{O}$$

arises as $T_n \mapsto a_n(g)$ for a normalized eigenform g in $S_2(\Gamma_0(N), K)$ whose associated newform is f. We shall write $\mathcal{P}$ for the kernel of this homomorphism; thus $\mathcal{P}$ is the preimage in $\mathbb{T}_\mathcal{O}$ of the ideal $\wp$ of $\mathbb{T}_\Sigma$. Note that $\mathcal{P}_\mathfrak{m}$ corresponds to $\wp$ under the isomorphism $\mathbb{T}_\mathfrak{m} \cong \mathbb{T}_\Sigma$.

Choose a number field K_0 containing the Fourier coefficients of g and let $\mathcal{O}_0$ denote the valuation ring $\mathcal{O} \cap K_0$. We choose the basis $\{x, y\}$ for $L_H[\wp]$ in the image of

$$H^1(X, \mathcal{O}_0)[\mathcal{P}_0] \subset H^1(X, \mathcal{O})[\mathcal{P}] \xrightarrow{\sim} H^1(X, \mathcal{O})_\mathfrak{m}[\mathcal{P}] = L_H[\wp],$$

where $\mathcal{P}_0 = \mathcal{P} \cap \mathbb{T}_{\mathcal{O}_0}$. Let $\mathcal{P}_\mathbb{C}$ denote the kernel of the homomorphism

$$\theta_g : \mathbb{T}_\mathbb{C} \to \mathbb{C}$$

associated to the eigenform g. The two-dimensional complex vector space

$$(H^1(X, \mathcal{O}_0)[\mathcal{P}_0]) \otimes_{\mathcal{O}_0} \mathbb{C} = H^1(X, \mathbb{C})[\mathcal{P}_\mathbb{C}]$$

has a canonical basis

$$\{\omega_g, \bar{\omega}_{g^c}\},$$

where we view these differential forms as cohomology classes and the superscript c indicates complex conjugation applied to the Fourier coefficients. (Recall that ω_g is the holomorphic differential on X defined by $\sum a_n(g)q^{n-1}dq$, so $\bar{\omega}_{g^c}$ is the antiholomorphic differential defined by $\sum a_n(g)\bar{q}^{n-1}d\bar{q}$ where $q = e^{2\pi i\tau}$.) Let A denote the matrix in $GL_2(\mathbb{C})$ such that

$$(\omega_g, \bar{\omega}_{g^c}) = (x, y)A.$$

Theorem 4.20 *With the above notation we have (up to sign)*

$$\langle x, y\rangle = (i\pi \det A)^{-1} N_f L(\mathrm{Symm}^2 f, 2) \prod_{p|N/N_f} c_p(f)),$$

where

$$c_p(f) = \begin{cases} p^{\delta_p} L_p(\mathrm{Symm}^2 f, 2)^{-1} & \text{if } p \neq \ell \\ (\alpha_\ell(f) - \beta_\ell(f))(1 - \alpha_\ell^{-2}(f)) & \text{if } p = \ell. \end{cases}$$

Proof: We have

$$\langle x, y\rangle_H \det A = \int_X \omega_g \wedge \bar{\omega}_{wg^c} = 8\pi^2 i \langle g, (wg)^c\rangle$$

where the last $\langle , \rangle$ denotes the Petersson inner product and we have used that $wg^c = (wg)^c$ for forms on $\Gamma_0(N)$. We then appeal to a formula of Shimura (see [Shi5] (2.5) and [Hil] (5.13)) to obtain

$$\langle g, (wg)^c\rangle = (48\pi)^{-1}[SL_2(\mathbb{Z}) : \Gamma_0(N)]\mathrm{res}_{s=2} D(g, wg, s).$$

By [Shi5] lemma 1, the Dirichlet series $D(g, h, s)$ is defined by $\sum a_n(g)a_n(h)n^{-s}$ and if g and h are normalized eigenforms then this has an Euler product expression in which the factors are

$$(1 - \alpha_p(g)\beta_p(g)\alpha_p(h)\beta_p(h)p^{-2s})(1 - \alpha_p(g)\alpha_p(h)p^{-s})^{-1} \cdot$$
$$(1 - \alpha_p(g)\beta_p(h)p^{-s})^{-1}(1 - \beta_p(g)\alpha_p(h)p^{-s})^{-1}(1 - \beta_p(g)\beta_p(h)p^{-s})^{-1},$$

(where the α_p's and β_p's are defined as they were for f). Using the recipe for obtaining g from f (see lemma 4.6) we find that if ℓ is not in Σ then

$$D(g, wg, s) = \pm D_\Sigma(f, f, s) \prod_{p|N/N_f} p^{-1},$$

(where the subscript Σ indicates that the primes in Σ are removed from the defining Euler product). If ℓ is in Σ, then we must also multiply by the factor

- $(1 - \ell^{-s})^{-1}$ if ℓ divides N_f;
- $(1 - \ell^{1-s})^{-1}(1 - \alpha_\ell^2(f)\ell^{-s})^{-1}(\beta_\ell(f) - \alpha_\ell^{2-s}(f))$ otherwise.

To obtain the formula in this last case, consider the eigenform

$$f_\alpha = f - \beta_\ell(f)f(\ell\tau)$$

on $\Gamma_0(N_f\ell)$. Since

$$w_{N_f\ell}f_\alpha = \pm(\ell f(\ell\tau) - \alpha_\ell^{-1}(f)f)$$

and

$$D(f_\alpha, \ell f(\ell\tau), s) = \alpha_\ell(f)\ell^{1-s}D(f_\alpha, f, s),$$

we get

$$D(f_\alpha, w_{N_f\ell}f_\alpha, s) = \pm\ell^{-1}(\beta_\ell(f) - \alpha_\ell(f)\ell^{2-s})D(f_\alpha, f, s)$$

and use the Euler product. The Euler product expression also gives

$$D_\Sigma(f, f, s) = \frac{\zeta_N(s-1)L_\Sigma(\mathrm{Symm}^2 f, s)}{\zeta_N(2s-2)},$$

where $\zeta_N(s)$ is the Riemann zeta function with the Euler factors removed at primes dividing N. Thus

$$\mathrm{res}_{s=2}D(g, wg, s) = \pm\frac{6}{\pi^2}\frac{NL_\Sigma(\mathrm{Symm}^2 f, 2)\prod_{p|N/N_f} p}{[SL_2(\mathbb{Z}) : \Gamma_0(N)]},$$

with the extra factor of $(1-\ell^{-2})^{-1}$ or

$$(\beta_\ell(f) - \alpha_\ell(f))(1-\beta_\ell^{-2}(f))^{-1}(1-\ell^{-1})^{-1}$$

if ℓ is in Σ. The theorem follows. □

Corollary 4.21 *With the above notation (in particular, $\Sigma \subset \Sigma_{\bar\rho}$), we have*

$$\frac{\ell^m N_f L_\Sigma(\mathrm{Symm}^2 f, 2)}{i\pi \det A} \in \mathcal{O}_0$$

is a generator of η_Σ, where

$$m = \begin{cases} 0, & \text{if } \ell \notin \Sigma, \\ 2, & \text{if } \ell \in \Sigma \text{ and } \ell | N_f, \\ 3, & \text{if } \ell \in \Sigma \text{ and } \ell \nmid N_f. \end{cases}$$

Remark 4.22 Without theorem 4.18 we obtain instead that η_Σ is contained in the ideal generated by the expression in the corollary.

Comparing η_Σ's: Suppose that $\Sigma' \subset \Sigma_{\bar\rho}$ is a finite set of primes containing Σ. In view of corollary 4.21, theorem 3.36 reduces to a comparison of $\det A$

and det A' where A' is defined using Σ' instead of Σ. We shall make the desired comparison using Wiles' generalization of a result of Ihara.

Applying proposition 4.7 to Σ', we have an isomorphism between $\mathbb{T}_{\Sigma'}$ and $\mathbb{T}'_{\mathfrak{m}'}$ where $\mathfrak{m}'$ is a certain maximal ideal of $\mathbb{T}'_{\mathcal{O}} = \mathbb{T}'_{\mathbf{Z}} \otimes \mathcal{O}$ with $\mathbb{T}'_{\mathbf{Z}}$ acting on $S_2(\Gamma_0(N_{\Sigma'}))$. We write $\mathcal{P}'$ for the preimage in $\mathbb{T}'_{\mathcal{O}}$ of $\wp'$. We let $N' = N_{\Sigma'}$, $X' = X_0(N')$, $J' = J_0(N')$ and define self-dual $\mathbb{T}'_{\mathfrak{m}'}$-modules $L'_{\mathcal{T}}$ and L'_H as above.

We now define a homomorphism

$$\phi_{\mathcal{T}} : L'_{\mathcal{T}} \to L_{\mathcal{T}}$$

of $\mathbb{T}'_{\mathfrak{m}'}$-modules in the case that $\Sigma' = \Sigma \cup \{p\}$ for some prime $p \in \Sigma_{\bar{\rho}} - \Sigma$. We first suppose that $p \neq \ell$, in which case $N' = Np^2$ and the surjective homomorphism of $\mathcal{O}$-algebras $\mathbb{T}'_{\mathfrak{m}'} \to \mathbb{T}_{\mathfrak{m}}$ (corresponding to $\mathbb{T}_{\Sigma'} \mapsto \mathbb{T}_{\Sigma}$) is defined by $T_r \mapsto T_r$ if $r \neq p$ and $T_p \mapsto 0$. The maps $\tau \mapsto \tau$, $\tau \mapsto p\tau$ and $\tau \mapsto p^2\tau$ on the upper half-plane induce maps $X' \to X$ which we denote α, β and γ. These give rise by functoriality to a map

$$\begin{array}{rcl} J' & \to & J \times J \times J \\ x & \mapsto & (\alpha_* x, \beta_* x, \gamma_* x) \end{array}$$

which induces an $\mathcal{O}$-linear homomorphism

$$\mathcal{T}_\ell(J') \otimes_{\mathbf{Z}_\ell} \mathcal{O} \to (\mathcal{T}_\ell(J) \otimes_{\mathbf{Z}_\ell} \mathcal{O})^3.$$

We compose with the projection to

$$L_{\mathcal{T}}^3 = (\mathcal{T}_\ell(J) \otimes_{\mathbf{Z}_\ell} \mathcal{O})^3_{\mathfrak{m}}$$

followed by the map $L_{\mathcal{T}}^3 \to L_{\mathcal{T}}$ defined by the matrix $(1, -p^{-1}T_p, p^{-1})$. The reason for this choice of matrix is that

$$g' = (\alpha^* - p^{-1}\beta^* T_p + p^{-1}\gamma^*)g \tag{4.4.3}$$

is the eigenform corresponding to $\mathcal{P}'$.

Lemma 4.23 *The composite*

$$\mathcal{T}_\ell(J') \otimes_{\mathbf{Z}_\ell} \mathcal{O} \to L_{\mathcal{T}} \tag{4.4.4}$$

is $\mathbb{T}'_{\mathcal{O}}$-linear where $L_{\mathcal{T}}$ is regarded as a $\mathbb{T}'_{\mathcal{O}}$-module via the surjection

$$\mathbb{T}'_{\mathcal{O}} \to \mathbb{T}'_{\mathfrak{m}'} \to \mathbb{T}_{\mathfrak{m}}.$$

We leave the proof to the reader after pointing out that the map on Hecke algebras can be rewritten as

$$\begin{array}{ccccccc} \mathbb{T}'_{\mathcal{O}} & \to & \mathbb{T}_{\mathcal{O}}[u]/(f(u)) & \to & \mathbb{T}_{\mathfrak{m}}[u]/(f(u)) & \to & \mathbb{T}_{\mathfrak{m}} \\ T_p & \mapsto & u & & u & \mapsto & 0, \end{array}$$

where $f(u) = u^3 - T_p u^2 + pu$.

It follows that (4.4.4) factors through $L'_\mathcal{T}$ and we thus obtain the desired homomorphism

$$\phi_\mathcal{T} : L'_\mathcal{T} \to L_\mathcal{T}$$

of $\mathbb{T}'_{\mathfrak{m}'}$-modules. Note that $\phi_\mathcal{T}$ is also the composite of the canonical splitting

$$L'_\mathcal{T} \hookrightarrow \mathcal{T}_\ell(J') \otimes_{\mathbf{Z}_\ell} \mathcal{O}$$

with (4.4.4).

The construction of $\phi_\mathcal{T}$ is similar for $p = \ell$ except that we have only two copies of J and the map $L^2_\mathcal{T} \to L_\mathcal{T}$ is given by the matrix $(1, -\alpha_\ell^{-1})$ where α_ℓ is the unit root of (3.3.2). For arbitrary Σ' satisfying $\Sigma \subset \Sigma' \subset \Sigma_{\bar{\rho}}$ we define $\phi_\mathcal{T}$ as a composite of the maps defined above. It is independent of the choice of ordering of $\Sigma' - \Sigma$. Recall that $\phi'_\mathcal{T}$ is used to denote the dual map $L_\mathcal{T} \to L'_\mathcal{T}$. We define $\phi_H : L'_H \to L_H$ as the adjoint of the map ϕ'_H which renders the diagram

$$\begin{array}{lccc} \phi'_H : & H^1(X, \mathcal{O})_\mathfrak{m} & \to & H^1(X', \mathcal{O})_{\mathfrak{m}'} \\ & \downarrow & & \downarrow \\ & \mathrm{Hom}_\mathcal{O}(L_\mathcal{T}, \mathcal{O}) & \to & \mathrm{Hom}_\mathcal{O}(L'_\mathcal{T}, \mathcal{O}) \end{array}$$

commutative (where the vertical maps are the natural isomorphisms and the bottom one is dual to $\phi_\mathcal{T}$).

The second crucial result whose discussion we postpone until the next section is the following generalization of a lemma of Ihara.

Lemma 4.24 *$\phi_\mathcal{T}$ and ϕ_H are surjective.*

Suppose again we are in the case $\Sigma' = \Sigma \cup \{p\}$ with $p \neq \ell$. One need only unravel the definition of ϕ'_H to see that the diagram

$$\begin{array}{ccc} H^1(X, \mathcal{O}_0)[\mathcal{P}_0] & \to & H^1(X', \mathcal{O}_0)[\mathcal{P}'_0] \\ \downarrow & & \downarrow \\ L_H & \overset{\phi'_H}{\to} & L'_H, \end{array}$$

commutes, where the top arrow is defined as

$$\alpha^* - p^{-1}\pi(T_p)\beta^* + p^{-1}\gamma^*.$$

Extending scalars to $\mathbb{C}$, this map sends ω_g to $\omega_{g'}$ and $\bar{\omega}_{g^c}$ to $\bar{\omega}_{g'^c}$ where g' is defined by (4.4.3). In the case $\Sigma' = \Sigma \cup \{\ell\}$, the same assertion holds if $\mathcal{O}_0$ is chosen so that it contains $\pi(\alpha_\ell) = a_\ell(g')$ and the top arrow is defined as

$$\alpha^* - \pi(\alpha_\ell)^{-1}\beta^*.$$

We conclude that if $\{x, y\}$ is a basis for $H^1(X, \mathcal{O}_0)[\mathcal{P}_0]$ and the matrices A and A' are defined using the bases $\{x, y\}$ for $H^1(X, \mathcal{O}_0)[\mathcal{P}_0]$ and $\{\phi'_H x, \phi'_H y\}$

for $H^1(X', \mathcal{O}_0)[\mathcal{P}'_0]$, then $A' = A$. Theorem 3.36 is now immediate from corollary 4.21. In fact, we have proved that

$$\eta_{\Sigma'} = \pi(\prod_{p\in\Sigma'-\Sigma} c_p)\eta_\Sigma.$$

$\square$

Remark 4.25 Note that to obtain the inclusion stated in theorem 3.36 it suffices to apply theorem 4.18 for Σ, rather than both Σ and Σ' (cf. remark 4.22).

Wiles' argument: The method of [W3] sec. 2.2, like that of Ribet in [R4], is more direct than the one given above but does not explicitly illustrate the relation with values of L-functions. The approach is simply to compute the composite $\phi_T\phi'_T$. It suffices to consider the case $\Sigma' = \Sigma \cup \{p\}$ and we suppose first that $p \neq \ell$. Let Δ denote the endomorphism of J^3 defined by the matrix

$$\Delta = \begin{pmatrix} \alpha_* \\ \beta_* \\ \gamma_* \end{pmatrix} w'_*(\alpha^*, \beta^*, \gamma^*)w_*.$$

Using the relations $\alpha w' = w\gamma$ and $\beta w' = w\beta$, we find that

$$\Delta = \begin{pmatrix} \alpha_*\gamma^* & \alpha_*\beta^* & \alpha_*\alpha^* \\ \beta_*\gamma^* & \beta_*\beta^* & \beta_*\alpha^* \\ \gamma_*\gamma^* & \gamma_*\beta^* & \gamma_*\alpha^* \end{pmatrix},$$

which can be computed for example by considering its effect on the cotangent space $S_2(\Gamma_0(N))^3$. The result is that

$$\Delta = \begin{pmatrix} T_p^2 - (p+1) & pT_p & p(p+1) \\ pT_p & p(p+1) & pT_p \\ p(p+1) & pT_p & T_p^2 - (p+1) \end{pmatrix},$$

which we note commutes with the action of $\mathbb{T}_\mathbf{Z}$ on J^3. One then checks that

$$\begin{aligned} \phi_T\phi'_T &= (1, -p^{-1}T_p, p^{-1})\ (\mathcal{T}_\ell(\Delta) \otimes_{\mathbf{Z}_\ell} \mathcal{O})_\mathfrak{m} \begin{pmatrix} 1 \\ -p^{-1}T_p \\ p^{-1} \end{pmatrix} \\ &= -p^{-2}(p-1)(T_p^2 - (p+1)^2). \end{aligned}$$

The case of $p = \ell$ is similar but simpler. One uses

$$\Delta = \begin{pmatrix} \alpha_* \\ \beta_* \end{pmatrix} w'_*(\alpha^*, \beta^*)w_* = \begin{pmatrix} T_\ell & \ell+1 \\ \ell+1 & T_\ell \end{pmatrix}$$

and gets

$$\phi_T\phi'_T = -(\alpha_\ell - \ell\alpha_\ell^{-1})(1 - \alpha_\ell^{-2}).$$

Thus in either case we find that $\phi_T\phi'_T$ acts on $L'_T[\mathcal{P}]$ by an element of $\mathcal{O} = \mathbb{T}_\mathfrak{m}/\mathcal{P}$ which is a unit times $\pi(c_p)$. Theorem 3.36 then follows from corollary 4.19 and lemma 4.24.

4.5 Homological results

In this section we sketch the proofs of theorem 4.18 and lemma 4.24, but shall often refer the reader to ch. 2 of [W3] for more details.

Multiplicities: We first consider theorem 4.18, generalizing a result proved by Mazur in sections II.14 and II.15 of [Mazl]. Recall that $L_{\mathcal{T}}$ (resp. L_H) is defined as $(\mathcal{T}_\ell(J_0(N))\otimes_{\mathcal{O}})_{\mathfrak{m}}$ (resp. $H^1(X_0(N),\mathcal{O})_{\mathfrak{m}}$) where $\mathfrak{m}$ is a maximal of the ideal of the Hecke algebra $\mathbb{T}_{\mathcal{O}}$ generated by the Hecke operators on $S_2(\Gamma_0(N))$. We are assuming moreover that $\rho_{\mathfrak{m}}$ is irreducible and that one of the following holds:

- ℓ does not divide N;
- ℓ^2 does not divide N and $T_\ell \notin \mathfrak{m}$.

We wish to prove that $L_{\mathcal{T}}$ and L_H are free over $\mathbb{T}_{\mathfrak{m}}$.

Since L_H and $L_{\mathcal{T}}$ are isomorphic as $\mathbb{T}_{\mathfrak{m}}$-modules, it suffices to prove that $L_{\mathcal{T}}$ is free. Furthermore, to prove that $L_{\mathcal{T}}$ is free, it suffices to prove that the localization of $\mathcal{T}_\ell(J_0(N))$ at $\mathfrak{m}\cap\mathbb{T}_{\mathbf{Z}_\ell}$ is free; i.e., we may replace $\mathcal{O}$ by $\mathbb{Z}_\ell$ and $\mathfrak{m}$ by its intersection with $\mathbb{T}_{\mathbf{Z}_\ell}$ and consider

$$\mathcal{T}_\ell(J_0(N))_{\mathfrak{m}}.$$

Since $\mathcal{T}_\ell(J_0(N))\otimes_{\mathbf{Z}_\ell}\mathbb{Q}_\ell$ is free is free of rank two over $\mathbb{T}_{\mathbf{Q}_\ell}$, it suffices to prove that

$$\dim_{\mathbb{T}/\mathfrak{m}}(L_{\mathcal{T}}/\mathfrak{m}L_{\mathcal{T}}) = 2.$$

Since

$$\mathcal{T}_\ell(J_0(N))/\ell\mathcal{T}_\ell(J_0(N)) \cong J_0(N)[\ell] \cong \mathrm{Hom}_{\mathbf{F}_\ell}(J_0(N)[\ell],\mathbb{F}_\ell)$$

as $\mathbb{T}_{\mathbf{F}_\ell}$-modules, it suffices to prove the following cases of [W3] thm. 2.1.

Theorem 4.26 *Let $\mathfrak{m}$ be a maximal ideal of $\mathbb{T}_{\mathbf{Z}_\ell}$. Suppose that $\rho_{\mathfrak{m}}$ is irreducible and that either*

(a) *ℓ does not divide N, or*

(b) *$\ell^2 \nmid N$ and $T_\ell \notin \mathfrak{m}$.*

Then

$$\dim_{\mathbb{T}_{\mathbf{Z}_\ell}/\mathfrak{m}} J_0(N)[\mathfrak{m}] = \dim_{\mathbb{T}_{\mathbf{Z}_\ell}/\mathfrak{m}}(J_0(N)[\ell]/\mathfrak{m}) = 2.$$

The proof of theorem 4.26 in case (b) requires more of the theory of group schemes and Néron models than we wish to delve into here. We shall therefore only explain the proof in the case (a) (which is [R5] thm. 5.2(b), see Ribet's paper for more details) and refer to [W3] sec. 2.1 for the general case. We remark however that the proof of theorem 3.42 appeals to theorem 3.36 only in the case $\Sigma = \emptyset$ (cf. remarks 4.22 and remark 4.25). Recall that ℓ

divides $N_\emptyset$ only if $\bar{\rho}$ is not good. So for the purposes of proving the Shimura-Taniyama conjecture for semistable elliptic curves, (b) can be replaced by the stronger hypothesis

(b') $\ell^2 \nmid N$ and $\rho_{\mathfrak{m}}|_{G_\ell}$ is not good.

The result in this case is due to Mazur and Ribet (the main result of [MRi]) and the proof is slightly easier than in the case of (b).

Returning to case (a), we appeal to a general property of the functor $\mathbb{D}$ of theorem 2.31 which follows from cor. 5.11 of [Oda].

Theorem 4.27 *Suppose that A is an abelian variety over $\mathbb{Z}_\ell$ (i.e., the Néron model of an abelian variety over $\mathbb{Q}_\ell$ with good reduction). There is a canonical and functorial isomorphism of vector spaces over $\mathbb{F}_\ell$*

$$\mathbb{D}(A(\bar{\mathbb{Q}}_\ell)[\ell]^*)^0 \cong \mathrm{Cot}_0(A/\mathbb{F}_\ell),$$

where Cot_0 denotes the cotangent space at the origin.

In the case that A is the Jacobian of a smooth proper curve X over $\mathbb{Z}_\ell$, we have also a canonical isomorphism

$$\mathrm{Cot}_0(A/\mathbb{F}_\ell) \cong H^0(X, \Omega^1_{X/\mathbb{F}_\ell})$$

and in the case $X = X_0(N)$ with $\ell \nmid N$, this provides an isomorphism of $\mathbb{T}_{\mathbb{F}_\ell}$-modules

$$\mathbb{D}(J_0(N)(\bar{\mathbb{Q}}_\ell)[\ell]^*)^0 \cong S_2(\Gamma_0(N), \mathbb{F}_\ell). \tag{4.5.1}$$

Consider now the action of $G_\mathbb{Q}$ on the points of

$$V = J_0(N)[\ell]^*[\mathfrak{m}] \cong (J_0(N)[\ell]/\mathfrak{m})^*.$$

By the argument in [Maz1] prop. II.14.2 or by the main result of [BLR], we know that every Jordan-Hölder constituent of the representation of $G_\mathbb{Q}$ on this $\mathbb{T}/\mathfrak{m}$-vector space is isomorphic to $\rho_\mathfrak{m}^* \cong \rho_\mathfrak{m}$. It follows that

$$\dim_{\mathbb{T}/\mathfrak{m}} \mathbb{D}(V) = 2 \dim_{\mathbb{T}/\mathfrak{m}} \mathbb{D}(V)^0$$

where we now regard V as a good G_ℓ-module. On the other hand (4.5.1) implies that

$$\mathbb{D}(V)^0 \cong S_2(\Gamma_0(N), \mathbb{F}_\ell)[\mathfrak{m}],$$

which is one-dimensional over $\mathbb{T}/\mathfrak{m}$ by the q-expansion principle (lemma 1.34 for example). We have now shown that $\mathbb{D}(V)$ is two-dimensional over $\mathbb{T}/\mathfrak{m}$, and it follows that so is V.

Ihara's lemma: We now sketch the proof of lemma 4.24, which proceeds by analyzing the behavior of the homology of modular curves under certain degeneracy maps. Note that it suffices to prove the lemma for $\phi_\mathcal{T}$.

If N and M are positive integers, then we let $\Gamma_1(N,M) = \Gamma_1(N) \cap \Gamma_0(M)$ and write $Y_1(N,M)$ (resp. $X_1(N,M)$) for the associated non-compactified (resp. compactified) modular curve. The key intermediate result is the following:

Lemma 4.28 *Suppose that N is a positive integer and p is a prime not dividing N.*

(a) *The map*

$$\begin{array}{ccc} H_1(X_1(N,p),\mathbb{Z}) & \to & H_1(X_1(N),\mathbb{Z})^2 \\ x & \mapsto & (\alpha_* x, \beta_* x) \end{array}$$

is surjective, where α is defined by $\tau \mapsto \tau$ and β by $\tau \mapsto p\tau$.

(b) *If $N > 3$ and ℓ is an odd prime different from p, then the sequence*

$$\begin{array}{ccccc} H_1(Y_1(Np,p^2),\mathbb{Z}_\ell) & \to & H_1(Y_1(Np),\mathbb{Z}_\ell)^2 & \to & H_1(Y_1(N),\mathbb{Z}_\ell) \\ x & \mapsto & (\alpha_{1*}x, \beta_{1*}x); \quad (y,z) & \mapsto & \beta_{2*}y - \alpha_{2*}z \end{array}$$

is exact, where the maps α_1 and α_2 are defined by $\tau \mapsto \tau$ and β_1 and β_2 by $\tau \mapsto p\tau$.

To prove the lemma, one first translates the statement into one about the homology (or cohomology) of the congruence subgroups involved. Part (a) of the lemma is due to Ihara [Ih], but see also the proof of thm. 4.1 of [R4]. Part (b) due to Wiles is more elementary and is established in the course of proving lemma 2.5 of [W3] (see the sequence (2.13)). The result stated there is in terms of cohomology rather than homology, but the one given here is immediate from it.

Remark 4.29 A method of Khare [Kh], sec. 2 using modular symbols yields an alternate proof of part (b) (after localization at a non-Eisenstein maximal ideal).

To deduce lemma 4.24 from lemma 4.28 we also need the following result:

Lemma 4.30 *Suppose that $\mathcal{O}$ is the ring of integers of a finite extension of $\mathbb{Q}_\ell$. Let M be a positive integer and H a subgroup of $(\mathbb{Z}/M\mathbb{Z})^\times$. Let $\tilde{\mathbb{T}}_\mathcal{O} = \tilde{\mathbb{T}}_\mathbf{Z} \otimes \mathcal{O}$ where $\tilde{\mathbb{T}}_\mathbf{Z}$ is the subring of endomorphisms of $M_2(\Gamma_1(M))$ generated by the operators $\langle d \rangle$ and T_r for primes $r \nmid M$. If $\mathfrak{n}$ is a non-Eisenstein maximal ideal then the localization at $\mathfrak{n}$ of the natural map*

(a) $H_1(Y_1(M),\mathcal{O}) \to H_1(X_1(M),\mathcal{O})$ *is an isomorphism;*

(b) $H_1(X_1(M),\mathcal{O}) \to H_1(X_H(M),\mathcal{O})$ *is surjective.*

To prove part (a), one checks that

$$H_1(Y_1(M),\mathbb{Z}) \to H_1(X_1(M),\mathbb{Z})$$

is surjective and that $T_r = r+1$ on the kernel for primes $r \equiv 1 \bmod M$ (cf. proposition 4.13, part (c)).

Part (b) is most easily proved by showing that $G_{\mathbf{Q}}$ acts trivially on the cokernel of the natural map

$$\mathcal{T}_\ell(J_1(M)) \to \mathcal{T}_\ell(J_H(M))$$

(using [LO], prop. 6 for example) and then applying lemma 4.12.

Finally we explain how to use lemmas 4.28 and 4.30 to prove lemma 4.24. It suffices to consider the case $\Sigma' = \Sigma \cup \{p\}$. We discuss only the more difficult case of $p \neq \ell$. We define $N = N_\Sigma$, $N' = Np^2$, $\mathfrak{m}$ and $\mathfrak{m}'$ as in the preceding section. Thus $\mathfrak{m}'$ is a maximal ideal of $\mathbb{T}'_{\mathcal{O}}$ where $\mathbb{T}'_{\mathbf{Z}}$ acts on $S_2(\Gamma_0(N'))$. All the maps we consider in the argument below will respect the action of the Hecke operators in $\tilde{\mathbb{T}}''_{\mathbf{Z}}$ where $\tilde{\mathbb{T}}''_{\mathbf{Z}}$ is the subring of endomorphisms of $M_2(\Gamma_1(Np,p^2))$ generated by the operators $\langle d\rangle$ and T_r for primes r not dividing Np. We let $\mathfrak{n}$ denote the preimage of $\mathfrak{m}'$ in $\tilde{\mathbb{T}}''_{\mathcal{O}} = \tilde{\mathbb{T}}''_{\mathbf{Z}} \otimes \mathcal{O}$. We first apply lemma 4.28(b) to show that

$$H_1(Y_1(Np,p^2),\mathcal{O})_{\mathfrak{n}} \to H_1(Y_1(Np),\mathcal{O})^2_{\mathfrak{n}} \to H_1(Y_1(N),\mathcal{O})_{\mathfrak{n}}$$

is exact. Next we apply lemma 4.30(a) to obtain the exactness of

$$H_1(X_1(Np,p^2),\mathcal{O})_{\mathfrak{n}} \to H_1(X_1(Np),\mathcal{O})^2_{\mathfrak{n}} \to H_1(X_1(N),\mathcal{O})_{\mathfrak{n}}.$$

Applying lemma 4.30(b) with $M = Np$, we find that

$$H_1(X_1(N,p^2),\mathcal{O})_{\mathfrak{n}} \to H_1(X_1(N,p),\mathcal{O})^2_{\mathfrak{n}} \to H_1(X_1(N),\mathcal{O})_{\mathfrak{n}}$$

is exact. We then apply lemma 4.28(a) to conclude that the map

$$\begin{array}{ccc} H_1(X_1(N,p^2),\mathcal{O})_{\mathfrak{n}} & \to & H_1(X_1(N),\mathcal{O})^3_{\mathfrak{n}} \\ x & \mapsto & (\alpha_* x, \beta_* x, \gamma_* x) \end{array}$$

is surjective where α, β and γ are defined respectively by $\tau \mapsto \tau$, $\tau \mapsto p\tau$ and $\tau \mapsto p^2\tau$. Finally, we use lemma 4.30(b) again (now with $M = N$) to get the surjectivity of

$$H_1(X_0(Np^2),\mathcal{O})_{\mathfrak{n}} \to H_1(X_0(N),\mathcal{O})^3_{\mathfrak{n}},$$

hence that of (4.4.4) upon localizing at $\mathfrak{m}'$.

5 Commutative algebra

In this section we collect some basic facts of commutative algebra that are used in the proof. We recall that $\mathcal{O}$ is the ring of integers in a finite extension K of $\mathbf{Q}_\ell$, and that $\mathcal{O}$ has residue field k. Let $\mathcal{C}_{\mathcal{O}}$ denote as in section 2.6 the category of complete noetherian local $\mathcal{O}$-algebras with residue field k.

5.1 Wiles' numerical criterion

In this section we state a numerical criterion discovered by Wiles for a map between two rings in $\mathcal{C}_{\mathcal{O}}$ to be an isomorphism.

Complete intersections: We say that a ring A in $\mathcal{C}_{\mathcal{O}}$ is *finite flat* if it is finitely generated and torsion free as an $\mathcal{O}$-module. A key ingredient in Wiles' isomorphism criterion is played by the concept of *complete intersection*, for which we give the following naive definition.

Definition 5.1 An object A in $\mathcal{C}_{\mathcal{O}}$ which is finite flat is called a *complete intersection* if it can be expressed as a quotient

$$A \simeq \mathcal{O}[[X_1, \dots, X_n]]/(f_1, \dots, f_n),$$

where there are as many relations as there are variables.

Remark 5.2 It is also true that if an object $\mathcal{O}[[Y_1, \dots, Y_r]]/J$ in $\mathcal{C}_{\mathcal{O}}$ which is finite flat is a complete intersection, then necessarily J can be generated by r elements. See for example [Mat], thm. 21.2 (and lemma 5.11 below). We will not use this fact in this chapter, although a special case is proved in the course of establishing lemma 5.30.

The category $\mathcal{C}_{\mathcal{O}}^{\bullet}$: The numerical criterion of Wiles is stated more naturally in terms of rings A in the category $\mathcal{C}_{\mathcal{O}}$ which are endowed with some *extra structure*: namely, a surjective $\mathcal{O}$-algebra homomorphism $\pi_A : A \longrightarrow \mathcal{O}$. Let $\mathcal{C}_{\mathcal{O}}^{\bullet}$ be the category whose objects are pairs (A, π_A), where A is an object of $\mathcal{C}_{\mathcal{O}}$ and $\pi_A : A \longrightarrow \mathcal{O}$ is a surjective $\mathcal{O}$-algebra homomorphism, also called the *augmentation map* attached to A. Morphisms in $\mathcal{C}_{\mathcal{O}}^{\bullet}$ are local ring homomorphisms which are compatible in the obvious way with the augmentation maps. By abuse of notation one often omits mentioning the augmentation map π_A when talking of objects in $\mathcal{C}_{\mathcal{O}}^{\bullet}$, and simply uses A to denote (A, π_A), when this causes no confusion. Objects of $\mathcal{C}_{\mathcal{O}}^{\bullet}$ will also be called *augmented rings*.

The invariants Φ_A and η_A: One associates to an augmented ring (A, π_A) two basic invariants:

$$\Phi_A = (\ker \pi_A)/(\ker \pi_A)^2;$$
$$\eta_A = \pi_A(\mathrm{Ann}_A \ker \pi_A).$$

Here $\mathrm{Ann}_A(I)$ denotes the annihilator ideal of the ideal I in A.

The invariant Φ_A can be thought of as a tangent space for the object A. (More precisely, it is the cotangent space of the scheme spec(A) at the point $\ker \pi_A$.) It is a finitely generated $\mathcal{O}$-module.

The invariant η_A seems less familiar at first sight. It is called the *congruence ideal.* (The reason for this terminology should become clearer shortly.) We are now ready to state Wiles' numerical criterion:

Theorem 5.3 *Let $\phi : R \longrightarrow T$ be a surjective morphism of augmented rings. Assume that T is finitely generated and torsion-free as an $\mathcal{O}$-module, and that $\eta_T \neq (0)$. (And hence, in particular, $\#(\mathcal{O}/\eta_T) < \infty$.) Then the following are equivalent:*

(a) *The inequality $\#\Phi_R \leq \#(\mathcal{O}/\eta_T)$ is satisfied.*

(b) *The equality $\#\Phi_R = \#(\mathcal{O}/\eta_T)$ is satisfied.*

(c) *The rings R and T are complete intersections, and the map $\phi : R \longrightarrow T$ is an isomorphism.*

Remark 5.4 The above theorem is slightly different from the one that appears in [W3], where it is assumed from the outset that the ring T is Gorenstein. In the form in which we state it above, the theorem is due to H. Lenstra [Len]. Our presentation follows Lenstra's very closely. For the original (and slightly different) point of view, the reader should consult the appendix of [W3].

Some examples: Before going further, it may be good to pause and consider some examples of objects of $\mathcal{C}_{\mathcal{O}}^{\bullet}$ and the invariants associated to them. While logically independent of the proof, the examples should help the reader develop some intuition. (For a systematic way to compute the tangent spaces Φ_A, see the paragraph at the end of section 5.2.)

Example 1:

$$\begin{aligned} A &= \{(a,b) \in \mathcal{O} \times \mathcal{O}, \quad a \equiv b \pmod{\lambda^n}\} \\ &\simeq \mathcal{O}[[T]]/(T(T-\lambda^n)), \qquad \text{with } \pi_A(a,b) := a. \\ &\Phi_A \simeq \mathcal{O}/\lambda^n\mathcal{O}, \qquad \eta_A = (\lambda^n). \end{aligned}$$

Example 2:

$$\begin{aligned} A &= \{(a,b,c) \in \mathcal{O} \times \mathcal{O} \times \mathcal{O}, a \equiv b \equiv c \pmod{\lambda}\} \\ &\simeq \mathcal{O}[[X,Y]]/(X(X-\lambda), Y(Y-\lambda), XY), \qquad \text{with } \pi_A(a,b,c) := a. \\ &\Phi_A \simeq \mathcal{O}/\lambda\mathcal{O} \times \mathcal{O}/\lambda\mathcal{O}, \qquad \eta_A = (\lambda). \end{aligned}$$

Example 3:

$$A = \mathcal{O}[[X]]/(X^2), \qquad \text{with } \pi_A(f) := f(0).$$

$$\Phi_A \simeq \mathcal{O}, \qquad \eta_A = 0.$$

Example 4:

$$\begin{aligned} A &= \left\{(a,b,c,d) \in \mathcal{O} \times \cdots \times \mathcal{O}, \begin{array}{ll} a \equiv b \equiv c \equiv d & \pmod{\lambda}, \\ a+d \equiv b+c & \pmod{\lambda^2} \end{array}\right\} \\ &\simeq \mathcal{O}[[X,Y]]/(X(X-\lambda), Y(Y-\lambda)), \qquad \text{with } \pi_A(a,b,c,d) := a. \end{aligned}$$

$$\Phi_A \simeq \mathcal{O}/\lambda\mathcal{O} \times \mathcal{O}/\lambda\mathcal{O}, \qquad \eta_A = (\lambda^2).$$

Example 5:

$$A = \mathcal{O}[[X_1, \dots, X_n]], \qquad \text{with } \pi_A(f) := f(0).$$

$$\Phi_A \simeq \mathcal{O}^n, \qquad \eta_A = (0).$$

Example 6: $(\lambda = \ell \equiv -1 \pmod 4, \quad \mathcal{O} = \mathbb{Z}_\ell)$.

$$\begin{aligned} A &= \{(a, b+ci) \in \mathbb{Z}_\ell \times \mathbb{Z}_\ell[i], \quad a \equiv b \pmod{\ell^2}, \quad c \equiv 0 \pmod{\ell}\} \\ &\simeq \mathbb{Z}_\ell[[X]]/(X(X^2+\ell^2)), \qquad \text{with } \pi_A(a, b+ci) := a. \end{aligned}$$

$$\Phi_A \simeq \mathbb{Z}/\ell^2\mathbb{Z}, \qquad \eta_A = (\ell^2).$$

Example 7:

$$A = \mathcal{O}[[T]]/(\lambda T) \simeq \mathcal{O} \oplus kT \oplus kT^2 \oplus \cdots, \qquad \text{with } \pi_A(f) = f(0).$$

$$\Phi_A \simeq k, \qquad \eta_A = (\lambda).$$

5.2 Basic properties of Φ_A and η_A

In this section we collect some of the basic properties of the invariants Φ_A and η_A, and prove the equivalence of (a) and (b) in theorem 5.3.

Behaviour of Φ_A under morphisms: The assignment $A \mapsto \Phi_A$ is a functor from the category $\mathcal{C}_\mathcal{O}^\bullet$ to the category of $\mathcal{O}$-modules; a morphism $A \longrightarrow B$ in $\mathcal{C}_\mathcal{O}^\bullet$ induces a homomorphism $\Phi_A \longrightarrow \Phi_B$ of $\mathcal{O}$-modules. Moreover, if $A \longrightarrow B$ is surjective, then so is the induced map on the tangent spaces. Therefore, when A maps surjectively onto B we have

$$\#\Phi_A \geq \#\Phi_B. \tag{5.2.1}$$

There is also a converse to this, which will be useful later:

Lemma 5.5 *If the homomorphism $\Phi_A \longrightarrow \Phi_B$ is surjective, then $A \longrightarrow B$ is also surjective.*

Proof: This follows from Nakayama's lemma. (Cf. [Ha], ch. II, sec. 7.4 and [Mat], th. 8.4.) □

Behaviour of η_A under (surjective) morphisms: Unlike the assignment $A \mapsto \Phi_A$, the assignment $A \longrightarrow \eta_A$ is not functorial, but it does have a nice behaviour under surjective morphisms: namely, if $\phi : A \longrightarrow B$ is surjective, then

$$\eta_A \subset \eta_B, \quad \text{i.e.,} \quad \#(\mathcal{O}/\eta_A) \geq \#(\mathcal{O}/\eta_B). \tag{5.2.2}$$

This is simply because in that case ϕ induces a map

$$\mathrm{Ann}_A \ker \pi_A \longrightarrow \mathrm{Ann}_B \ker \pi_B.$$

Relation between the invariants Φ_A and η_A: In general, we have the following inequality:

$$\#\Phi_A \geq \#(\mathcal{O}/\eta_A). \tag{5.2.3}$$

The key behind proving this identity is to interpret $\#\Phi_A$ in terms of Fitting ideals.

Digression on Fitting ideals: If R is a ring (in $\mathcal{C}_{\mathcal{O}}$, say) and M is a finitely generated R-module, we express M as a quotient of R^n for some n:

$$0 \longrightarrow M' \longrightarrow R^n \longrightarrow M \longrightarrow 0. \tag{5.2.4}$$

The *Fitting ideal* of M, denoted $\mathrm{Fitt}_R(M)$, is the ideal of R generated by the determinants $\det(v_1, \ldots, v_n)$, where the vectors $v_i \in R^n$ range over all possible choices of elements of $M' \subset R^n$. One checks that this ideal does not depend on the choice of exact sequence (5.2.4), and hence is an invariant of the R-module M. For example, if M is a finitely generated $\mathcal{O}$-module, we may write

$$M = \mathcal{O}^r \oplus \mathcal{O}/(\lambda^{n_1}) \oplus \mathcal{O}/(\lambda^{n_2}) \oplus \cdots \oplus \mathcal{O}/(\lambda^{n_k}),$$

with $n_1 \geq n_2 \geq \cdots \geq n_k$, and the sequence n_i is completely determined by M. The Fitting ideal $\mathrm{Fitt}_{\mathcal{O}}(M)$ is then the ideal of $\mathcal{O}$ generated by $\lambda^{n_1+\cdots+n_k}$, if $r = 0$, and is the 0-ideal if $r > 0$. Note in particular that, if M is a finite $\mathcal{O}$-module, then

$$\#M = \#(\mathcal{O}/\mathrm{Fitt}_{\mathcal{O}}(M)). \tag{5.2.5}$$

Furthermore, if M is any R-module, it follows directly from the definition that

$$\mathrm{Fitt}_R(M) \subset \mathrm{Ann}_R(M). \tag{5.2.6}$$

Fitting ideals behave well under tensor products: in particular, if M is a finitely generated A-module, where A is an object in $\mathcal{C}^{\bullet}_{\mathcal{O}}$, then:

$$\pi_A(\mathrm{Fitt}_A(M)) = \mathrm{Fitt}_{\mathcal{O}}(M \otimes_A \mathcal{O}), \tag{5.2.7}$$

where the tensor product is taken with respect to the augmentation map π_A. For more details and references on the Fitting ideal, see [Len] for example.

Now we are ready to prove equation (5.2.3). For, noting that $\Phi_A = \ker \pi_A \otimes_A \mathcal{O}$, where the tensor product is taken with respect to π_A, and applying equation (5.2.7) with $M = \ker \pi_A$, we have:

$$\mathrm{Fitt}_{\mathcal{O}}(\Phi_A) = \pi_A(\mathrm{Fitt}_A(\ker \pi_A)) \subset \pi_A(\mathrm{Ann}_A \ker \pi_A) = \eta_A, \tag{5.2.8}$$

where the containment follows from equation (5.2.6). Now the inequality (5.2.3) follows by combining (5.2.5) and (5.2.8).

As a consequence, we have

Corollary 5.6 *The statements* (*a*) *and* (*b*) *in theorem 5.3 are equivalent.*

Proof. If $R \longrightarrow T$ is a surjective map of augmented rings, then $\#\Phi_R \geq \#\Phi_T$, by equation (5.2.1). But equation (5.2.3) gives the inequality $\#\Phi_T \geq \#(\mathcal{O}/\eta_T)$. Hence the inequality $\#\Phi_R \geq \#(\mathcal{O}/\eta_T)$ always holds, so that (*a*) implies (*b*) in theorem 5.3. The reverse implication is clear. □

Remark 5.7 (Computing the tangent spaces Φ_A): Any object (A, π_A) in $\mathcal{C}^{\bullet}_{\mathcal{O}}$ can be expressed as a quotient of the object $U = \mathcal{O}[[X_1, \dots, X_n]]$ of example 5 with augmentation map given by $\pi_U(f) = f(0)$. Indeed, one can take $a_1, \dots, a_n$ to be A-module generators of the finitely generated A-module $\ker \pi_A$, and obtain the desired quotient map by sending X_i to a_i.

The tangent space Φ_U of U is a free $\mathcal{O}$-module of rank n which can be written down canonically as

$$\mathcal{O}X_1 \oplus \mathcal{O}X_2 \oplus \cdots \oplus \mathcal{O}X_n,$$

the natural map from $(\ker \pi_U)$ being simply the map which sends a power series $f \in U$ with no constant term to its degree 1 term, which we will denote by $\bar{f}$.

If A is expressed as a quotient $U/(f_1, \dots, f_r)$, then one has

$$\Phi_A = \Phi_U/(\bar{f}_1, \dots, \bar{f}_r). \tag{5.2.9}$$

5.3 Complete intersections and the Gorenstein condition

In this section we show that complete intersections satisfy a Gorenstein condition, and that (*c*) implies (*a*) and (*b*) in the statement of Wiles isomorphism criterion (theorem 5.3).

Definition 5.8 An object A in $\mathcal{C}_{\mathcal{O}}$ which is finite flat is said to be *Gorenstein* if

$$\mathrm{Hom}_{\mathcal{O}}(A, \mathcal{O}) \simeq A \quad \text{as } A\text{-modules.}$$

Proposition 5.9 *Suppose A in $\mathcal{C}_{\mathcal{O}}$ is finite flat. If A is a complete intersection, then A is Gorenstein.*

The remainder of this section will be devoted to proving proposition 5.9. Since the proof is a bit long and involved, and the concepts it uses are not used elsewhere, the reader is advised on a first reading to take it on faith and skip to the next section. A more direct proof of proposition 5.9 which bypasses the arguments of this section is explained in [Len].

We let A be a ring which is finite flat, and is a complete intersection. (Hence, A can be written as $\mathcal{O}[[X_1, \ldots, X_n]]/(f_1, \ldots, f_n)$.) We assume that the augmentation map for A is induced from the map on $\mathcal{O}[[X_1, \ldots, X_n]]$ sending f to $f(0)$. This implies that $f_i(0) = 0$, i.e., the f_i have no constant term.

We recall some definitions from commutative algebra that we will need. An ideal I of a local ring R is said to be ***primary*** if $I \neq R$ and every zero divisor in R/I is nilpotent. If $(x_1, \ldots, x_n)$ generates a primary ideal of R, and $n = \dim R$, then $(x_1, \ldots, x_n)$ is called a ***system of parameters*** for R.

Lemma 5.10 *The sequence $(f_1, \ldots, f_n, \lambda)$ is a system of parameters for $U = \mathcal{O}[[X_1, \ldots, X_n]]$.*

Proof: The quotient ring $U/(f_1, \ldots, f_n, \lambda)$ is local and is finitely generated as a k-vector space; therefore every element in its maximal ideal is nilpotent. □

A sequence $(x_1, \ldots, x_n)$ in a ring R is said to be a ***regular sequence*** if x_i is not a zero-divisor in $R/(x_1, \ldots, x_{i-1})$ for $i = 1, \ldots, n$.

Lemma 5.11 *The sequence $(f_1, \ldots, f_n)$ is a regular sequence for U.*

Proof: The ring U is Cohen Macaulay, since $\lambda, X_1, \ldots, X_n$ is a system of parameters of U which is also a regular sequence. Hence, by theorem 17.4 (iii) of [Mat], $(f_1, \ldots, f_n, \lambda)$ is a regular sequence in U. A fortiori, the sequence $(f_1, \ldots, f_n)$ is also a regular sequence. □

The proofs of lemma 5.10 and 5.11 use only the fact that A is finitely generated as an $\mathcal{O}$-module, and not that A is flat. As a corollary of this proof, we therefore have:

Corollary 5.12 *If R is an object of $\mathcal{C}^\bullet_\mathcal{O}$ which is finitely generated as an $\mathcal{O}$-module, and $R \simeq \mathcal{O}[[X_1, \ldots, X_n]]/(f_1, \ldots, f_n)$, then R is also flat, and hence is a complete intersection.*

To go further, we will introduce the Koszul complex

$$K(\underline{x}, R) := \oplus_{p=0}^{n} K_p(\underline{x}, R)$$

associated to a local ring R and a sequence $\underline{x} = (x_1, \ldots, x_n)$ of elements in its maximal ideal. For more details on the Koszul complex and its relation to regular sequences, the reader can consult [Mat], especially §16, or [Bour] X, or [Se3]. This complex is defined to be the free graded differential algebra generated by symbols $u_1, \ldots, u_n$:

$$K_p(\underline{x}, R) := \oplus_{i_1 < i_2 < \cdots < i_p} R \cdot u_{i_1} \wedge \cdots \wedge u_{i_p},$$

with differential $d : K_p \longrightarrow K_{p-1}$ defined by

$$d(u_{i_1} \wedge \cdots \wedge u_{i_p}) = \sum_{t=1}^{p} (-1)^t x_t \cdot u_{i_1} \wedge \cdots \wedge u_{i_{t-1}} \wedge u_{i_{t+1}} \wedge \cdots \wedge u_{i_p}.$$

We denote by $H_p(\underline{x}; R)$ the homology groups of this complex. We record here the main properties of this complex that we will use.

Proposition 5.13 (a) $H_0(\underline{x}; R) = R/(\underline{x})$.

(b) *There is a long exact homology sequence*

$$\cdots \longrightarrow H_p(\underline{x}; R) \longrightarrow H_p(\underline{x}, x_{n+1}; R) \longrightarrow H_{p-1}(\underline{x}; R) \stackrel{\pm x_{n+1}}{\longrightarrow}$$
$$H_{p-1}(\underline{x}; R) \longrightarrow H_{p-1}(\underline{x}, x_{n+1}; R) \longrightarrow H_{p-2}(\underline{x}; R) \longrightarrow \cdots$$

(c) *$H_p(\underline{x}; R)$ is annihilated by the ideal $(\underline{x})$, i.e., it has a natural $R/(\underline{x})$-module structure.*

(d) *If $\underline{x}$ is a regular sequence, then $H_p(\underline{x}; R) = 0$ for all $p > 0$ (i.e., the complex $K_p(\underline{x}; R)$ is a free resolution of $R/(\underline{x})$.)*

Proof: The first assertion follows directly from the definition. For (*b*) and (*c*), see [Mat], th. 16.4. The assertion (*d*) can be proved by a direct induction argument on n, using the long exact homology sequence: For $p > 1$, this sequence becomes

$$0 \longrightarrow H_p(\underline{x}, x_{n+1}; R) \longrightarrow 0,$$

and for $p = 1$, it is

$$0 \longrightarrow H_1(\underline{x}, x_{n+1}; R) \longrightarrow H_0(\underline{x}; R) \stackrel{x_{n+1}}{\longrightarrow} H_0(\underline{x}; R).$$

But the assumption that $\underline{x}, x_{n+1}$ is a regular sequence means that multiplication by x_{n+1} is injective on $H_0(\underline{x}; R) = R/(\underline{x})$. Hence, the assertion (*d*) follows. □

Now, we turn to the proof of proposition 5.9, following a method of Tate which is explained in the appendix of [MRo]. For any ring R, write $R[[\underline{X}]] := R[[X_1, \ldots, X_n]]$. Let $a_1, \ldots, a_n$ be the images in A of $X_1, \ldots, X_n$ by the natural map

$$\alpha : \mathcal{O}[[\underline{X}]] \longrightarrow A = \mathcal{O}[[\underline{X}]]/(f_1, \ldots, f_n),$$

and let

$$\beta : A[[\underline{X}]] \longrightarrow A$$

be the natural map which sends X_i to a_i. The sequence $(g_i) = (X_i - a_i)$ generates the kernel of β. Since the f_i, viewed as polynomials in $A[[\underline{X}]]$, also belong to $\ker \beta$, we have:

$$(f_1, \ldots, f_n) = (g_1, \ldots, g_n)M, \tag{5.3.1}$$

where M is an $n \times n$ matrix with coefficients in $A[[\underline{X}]]$. Let

$$D = \det(M) \in A[[\underline{X}]].$$

Our goal is to construct an A-module isomorphism

$$\mathrm{Hom}_{\mathcal{O}}(A, \mathcal{O}) \longrightarrow A.$$

We begin by constructing a surjective ($\mathcal{O}$-linear) map

$$\mathrm{Hom}_{\mathcal{O}[[\underline{X}]]}(A[[\underline{X}]], \mathcal{O}[[\underline{X}]]) \longrightarrow A.$$

Lemma 5.14 *(Tate): The function $\Phi(f) = \alpha(f(D))$ induces an isomorphism of $\mathcal{O}[[\underline{X}]]$-modules*

$$\mathrm{Hom}_{\mathcal{O}[[\underline{X}]]}(A[[\underline{X}]], \mathcal{O}[[\underline{X}]])/(g_1, \dots, g_n) \longrightarrow A.$$

Proof: By lemma 5.11, the sequence $(\underline{f}) = (f_1, \dots, f_n)$ is a regular $\mathcal{O}[[\underline{X}]]$-sequence. One can see directly from the definition that the sequence $(\underline{g}) = (g_i) = (X_i - a_i)$ is a regular $A[[\underline{X}]]$-sequence. Let $K(\underline{f})$ and $K(\underline{g})$ be the Koszul complexes associated to these two sequences. It follows from proposition 5.13 that the Koszul complex $K(\underline{f})$ is a resolution of A by free $\mathcal{O}[[\underline{X}]]$-modules, and the Koszul complex $K(\underline{g})$ is a resolution of A by free $A[[\underline{X}]]$-modules, and hence a fortiori, by free $\mathcal{O}[[\underline{X}]]$-modules. We define a map $\Phi : K(\underline{f}) \longrightarrow K(\underline{g})$ of complexes by letting

$$\Phi_0 : K_0(\underline{f}) \longrightarrow K_0(\underline{g})$$

be the natural inclusion of $\mathcal{O}[[\underline{X}]]$ into $A[[\underline{X}]]$, and letting

$$\Phi_1 : K_1(\underline{f}) \longrightarrow K_1(\underline{g})$$

be the map defined by

$$(\Phi_1(u_1), \dots, \Phi_1(u_n)) = (v_1, \dots, v_n)M,$$

and extending it by skew-linearity to a map of exterior algebras. One can check that the resulting map Φ is a morphism of complexes which induces the identity map $A \longrightarrow A$, and satisfies

$$\Phi_n(u_1 \wedge \cdots \wedge u_n) = D \cdot v_1 \wedge \cdots \wedge v_n.$$

Applying the functor $\mathrm{Hom}_{\mathcal{O}[[\underline{X}]]}(-, \mathcal{O}[[\underline{X}]])$ to these two free resolutions, and taking the homology of the resulting complexes, we find that since Φ is a homotopy equivalence of complexes it induces an isomorphism on the cohomology, and in particular, on the nth cohomology:

$$\Phi_n : \mathrm{Hom}_{\mathcal{O}[[\underline{X}]]}(A[[\underline{X}]], \mathcal{O}[[\underline{X}]])/(g_1, \dots, g_n) \xrightarrow{\simeq}$$

$$\mathrm{Hom}_{\mathcal{O}[[\underline{X}]]}(\mathcal{O}[[\underline{X}]], \mathcal{O}[[\underline{X}]])/(f_1, \dots, f_n) \simeq A,$$

which is given explicitly by the formula:

$$\Phi_n(f) = \alpha(f(D)).$$

□

We finally come to the proof of proposition 5.9, which we can state in a more precise form.

Lemma 5.15 *The map* $\Psi : \mathrm{Hom}_{\mathcal{O}}(A, \mathcal{O}) \longrightarrow A$ *defined by* $\Psi(f) = \alpha(\tilde{f}(D))$, *where* $\tilde{f} : A[[\underline{X}]] \longrightarrow \mathcal{O}[[\underline{X}]]$ *is the base change of* f, *is an* A*-module isomorphism, and hence,* A *is Gorenstein.*

Proof: The key point is to show that Ψ is A-linear. By definition, if

$$a = \alpha(a') \in A, \text{ with } a' \in \mathcal{O}[[\underline{X}]],$$

then

$$\Psi(af) = \alpha(\tilde{f}(aD)) = \alpha(\tilde{f}((a - a')D)) + \alpha(\tilde{f}(a'D)).$$

Since $a - a' \in \ker \beta$, it can be written as an $A[[\underline{X}]]$-linear combination of the g_i. By multiplying the relation

$$(f_1, \dots, f_n) = (g_1, \dots, g_n)M$$

by the matrix $D \cdot M^{-1} \in M_n(A[[\underline{X}]])$, one sees that the Dg_i can be written as $A[[\underline{X}]]$-linear combinations of the f_i's. Hence, so can the expression $(a-a')D$. By the $\mathcal{O}[[\underline{X}]]$-linearity of $\tilde{f}$, and the fact that each f_i belongs to $\ker \alpha$, it follows that

$$\alpha(\tilde{f}((a - a')D)) = 0.$$

Therefore,

$$\Psi(af) = \alpha(a'\tilde{f}(D)) = a\Psi(f).$$

This shows that Ψ is A-linear. To show that Ψ is surjective, observe that if $f_1, \dots, f_r$ is a $\mathcal{O}$-basis of $\mathrm{Hom}_{\mathcal{O}}(A, \mathcal{O})$, then $\tilde{f}_1, \dots, \tilde{f}_r$ is a $\mathcal{O}[[\underline{X}]]$-basis for $\mathrm{Hom}_{\mathcal{O}[[\underline{X}]]}(A[[\underline{X}]], \mathcal{O}[[\underline{X}]])$. Hence, for all $a \in A$, there exist $p_1, \dots, p_r$ such that

$$\Phi_n(p_1 \tilde{f}_1 + \cdots + p_r \tilde{f}_r) = a.$$

But this means that

$$\Psi(\alpha(p_1) f_1 + \cdots + \alpha(p_r) f_r) = a,$$

so that Ψ is surjective. Finally, since $\mathrm{Hom}_{\mathcal{O}}(A, \mathcal{O})$ and A are free $\mathcal{O}$-modules of the same finite rank, and Ψ is surjective, it must also be injective. Hence Ψ is an isomorphism, as was to be shown. □

Example 5.16 Let $A \subset \mathcal{O} \times \mathcal{O}$ be the ring of example 1 in section 5.1. Then, $f = T^2 - \lambda^n T$, and $g = T - (0, \lambda^n)$. Hence, $f = (T - (\lambda^n, 0))g$, so that $D = T - (\lambda^n, 0)$. It follows that

$$\Phi(h) = \alpha(\tilde{h}(T - (\lambda^n, 0))) = \alpha(Th(1,1) - h(\lambda^n, 0)) = (0, \lambda^n)h(1,1) - h(\lambda^n, 0).$$

The reader can check directly that Φ is indeed an A-linear isomorphism from the A-module $\mathrm{Hom}_{\mathcal{O}}(A, \mathcal{O})$ to A.

Example 5.17 Let $A = \mathcal{O}[\epsilon]/(\epsilon^2)$ be the ring of example 3 in section 5.1. Then, $f = T^2$, and $g = T - \epsilon$. Hence, $f = (T + \epsilon)g$, so that $D = T + \epsilon$. It follows that

$$\Phi(f) = \alpha(\tilde{f}(T + \epsilon)) = \alpha(Tf(1) + f(\epsilon)) = \epsilon f(1) + f(\epsilon).$$

Exercise 5.18 Write down explicitly an isomorphism $A \tilde{\to} \mathrm{Hom}_{\mathcal{O}}(A, \mathcal{O})$ for the complete intersection $A = \mathcal{O}[[X, Y]]/(X(X - \lambda), Y(Y - \lambda))$ of example 4 of section 5.1.

5.4 The Congruence ideal for complete intersections

Let A be an object of $\mathcal{C}_{\mathcal{O}}^{\bullet}$, which is finite flat and is a complete intersection as in the previous section, so that $A \simeq \mathcal{O}[[X_1, \ldots, X_n]]/(f_1, \ldots, f_n)$.

Using the result of the previous section, we will give an explicit formula for computing η_A in this case, and prove that (*c*) implies (*b*) in theorem 5.3.

Let $A^\vee := \mathrm{Hom}_{\mathcal{O}}(A, \mathcal{O})$, and let $\pi_A^\vee : \mathcal{O}^\vee \longrightarrow A^\vee$ be the dual map. From the Gorenstein property of A, we may identify $A^\vee$ with A (as A-modules). Fix any identification $\Psi : A^\vee \xrightarrow{\simeq} A$. (Any two such differ by a unit in A.) One checks that

$$\Psi\pi_A^\vee(\mathcal{O}^\vee) = \mathrm{Ann}_A \ker \pi_A.$$

Hence, η_A is the image of the map $\pi_A \Psi \pi_A^\vee$. By using the explicit construction of Ψ given in lemma 5.15, we find:

$$\pi_A \Psi \pi_A^\vee(\mathcal{O}^\vee) = \pi\alpha\tilde{\pi}(D),$$

where D is the determinant defined in section 5.3. By a direct calculation using equation (5.3.1), one sees that the right hand side is equal to $\det(\partial f_i/\partial X_j(0))$. Hence we have shown:

Proposition 5.19 *Suppose that A is an object of $\mathcal{C}_{\mathcal{O}}^{\bullet}$ which is finite flat. If $A = \mathcal{O}[[X_1, \ldots, X_n]]/(f_1, \ldots, f_n)$ is a complete intersection, then*

$$\eta_A = (\det(\partial f_i/\partial X_j(0))).$$

This proposition implies:

Corollary 5.20 *Let A in $\mathcal{C}_{\mathcal{O}}^{\bullet}$ be finite flat. If A is a complete intersection, then $\#\Phi_A = \#(\mathcal{O}/\eta_A)$. Hence, the statement (c) implies statement (b) in theorem 5.3.*

Proof: The equation (5.2.9) of remark 5.7 implies that

$$\#\Phi_A = \#\mathcal{O}/(\det(\partial f_i/\partial X_j(0))).$$

The result follows. □

5.5 Isomorphism theorems

The usefulness of the notion of complete intersections comes from the following two (vaguely stated) principles:
1. Isomorphisms to complete intersections can often be recognized by looking at their effects on the tangent spaces.
2. Isomorphisms from complete intersections can often be recognized by looking at their effects on the invariants η.

These vague principles are made precise in theorems 5.21 and 5.24 respectively.

Theorem 5.21 *Let $\phi : A \longrightarrow B$ be a surjective morphism of augmented rings, with B a (finite, flat) complete intersection. If ϕ induces an isomorphism from Φ_A to Φ_B, and these modules are finite, then ϕ is an isomorphism.*

Remark 5.22 Let $A = \mathcal{O}[[X,Y]]/(X(X-\lambda), Y(Y-\lambda))$ be the ring of example 4, let $B = \mathcal{O}[[X,Y]]/(X(X-\lambda), Y(Y-\lambda), XY)$ be the ring of example 2, and let $\phi : A \longrightarrow B$ be the natural projection. The map ϕ induces an isomorphism $\Phi_A \longrightarrow \Phi_B$, even though ϕ is not an isomorphism. The assumption that B is a complete intersection is crucial for concluding that ϕ is an isomorphism.

Remark 5.23 The natural map

$$\mathcal{O}[[X]]/(X^3) \longrightarrow \mathcal{O}[[X]]/(X^2)$$

is a surjective morphism inducing an isomorphism on tangent spaces, and the target ring is a complete intersection. Yet this map is *not* an isomorphism. This shows that the assumption on the finiteness of the tangent spaces cannot be dispensed with.

Proof of theorem 5.21: Recall that $U = \mathcal{O}[[X_1, \dots, X_n]]$ is the augmented ring of example 5 of section 5.1. Let

$$\nu_B : U \longrightarrow B$$

be a surjective morphism of augmented rings with $\ker \nu_B = (f_1, \ldots, f_n)$. Let $b_1, \ldots, b_n \in \ker \pi_B$ denote the images of $X_1, \ldots, X_n$ by ν_B, and let $a_1, \ldots, a_n \in \ker \pi_A$ denote inverse images of $b_1, \ldots, b_n$ by ϕ. Since ϕ is an isomorphism on tangent spaces, the elements a_i generate $(\ker \pi_A)/(\ker \pi_A)^2$. Hence the morphism

$$\nu_A : \mathcal{O}[[X_1, \ldots, X_n]] \longrightarrow A$$

defined by $\nu_A(X_i) = a_i$ induces a surjection $\Phi_U \longrightarrow \Phi_A$, and so it is surjective, by lemma 5.5.

We claim that $\ker \nu_B$ is contained in $\ker \nu_A$ (and hence, $\ker \nu_B = \ker \nu_A$). For, let $g_1, \ldots, g_n$ be elements of $\ker \nu_A$ whose linear terms $\bar{g}_1, \ldots, \bar{g}_n$ generate the kernel of

$$\bar{\nu}_A : \Phi_U \longrightarrow \Phi_A.$$

Since $\ker \nu_A \subset \ker \nu_B$, it follows that there exists an $n \times n$ matrix $M \in M_n(U)$ with entries in U such that

$$(g_1, \ldots, g_n) = (f_1, \ldots, f_n) M.$$

Let $\bar{M}$ be the matrix of constant terms of the matrix M. Then we have

$$(\bar{g}_1, \ldots, \bar{g}_n) = (\bar{f}_1, \ldots, \bar{f}_n) \bar{M}.$$

Since $(\bar{g}_1, \ldots, \bar{g}_n)$ and $(\bar{f}_1, \ldots, \bar{f}_n)$ generate the same submodules of rank n and finite index in Φ_U, it follows that $\det \bar{M}$ is a unit in $\mathcal{O}$. Hence, M is invertible, and therefore the f_i can be expressed as a U-linear combination of the g_j. This implies that $\ker \nu_B \subset \ker \nu_A$. Now we see that $\nu_A \nu_B^{-1}$ gives a well-defined inverse to ϕ, so that ϕ is an isomorphism. □

Theorem 5.24 *Let $\phi : A \longrightarrow B$ be a surjective morphism of augmented rings. Suppose that A and B are finite flat, and that A a complete intersection. If $\eta_A = \eta_B \neq 0$, then ϕ is an isomorphism.*

Remark 5.25 The torsion-freeness assumption on B is essential: if n is large enough, then $B = A/(\ker \pi_A)^n$ satisfies $\eta_A = \eta_B$, although the natural map $A \longrightarrow B$ is not injective when $A \neq \mathcal{O}$.

Proof: By proposition 5.9, A is Gorenstein, i.e., it satisfies

$$A^\vee := \mathrm{Hom}_{\mathcal{O}}(A, \mathcal{O}) \simeq A \text{ as } A\text{-modules.}$$

Now, we observe that

$$\ker \pi_A \cap \mathrm{Ann}_A \ker \pi_A = 0, \tag{5.5.1}$$

and likewise for B. For, let x be a non-zero element of η_A, and let $x' \in \mathrm{Ann}_A \ker \pi_A$ satisfy $\pi_A(x') = x$. For all $a \in \ker \pi_A \cap \mathrm{Ann}_A \ker \pi_A$, we have

$$0 = a(x - x') = ax,$$

the first equality because a belongs to $\mathrm{Ann}_A \ker \pi_A$ and $(x - x')$ belongs to $\ker \pi_A$, the second equality because a belongs to $\ker \pi_A$ and x' belongs to $\mathrm{Ann}_A \ker \pi_A$. Hence a belongs to the $\mathcal{O}$-torsion submodule of A, and therefore is 0.

It follows from (5.5.1) that the homomorphism π_A (resp. π_B) induces an isomorphism from $\mathrm{Ann}_A \ker \pi_A$ (resp. $\mathrm{Ann}_B \ker \pi_B$) to η_A (resp. η_B). Since η_A is isomorphic to η_B, it follows that ϕ induces an isomorphism from $\mathrm{Ann}_A \ker \pi_A$ to $\mathrm{Ann}_B \ker \pi_B$, i.e.,

$$\phi \mathrm{Ann}_A \ker \pi_A = \mathrm{Ann}_B \ker \pi_B.$$

From (5.5.1) it also follows a fortiori that

$$\ker \phi \cap \mathrm{Ann}_A \ker \pi_A = 0,$$

hence there is an exact sequence of A-modules:

$$0 \longrightarrow \ker \phi \oplus \mathrm{Ann}_A \ker \pi_A \longrightarrow A. \tag{5.5.2}$$

The cokernel of the last map is

$$A/(\ker \phi \oplus \mathrm{Ann}_A \ker \pi_A) \simeq B/(\phi \mathrm{Ann}_A \ker \pi_A) \simeq B/(\mathrm{Ann}_B \ker \pi_B),$$

which is torsion-free, since there is a natural injection

$$B/(\mathrm{Ann}_B \ker \pi_B) \hookrightarrow \mathrm{End}_{\mathcal{O}}(\ker \pi_B).$$

Hence, the exact sequence (5.5.2) splits over $\mathcal{O}$. Taking $\mathcal{O}$ duals in (5.5.2) and using the Gorenstein condition for A, we thus get an exact sequence of A-modules:

$$A \longrightarrow (\ker \phi)^\vee \oplus (\mathrm{Ann}_A \ker \pi_A)^\vee \longrightarrow 0.$$

Applying the functor $- \otimes_A k$ (relative to the map $A \longrightarrow k$), we find

$$1 = \dim_k(A \otimes_A k) \geq \dim_k((\ker \phi)^\vee \otimes_A k) + \dim_k((\mathrm{Ann}_A \ker \pi_A)^\vee \otimes_A k).$$

Since $\eta_A \neq 0$, it follows that $(\mathrm{Ann}_A \ker \pi_A)^\vee \otimes_A k \neq 0$, and hence we must have

$$(\ker \phi)^\vee \otimes_A k = 0.$$

Therefore by Nakayama's lemma and duality, $\ker \phi = 0$, which proves the theorem. □

5.6 A resolution lemma

It turns out that objects in $\mathcal{C}_{\mathcal{O}}^{\bullet}$ can be "resolved" (in a weak sense) by a complete intersection, namely,

Theorem 5.26 *Let A be an augmented ring which is finite flat over $\mathcal{O}$. Then there is a morphism $\tilde{A} \longrightarrow A$ in the category $\mathcal{C}_{\mathcal{O}}^{\bullet}$ such that:*

(a) *the ring $\tilde{A}$ is finite flat over $\mathcal{O}$ and is a complete intersection;*

(b) *the map $\tilde{A} \longrightarrow A$ induces an isomorphism $\Phi_{\tilde{A}} \longrightarrow \Phi_A$.*

Proof: Write A as a quotient of $U = \mathcal{O}[[X_1, \dots, X_n]]$ (with $\pi_U : U \longrightarrow \mathcal{O}$ the map which sends each X_i to 0.) Let $f_1, \dots, f_n$ be elements in the kernel of the natural map $U \longrightarrow A$, such that $\bar{f}_1, \dots, \bar{f}_n$ generate the kernel of $\Phi_U \longrightarrow \Phi_A$. Now letting

$$\tilde{A} = U/(f_1, \dots, f_n)$$

would give the desired ring $\tilde{A}$, provided $\tilde{A}$ is finitely generated: indeed, the flatness of $\tilde{A}$ would follow from corollary 5.12.

Thus we need to show that the f_i can be chosen so that $\tilde{A}$ is finitely generated. Let $a_1, \dots, a_n$ be $\mathcal{O}$-module generators of the finitely generated module $\ker \pi_A$, and define a homomorphism ϕ from the polynomial ring

$$V = \mathcal{O}[X_1, \dots, X_n]$$

to A by sending X_j to a_j. Clearly ϕ is surjective. Let $f_1, \dots, f_n$ be elements of $\ker \phi$ chosen as above, and let m denote their maximal degree. Since the elements a_i^2 belong to $\ker \pi_A$, we may write

$$a_i^2 = h_i(a_1, \dots, a_n),$$

where $h_i(X_1, \dots, X_n)$ is a linear polynomial. Now, replacing the relations f_i by the relations

$$f_i + X_i^m h_i - X_i^{m+2},$$

we find that the ring $V/(f_1, \dots, f_n)$ is a finitely generated $\mathcal{O}$-module: it can be generated by the images of the monomials of degree $\leq n(m+1)$, since the relations allow us to rewrite any monomial of higher degree in terms of ones of lower degree. Completing at the ideal $(\lambda, X_1, \dots, X_n)$, we find that the ring

$$\tilde{A} = U/(f_1, \dots, f_n)$$

has the desired properties: the natural homomorphism from $\tilde{A}$ to A induces an isomorphism on the tangent spaces, since the linear terms of the f_i generate the kernel of the induced map $\Phi_U \longrightarrow \Phi_A$ on the tangent spaces, and $\tilde{A}$ is a finitely generated $\mathcal{O}$-module, since $V/(f_1, \dots, f_n)$ is. □

5.7 A criterion for complete intersections

The results we have accumulated so far allow us to give an important criterion for an object A to be a complete intersection:

Theorem 5.27 *Let A be an augmented ring which is a finitely generated torsion-free $\mathcal{O}$-module. If $\#\Phi_A \leq \#(\mathcal{O}/\eta_A) < \infty$, then A is a complete intersection.*

Proof: Let $\phi : \tilde{A} \longrightarrow A$ be the surjective morphism given by the resolution theorem (theorem 5.26). Then we have

$$\#(\mathcal{O}/\eta_A) \geq (\#\Phi_A) = (\#\Phi_{\tilde{A}}) \geq \#(\mathcal{O}/\eta_{\tilde{A}}),$$

where the first inequality is by assumption, the second by the choice of $\tilde{A}$, and the third is by the equation (5.2.3). On the other hand, by equation (5.2.2), we have

$$\#(\mathcal{O}/\eta_{\tilde{A}}) \geq \#(\mathcal{O}/\eta_A).$$

It follows that

$$\eta_A = \eta_{\tilde{A}},$$

so that ϕ is an isomorphism by theorem 5.24. It follows that A is a complete intersection. □

5.8 Proof of Wiles' numerical criterion

Theorem 5.28 *Let R and T be augmented rings such that T is a finitely generated torsion-free $\mathcal{O}$-module, and let $\phi : R \longrightarrow T$ be a surjective morphism. If*

$$\#\Phi_R \leq \#(\mathcal{O}/\eta_T) < \infty,$$

then R and T are complete intersections, and ϕ is an isomorphism.

Proof: We have:

$$\#(\mathcal{O}/\eta_T) \leq \#\Phi_T \leq \#\Phi_R \leq \#(\mathcal{O}/\eta_T),$$

where the first inequality is by equation (5.2.3), the second follows from the surjectivity of ϕ, and the third is by hypothesis. Therefore,

$$\#\Phi_T = \#(\mathcal{O}/\eta_T),$$

and hence T is a complete intersection by theorem 5.27. Since the orders of Φ_R and Φ_T are the same, ϕ induces an isomorphism between them. Hence ϕ is an isomorphism $R \longrightarrow T$, by theorem 5.21. This completes the proof. □

Theorem 5.28 shows that the statements (*a*) and (*b*) in theorem 5.3 imply the statement (*c*). Combining corollaries 5.6, 5.20, and theorem 5.28 completes the proof of theorem 5.3.

5.9 A reduction to characteristic ℓ

Let $\mathcal{C}_k$ be the category of complete local noetherian k-algebras with residue field k. Again all morphisms are assumed to be local. There is a natural functor $A \mapsto \bar{A}$ from $\mathcal{C}_{\mathcal{O}}$ to $\mathcal{C}_k$ which send A to $\bar{A} := A/\lambda$.

We say that an object A of $\mathcal{C}_k$ which is finite dimensional as a k-vector space is a complete intersection if it is isomorphic to a quotient

$$A = k[[X_1, \dots, X_r]]/(f_1, \dots, f_r).$$

Note that if an object A of $\mathcal{C}_{\mathcal{O}}$ is a complete intersection, then $\bar{A}$ is a complete intersection in $\mathcal{C}_k$. As a partial converse, we have:

Lemma 5.29 *Suppose that $R \mapsto T$ is a map in the category $\mathcal{C}_{\mathcal{O}}$, and that T is finitely generated and free as an $\mathcal{O}$-module. Then $R \longrightarrow T$ is an isomorphism of complete intersections, if and only if $\bar{R} \longrightarrow \bar{T}$ is.*

Proof: This is an exercise and is left to the reader. □

We now come to the proof of lemma 3.39 of section 3.4:

Lemma 5.30 *Suppose that $K \subset K'$ are local fields with rings of integers $\mathcal{O} \subset \mathcal{O}'$ and that A is an object of $\mathcal{C}_{\mathcal{O}}$ which is finitely generated and free as an $\mathcal{O}$-module. Then A is a complete intersection if and only if $A \otimes_{\mathcal{O}} \mathcal{O}'$ is.*

Proof: One implication is clear. Let k and k' be the residue fields of $\mathcal{O}$ and $\mathcal{O}'$ respectively. By lemma 5.29 it is enough to prove that, if R is an object of $\mathcal{C}_k$ which is finite dimensional as a k-vector space, then

$$R' = R \otimes_k k' \text{ is a complete intersection} \quad \Rightarrow \quad R \text{ is a complete intersection.}$$

Let $\mathfrak{m}$ and $\mathfrak{m}'$ denote the maximal ideals of R and R' respectively.

By assumption, we have $R' = k'[[Y_1, \ldots, Y_r]]/J$, where the ideal J can be generated by r elements. We can assume without loss of generality (by adding extra variables and relations if necessary) that the images of $Y_1, \ldots Y_r$ generate $\mathfrak{m}'$ as a k'-vector space. Now, let

$$\phi : k[[X_1, \ldots, X_r]] \longrightarrow R$$

be a ring homomorphism, such that the images of $X_1, \ldots, X_r$ generate $\mathfrak{m}$ as a k-vector space, and let

$$\phi' : k'[[X_1, \ldots, X_r]] \longrightarrow R'$$

be the extension of scalars. Let I and $I' = I \otimes_k k'$ be the kernels of these two maps. We claim that I' can be generated by r elements. (In fact, this is also true without the assumption that the images of $X_1, \ldots, X_r$ generate $\mathfrak{m}'$ as a k' vector space, although we use this assumption in the proof below. Cf. remark 5.2.) To see this, choose an isomorphism of k'-vector spaces $k'X_1 \oplus \cdots \oplus k'X_r \longrightarrow k'Y_1 \oplus \cdots \oplus k'Y_r$ such that

$$\begin{array}{ccc} k'X_1 \oplus \cdots \oplus k'X_r & \longrightarrow & k'Y_1 \oplus \cdots \oplus k'Y_r \\ \downarrow & & \downarrow \\ \mathfrak{m}' & = & \mathfrak{m}' \end{array}$$

commutes. Extending this map to an isomorphism of k'-algebras,

$$\nu : k'[[X_1, \ldots, X_r]] \longrightarrow k'[[Y_1, \ldots, Y_r]],$$

one sees that $I' = \nu^{-1}(J)$, and hence can be generated by r elements, as claimed.

In particular we have $\dim_{k'}(I'/\mathfrak{m}'I') \le r$, and

$$\dim_k(I/\mathfrak{m}I) = \dim_{k'}((I/\mathfrak{m}I) \otimes_k k') \le \dim_{k'}(I'/\mathfrak{m}'I') \le r.$$

Nakayama's lemma now implies that I can be generated by r elements, and hence R is a complete intersection. This proves the lemma. □

5.10 J-structures

We now turn to the proof of theorem 3.41. In view of the last section we will work in characteristic ℓ. Thus let $\pi : R \twoheadrightarrow T$ be a surjective morphism in the category $\mathcal{C}_k$, where R and T are finite dimensional as k-vector spaces. Let r be a non-negative integer. If $J \lhd k[[S_1, ..., S_r]]$ and $J \subset (S_1, ..., S_r)$ then by a strong J-structure we shall mean a commutative diagram in $\mathcal{C}_k$

$$\begin{array}{ccccc} & & k[[S_1,...,S_r]] & & \\ & & \downarrow & & \\ k[[X_1,...,X_r]] & \twoheadrightarrow & R' & \twoheadrightarrow & T' \\ & & \downarrow & & \downarrow \\ & & R & \twoheadrightarrow & T, \end{array}$$

such that

(a) $T'/(S_1, ..., S_r)T' \xrightarrow{\sim} T$ and $R'/(S_1, ..., S_r)R' \twoheadrightarrow R$,

(b) for each ideal $I \supset J$, $I = \ker(k[[S_1, ..., S_r]] \to T'/I)$,

(c) $J = \ker(k[[S_1, ..., S_r]] \to T')$ and $R' \hookrightarrow T' \oplus R$.

We refer to these as *strong* J-structures because we have slightly altered the conditions from section 3.4 for technical convenience in the rest of this section.

We will prove the following result.

Theorem 5.31 *Suppose there exist a sequence of ideals $J_n \lhd k[[S_1, ..., S_r]]$ such that $J_0 = (S_1, ..., S_r)$, $J_n \supset J_{n+1}$, $\bigcap_n J_n = (0)$ and for each n there exists a strong J_n-structure. Then $R \xrightarrow{\sim} T$ and these rings are complete intersections.*

Before proving theorem 5.31 we shall explain how to deduce theorem 3.41 from it. If J is an ideal of $\mathcal{O}[[S_1, ..., S_r]]$ we will use $\bar{J}$ to denote its image in $k[[S_1, ..., S_r]]$. If $\mathcal{S}$ given by

$$\begin{array}{ccccc} & & \mathcal{O}[[S_1,...,S_r]] & & \\ & & \downarrow & & \\ \mathcal{O}[[X_1,...,X_r]] & \twoheadrightarrow & R' & \twoheadrightarrow & T' \\ & & \downarrow & & \downarrow \\ & & R & \twoheadrightarrow & T, \end{array}$$

is a J-structure for $R \to T$, let $\mathfrak{a} = \ker(R \longrightarrow T)$, and let R'' denote the ring $\mathrm{Im}(R' \to (R/\mathfrak{m}_R\mathfrak{a} \oplus (T'/J)) \otimes_{\mathcal{O}} k)$. Then $\bar{\mathcal{S}}$ given by

$$\begin{array}{ccccc} & & k[[S_1,...,S_r]] & & \\ & & \downarrow & & \\ k[[X_1,...,X_r]] & \twoheadrightarrow & R'' & \twoheadrightarrow & (T'/J)\otimes_{\mathcal{O}} k \\ & & \downarrow & & \downarrow \\ & & (R/\mathfrak{m}_R\mathfrak{a})\otimes_{\mathcal{O}} k & \twoheadrightarrow & T\otimes_{\mathcal{O}} k, \end{array}$$

is a strong $\bar{J}$-structure for $(R/\mathfrak{m}_R\mathfrak{a}) \otimes_{\mathcal{O}} k \to T \otimes_{\mathcal{O}} k$. If J_n is a sequence of ideals as in theorem 3.41 then $\bar{J}_n$ is a sequence of nested ideals with $\bar{J}_0 = (S_1, ..., S_r)$ and $\bigcap_n \bar{J}_n = (0)$. Moreover if for each n there is a J_n-structure $\mathcal{S}_n$ for $R \to T$ then for each n there is a strong $\bar{J}_n$-structure $\bar{\mathcal{S}}_n$ for

$$(R/\mathfrak{m}_R\mathfrak{a}) \otimes_{\mathcal{O}} k \to T \otimes_{\mathcal{O}} k.$$

Then by theorem 5.31 we see that this map is an isomorphism of complete intersections. Lemma 5.29 shows that theorem 3.41 follows. □

Before returning to the proof of theorem 5.31, let us first make some remarks about strong J-structures.

- The set of strong J structures for all ideals $J \subset (S_1, ..., S_r) \subset k[[S_1, ..., S_r]]$ forms a category, with the obvious notion of morphism.

- If $\mathcal{S}$ is a strong J-structure and if $(S_1, ..., S_r) \supset J' \supset J$ then there is a natural J'-structure $\mathcal{S}$ mod J' obtained by replacing T' by T'/J' and R' by the image of $R' \to R \oplus (T'/J')$.

- If R is a finite dimensional k-vector space and if J has finite index in $k[[S_1, ..., S_r]]$ then there are only finitely many isomorphism classes of strong J-structure. (This follows because we can bound the order of R' in any J-structure. Explicitly we must have $\#R' \leq (\#R)(\#k[[S_1, ..., S_r]]/J)^{\dim_k T}$.)

Lemma 5.32 *Suppose that R is a finite dimensional k vector space. Suppose also that $\{J_n\}$ is a nested (decreasing) sequence of ideals and that $J = \bigcap_n J_n$. If for each n a strong J_n structure exists then a strong J structure exists.*

Proof: We may suppose that each J_n has finite index in $k[[S_1, ..., S_r]]$. Let $\mathcal{S}_n$ denote a strong J_n-structure. Let $\mathcal{S}_{n,m} = \mathcal{S}_n$ mod J_m if $m \leq n$. Because there are only finitely many isomorphism classes of strong J_m structure, we may recursively choose integers $n(m)$ such that

- $\mathcal{S}_{n(m),m} \cong \mathcal{S}_{n,m}$ for infinitely many n,
- if $m > 1$ then $\mathcal{S}_{n(m),m-1} \cong \mathcal{S}_{n(m-1),m-1}$.

Let $\mathcal{S}'_m = \mathcal{S}_{n(m),m}$. Then $\mathcal{S}'_m$ is a strong J_m structure and if $m \geq m_1$ then $\mathcal{S}'_m$ mod $J_{m_1} \cong \mathcal{S}'_{m_1}$. One checks that $\mathcal{S} = \varprojlim \mathcal{S}'_m$ is the desired strong J-structure. □

Lemma 5.33 *Suppose that a strong* (0) *structure exists. Then the map* $R \longrightarrow T$ *is an isomorphism, and these rings are complete intersections.*

Proof: Because $k[[S_1, ..., S_r]] \hookrightarrow T'$ (and T' is a finitely generated $k[[S_1, ..., S_r]]$-module by Nakayama's lemma, cf. [Mat], thm. 8.4) we see that the Krull dimension of T' is at least r. On the other hand $k[[X_1, ..., X_r]] \twoheadrightarrow T'$ and so by Krull's principal ideal theorem this map must be an isomorphism. Thus

$$k[[X_1, ..., X_r]] \xrightarrow{\sim} R' \xrightarrow{\sim} T'.$$

Hence we have that

$$k[[X_1, ..., X_r]]/(S_1, ..., S_r) \xrightarrow{\sim} R'/(S_1, ..., S_r) \xrightarrow{\sim} T,$$

and the lemma follows. □

Theorem 5.31 follows at once from these two lemmas. □

Bibliography

[Ant2] W. Kuyk, P. Deligne, eds., Modular Functions of One Variable II, Lecture Notes in Math. **349**, Springer-Verlag, New York, Berlin, Heidelberg, 1973.

[Ant3] W. Kuyk, J.-P. Serre, eds., Modular Functions of One Variable III, Lecture Notes in Math. **350**, Springer-Verlag, New York, Berlin, Heidelberg, 1973.

[Ant4] B. Birch, W. Kuyk, eds., Modular Functions of One Variable IV, Lecture Notes in Math. **476**, Springer-Verlag, New York, Berlin, Heidelberg, 1975.

[Ant5] J.-P. Serre, D. Zagier, eds., Modular Functions of One Variable V, Lecture Notes in Math. **601**, Springer-Verlag, New York, Berlin, Heidelberg, 1977.

[AL] A.O.L. Atkin and J. Lehner, *Hecke operators on* $\Gamma_0(m)$, Math. Ann. **185** (1970), 134–160.

[BCEM] J. Buhler, R. Crandall, R. Ernvall, T. Metsänkylä, *Irregular primes and cyclotomic invariants to four million*, Math. Comp. **61** (1993), 151–153.

[BK] S. Bloch, K. Kato, *L-functions and Tamagawa numbers of motives*, in: The Grothendieck Festschrift I, Progress in Math. **86**, Birkhäuser, Boston, Basel, Berlin, 1990, pp. 333–400.

[BLR] N. Boston, H. Lenstra, K. Ribet, *Quotients of group rings arising from two-dimensional representations*, C. R. Acad. Sc. **t312** (1991), 323-328.

[BM] N. Boston, B. Mazur, *Explicit universal deformations of Galois representations*, Algebraic Number Theory, 1–21, Adv. Stud. Pure Math. **17**, Acad. Press, Boston MA, 1989.

[Bour] N. Bourbaki, Algèbre (Éléments de Mathématiques, Fasc. XXIII), Hermann, Paris, 1958.

[Ca1] H. Carayol, *Sur les représentations p-adiques associées aux formes modulaires de Hilbert*, Ann. Sci. Ec. Norm. Super. **19** (1986), 409–468.

[Ca2] H. Carayol, *Sur les représentations galoisiennes modulo ℓ attachées aux formes modulaires*, Duke Math. J. **59** (1989), 785–801.

[Ca3] H. Carayol, *Formes modulaires et représentations galoisiennes à valeurs dans un anneau local complet*, in: p-adic Monodromy and the Birch-Swinnerton-Dyer Conjecture (B. Mazur and G. Stevens, eds.), Contemporary Math. **165**, 1994, pp. 213–237.

[Co] J. Coates, S.T. Yau, eds., Elliptic Curves, Modular Forms and Fermat's Last Theorem, International Press, Cambridge, 1995.

[CPS] E. Cline, B. Parshall, L. Scott, *Cohomology of finite groups of Lie type I*, Publ. Math. IHES **45** (1975), 169–191.

[Cr] J. Cremona, Algorithms for Modular Elliptic Curves, Cambridge Univ. Press, Cambridge, 1992.

[CR] C.W. Curtis, I. Reiner, Representation Theory of Finite Groups and Associative Algebras, Wiley Interscience, New York, 1962.

[CS] G. Cornell and J.H. Silverman, eds., Arithmetic Geometry, Springer-Verlag, New York, Berlin, Heidelberg, 1986.

[CW] J. Coates, A. Wiles, *On the conjecture of Birch and Swinnerton-Dyer*, Inv. Math. **39** (1977), 223–251.

[Da1] H. Darmon, *Serre's conjectures*, in [Mu2], pp. 135–153.

[Da2] H. Darmon, *The Shimura-Taniyama Conjecture, (d'après Wiles)*, (in Russian) Uspekhi Mat. Nauk 50 (1995), no. 3(303), 33–82. (English version to appear in Russian Math Surveys)

[De] P. Deligne, *Formes modulaires et représentations ℓ-adiques*, in: Lecture Notes in Math. **179**, Springer-Verlag, New York, Berlin, Heidelberg, 1971, pp. 139–172.

[DR] P. Deligne, M. Rapoport, *Les schémas de modules de courbes elliptiques*, in [Ant2], pp. 143–316.

[dS] E. de Shalit, *On certain Galois representations related to the modular curve $X_1(p)$*, to appear in Compositio Math.

[dSL] B. de Smit, H.W. Lenstra, *Explicit construction of universal deformation rings*, to appear in the Proceedings of the Conference on Fermat's Last Theorem, Boston University, August 9–18, 1995.

[DS] P. Deligne, J.-P. Serre, *Formes modulaires de poids* 1, Ann. Sci. Ec. Norm. Sup. **7** (1974), 507–530.

[Di1] F. Diamond, *The refined conjecture of Serre*, in [Co], pp. 22–37.

[Di2] F. Diamond, *On deformation rings and Hecke rings*, to appear in Annals of Math.

[Dic1] L.E. Dickson, History of the Theory of Numbers, Vol. II, Chelsea Publ. Co., New York, 1971.

[Dic2] L.E. Dickson, Linear Groups with an Exposition of the Galois Field Theory, Teubner, Leipzig, 1901.

[Dir] P.G.L. Dirichlet, *Mémoire sur l'impossibilité de quelques équations indéterminées du cinquième degré*, Jour. für Math. (Crelle) **3** (1828), 354–375.

[DI] F. Diamond and J. Im, *Modular forms and modular curves*, in [Mu2], pp. 39–133.

[DK] F. Diamond and K. Kramer, *Modularity of a family of elliptic curves* Math. Research Letters **2** (1995), 299–305.

[DO] K. Doi, M. Ohta, *On some congruences between cusp forms on* $\Gamma_0(N)$, in [Ant5], pp. 91–106.

[Dr] V.G. Drinfeld, *Elliptic modules*, (Russian) Math Sbornik **94** (1974), 594–627. (English translation: Math. USSR, Sbornik **23** (1973).)

[DT1] F. Diamond and R. Taylor, *Non-optimal levels of mod* ℓ *modular representations*, Invent. Math. **115** (1994) 435–462.

[DT2] F. Diamond and R. Taylor, *Lifting modular mod* ℓ *representations*, Duke Math. J. **74** (1994) 253–269.

[Edi] B. Edixhoven, *The weight in Serre's conjectures on modular forms*, Invent. Math. **109** (1992), 563–594.

[Edw] H.M. Edwards, Fermat's Last Theorem: A genetic introduction to algebraic number theory, Graduate Texts in Math. **50**, Springer-Verlag, New York, Berlin, Heidelberg, 1977.

[Ei] M. Eichler, *Quaternäre quadratische Formen und die Riemannsche Vermutung für die Kongruenzzetafunktion*, Arch. Math. **5** (1954) 355–366.

[Fa] G. Faltings, *Endlichkeitssätze für abelsche Varietäten über Zahlkörpern*, Inv. Math. **73** (1983), 349–366. (English translation in [CS].)

[Fl1] M. Flach, *A finiteness theorem for the symmetric square of an elliptic curve*, Inv. Math. **109** (1992), 307–327.

[Fl2] M. Flach, *On the degree of modular parametrizations*, Séminaire de théorie des nombres, Paris, 1991-92, 23-36; Prog. in Math. **116**, Birkhäuser, Boston, MA 1993.

[FL] J.-M. Fontaine and G. Laffaille, *Construction de représentations p-adiques*, Ann. Sci. Ec. Norm. Super. **15** (1982), 547–608.

[FM] J.-M. Fontaine and B. Mazur, *Geometric Galois representations*, in [Co], 41–78.

[Fo1] J.-M. Fontaine, *Groupes finis commutatifs sur les vecteurs de Witt*, C. R. Acad. Sc. **280** (1975), 1423–1425.

[Fo2] J.-M. Fontaine, *Groupes p-divisibles sur les corps locaux*, Astérisque **47–48** (1977).

[For] O. Forster, Lecures on Riemann Surfaces, Springer-Verlag, New York, Berlin, Heidelberg, 1981.

[Fr] G. Frey, *Links between solutions of* $A - B = C$ *and elliptic curves*, in: Number Theory, Ulm 1987, Proceedings, Lecture Notes in Math. **1380**, Springer-Verlag, New York, Berlin, Heidelberg, 1989, pp. 31–62.

[Gra] A. Granville, *The set of exponents for which Fermat's Last Theorem is true, has density one*, Comptes Rendus de l'Académie des Sciences du Canada **7** (1985), 55-60.

[Gre] R. Greenberg, *Iwasawa theory for p-adic representations*, in: Algebraic Number Theory, in honor of K. Iwasawa, Adv. Studies in Pure Math. **17**, Academic Press, San Diego, 1989, pp. 97–137.

[Gro] B.H. Gross, *A tameness criterion for Galois representations associated to modular forms* mod$p$, Duke Math. J. **61** (1990), 445–517.

[Ha] R. Hartshorne, Algebraic Geometry, Graduate Texts in Math. **52**, Springer-Verlag, New York, Berlin, Heidelberg, 1977.

[HB] D.R. Heath-Brown, *Fermat's Last Theorem for "almost all" exponents*, Bull. London. Math. Soc. 17 (1985), 15-16

[He] Y. Hellegouarch, *Points d'ordre* $2p^h$ *sur les courbes elliptiques*, Acta Arith. **26** (1974/75) 253–263.

[Hi1] H. Hida, *Congruences of cusp forms and special values of their zeta functions*, Inv. Math. **63** (1981), 225–261.

[Hi2] H. Hida, *A p-adic measure attached to the zeta functions associated with two elliptic modular forms. I*, Inv. Math. **79** (1985), 159–195.

[Hi3] H. Hida, *Galois representations into $GL_2(\mathbb{Z}_p[[X]])$ attached to ordinary cusp forms,* Inv. Math. **85** (1986), 545–613.

[Hu] B. Huppert, Endliche Gruppen I, Grundlehren Math. Wiss. **134**, Springer-Verlag, New York, Berlin, Heidelberg, 1983.

[Ig] J. Igusa, *Kroneckerian model of fields of elliptic modular functions*, Amer. J. Math. **81**, 561–577 (1959).

[Ih] Y. Ihara, *On modular curves over finite fields*, in: Proc. Intl. Coll. on Discrete Subgroups of Lie Groups and Applications to Moduli, Bombay, 1973, pp. 161–202.

[Kam] S. Kamienny, *Torsion points on elliptic curves over fields of higher degree*, Inter. Math. Res. Not. **6**, 1992, 129–133.

[Kat] N.M. Katz, *p-adic properties of modular schemes and modular forms*, in [Ant3], pp. 70–189.

[Kh] C. Khare, *Congruences between cusp forms: the (p,p) case*, to appear in Duke Math. J.

[Ki] F. Kirwan, Complex Algebraic Curves, LMS Student Texts **23**, Cambridge Univ. Press, Cambridge, 1993.

[KL] D.S. Kubert, S. Lang, Modular Units. Springer-Verlag, 1981.

[KM] N.M. Katz, and B. Mazur, Arithmetic Moduli of Elliptic Curves, Annals of Math. Studies **108**, Princeton Univ. Press, Princeton, 1985.

[Kn] A.W. Knapp, Elliptic Curves, Princeton Univ. Press, Princeton, 1992.

[Koc] H. Koch, Galoissche Theorie der p-Erweiterungen, Springer-Verlag, New York, Berlin, Heidelberg, 1970.

[Kol] V.A. Kolyvagin, *Finiteness of $E(\mathbb{Q})$ and sha$(E/\mathbb{Q})$ for a subclass of Weil curves*, Izv. Akad. Nauk. SSSR Ser. Mat. **52** (3) (1988), 522–540. (English translation: Math. USSR, Izvestiya **32** (1989).)

[Ku] E. Kunz, Introduction to Commutative Algebra and Algebraic Geometry, Birkhäuser, Boston, Basel, Berlin, 1985.

[La1] S. Lang, Algebraic Number Theory, Addison-Wesley, Reading, MA 1970.

[La2] S. Lang, Introduction to Modular Forms, Grundlehren Math. Wiss. **222**, Springer-Verlag, New York, Berlin, Heidelberg, 1976.

[Leg] A.M. Legendre, *Sur quelques objets d'analyse indéterminée et particulièrement sur le théorème de Fermat*, Mém. Acad. R. Sc. de l'Institut de France **6**, Paris, 1827.

[Len] H.W. Lenstra, *Complete intersections and Gorenstein rings*, in [Co], pp. 99–109.

[Liv] R. Livné, *On the conductors of mod ℓ representations coming from modular forms*, J. Number Theory **31** (1989), 133–141.

[Ll1] R.P. Langlands, *Modular forms and ℓ-adic representations*, in [Ant2], pp. 361–500.

[Ll2] R.P. Langlands, Base Change for $GL(2)$, Annals of Math. Studies **96**, Princeton Univ. Press, Princeton, 1980.

[LO] S. Ling, J. Oesterlé, *The Shimura subgroup of $J_0(N)$*, Astérisque **196–197** (1991), 171–203.

[Mat] H. Matsumura, Commutative Ring Theory, Cambridge Studies in Adv. Math. **8**, Cambridge Univ. Press, Cambridge, 1986.

[Maz1] B. Mazur, *Modular curves and the Eisenstein ideal*, Publ. Math. IHES **47** (1977), 33–186.

[Maz2] B. Mazur, *Rational isogenies of prime degree*, Inv. Math. **44** (1978), 129–162.

[Maz3] B. Mazur, *Deforming Galois representations*, in: Galois groups over $\mathbb{Q}$, Y. Ihara, K. Ribet, J-P. Serre, eds., MSRI Publ. **16**, Springer-Verlag, New York, Berlin, Heidelberg, 1989, pp. 385–437.

[Mc] W. McCallum, *On the method of Coleman and Chabauty*, Math. Ann. **299** (1994), 565–596.

[Mer] L. Merel, *Bornes pour la torsion des courbes elliptiques sur les corps de nombres*, Inventiones Math., to appear.

[Mes] J-F. Mestre, *Constructions polynomiales et théorie de Galois*, Proceedings of the ICM, Zurich, August 1994.

[Mi] J.S. Milne, Arithmetic Duality Theorems, Perspectives in Math., Academic Press, San Diego, 1986.

[MRi] B. Mazur, K. Ribet, *Two-dimensional representations in the arithmetic of modular curves*, Astérisque **196–197** (1991), 215–255.

[MRo] B. Mazur, L. Roberts, *Local Euler characteristics*, Inv. Math. **9** (1970), 201–234.

[MT] B. Mazur, J. Tilouine, *Représentations galoisiennes, différentielles de Kähler et "conjectures principales"*, Publ. Math. IHES **71** (1990), 65–103.

[Mu1] V.K. Murty, Introduction to Abelian Varieties, CRM Monograph Series **3**, AMS Publ., Providence, 1993.

[Mu2] V.K. Murty, ed., Seminar on Fermat's Last Theorem, CMS Conference Proceedings **17**, AMS Publ., Providence, 1995.

[MW] B. Mazur, A. Wiles, *Class fields of abelian extensions of* $\mathbb{Q}$, Inv. Math. **76** (1984), 179–330.

[Na] K. Nagao, *An example of an elliptic curve over* $\mathbb{Q}(T)$ *with rank* ≥ 13, Proc. of The Japan Acad., **70**, 1994, 152-153.

[Oda] T. Oda, *The first De Rham cohomology group and Dieudonné modules*, Ann. Sci. Ec. Norm. Super. **2** (1969), 63–135.

[Oe] J. Oesterlé, *Nouvelles aproches du "théorème" de Fermat"*, Séminaire Bourbaki no. 694 (1987-88), Astérisque **161–162** (1988), 165–186.

[Ogg] A. Ogg, Modular Forms and Dirichlet Series, Benjamin, New York, 1969.

[Oo] F. Oort, Commutative Group Schemes, Lecture Notes in Math. **15**, Springer-Verlag, New York, Berlin, Heidelberg, 1966.

[Ram] R. Ramakrishna, *On a variation of Mazur's deformation functor*, Compositio Math. **87** (1993), 269–286.

[Ray] M. Raynaud, *Schémas en groupes de type* $(p, p, ..., p)$, Bull. Soc. Math. France **102** (1974), 241–280.

[R1] K. Ribet, *A modular construction of unramified p-extensions of* $\mathbb{Q}(\mu_p)$, Inv. Math. **34**, 151-162.

[R2] K. Ribet, *Twists of modular forms and endomorphisms of abelian varieties*, Math. Ann. **253** (1980), 43-62.

[R3] K.A. Ribet, *The* ℓ*-adic representations attached to an eigenform with Nebentypus: a survey*, in [Ant5], pp. 17–52.

[R4] K.A. Ribet, *Congruence relations between modular forms*, Proc. ICM, 1983, 503–514.

[R5] K.A. Ribet, *On modular representations of* Gal $(\bar{\mathbb{Q}}/\mathbb{Q})$ *arising from modular forms*, Inv. Math. **100** (1990), 431–476.

[R6] K.A. Ribet, *Report on mod ℓ representations of* Gal $(\bar{\mathbb{Q}}/\mathbb{Q})$, in: Motives, Proc. Symp. in Pure Math. **55** (2), 1994, pp. 639–676.

[R7] K.A. Ribet, *Abelian varieties over $\mathbb{Q}$ and modular forms*, 1992 Proceedings of KAIST Mathematics Workshop, Korea Advanced Institute of Science and Technology, Taejon, 1992, pp. 53–79.

[R8] K.A. Ribet, *Galois representations and modular forms*, Bull. AMS. **32** (1995), 375–402.

[Ri] P. Ribenboim, 13 Lectures on Fermat's Last Theorem, Springer-Verlag, New York, Berlin, Heidelberg, 1979.

[Sa] T. Saito, *Conductor, discriminant and the Noether formula for arithmetic surfaces*, Duke Math. J. **57** (1988), 151–173.

[Se1] J.-P. Serre, Cohomologie Galoisienne, Lecture Notes in Math. **5**, Springer-Verlag, New York, Berlin, Heidelberg, 1964.

[Se2] J.-P. Serre, Corps Locaux, Hermann, Paris, 1962.

[Se3] J.-P. Serre, Algèbre Locale, Multiplicités, Cours au Collège de France, 1957-58, Lecture Notes in Math. **11**, Springer-Verlag, New York, Berlin, Heidelberg, 1965.

[Se4] J.-P. Serre, A Course in Arithmetic, Graduate Texts in Math. **7**, Springer-Verlag, New York, Berlin, Heidelberg, 1973.

[Se5] J.-P. Serre, Abelian ℓ-adic Representations and Elliptic Curves, Benjamin, New York, 1968.

[Se6] J.-P. Serre, *Propriétés galoisiennes des points d'ordre fini des courbes elliptiques*, Inv. Math. **15** (1972), 259–331.

[Se7] J.-P. Serre, *Sur les représentations modulaires de degré* 2 *de* Gal $(\bar{\mathbb{Q}}/\mathbb{Q})$, Duke Math. J. **54** (1987), 179–230.

[Sha] S. Shatz, *Group schemes, formal groups, and p-divisible groups*, in [CS], pp. 29–78.

[Shi1] G. Shimura, *Correspondances modulaires et les fonctions ζ de courbes algébriques*, J. Math. Soc. Japan **10** (1958), 1–28.

[Shi2] G. Shimura, Introduction to the Arithmetic Theory of Automorphic Functions, Princeton Univ. Press, Princeton, 1971.

[Shi3] G. Shimura, *On the factors of the jacobian variety of a modular function field*, J. Math. Soc. Japan **25** (1973), 523–544.

[Shi4] G. Shimura, *On the holomorphy of certain Dirichlet series*, Proc. LMS (3) **31** (1975), 79–98.

[Shi5] G. Shimura, *The special values of the zeta functions associated with cusp forms*, Comm. Pure Appl. Math. **29** (1976), 783–804.

[Si1] J. Silverman, The Arithmetic of Elliptic Curves, Graduate Texts in Math. **106**, Springer-Verlag, New York, Berlin, Heidelberg, 1986.

[Si2] J. Silverman, Advanced Topics in the Arithmetic of Elliptic Curves, Graduate Texts in Math. **151**, Springer-Verlag, New York, Berlin, Heidelberg, 1994.

[St] J. Sturm, *Special values of zeta functions, and Eisenstein series of half integral weight*, Amer. J. Math. **102** (1980), 219–240.

[ST] J.-P. Serre, J.T. Tate, *Good reduction of abelian varieties*, Annals of Math. **88** (1968), 492–517.

[Su] J. Suzuki, *On the generalized Wieferich criterion*, Proc. Japan Acad. **70** (1994), 230-234.

[Ta] J.T. Tate, *Algorithm for determining the type of a singular fiber in an elliptic pencil*, in [Ant4], pp. 33–52.

[Te] G. Terjanian, *Sur l'équation* $x^{2p} + y^{2p} = z^{2p}$, C.R. Acad. Sci. Paris, **285** (1977) 973-975.

[Th] F. Thaine, *On the ideal class groups of real abelian number fields*, Annals of Math. **128** (1988), 1–18.

[Tu] J. Tunnell, *Artin's conjecture for representations of octahedral type*, Bull. AMS **5** (1981), 173–175.

[TW] R. Taylor, A. Wiles, *Ring theoretic properties of certain Hecke algebras*, Annals of Math. **141** (1995), 553–572.

[Wa] L. Washington, Introduction to Cyclotomic Fields, Graduate Texts in Math. **83**, Springer-Verlag, New York, Berlin, Heidelberg, 1982.

[We1] A. Weil, Variétés Abéliennes et Courbes Algébriques, Hermann, Paris, 1948.

[We2] A. Weil, *Über die Bestimmung Dirichletscher Reihen durch Funktionalgleichungen*, Math. Ann. **168** (1967), 165–172.

[W1] A. Wiles, *On p-adic representations for totally real fields*, Annals of Math. **123** (1986), 407–456.

[W2] A. Wiles, *On ordinary* λ*-adic representations associated to modular forms*, Inv. Math. **94** (1988), 529–573.

[W3] A. Wiles, *Modular elliptic curves and Fermat's Last Theorem*, Annals of Math. **141** (1995), 443–551.

Renormalization ideas in conformal dynamics

Mikhail Lyubich

Contents

1 Introduction

How to look at a dynamical system f at a small scale? You should take a small piece of the phase space, consider the *first return map* to this piece, and then rescale it to "the original size". The new dynamical system is called the *renormalization* Rf of the original one. It may happen that Rf looks "similar" to f, and then you can try to repeat this procedure, and construct the second renormalization $R^2 f$, etc. Asymptotic properties of this sequence of renormalizations reflect micro-structure of the original system. For example, convergence of the sequence $R^n f$ to a map f_* independent of f (from some class of similar maps) means that all maps of this class have in small scales a universal geometry represented by f_*.

A striking phenomenon of this kind is the Feigenbaum-Coullet-Tresser Universality Law ([CT, F], see [McM1], §6). It deals with the class of sufficiently smooth unimodal maps of an interval I with the critical point 0 of a given type $|x|^d$ ("unimodal" means: "with one critical point"). Under some combinatorial assumptions on the positions of the first four iterates of the critical point, the interval $J = [-f^2 0, f^2 0]$ turns out to be invariant under f^2. Moreover $f^2|J$ is again a unimodal map of the same class. Rescaling J to the original size, we obtain the "doubling renormalization" Rf of f. A map f of such kind can be called "renormalizable". If it happens that this procedure can be repeated, we have twice renormalizable maps, etc. The Universality Law asserts that the renormalizations $R^n f$ of infinitely renormalizable maps converge to a map f_* independent of f. Thus all infinitely renormalizable unimodal maps with a given type of the critical point have asymptotically the same geometry in small scales. A similar picture is observed not only for the doubling renormalization but for other periods as well.

We have here a kind of the rigidity phenomenon: Combinatorics of an object determines its geometry. Compare it with the Rigidity Conjecture discussed by McMullen [McM1]. The latter is concerned with a finitely dimensional family of globally defined objects, rational maps. The rigidity conclusion is also global: the geometry of the whole Julia set is determined by combinatorics. In the Feigenbaum-Coullet-Tresser situation we deal with an infinitely dimensional family of "partially defined" maps (think of polynomial-like maps). The Julia set is not rigid any more (as one can vary the multipliers of periodic points), but its most important part (the post-critical set) is still rigid!

The Universality Law was backed by numerous computer experiments and then by a computer assisted proof (Lanford [La]) - see the book [CE] and the survey [VSK] for that stage of events). In mid 80th Sullivan suggested a program of conceptual understanding of this phenomenon [S1-S3]. It included three big steps:

- Geometric *a priori* bounds;
- From geometric bounds to quasi-rigidity and further to global rigidity of polynomials;
- Contracting property of the renormalization transformation R with respect to an appropriate Teichmüller metric.

The first step of this program motivated by the work of Douady and Hubbard on polynomial-like maps [DH2] proved to be a hard analytical issue. Sullivan resolved it for real infinitely renormalizable maps of "bounded type". In §4 we will discuss, along with this work, the further development which settled the problem for all real quadratic maps and many complex ones [L5, GS2, LS, LY].

The second step has been resolved by a nice geometric argument based on the theory of quasi-conformal maps (see §5). The ideas for this part introduced into dynamics by Sullivan and Thurston can be tracked back to the Mostow Rigidity.

For the last step, Sullivan developed a sophisticated Teichmüller theory of "Riemann surface laminations" (see [S3] and the book of de Melo & van Stiren [MvS]). A different approach was suggested by McMullen who introduced a global dynamical object called a *tower*, and reduced the universality law to the quasi-conformal rigidity of towers [McM3].

What we have described above is the universality phenomenon in the *dynamical* plane. Not less intriguing is the *parameter* universality, which was actually discovered first by Feigenbaum, Coullet and Tresser. They gave an explanation of this phenomenon based upon conjectural hyperbolicity of the renormalization transformation R at the fixed point f_*. Proofs of this conjecture for period doubling case were given by Lanford [La] and Eckmann-Epstein [EE]. Recently the author proved it for all real combinatorial types [L8]. The proof is based upon a Rigidity Theorem for quadratic-like maps with *a priori bounds* [L7]. We will discuss this issue in §6.

We see that the global rigidity problem is an intimate part of the universality phenomenon. There has been recently several big breakthroughs in this problem which first looked complementary to the renormalization theory, but then were linked to it. The combinatorial game called "puzzle" appeared in the work of Branner, Hubbard and Yoccoz [BH, H] and allowed one to settle the rigidity problem for all maps with one non-escaping critical point of quadratic type, which are "at most finitely renormalizable" (see §4.4). Further contribution to the rigidity problem has been made by McMullen [McM2], Swiatek [Sw] and the author [L5]-[L7], which, in particular, settled it for real quadratic maps.

The notion which links "non-renormalizable" and "infinitely renormalizable" cases is "generalized renormalization". It allows one to embed the non-renormalizable maps into the renormalization theory, and to handle a number of geometric and measure-theoretical problems of real and complex dynamics [LM], [L3]-[L5], [SN]. We will particularly emphasize renormalization in the family of Fibonacci maps. A new curious phenomenon enlightened by this family is dependence of geometric and measure-theoretic properties of the map on the degree. In particular, the quadratic Fibonacci map has the Julia set of measure zero (Lyubich-Shishikura [L3]), while the maps of sufficiently high degree have positive measure Julia sets (Nowicki-van Strien [SN]).

This paper is linked to McMullen's paper [McM1] in this volume: concepts and results discussed in [McM1] may be used here without extra comments.

2 Combinatorics of complex unimodal maps

2.1 Renormalization in the sense of Douady and Hubbard

Polynomial-like maps were introduced by Douady and Hubbard in order to explain partial self-similarity of the Mandelbrot set. It is important to realize that a "polynomial-like map" actually means a *germ* near the filled Julia sets, so that there is a flexibility in the choice of the domain and range. Thus referring to a *conjugacy* between two polynomial-like maps we mean conjugacy near the filled Julia sets. In particular, the germ of a polynomial f near its filled Julia set $K(f)$ is polynomial-like. In this sense polynomials are also considered as polynomial-like maps. Polynomial-like maps with a single critical point (maybe degenerate) will be called *unimodal*, or *complex unimodal maps*.

Besides topological/quasi-conformal/conformal/affine categories of conjugacies, there is one more category called *hybrid*. Two polynomial-like maps are hybrid equivalent if they are conjugate by a quasi-conformal map h such that $\bar\partial h = 0$ almost everywhere on the filled Julia set. Let $\mathcal{H}(f)$ denote the hybrid class of a polynomial-like map f modulo conformal equivalence. The following basic result explains the importance of the hybrid category:

Straightening Theorem [DH2]. *Every hybrid class $\mathcal{H}(f)$ contains a polynomial. This polynomial is unique (modulo affine conjugacy) provided the Julia set $J(f)$ is connected.*

Sullivan views these hybrid classes as infinitely dimensional Teichmüller spaces [S1]. The Teichmüller pseudo-metric on this space is defined as follows:

$$\mathrm{dist}_T(f,g) = \inf \log K_h,$$

where h runs over the hybrid conjugacies between f an g, and K_h stands for its dilatation. It is not obvious but turns out to be true that this pseudo-metric is actually a metric provided $J(f)$ is connected. Note that unlike the classical situation, by the Straightening Theorem this Teichmüller space has a *preferred point.*

Let M_d be the connectedness locus of the family $z \mapsto z^d + c$, that is, the set of parameter values c for which the Julia set $J(z^d + c)$ is connected. Let Γ_n be the cyclic group of rotations $z \mapsto e^{2\pi i m/n} z$ of order n. By the Straightening Theorem, for any complex unimodal map f of degree d with connected Julia set, the hybrid class $\mathcal{H}(f)$ contains a unique polynomial $z \mapsto z^d + c(f)$ (modulo conjugacy by rotations $\gamma \in \Gamma_{d-1}$), where $c(f)$ belongs to M_d. In particular, hybrid classes of quadratic-like maps with connected Julia set are labeled by the points of the Mandelbrot set $M \equiv M_2$.

Renormalization (in the sense of Douady and Hubbard) of a complex unimodal map f means extracting from it a complex unimodal map $Rf = f^p$ of the same degree (see [McM1], §6). If this is possible, the map is usually called *renormalizable.* Notice that Rf is not a polynomial even when f is, so that the renormalization procedure automatically leads to the class of polynomial-like maps.

However this procedure respects the hybrid equivalence, so that it induces a map σ_d from a part of the connectedness locus M_d/Γ_{d-1}, where the map is renormalizable, into M_d/Γ_{d-1}. Douady and Hubbard have proved [D, DH2] that in degree two this map gives a homeomorphism of appropriate pieces of the Mandelbrot set onto the whole M. These pieces are exactly "small copies of M".

Different copies specify different "combinatorial types" of the renormalization. Let us consider the family $\mathcal{M}$ of *maximal* copies of M, that is, the copies which are not contained in any other copies. We have a map $\sigma \equiv \sigma_2 : \cup_{M'_l \in \mathcal{M}} M'_l \to M$ from the union of these copies onto M. If a parameter value $c \in M$ is infinitely renormalizable then we can apply this map infinitely many time. Let us keep track of the combinatorial types of the corresponding renormalizations by looking how the trajectory $c, \sigma c, \sigma^2 c, \ldots$ travels through the copies: let $\sigma^n c \in M'_{l(n)}$, $n = 0, 1, \ldots$. Let us call the sequence $\tau(f) = \{l(0), l(1), \ldots\}$ the *combinatorial type* of f. The combinatorial class $Com(f)$ of an infinitely renormalizable quadratic-like map is the set of maps with the same combinatorial type. (One can show that this definition fits to the definition in terms of rational laminations, see [McM1], §5).

2.2 Generalized renormalization

So, there are "non-renormalizable" maps. How does it fit to the general idea of renormalization indicated in the introduction? The answer is hidden in the word "similar": renormalization Rf is supposed to be similar to f. In the above discussion the criterion for this similarity was the polynomial-like property in the sense of Douady and Hubbard. But why should we stick to it? It turns out that there is a fruitful extension of the class of polynomial-like maps which allows us to apply the renormalization ideas to "non-renormalizable" maps as well.

Let U_i be a family of disjoint topological disks compactly contained in another topological disk V. A generalized polynomial-like map $f : \cup U_i \to V$ is a branched covering which is univalent on all U_i except at most finitely many. If such a map has a single critical point, it is called (generalized) unimodal. If this point is non-degenerate, f is also called a (generalized) quadratic-like map. The filled Julia set $K(f)$ is again defined as the set of non-escaping points, and the Julia set $J(f)$ is defined as its boundary. In this setting, we have the following Straightening Theorem: *Any generalized polynomial-like map is hybrid equivalent to a polynomial with the same number of non-escaping critical points.*

We can now try to renormalize a complex unimodal polynomial $z \mapsto z^d + c$ in the class of generalized complex unimodal maps with non-escaping critical point. We will see that this is indeed possible for all "combinatorially recurrent" polynomials. However it requires a careful selection of the disk V. Indeed, if you take a random disk V and pull it back along an orbit $z \in V, fz, \dots, f^n z \in V$, you may well obtain a domain U which intersects V in a crazy way. We will discuss two good ways to select the domain: as a Yoccoz puzzle piece and (for real maps) just as a Euclidean disk.

2.3 Yoccoz puzzle

The puzzle provides us with a family of topological disks which always intersect nicely. The idea is to cut a neighborhood of the filled Julia set by a *forward invariant* family of curves, and then pull the corresponding domains back. A nice selection of the neighborhood is a topological disk D bounded by some equipotential E (note that fE encloses E). A nice selection of the cuts is the union of several rational external rays R_i (see [McM1], §5 for the definition).

(The most popular choice in the quadratic case is the following. Let $f = P_c : z \mapsto z^2 + c$, with $c \in M$ but outside the main cardioid of the Mandelbrot set. Such a quadratic has a fixed point α which is a landing point of *more than one* external rays R_i cyclically permuted by f.)

So assume that the rays R_i divide D into the pieces $Y_i^{(0)}$, "puzzle pieces of depth 0". The puzzle pieces of depth n are defined as the closures of the components of $f^{-n}\,\mathrm{int}\,Y_i^{(0)}$. The pieces of depth n form a tiling $\mathcal{T}_n$ of the disk $D_n = f^{-n}D$. Moreover $\mathcal{T}_n$ is a refinement of $\mathcal{T}_{n-1}|D_n$. The pieces containing 0 are called *critical.* We will either label the critical pieces with subscript 0, or skip the subscript all together.

Thus any two puzzle pieces are either nested or have disjoint interiors, and moreover the image of any puzzle piece of depth $n > 1$ is a puzzle piece of depth $n-1$. These two obvious facts express the extremely useful *Markov property* of the family of puzzle pieces. It prevents the intersection troubles mentioned above, and allows us to carry out a generalized renormalization procedure.

2.4 Construction of $T^n f$

Let $\mathcal{O} = \mathcal{O}(f)$ denote the *post-critical set,* that is, the closure of the orbit $\{f^n 0\}_{n=0}^{\infty}$ of the critical point.

Lemma 2.1 *Let $V = Y^{(n)}$ be a critical puzzle piece for a unimodal polynomial $f : z \mapsto z^d + c$ such that:*

- *V is compactly contained in the previous critical puzzle piece $Y^{(n-1)}$;*
- *The critical point returns infinitely many times to V.*

Then f admits a generalized unimodal renormalization $T_V f : \cup U_i \to V$ with range V.

Proof. Given a point $z \in V \cap \mathcal{O}$ which returns back to $\mathrm{int}\,V$, let $r(z)$ denote the first return time. Let us consider the pull-back $U(z)$ of V along the orbit $z, fz, \ldots, f^{r(z)}z$ (that is, the puzzle piece containing z which is mapped under $f^{r(z)}$ onto V). It follows from the Markov property that all puzzle pieces $U(z)$ are contained in V, and any two of them either coincide or have disjoint interiors. Moreover, our first assumption implies that they are compactly contained in V. The map $g : \cup U(z) \to V$ defined as $g|U(z) = f^{r(z)}$ is the desired renormalization. □

Remarks. 1. Yoccoz showed that a non-renormalizable (in the sense of Douady & Hubbard) unimodal polynomial with all periodic points repelling always has a puzzle piece satisfying the first assumption. The second assumption is minor as polynomials with non-recurrent critical point can be easily treated.

2. In most interesting cases the domain of $T_V f$ consists only of finitely many pieces $U(z)$ (see, e.g., the Fibonacci maps below). This is the situation when the renormalization philosophy becomes really valuable.

By repeating the above construction we can now construct a sequence of generalized renormalizations

$$T^n f \equiv g_n : \bigcup_i V_i^n \to V_0^{n-1} \equiv V^{n-1}. \tag{2.1}$$

Namely, starting with a critical puzzle piece $V^0 \equiv V_0^0$, let us inductively define g_n as $T_{V^{n-1}}f$. The sequence of puzzle pieces

$$V^0 \supset V^1 \supset \dots \tag{2.2}$$

is called the ***principal nest***. Understanding of this nest is the key to the full dynamical picture.

The maps $g_n : V^n \to V^{n-1}$ are quadratic-like in the sense of Douady & Hubbard. Note however that not all of them are different (remember that a polynomial-like map means a germ). Let $\{n(k)\}$ be the sequence of levels where new quadratic-like maps $g_n|V^n$ are created. (These levels are characterized by the property that the critical point returns to V^{n-1} later than to V^{n-2}). This sequence is finite if and only if the map f is renormalizable in the sense of Douady & Hubbard. (Indeed finiteness of this sequence means that one of the maps $g_n|V^n$ has a non-escaping critical point, so that it gives a renormalization of f.) Let us define the ***height*** $\chi(f)$ as the length of the sequence $\{n(k)\}$. This combinatorial parameter has a big impact on the geometry of the map.

2.5 Combinatorics of the Fibonacci maps

For every even degree d, there is a remarkable non-renormalizable map $F_d : z^d + c_d$ called Fibonacci, $c_d \in \mathbb{R}$. Such a map is combinatorially determined by the property that the closest returns of the critical point to itself occur at the Fibonacci moments. It is extremal in many respects which makes it a good candidate for different interesting properties and, on the other hand, a test example to work out general results.

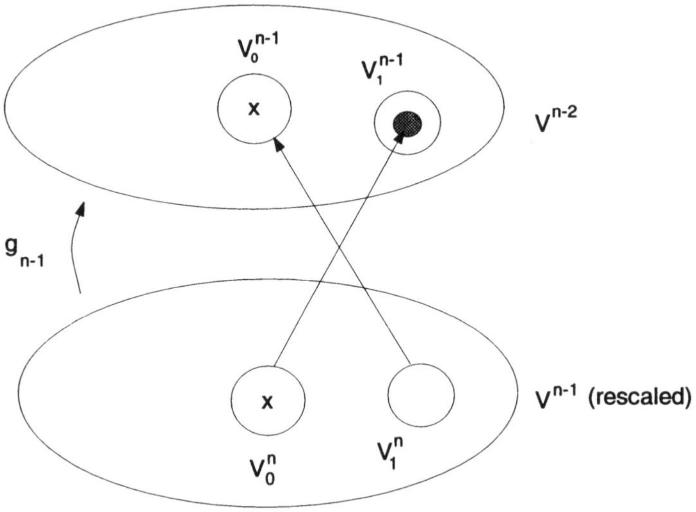

Figure 1. Fibonacci renormalization scheme.

The most efficient way to understand combinatorics of the Fibonacci maps is given by the generalized renormalization (2.1). It turns out that the domain of $g_n = T^n F_d$ consists of only *two* puzzle pieces, so that

$$g_n : V_0^n \cup V_1^n \to V_0^{n-1}, \tag{2.3}$$

where $g_n : V_0^n \to V_0^{n-1}$ is a d-to-1 covering, while $g_n : V_1^n \to V_0^{n-1}$ is univalent. Moreover, $g_n(0) \in V_1^n$, while $g_n(g_n 0) \in V_0^n$. This is in a sense the fastest possible recurrence of the critical point.

Figure 1 shows how to pass from one renormalization level of the Fibonacci map to the next. Even if you never heard about Fibonacci maps, the generalized renormalization analysis would inevitably lead you to this scheme.

Remark. Fibonacci maps appeared independently in several works: Branner & Hubbard [BH], Hofbauer & Keller [HK], Shibayama [Sh]... (see [LM] for more detailed account). The above renormalization scheme for these maps was suggested in [LM].

3 Parameter plane vs dynamical plane

3.1 Parameter rays vs dynamical rays

There is a remarkable similarity between the dynamical and parameter planes of conformal dynamical systems (actually it goes beyond conformal setting). Properties in the dynamical plane is usually reflected in the parameter plane. For example, little Julia sets in the dynamical plane reflect themselves as little copies of the Mandelbrot set in the parameter plane, see §2.1. As Douady put it: "You plow in the dynamical plane and then harvest in the parameter plane". There are different ways to pick this harvest. Historically the first one was based on a relation between the Riemann mapping

$$\phi_M : \mathbb{C} \setminus M \to \mathbb{C} \setminus \mathbb{D}$$

of the complement of the Mandelbrot set and the corresponding mappings in the dynamical plane [DH1].

Any quadratic polynomial $f = P_c : z \mapsto z^2 + c$ near infinity "looks like" z^2. The precise statement is that it is analytically conjugate to $z \mapsto z^2$, namely there is a conformal map $\phi = \phi_c$ near ∞, fixing ∞, tangent to id at ∞, and such that $\phi(fz) = (fz)^2$. There is a classical explicit formula, due to Böttcher, for this map, namely

$$\phi(z) = \lim_{n\to\infty} (f^n z)^{1/2^n}. \tag{3.4}$$

By means of this formula the map can be extended to larger domains until they hit the critical point 0. In the case of connected Julia set this gives the Riemann mapping $\phi : D(\infty) \to \mathbb{C} \setminus \mathbb{D}$ of the whole basin of infinity onto the complement of the unit disk. In the disconnected case it maps the complement of the "figure 8" onto the complement of some disk of radius $R > 1$.

In the parameter plane one can write down the similar formula, only instead of iterates of a single polynomial one should consider the sequence of polynomials $Q_n(c) = Q_{n-1}(c)^2 + c$, $Q_0(c) \equiv 0$, which keeps track of the orbit of the critical point

$$0 \mapsto c \mapsto c + c^2 \mapsto (c + c^2)^2 + c \mapsto \dots$$

Namely

$$\phi_M(c) = \lim_{n\to\infty} (Q_n c)^{1/2^n} \tag{3.5}$$

This yields connectivity of the Mandelbrot set together with the following remarkable formula for the corresponding Riemann mapping (Douady-Hubbard, Sibony [DH1]):

$$\phi_M(c) = \phi_c(c), \quad c \in \mathbb{C} \setminus M. \tag{3.6}$$

So a point $c \in \mathbb{C} \setminus M$ has a double personality: as a parameter value and as the critical value for the corresponding polynomial P_c. Both personalities can be identified by their uniformizing coordinates: the external angles and equipotential levels. Formula (3.6) says us that these identifications coincide!

3.2 Combinatorial classes

In [McM1] dynamical rational laminations $\lambda_{\mathbb{Q}}(P_c)$ are defined. We can similarly define a parameter rational lamination $\lambda_{\mathbb{Q}}(M)$ which describes how the rational rays in $\mathbb{C}\setminus M$ land. Douady & Hubbard [DH1] gave a full combinatorial description of this lamination. In particular, they proved using (3.6) that rational rays with odd denominators land at parabolic points, while rational rays with even denominators land at post-critically finite points (also called Misiurewicz).

Moreover formula (3.6) shows that the dynamical lamination bifurcates exactly at the moments when c crosses the parameter rational rays. Thus combinatorial classes in the quadratic family can be defined as the pieces on which the parameter rational rays partition the Mandelbrot set. Notice that this definition gives an extension of the notion of a combinatorial class to the maps with indifferent cycles as well (compare [McM1], Theorem 6.1). This also explains the relation between combinatorial rigidity and local connectivity of M (the last property is usually abbreviated as MLC).

Indeed, let us show that combinatorial rigidity implies MLC. Assume that the combinatorial class of $c \in M$ is a single point $\{c\}$. Then there is a nest of topological disks $D_1 \supset D_2 \supset \dots$ bounded by rational parameter external rays and shrinking to c. The complement $M \setminus D_k$ splits into finitely many disjoint parts. It follows that the intersection $M \cap D_k$ is connected (for otherwise the whole Mandelbrot set would be disconnected), and local connectivity at c follows.

The reverse property requires a finer combinatorial analysis similar to the proof that MLC implies density of expanding maps (see [DH1, Sch]).

3.3 Parapuzzle

The dynamical puzzle constructed in the previous section can also be transferred to the parameter plane. First splitting is given by the combinatorial rotation number of the fixed point α. As we mentioned before, there are finitely many external rays landing at α which are cyclically permuted by dynamics. The rotation number q/p of this permutation is called the *combinatorial rotation number of* α.

To determine this number looking at the parameter plane, you should do the following. Take a bifurcation point b on the main cardioid of the Mandelbrot set. The polynomial P_b has a parabolic point α_b with rotation number $e^{2\pi i q/p}$. The component of $M \setminus \{b\}$ which does not intersect the main cardioid is called *q/p-limb* $L_{q/p}$ of M. It turns out that the combinatorial rotation number of α_c is equal to q/p if and only if $c \in L_{q/p}$.

To construct further the nest of parapuzzle pieces $Z^{(n)}(c)$ about a parameter value $c \in M$ we should consider the corresponding dynamical nest $Y^{(n)}(c)$ about the critical value c. By definition, the parameter piece $Z^{(n)}(c)$ is bounded by the external rays and equipotentials of the same arguments and levels as the dynamical pieces $Y^{(n)}(c)$. By (3.6), all quadratics within the parameter piece have the same combinatorics up to depth n. By the discussion of §3.2, combinatorial rigidity of of a non-renormalizable polynomial P_c amounts to shrinking of these parameter pieces to c.

In particular, we have a ***principal parapuzzle nest*** $\{W^n(c)\}$ corresponding to the principal nest $\{V^n(c)\}$. The generalized renormalizations $T^n P_b$ have the domains with the same combinatorics when b ranges over $W^{(n)}(c)$.

4 Geometric bounds

4.1 Compactness.

Convergence = pre-compactness + uniqueness of a limit point. This triviality often helps to understand better a nature of a specific deep problem we deal with.

If we are after convergence of the sequence of renormalizations $R^n f$, we should first try to prove its pre-compactness or, at least, boundedness in some metric. Both approaches turn out to be fruitful, and both amount to the same analytical issue, namely complex *a priori* bounds.

Let $\mathcal{PL}_d$ denote the space of complex unimodal maps of degree d up to conformal equivalence (normalized so that 0 is the critical point). This space can be supplied with a *Carathéodory topology.* Convergence of a sequence $f_n : U_n \to V_n$ to $f : U \to V$ in this topology means Carathéodory convergence of pointed domains $(U_n, 0)$ and ranges $(V_n, 0)$ to $(U, 0)$ and $(V, 0)$ respectively, and compact-open convergence of the corresponding maps (all after appropriate choice of representatives of conformal classes).

Let $\mathcal{PL}_d(\mu)$ denote the subspace of $\mathcal{PL}_d$ consisting of maps which have a fundamental annulus of modulus at least $\mu > 0$ (we will also express this by saying that $\mathrm{mod}\, f \geq \mu$). A normality argument yields the following fact (see [McM2]):

Lemma 4.1 *For any $\mu > 0$, the space $\mathcal{PL}_d(\mu)$ is compact.*

Thus compactness of the sequence of renormalized maps $R^n f$ amounts to complex *a priori bounds* for these maps: $\mathrm{mod}(R^n f) \geq \mu$, $n = 0, 1, \ldots$, for some $\mu > 0$.

On the other hand, in §2.1 we introduced the Teichmüller metric on the hybrid classes $\mathcal{H}(f)$ of polynomial-like maps. Let $\mathcal{H}(f, \mu)$ denote the subspace of the hybrid class consisting of maps with $\mathrm{mod}(f) \geq \mu$. Since the dilatation of the straightening map depends only on the modulus of the fundamental annulus, we have the following fact:

Lemma 4.2 *The set $\mathcal{H}(f, \mu)$ has a bounded diameter in the Teichmüller metric.*

So both approaches lead us to the problem of *a priori* bounds.

4.2 Real bounds

Before passing to the complex plane let us analyze the situation on the real line. For an infinitely renormalizable real unimodal map f the postcritical set $\mathcal{O}$ is a Cantor set with the following structure:

$$\mathcal{O} = \bigcap_{n=0}^{\infty} \bigcup_{k=0}^{q_n - 1} I_k^n,$$

where $I_0^1 \supset I_0^2 \supset \ldots \ni 0$, and for any n the intervals I_i^n are cyclically permuted by f. The *real* n-fold renormalization $R^n f$ is just $f^{q_n}|I_0^n$. (Like in the complex situation, there is some flexibility in the choice of intervals I_0^n). The ratios $p_n = q_{n-1}/q_n$ are called *the relative periods.* Every interval of level $n-1$ contains exactly p_n intervals of the next level n. The Feigenbaum-Coullet-Tresser situation discussed in the introduction corresponds to the doubling on all levels: $p_n = 2$, $n = 1, 2, \ldots$.

One says that f has a *bounded combinatorics* if the relative periods are uniformly bounded. By the *gaps* of level n we mean the connected components of $I_k^{n-1} \setminus I_j^n$. One says that the Cantor set $\mathcal{O}$ has bounded geometry if for any I_k^{n-1}, all intervals and all gaps of level n belonging to I_k^{n-1} are commensurable (with a constant independent of level n and the interval I_k^{n-1}). Cantor subsets of $\mathbb{R}$ with bounded geometry have Hausdorff dimension strictly in between 0 and 1.

Theorem 4.3 (see [G, BL, S2]) *Let f be an infinitely renormalizable real unimodal map. Then:*

- *There is an absolute $\delta > 0$ and intervals $0 \in S^n \subset T^n$ such that $f^{q_n} : (S^n, \partial S^n) \to (T^n, \partial T^n)$ is a unimodal map, and $|T^n| \geq (1+\delta)|S^n|$;*
- *The real renormalizations $R^n f$ form a pre-compact family in C^1 topology;*
- *If f has bounded combinatorics then the post-critical set $\mathcal{O}$ has bounded geometry.*

4.3 Sullivan's bounds

Sullivan's idea is to complexify the real bounds using the following hyperbolic disks. Let $I \subset \mathbb{R}$ be an interval. Let us consider the complex plane with two slits, $D_0(I) = \mathbb{C} \setminus (\mathbb{R} \setminus I)$. It is conformally equivalent to the unit disk $\mathbb{D}$ and thus can be supplied with the hyperbolic metric. By symmetry, I is a hyperbolic geodesic in this metric. It is easy to check that the hyperbolic r-neighborhood of this geodesic is the union $D_\theta(I)$ of two symmetric segments of Euclidean disks which meet the real line at angle $\theta = \theta(r)$ (see Figure 2). As analytic maps contract the Poincaré metric, we have:

Schwarz lemma. *Let $\phi : D_0(J) \to D_0(J')$ be an analytic map which transforms the interval J into J'. Then for any $\theta \in (0, \pi)$, $\phi(D_\theta(J)) \subset D_\theta(J')$.*

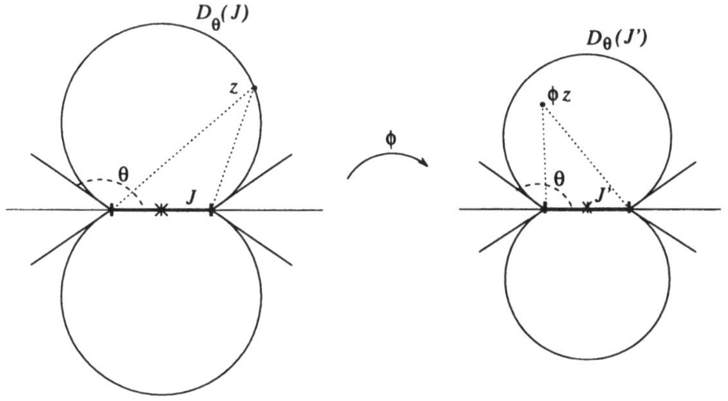

Figure 2. Poincarè disks.

Let us now have a renormalizable map $f \in \mathcal{PL}_d$ preserving the real line, with renormalization $Rf = f^p$. By Theorem 4.3, there is an orbit of intervals

$$J_0 \to J_1 \to \dots \to J_p \supset J_0 \tag{4.7}$$

with the following properties: $f : J_0 \to J_1$ is unimodal with $f(\partial J_0) \subset \partial J_1$ (a real version of a double covering), while all $f : J_k \to J_{k+1}$ are diffeomorphisms for $k = 1, \dots, p-1$. Moreover, $J_p \supset (1+\delta)J_0$ with an absolute $\delta > 0$.

Let us now take a Poincaré disk $V_p = D_\theta(I_p)$, and pull it back along the orbit (4.7): let V_k be the component of $f^{-1}V_{k+1}$ containing J_k. By the Schwarz lemma $V_k \subset D_\theta(I_k)$ for all $k = 1, 2 \dots, p$, but not necessarily for $k = 0$.

This little phenomenon is a source of big troubles. The way Sullivan settles it is the following: For the maps of bounded type, he first proves the so called Sector Lemma asserting that the pull-back of the *whole slit complex* plane $D_0(I_p)$ (think of it as the Poincare disk of infinite radius) is contained in the union of two symmetric ϵ-sectors based on I_0, with some $\epsilon > 0$ dependent only on the combinatorial bounds on f. It follows that for sufficiently small θ, the pull-back of $D_\theta(I_p)$ under f^p is contained well inside itself. This gives complex *a priori* bounds for infinitely renormalizable real maps of bounded type. As the Sector Lemma fails for unbounded combinatorics, this case requires a different treatment.

4.4 Divergence property

A totally different methods to estimate geometric moduli have been developed in the framework of the puzzle. The first results of this kind appeared in the works of Branner & Hubbard [BH] and Yoccoz (see [H, M2]) on dynamics of polynomials with one non-escaping critical point of quadratic type. We will state the results in the quadratic case. Recall that $Y^{(n)}(z)$ stands for the Yoccoz puzzle pieces of depth n containing z, while $Z_n(c)$ stands for the nest of parapuzzle pieces about c (see §2.3).

Theorem 4.4 *Let f be a non-renormalizable quadratic polynomial with all periodic points repelling. Then for any $z \in J(f)$,*

$$\sum \operatorname{mod}(Y^{(n)}(z) \setminus Y^{(n+1)}(z)) = \infty.$$

Hence $\operatorname{diam}(Y^{(n)}(z)) \to 0$ *as* $n \to \infty$.

The last conclusion follows from the Grötzcsh inequality.

Quantifying in ingenious way the ideas outlined in §3, Yoccoz has transferred the last result to the parameter plane:

Theorem 4.5 *Let $c \in M$ be a non-renormalizable parameter value. Then*

$$\sum \operatorname{mod}(Z_n(c) \setminus Z_{n+1}(c)) = \infty.$$

Hence the Mandelbrot set is locally connected at c and the quadratic P_c is combinatorially rigid.

These results are easily extended to at most finitely renormalizable quadratic polynomials (with an appropriate choice of the puzzle), but they are not enough for infinitely renormalizable maps. Also, even for non-renormalizable maps, many geometric issues need better bounds, which will be discussed next.

4.5 Growth of the principal moduli

Given a point $c \in \mathbb{C}$, let $d(c)$ stand for the distance from c to the union of the expanding domain of the Mandelbrot set bounded by the main cardioid and all expanding components attached to it. With the notations of §3.3 for the principal nest, we have the following estimate:

Theorem 4.6 ([L5, L6]) *Let P_c be a quadratic polynomial with $c \in M$ and $d(c) \geq \epsilon > 0$. Then*

$$\operatorname{mod}(V^{n(k)-1} \setminus V^{n(k)}) \geq C(\epsilon)k.$$

Remark. For real maps a related result was independently obtained by Graczyk & Swiatek [GS1].

Theorem 4.6 implies that infinitely renormalizable maps with sufficiently big height on all levels have big moduli:

Corollary 4.7 *Let P_c be an infinitely renormalizable quadratic polynomial. Let $c_n \in M$ label the hybrid class of the renormalization $R^n f$, while χ_n stand for its height. Assume that*

- $d(c_n) \geq \epsilon > 0$, $n = 0, 1, \dots$;
- $\chi_n \geq \chi$.
  Then $\operatorname{mod}(R^n f) \geq \mu(\epsilon, \chi)$, $n = 0, 1 \dots$, *where for any given* $\epsilon > 0$, $\mu(\chi, \epsilon) \to \infty$ *as* $\chi \to \infty$.

Thus the renormalizations $R^n f$ are becoming purely quadratic as the height χ grows. Let us now state the parameter counterpart of these results (with notations of §3.3 for the principal parapuzzle nest):

Theorem 4.8 ([L9, W]) *Let M' be a maximal copy of the Mandelbrot set with $d(c) \geq \epsilon > 0$ for all $c \in M'$. Then for $c \in M'$,*

$$\operatorname{mod}(W^{n(k)-1}(c) \setminus W^{n(k)}(c)) \geq C(\epsilon)k.$$

Corollary 4.9 *Let the assumptions of Corollary 4.7 be satisfied throughout an infinitely renormalizable combinatorial class $Com(c) = \cap M_n$, where M_n is the corresponding nest of the little Mandelbrot copies. Then the M_n exponentially shrink. In particular $Com(c) = \{c\}$, so that P_c is combinatorially rigid.*

The way we derive these parameter results from the dynamical ones is different from the Yoccoz's method. It is based on the theory of holomorphic motions (see §5.2): the transversal quasi-conformal structure is the key which allows us to compare dynamical and parameter moduli.

4.6 Geometry of the quadratic Fibonacci puzzle.

Let us illustrate the above geometric results in the Fibonacci case. Let R_i^n denote the maximal annulus in $V^{n-1} \setminus (V_0^n \cup V_1^n)$ which goes around V_i^n but does not go around V_{1-i}^n (see Figure 1). Let us consider the following asymmetric combination:

$$\sigma_n = \operatorname{mod} V_0^n + \frac{1}{2} \operatorname{mod} V_1^n.$$

Pulling these annuli back according to the Fibonacci scheme, one can see that $\sigma_{n+1} \geq \sigma_n$. This yields the bounds $\operatorname{mod}(V^{n-1} \setminus V^n) \geq \epsilon > 0$, $n = 0, 1, \ldots$, but not yet the growth of the moduli.

To prove the growth one needs to analyze the positions of the V_i^n in V^{n-1}, and the shapes of these puzzle pieces. What, after all, makes the moduli grow is *pinching* of the puzzle pieces. Carring this renormalization analysis further, we can find exactly what is the asymptotic shape of the quadratic Fibonacci puzzle pieces: It is just the filled Julia set of $z \mapsto z^2 - 1$! (see Figure 3)

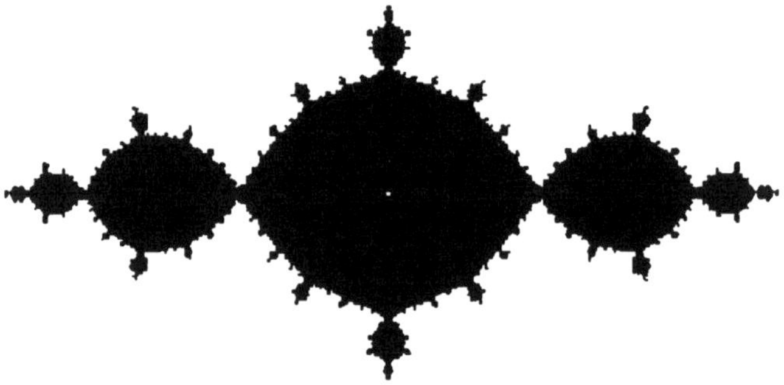

Figure 3. Degree two Fibonacci puzzle piece (made by B. Yarrington).

To explain this phenomenon, let us consider the triples of points $t_n = \{0, g_n 0, a_n\}$, where a_n is an appropriately chosen point of $\partial V^n \cap \mathbb{R}$. Then t_{n+1} is the pull-back of t_n by the map $g_n : V^n \to V^{n-1}$, which is exponentially close to a quadratic map (according to Theorem 4.6). This pull-back coincides (up to an exponentially small error) with the Thurston transformation τ in the Teichmüller space of thrice punctured planes (see [McM1] and the discussion in the next section). As τ is contracting, t_n converge to its fixed point, which corresponds to the superattracting period two cycle $0 \mapsto -1 \mapsto 0$ of $z \mapsto z^2 - 1$.

Let us finally mention the following consequence of the above discussion: the post-critical set $\mathcal{O}$ of the quadratic Fibonacci map is a Cantor set with exponentially decaying geometry [LM] (that is, the intervals of the next level are exponentially small as compared with the intervals of the previous level). In particular, this set has zero Hausdorff dimension. This is quite different from the bounded geometry of the Feigenbaum attractors (see §4.2). What is more surprising that this is also different from the geometry of the post-critical sets for higher degree Fibonacci maps. This curious phenomenon will be discussed in §4.8.

4.7 Complex bounds for real quadratics

Notice that among Poincaré disks introduced in §4.3 there is one especially nice, namely the Euclidean disk $D(I) \equiv D_{\pi/2}(I)$. What if to try to create a (generalized) polynomial-like map by pulling it back?

Let us have an orbit of intervals (4.7) (not-necessarily corresponding to the renormalization level). Take the Euclidean disk $D(J_p)$ and pull it back along this orbit. We will obtain a sequence of pull-backs $V_k \subset D(I_k)$, $k = p, p-1, \dots, 1$. To settle the trouble with the last square root pull-back indicated in §4.3, we need some control of the position of the critical value $f0$ in J_1. If this position is sufficiently "high" (that is, $|fJ_0|/|J_1|$ is sufficiently big), we are fine; otherwise $V_0 \not\subset D(I_0)$.

The appropriate estimate was made first for the Fibonacci maps [LM], and then extended to much wider range of combinatorial patterns. Let f be a renormalization $R^m P_c$ of a quadratic polynomial. Recall that $\chi(f)$ stands for the height of f (see §2.4).

Lemma 4.10 **([L4])** *There are absolute* $\epsilon_* > 0$ *and* χ_* *with the following property. If* $\chi(f) \geq \chi_*$, *then* f *admits a generalized renormalization* $g_n : \cup U_i \to D(I)$ *with a definite modulus:* $\mathrm{mod}(D(I) \setminus U_0) \geq \epsilon_* > 0$.

After a generalized polynomial-like map is created, according to Theorem 4.6 the moduli start to grow, so that we have:

Corollary 4.11 *If* f *is renormalizable in the sense of Douady and Hubbard then* $\mathrm{mod}(Rf) \geq \mu(\chi(f))$ *where* $\mu(\chi) \to \infty$ *as* $\chi \to \infty$.

Remark. The "height" in the above results can be replaced by a finer combinatorial parameter called "essential period" (see [LY] for the definition). If the essential period is bounded, the true period can be big only because some of the maps $g_n : V^n \to V^{n-1}$ are combinatorially close to $z \mapsto z^2 + 1/4$. This parameter is responsible for the dichotomy between "decaying" and "essentially bounded" geometry.

The combinatorial condition of Corollary 4.11 is a kind of complementary to Sullivan's bounded type. A gap between these two results has been recently filled in [LY], by an appropriate extension of Sullivan's sector lemma.

Theorem 4.12 *Any infinitely renormalizable real quadratic polynomial has complex a priori bounds.*

Different proofs of this result have been independently given by Graczyk & Swiatek [GS2] and Levin & van Strien [LS]. The latter work also treats the higher degree unimodal case $x \mapsto x^d + c$.

4.8 Higher degree Fibonacci maps

The geometry of higher degree Fibonacci maps F_d is quite different from the quadratic case, and more similar to the geometry of infinitely renormalizable maps of bounded type:

Lemma 4.13 *The post-critical set $\mathcal{O}$ for higher degree Fibonacci maps ($d > 2$) has bounded geometry. There are a priori bounds for generalized polynomial-like renormalizations $T^n f$.*

The latter property means that there is an $\epsilon > 0$, a choice of the domains of the generalized renormalizations $T^n F_d : V_0^n \cup V_1^n \to V^{n-1}$ (see §2.5) and annuli $R_i^n \subset V^{n-1} \setminus (V_0^n \cup V_1^n)$ defined in §4.6 such that $\mathrm{mod}(R_i^n) \geq \epsilon > 0$. (The former property shows that the $\mathrm{mod}(R_i^n)$ are also bounded from above). These *a priori* bounds for Fibonacci maps of higher degree are obtained by pulling back Euclidean disks in the same way as in the quadratic case [LM].

The reason why higher degrees differ degree 2 can be roughly seen in the following way. Let $I^n = V^n \cap \mathbb{R}$ stand for the real traces of the principal puzzle pieces. Let $\lambda_n = |I^n|/|I^{n-1}|$ be the corresponding *scaling factors.* Then one can write a recurrent relation between these scaling factors, which looks (up to bounded factors) like this

$$\lambda_{n+1}^d \approx \frac{1}{d}\lambda_n\lambda_{n-1}.$$

Pretending that this relation is precise, we see that its solutions decay to 0 for $d \leq 2$, and stay bounded away from 0 for $d > 2$.

Remark. This difference between Fibonacci maps of degree two and higher degrees was first pointed out in [LM]. Curiously there is a similar phenomenon for quite a different class of maps (circle maps with flat spot) which had been earlier studied by Tangerman & Veerman [TV].

5 Rigidity

In this section we will continue McMullen's discussion of the rigidity problem, see [McM1], §§2,5.

5.1 Deformation spaces

Action of a rational function on the Fatou set produces a Riemann surface $S(f)$ with affine foliation on some components. As described in [McM1], §2, there is a way to deform a rational function f by deforming its Riemann surface $S(f)$ (respecting the affine foliation). Namely, a conformal structure on $S(f)$ compatible with the affine structure on the leaves can be lifted to an f-invariant measurable structure μ on the Riemann sphere with bounded dilatation (on the Julia set μ coincides with the standard structure σ). By the Measurable Riemann Mapping Theorem (see the Appendix), there a quasi-conformal map $h_\mu : \bar{\mathbb{C}} \to \bar{\mathbb{C}}$ such that $(h_\mu)_*(\mu) = \sigma$. Then $f_\mu = h_\mu \circ f \circ h_\mu^{-1}$ is a new rational function (defined up to conformal equivalence). Let $\mathcal{D}ef(f)$ stand for the space of functions (modulo conformal equivalence) which can be obtained in such a way. This space is parametrized by the Teichmüller space of $S(f)$.

Let $\mathcal{M}$ be an analytic family of rational functions modulo conformal equivalence (examples to keep in mind: a family of rational functions of degree d, a family of polynomials of degree d, a family of complex unimodal maps $z \mapsto z^d + c$ modulo the cyclic group Γ_{d-1}) . Let

$$\mathcal{C}om_{\mathcal{M}}(f) \supset \mathcal{T}op_{\mathcal{M}}(f) \supset \mathcal{QC}_{\mathcal{M}}(f)$$

denote respectively the combinatorial, topological and quasi-conformal classes of f in this family modulo conformal equivalence (actually combinatorial classes are so far defined for polynomials only). We skip the label $\mathcal{M}$ unless it may lead to confusion.

Deformation Conjecture. *For any f,*

- $\mathcal{C}om(f) = \mathcal{T}op(f)$ *whenever* $\mathcal{C}om(f)$ *is defined;*
- $\mathcal{T}op(f) = \mathcal{QC}(f)$*;*
- $\mathcal{QC}(f) = \mathcal{D}ef(f)$*, except for the Lattès examples.*

The last statement is equivalent to the absence of invariant line fields on the Julia set. By [MSS], this would imply density of expanding maps among rational maps. In the case when $\mathcal{D}ef(f) = \{f\}$ (that is, the Riemann surface $S(f)$ is rigid), the Deformation Conjecture turns into the *Rigidity Conjecture.* It can be refined as the *combinatorial/topological/quasi-conformal rigidity* conjecture which would assert that the corresponding class is a single map.

5.2 Structural stability

The maps belonging to open topological classes are called *structurally stable.* In early 80th the following advance towards the topological classification of rational maps was achieved (see [MSS, ST], and also [L1] for a part of this result):

Theorem 5.1 (Structural stability) *Structurally stable maps are dense in any holomorphic family. Connected components of structurally stable maps represent open quasi-conformal classes.*

The proof of this result based on the notion of a "holomorphic motion" is at least as important as the result itself. A holomorphic motion is a family $h_\lambda : X \to \mathbb{C}$ of injections of a set $X \subset \mathbb{C}$ holomorphically depending on λ (ranging withing some analytic manifold with a reference point λ_0) and such that $h_{\lambda_0} = \mathrm{id}$. A great property of a holomorphic motion (called "the λ-lemma") is that it automatically admits an extension to a holomorphic motion $h_\lambda : \bar{\mathbb{C}} \to \bar{\mathbb{C}}$ of the whole Riemann sphere, and that the maps of this motion are automatically quasi-conformal.

To prove Theorem 5.1 one constructs a holomorphic motion conjugating a function f_0 and a nearby function f. One can begin the construction of the motion with repelling periodic points, then extend it by the λ-lemma to the Julia set, then go to little neighborhoods of attracting cycles and spread the motion by dynamics onto the whole attracting basin, etc. There are obstructions for this construction, like parabolic cycles or coincidence of the grand orbits of two critical points, but one can show that they don't occur on a dense set of maps.

5.3 Unimodal families

Let us refine the above discussion in the unimodal case $P_c : z \mapsto z^d + c$. First note that in this case the set of expanding components (also called "hyperbolic") is the complement $\mathbb{C} \setminus M_d$ of the connectedness locus union the *expanding* components of $\operatorname{int} M_d$. For c belonging to an expanding component H, the polynomial P_c has an attracting cycle α_c. The *center* of the expanding component is the parameter value c_H for which the attracting cycle becomes superattracting (by a theorem of Douady and Hubbard [DH1], the center is indeed unique in any expanding component). Non-expanding components of $\operatorname{int} M_d$ are called *queer.* The maps in queer components must have invariant line fields on the Julia set. The set of structurally stable maps is the union of $\mathbb{C} \setminus M_d$, punctured expanding components $H \setminus \{c_H\}$ and queer components.

Now, we have the following quasi-conformal classification of the unimodal families

Theorem 5.2 (Quasi-conformal classification) *Quasi-conformal classes of the complex unimodal maps $z \mapsto z^d + c$ (modulo the cyclic group Γ_{d-1}) are the following:*

- $(\mathbb{C} \setminus M_d)/\Gamma_{d-1}$;
- *punctured hyperbolic components $H \setminus \{c_H\}$ of* $\operatorname{int} M_d/\Gamma_{d-1}$;
- *queer components;*
- *single points.*

Thus all maps P_c with $c \in \partial M_d$ are quasi-conformally rigid. To complete the proof of the Quasi-conformal Rigidity Conjecture, one should prove that there are no queer components.

Proof of Theorem 5.2. Note that the Measurable Riemann Mapping Theorem implies that quasi-conformal classes are always connected, and in the unimodal family they are either open, or single points. Indeed, let f and $\tilde{f}$ be two quasi-conformally equivalent maps, and h be the corresponding conjugacy. Then h induces an f-invariant conformal structure $h^*\sigma$ represented by the Beltrami differential $\mu = \bar{\partial} h d\bar{z}/\partial h dz$ with $\|\mu\|_\infty < 1$. But then we actually have a complex one-parameter family of f-invariant conformal structures corresponding to Beltrami differentials $\lambda\mu$ with $|\lambda| < 1/\|\mu\|_\infty$. By the Measurable Riemann Mapping Theorem, there is a family of quasi-conformal maps h_λ such that $(h_\lambda)_*(\lambda\mu) = \sigma$. The family of maps $f_\lambda = h_\lambda \circ f \circ h_\lambda^{-1}$ is an analytic disk in the quasi-conformal class $\mathcal{QC}(f)$ containing $\tilde{f}$. Moreover, in the unimodal case, this disk gives a neighborhood of f contained in $\mathcal{QC}(f)$. □

The following remark shows that in the unimodal families the first two parts of the Deformation Conjecture yield the last one:

Lemma 5.3 *Assume that for some $c \in M_d$, $\mathcal{C}om(c) = \mathcal{QC}(c)$. Then c is combinatorially rigid.*

Proof. Indeed the combinatorial classes are closed as the intersections of parapuzzle pieces, while the quasi-conformal classes are either open, or single points. Thus if two classes coincide, they must be a single point.

5.4 Bounds and line fields

Queer components are always associated with the invariant line fields on the Julia set. Thus to complete quasi-conformal classification of the unimodal families (and to prove density of expanding components) we need to show that line fields don't exist: a nice relation between the rigidity problem and ergodic theory. In turn, the latter problem has been reduced by McMullen to the problem of *a priori* bounds:

Theorem 5.4 ([McM2]) *Let f be an infinitely renormalizable complex unimodal map with a priori bounds: $R^n f \geq \epsilon > 0$. Then there are no invariant line fields on the Julia set $J(f)$.*

The idea going back to the works of Mostow and Sullivan in Kleinian groups is to dynamically blow up the invariant line field near a density point where it is almost constant. This will show that the field is a.e. compatible with a real analytic one, which easily leads to a contradiction.

To carry out the "blow up" procedure, one needs to know that dynamics is fairly expanding. McMullen managed to exploit a quite modest amount of expansion following from *a priori* bounds: In the fundamental annuli of the $R^n f$, the map is uniformly expanding with respect to the hyperbolic metric in $\mathbb{C} \setminus \mathcal{O}$.

5.5 Pull-back argument for bounded geometry

Lemma 5.3, or Theorem 5.4 reduce the rigidity problem (under appropriate circumstances) to a construction of a quasi-conformal conjugacy between two combinatorially equivalent maps f and $\tilde{f}$. The main method to carry this out is called "the pull-back argument". It was originated (at least in the dynamical setting) in the work of Thurston on post-critically finite maps (see [Th, DH3] and discussion in [McM1],§5). The idea is to start with a quasi-conformal map of a right homotopy type which preserves some dynamical data, to lift it up by iterates of f and $\tilde{f}$, and to obtain a quasi-conformal (or even conformal) conjugacy in the limit. However you need some luck to carry this procedure out: the respected dynamical data you start with should allow you to go through an infinite lifting procedure. In the simplest cases this data is just dynamics on the post-critical set.

Assume, for instance, you wish to show that two topologically equivalent post-critically finite maps are quasi-conformally equivalent. Start with any K-quasi-conformal map h_0 which conjugates f and $\tilde{f}$ on their post-critical sets $\mathcal{O}$ and $\tilde{\mathcal{O}}$, and homotopic to a topological conjugacy rel $\mathcal{O}$. Then h_0 can be lifted to a map h_1 homotopic to h_0 rel Ω. Moreover this map is also K-quasi-conformal since f and $\tilde{f}$ are analytic. Hence you can lift it again, etc.

By interpreting this procedure as iterates of a contracting transformation in the Teichmüller space of punctured spheres, Thurston proved that h_i converge, unless f is a Lattès example. The limit map is a quasi-conformal conjugacy between f and $\tilde{f}$.

For more complicated combinatorics, a problem arises at the very beginning of the procedure: Why is there a quasi-conformal map which conjugates f and $\tilde{f}$ on their post-critical sets? Such a fact depends on the geometry of the post-critical set, which thus becomes crucial for the rigidity problem. Real bounds of §4.2 allow one to handle the problem in the real infinitely renormalizable case of bounded type. Indeed, by Theorem 4.3 the post-critical sets have bounded geometry in this case. Then their complements can be constructed by gluing standard pairs of pants (a standard pair of pants is a round disk with several round disks removed) with bounded geometry. Then there is a K-quasi-conformal map between the respective pairs of pants which is affine on the boundary circles and orientation preserving on the real line (with a uniform K). By the Gluing Lemma from the Appendix, the complements of the post-critical sets are K-quasi-conformally equivalent, with the same K. Applying the Gluing Lemma again (remember that our Cantor sets lie on the real line), we obtain the desired quasi-conformal map h_0 to start with.

Now, the pull-back argument allows us to turn h_0 into a quasi-conformal conjugacy. Indeed, let us extend h_0 to a quasi-conformal map on the whole complex plane in such a way that it conjugates f to $\tilde{f}$ outside some equipotentials E and $\tilde{E}$ (remember that both maps are conformally equivalent to $z \mapsto z^d$ outside the Julia set). This map can be lifted to a map h_1 homotopic to h_0 rel the post-critical sets. Moreover, h_1 is K-quasi-conformal with the same dilatation K as h_0. Similarly h_1 can be lifted to a K-quasi-conformal map h_2, etc. (as in the post-critically finite case). By the Compactness Lemma from the Appendix, we can select a subsequence $h_{n(i)}$ uniformly converging to a K-quasi-conformal map h. Outside the Julia sets this map conformally conjugates f and $\tilde{f}$. As the Julia set of a polynomial with all periodic points repelling is nowhere dense, h conjugates f to $\tilde{f}$ on the whole plane, and the construction is completed.

Let us summarize the above discussion in the following rigidity result:

Proposition 5.5 (see [MvS, S2]) *For any infinitely renormalizable bounded combinatorial type τ, there is at most one real quadratic polynomial of type τ.*

5.6 Pull-back argument for decaying geometry

To fix the idea, let us consider a quadratic Fibonacci map $f \equiv F_2$ (we pretend that we yet don't know that such a map is unique). For this map we have a sequence of generalized renormalizations $g_n : V_0^n \cup V_1^n \to V_0^{n-1}$ with linearly increasing moduli $\text{mod}(V^{n-1} \setminus V^n)$ (see Theorem 4.6). So the pairs of pants $V_0^{n-1} \setminus (V_0^n \cup V_1^n)$ don't have bounded geometry. However we will check that the corresponding pairs of pants stay bounded "Teichmüller distance away", that is, they are K-quasi-conformal equivalent with a uniform K.

We will mark the objects corresponding to $\tilde{f}$ with tilde. Note that all puzzle pieces come together with the boundary parametrization, induced e.g., by the Böttcher coordinate in the complement of the Julia set. Let us have a K-quasi-conformal map

$$h_n : (V^{n-1}, V_0^n, V_1^n) \to (\tilde{V}^{n-1}, \tilde{V}_0^n, \tilde{V}_1^n),$$

respecting the boundary parametrization of the pieces. We would like to lift this map to a quasi-conformal map

$$h_{n+1} : (V^n, V_0^{n+1}, V_1^{n+1}) \to (\tilde{V}^n, \tilde{V}_0^{n+1}, \tilde{V}_1^{n+1})$$

with the same property. What causes a problem is that h_n does not carry the critical values $v_n = g_n(0)$ to $\tilde{v}_n = \tilde{g}_n(0)$. However, as $\text{mod}(V_0^{n-1} \setminus V_1^n)$ is linearly big, $h_n(v_n)$ is exponentially close to $\tilde{v}_n$ in the hyperbolic metric of $\tilde{V}^{n-1}$.

By lifting h_n to the non-central puzzle pieces $V_1^n \to \tilde{V}_1^n$ via the univalent maps $g_n : V_1^n \to V^{n-1}$ and $g_n : \tilde{V}_1^n \to \tilde{V}^{n-1}$, we obtain a K-quasi-conformal map $\hat{h}_n : V^{n-1} \to \tilde{V}^{n-1}$ matching with h_n on $V^{n-1} \setminus V_1^n$, with even better property: $\hat{h}_n(v_n)$ is exponentially close to $\tilde{v}_n$ in the hyperbolic metric of $\tilde{V}_1^n$.

Now we can replace $\hat{h}_n$ by another map H_n matching with it on $V^{n-1} \setminus V_1^n$, respecting the critical values and having dilatation K(1+exp small term). This map can be already lifted to V^n. It needs not yet respect boundary parametrization of V_i^{n+1} but one more repetition of the pull-back procedure will do the job.

Repeating this procedure we will construct a quasi-conformal equivalence between the pairs of pants of all levels with uniformly bounded dilatation (as the dilatation increases by exponentially small amount on every step, it stays bounded). Spreading it around the post-critical set, we conclude that the post-critical sets of two Fibonacci quadratics are quasi-conformally equivalent in the right homotopy class (respecting dynamics on the sets). Now the pull-back argument described in §5.5 turns this quasi-conformal map to a quasi-conformal conjugacy on the whole plane.

This argument can be carried out for all non-renormalizable quadratics which gives a different proof of Theorem 4.5. In the infinitely renormalizable case one needs complex *a priori* bounds in order to start this argument from scratch on every renormalization level. This leads to the following Rigidity Theorem (compare Corollary 4.9).

Theorem 5.6 **([L5, L7])** *Let P_c be an infinitely renormalizable quadratic polynomial. Let $c_n \in M$ label the hybrid classes of its renormalizations $R^n f$. Let us consider the following two properties.*

- $d(c_n) \geq \epsilon > 0$, $n = 0, 1, \ldots$ *(where $d(c)$ is defined in §4.5);*
- *P_c has a priori bounds:* $\mathrm{mod}(R^n P_c) \geq \epsilon$, $n = 0, 1, \ldots$
Every combinatorial class contains at most one quadratic polynomial with these two properties.

5.7 Rigidity of real maps

Theorem 5.4 and Theorem 4.12 yield:

Theorem 5.7 **([McM2])** *Non-expanding real quadratic polynomials are quasi-conformally rigid.*

(Actually McMullen derived this result from the real bounds, as the complex bounds were not available at that time).

Theorem 4.12 and Theorem 5.6 yield a stronger conclusion:

Theorem 5.8 *Combinatorial classes of non-expanding real quadratic polynomial intersect the real line in single points.*

Corollary 5.9 *Expanding real quadratics are dense in the family $z^2 + c$, $c \in \mathbb{R}$.*

The last two results were first announced by Swiatek [Sw] who approached them from the point of view of real dynamics. The above proof follows [L5, L7].

6 Universality

Let us now consider a combinatorial class $\mathcal{C}^{\mathbb{R}} \equiv Com^{\mathbb{R}}(M')$ of infinitely renormalizable real unimodal maps f which admit polynomial-like extensions to the complex plane and have a stationary combinatorial type $\tau(f) = (M', M', \dots)$. The Universality Law asserts that in this class there is a unique R-invariant map f_*, and the renormalizations of all other maps $f \in \mathcal{C}^{\mathbb{R}}$ converge to f_*. Moreover, the renormalization operator is hyperbolic at f_*, with one dimensional unstable manifold. We will sketch two approaches to the construction of the fixed point f_* and the stable manifold (due to Sullivan and McMullen), and then the author's approach to the unstable direction.

6.1 Sullivan's Contraction Lemma

By Proposition 5.5, all maps of $\mathcal{C}^{\mathbb{R}}$ are hybrid equivalent, $\mathcal{C}^{\mathbb{R}} \subset \mathcal{H}(f)$, so that this space can be supplied with the Teichmüller metric (see §2.1). If two maps f and $\tilde{f}$ are hybrid conjugate by a quasi-conformal map h then their renormalizations Rf and $R\tilde{f}$ are conjugate by a restriction of this map. It follows that the renormalization transformation R is contracting with respect to the Teichmüller metric.

This is not, though, enough to conclude that R has a globally attracting fixed point: To this end one needs a definite contraction. A result of this kind proved by Sullivan [MvS, S2, S3] is the following:

Lemma 6.1 (Contraction) *There exists a $\lambda \in (0,1)$ with the following property. For any two maps $f, g \in \mathcal{C}^{\mathbb{R}}$ there is an n such that*

$$\mathrm{dist}_T(R^n f, R^n g) \le \lambda\, \mathrm{dist}_T(f, g).$$

It follows that there is at most one limit map f_* for any orbit $\{R^m f\}$, which thus must be R-invariant. Moreover, this point is independent of f. On the other hand, due to ***a priori bounds*** and Compactness Lemma 4.1, any orbit has at least one accumulation point. It follows that there is a unique fixed point $f_* \in \mathcal{C}^{\mathbb{R}}$ which attracts all $f \in \mathcal{C}^{\mathbb{R}}$.

Sullivan's proof of the above Contraction Lemma uses the full scale machinery of the Teichmüller theory extended to objects called "Riemann surface laminations" [MvS, S2, S3].

6.2 McMullen's towers

Let f be an infinitely renormalizable unimodal map with complex ***a priori bounds***. By Lemma 4.1, the orbit $\{R^n f\}$ is Carathéodory compact. Let Λ be the set of limit points of this orbit. Then the restriction $f|\Lambda$ is invertible, so that for every $g \in \Lambda$, there is a two sided R-orbit $\hat{g}$

$$\ldots \mapsto g_{-1} \mapsto g \mapsto g_1 \mapsto \ldots \tag{6.8}$$

With appropriate normalization, such an orbit can be realized as a kind of multi-valued conformal dynamical system called "tower": given a point $z \in \mathbb{C}$, you can apply to it infinitely many maps of (6.8) and all their admissible compositions. The Julia set $J(\hat{g})$ of the tower is defined as $\mathrm{cl}(\bigcup J(g_n))$. With these concepts in hands, McMullen globalized his Rigidity Theorem 5.4:

Theorem 6.2 (Towers rigidity [McM3]) *Let f be an infinitely renormalizable map with a priori bounds, and $\hat{g}$ an associated tower. Then $J(\hat{g}) = \mathbb{C}$, and there are no invariant line fields on $J(\hat{g})$.*

So, *a priori* bounds imply pre-compactness of the orbit $\{R^n f\}$, combinatorial rigidity of the straightened maps, and rigidity of limit towers. Altogether these yield convergence: $R^n f \to f_*$. Indeed, let $g, g' \in \Lambda$ be two limit maps, and $\hat{g}, \hat{g}'$ be the corresponding towers. By the combinatorial rigidity (Proposition 5.5), $\hat{g}$ and $\hat{g}'$ are quasi-conformally equivalent. Hence by Theorem 6.2, they are conformally equivalent. Thus g and g' represent the same point in the hybrid class $\mathcal{H}(f)$, so that $\Lambda = \{g\}$, and convergence of $\{R^n f\}$ follows.

Carrying further these ideas McMullen has proven exponential convergence:

Theorem 6.3 (Exponential convergence [McM3]) *For any $f \in \mathcal{H}(f_*)$, the orbit $R^n f$ converges to the fixed point f_* exponentially fast in the Carathéodory topology.*

Thus the hybrid class $\mathcal{H}(f)$ is contained in the strong stable manifold $W^s(f_*)$ of the renormalization fixed point.

The last theorem has been independently announced by Jeremy Kahn.

6.3 Unstable direction

To complete the renormalization picture, we need to analyze the unstable direction of the renormalization operator (see Figure 4). Our analysis is based upon the Rigidity Theorem 5.6 and the following general lemma:

Lemma 6.4 (Small orbits) *Let B be a complex Banach space, $R : (U, 0) \to (U', 0)$ be an analytic map in a neighborhood of 0, $L = DR(0)$, and U' is compact. Let $\mathrm{spec}(L) = \mathrm{spec}^s(L) \cup \mathrm{spec}^n(L)$ where the $\mathrm{spec}^s(L)$ is contained in a disk of radius $r < 1$, while $\mathrm{spec}^n(L)$ lies in the unit circle. Let W^s be the strong stable manifold of R (corresponding to $\mathrm{spec}^s(L)$). Then R has small orbits outside W^s, that is, for any neighborhood $V \ni 0$, there is a point $f \notin W^s$ whose forward orbit $\{R^m f\}_{m=0}^{\infty}$ is contained in V.*

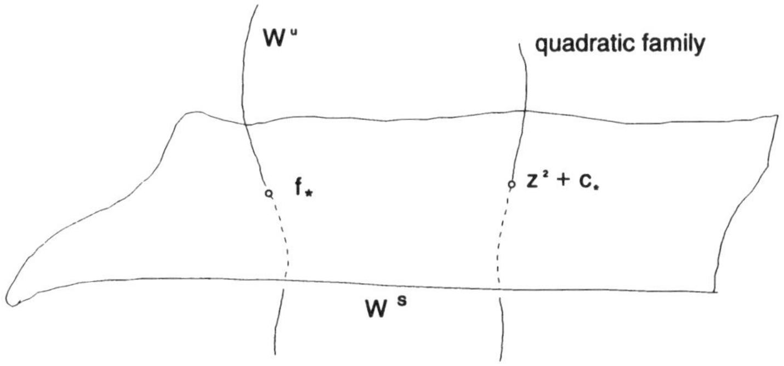

Figure 4. Renormalization picture.

Theorem 6.5 (Hyperbolicity [L8]) *The renormalization operator R is hyperbolic at the fixed point f_* with one dimensional unstable manifold.*

Proof. Clearly $W^s(f_*)$ cannot be open in the space of quadratic-like maps, since it intersects the quadratic family $z \mapsto z^2 + c$ at a single point on the real line (by Proposition 5.5).

If $\operatorname{codim} W^s(f_*) > 1$ then there would be a codimension 2 complex analytic submanifold X (in the space of quadratic-like maps) transversal to $W^s(f_*)$. Then the straightening map χ on $X \setminus \{f_*\}$ would omit the parameter value c_* representing the hybrid class of f_*. But one can see (using an index argument) that every nearby complex one dimensional submanifold $Y \subset X$ contains maps f with $\chi(f) = c_*$.

To prove hyperbolicity of R at f_*, let us assume by contradiction that $\operatorname{spec}(DR(f_*))$ is contained in the closed unit disk. Then by the Small Orbits Lemma, there is a quadratic-like map $f \notin W^s(f_*)$ such that the whole orbit $R^n f$ stays near f_*. But this means that f is an infinitely renormalizable map with *a priori* bounds. By Theorem 5.6, $f \in \mathcal{H}(f_*) \subset W^s(f_*)$, which is a contradiction. □

Remarks. 1. Note that even for real combinatorics the proof of Theorem 6.5 is complex is nature as Lemma 6.4 fails over reals. Moreover, the proof works equally well for any complex situation, provided R has a fixed point f_*.

2. For bounded type combinatorics, the renormalization operator has a hyperbolic invariant set instead of a single point.

6.4 Milnor's Hairiness Conjecture

In [M1] Milnor conjectured that the Mandelbrot set is becoming dense in small scales near the Feigenbaum-Coullet-Tresser point c_*. In other words, the rescalings of M by $T_\lambda : c \mapsto \lambda(c - c_*) + c_*$ tend (in the Hausdorff metric) to the whole complex plane $\mathbb{C}$ as $\lambda \to \infty$.

The dynamical counterpart of this conjecture asserts that the rescalings of the corresponding Julia set $J(P_{c_*})$ about 0 are becoming dense. This follows from the first part of McMullen's Theorem 6.2.

Now we can transfer this result to the parameter plane in the following way. The unstable manifold $W^u \equiv W^u(f_*)$ is a complex one-parameter family of quadratic-like maps, so that we can consider the Mandelbrot set $M^u \subset W^u$. For any $f \in W^u$, the inverse orbit $f_{-n} = R^{-n}f \to f_*$ can be viewed as a one-sided tower $\hat{f}$ with *a priori* bounds. If $\bigcup R^n M^u$ were not dense in W^u then the tower $\hat{f}$ would admit a holomorphic motion over the region $W^u \setminus \bigcup R^n M^u$. But this would contradict to the quasi-conformal rigidity of towers (an extended version of Theorem 6.2).

Thus $\bigcup R^n M^u$ is dense in W^u, so that M^u is hairy. Now one can transfer hairiness of M^u into the hairiness of the genuine Mandelbrot set M by means of holonomy from W^u to the quadratic family along the hybrid classes. This holonomy is transversally quasi-conformal since it can be viewed as a holomorphic motion.

Figure 5. Degree six Fibonacci puzzle piece (made by Scott Sutherland).

6.5 Universality law for higher degree Fibonacci maps

Generalized renormalization and *a priori* bounds of Lemma 4.13 allow us to carry the above discussion for the Fibonacci maps of even degree $d > 2$. For such a d, let $\mathcal{F}_d$ stand for the combinatorial class of polynomial-like Fibonacci maps f preserving the real line.

Theorem 6.6 *For any Fibonacci map $f \in \mathcal{F}_d$ of even degree $d > 2$, the generalized renormalizations $T^n f$ converge to a cycle $\{f_1, f_2\}$ of order two depending only on d.*

The combinatorial difference between f_1 and f_2 is that the restrictions of these maps on the corresponding non-critical puzzle pieces V_1^n have opposite orientation on the real line. The picture of the principal nest for degree 6 Fibonacci map below in Figure 5 shows that all puzzle pieces have approximately the same shape.

As in the quadratic case, these puzzle pieces have asymptotical shapes of the Julia set of an appropriate polynomial-like map, which explains all the pinchings visible at the picture. Unlike the quadratic case though, this polynomial-like map is not a genuine polynomial.

7 Measure of Julia sets

In this last section we will refine a bit McMullen's discussion of the measure problem (see Theorems 5.8 and 6.5 in [McM1]). It is intimately related to renormalization and rigidity.

7.1 Measure-theoretic attractors

The global dynamics of a rational map depends first of all on the structure of the post-critical set. One of the results of such a kind is that the post-critical set is a "measure-theoretic attractor" for the dynamics on the Julia set:

Lemma 7.1 ([L2]) *For Lebesgue almost all $z \in J(f)$, either* $\mathrm{orb}(z)$ *is dense in $\mathbb{C}$, or $\omega(z) \subset \mathcal{O}$.*

In the second case, conjecturally there are at most finitely many measure-theoretic attractors A_i such that the orbit of almost every z converge to one of these attractors: $\omega(z) = A_i$. The basins of these attractors are going to be ergodic components $f|J(f)$. Such results also depend on the geometric bounds, and have been resolved in the cases when the geometry is under a good control.

7.2 Quadratic case

The divergence property of Branner and Hubbard yields that the Julia set of the corresponding cubic polynomial is a Cantor set of zero measure. (The latter consequence was pointed out by McMullen, see [BH]). The passage from this result to the measure zero result for quadratic maps (Lyubich-Shishikura [L2]) can be made by means of the generalized renormalization

Tf. Indeed, by McMullen's argument, the Julia set of Tf has zero measure. Then by Lemma 7.1 the big Julia set $J(f)$ must have zero measure as well.

7.3 Higher degree case

In particular, the Julia set of the quadratic Fibonacci map F_2 has zero measure. What about higher degree Fibonacci maps $f = F_d$? There is an approach to this problem based on the renormalization theory and a random walk construction. Given the principal nest $V^0 \supset V^1 \supset \dots$, let us think of the annuli $A^n = V^{n-1} \setminus V^n$ as the states of the random walk with the transitions induced by $g_n|A^n$. Then drift of the random walk orbits to the higher levels means that the f-orbits converge with positive probability to the post-critical set: In this case the Julia set $J(f)$ has positive measure. The computer experiment carried out by the author jointly with Scott Sutherland suggested that this is indeed the case for $d = 32$.

G. Keller stated a rigorous lemma about random walks which is applicable to such kind of situations, while T.Nowicki & S. van Strien [SN] gave necessary geometric estimates to show that for sufficiently big d, there is a drift to higher levels. Altogether this gives the first example of a polynomial with the Julia set of positive measure.

8 Appendix: Quasi-conformal maps

Quasi-conformal maps play an outstanding role in conformal dynamics (see §2 of [McM1]). They are sufficiently regular to be a subject of analysis, and, on the other hand, so irregular that produce fractal geometric objects (e.g., Jordan curves with Hausdorff dimension greater than 1). By definition, a homeomorphism $h : U \to V$, where $U, V \subset \mathbb{C}$, is called quasi-conformal if it has locally integrable distributional derivatives ∂h, $\bar\partial h$, and $|\bar\partial h/\partial h| \leq k < 1$ almost everywhere. As this local definition is conformally invariant, one can define quasi-conformal homeomorphisms between Riemann surfaces.

One can associate to a quasi-conformal map an analytic object called Beltrami differential, namely

$$\mu = \frac{\bar\partial h}{\partial h}\frac{d\bar z}{dz},$$

with $\|\mu\|_\infty < 1$. The corresponding geometric object is a measurable family of infinitesimal ellipses (defined up to dilation), pull-backs by h_* of the field of infinitesimal circles. The eccentricities of these ellipses are ruled by $|\mu|$, and are uniformly bounded almost everywhere, while the orientation of the ellipses is ruled by the $\arg \mu$. The dilatation $K_h = (1 + \|\mu\|_\infty)/(1 - \|\mu\|_\infty)$ of h is the essential supremum of the eccentricities of these ellipses. A quasi-conformal map is called K-quasi-conformal if $K_h \leq K$.

One of the most remarkable facts of analysis is that the above statements can be reversed: Any Beltrami differential with $\|\mu_\infty\| < 1$ (a measurable field of ellipses with essentially bounded eccentricities) is locally generated by a quasi-conformal map, unique up to post-composition with an analytic map. Thus such a Beltrami differential on a Riemann surface S induces a conformal structure quasi-conformally equivalent to the original structure of S. Together with the Riemann mapping theorem this leads to the following statement:

Measurable Riemann Mapping Theorem. *Let μ be a Beltrami differential on $\bar{\mathbb{C}}$ with $\|\mu_\infty\| < 1$. Then there is a quasi-conformal map $h : \bar{\mathbb{C}} \to \bar{\mathbb{C}}$ which solves the Beltrami equation: $\bar\partial h/\partial h = \mu$ a.e.*

In what follows by a conformal structure we will mean a structure associated to measurable Beltrami differentials μ with $\|\mu\|_\infty < 1$. We will denote by σ the standard structure corresponding to zero Beltrami differential.

Another fundamental property of the space of quasi-conformal maps is compactness:

Compactness Lemma. *The space of K-quasi-conformal maps $h : \mathbb{C} \to \mathbb{C}$ normalized by $h(0) = 0$ and $h(1) = 1$ is compact in the uniform topology on the Riemann sphere.*

The following gluing property is also important:

Gluing Lemma. *Let us have two disjoint domains D_1 and D_2 with a smooth piece γ of their common boundary. Let $D = D_1 \cup D_2 \cup \gamma$. If $h : D \to \mathbb{C}$ is a homeomorphism such that $h|D_i$ is K-quasi-conformal, then h is K-quasi-conformal.*

One of Sullivan's leading ideas was the idea of the Teichmüller metric on the space of deformations of a conformal dynamical systems. The prototype for this metric is the classical Teichmüller metric on the space of marked Riemann surfaces. The distance $\text{dist}(S_1, S_2)$ between two marked Riemann surfaces is defined as the infimum of the dilatations K_h, where $h : S_1 \to S_2$ ranges over quasi-conformal homeomorphisms in the homotopy class respecting the marking.

A basic references on quasi-conformal maps is [A].

Bibliography

[A] L. Ahlfors. Lectures on quasi-conformal maps. Van Nostrand Co, 1966.

[BL] A. Blokh & M. Lyubich. Measure of solenoidal attractors of unimodal maps of the segment. Math. Notes., v. 48 (1990), 1085-1990.

[D] A. Douady. Chirurgie sur les applications holomorphes. In: "Proc. ICM", Berkeley, 1986, p. 724-738.

[BH] B. Branner & J.H. Hubbard. The iteration of cubic polynomials, Part II, Acta Math. v. 169 (1992), 229-325.

[CE] P. Collet & J.-P. Eckmann. Iterated maps on the interval as dynamical systems. Progress in Physics, Birkhäuser, 1980.

[CT] P. Coullet & C. Tresser. Itération d'endomorphismes et groupe de renormalisation. J. Phys. Colloque C 539, C5-25 (1978).

[DH1] A. Douady & J.H. Hubbard. Étude dynamique des polynômes complexes. Publication Mathematiques d'Orsay, 84-02 and 85-04.

[DH2] A. Douady & J.H. Hubbard. On the dynamics of polynomial-like maps. Ann. Sc. Éc. Norm. Sup., v. 18 (1985), 287-343.

[DH3] A. Douady & J.H. Hubbard, A proof of Thurston's topological characterization of rational functions. Acta Math., 1994.

[EE] J.-P. Eckmann & H. Epstein. Bounds on the unstable eigenvalue for period doubling. Comm. Math. Phys., v. 128 (1990), 427-435.

[F] M.J. Feigenbaum. Quantitative universality for a class of non-linear transformations. J. Stat. Phys., v. 19 (1978), 25-52.

[G] J. Guckenheimer. Limit sets of S-unimodal maps with zero entropy. Comm. Math. Phys., v. 110 (1987), 655-659.

[GS1] J. Graczyk & G. Swiatek. Induced expansion for quadratic polynomials. Preprint IMS at Stony Brook, # 1993/8.

[GS2] J. Graczyk & G. Swiatek. Polynomial-like property for real quadratic polynomials. Preprint 1995.

[H] J.H. Hubbard. Local connectivity of Julia sets and bifurcation loci: three theorems of J.-C. Yoccoz. In: "Topological Methods in Modern Mathematics, A Symposium in Honor of John Milnor's 60th Birthday", Publish or Perish, 1993.

[HJ] J. Hu & Y. Jiang. The Julia set of the Feigenbaum quadratic polynomial is locally connected. Preprint 1993.

[HK] F. Hofbauer and G. Keller. Some remarks on recent results about S-unimodal maps. Ann. Inst. Henry Poincaré, v. 53 (1990), 413-425.

[LS] G. Levin, S. van Strien. Local connectivity of Julia sets of real polynomials, Preprint IMS at Stony Brook, 1995/5.

[La] O.E. Lanford. A computer assisted proof of the Feigenbaum conjectures. Bull. Amer. Math. Soc., v.6 (1982), 427-434.

[L1] M. Lyubich. Some typical properties of the dynamics of rational maps, Russian Math. Surveys, v. 38:5 (1983), 154-155.

[L2] M. Lyubich. On typical behavior of the trajectories of a rational mapping of the sphere. Soviet Math. Dokl., v. 27 (1983), 22-25.

[L3] M. Lyubich. On the Lebesgue measure of the Julia set of a quadratic polynomial, Preprint IMS at Stony Brook, #1991/10.

[L4] M. Lyubich. Combinatorics, geometry and attractors of quasi-quadratic maps. Ann. Math, v.140 (1994), 347-404.

[L5] M. Lyubich. Geometry of quadratic polynomials: moduli, rigidity and local connectivity. Preprint IMS at Stony Brook #1993/9.

[L6] M. Lyubich. Dynamics of quadratic polynomials, I. Combinatorics and geometry of the Yoccoz puzzle. Preprint MSRI #26, 1995.

[L7] M. Lyubich. Dynamics of quadratic polynomials, II. Rigidity. Preprint IMS at Stony Brook, # 1995/14.

[L8] M. Lyubich. Feigenbaum-Coullet-Tresser Universality and Milnor's Hairiness Conjecture. Manuscript in preparation, 1995.

[L9] M. Lyubich. Dynamics of quadratic polynomials. Parapuzzle geometry. Manuscript in preparation, 1995.

[LM] M. Lyubich. & J. Milnor. The Fibonacci unimodal map, Journal of AMS, v. 6 (1993), 425-457.

[LY] M. Lyubich & M. Yampolsky. Complex bounds for real maps. Preprint MSRI # 34, 1995.

[M1] J. Milnor. Self-similarity and hairiness in the Mandelbrot set, "Computers in geometry and topology", Lect. Notes in Pure Appl Math, **114** (1989), 211-257.

[M2] J. Milnor. Local connectivity of Julia sets: expository lectures. Preprint IMS at Stony Brook #1992/11.

[MT] J. Milnor & W. Thurston. On iterated maps of the interval. "Dynamical Systems", Proc. U. Md., 1986-87, ed. J. Alexander, Lect. Notes Math., v. 1342 (1988), 465-563.

[McM1] C. McMullen. The classification of conformal dynamical systems. Paper in this volume.

[McM2] C. McMullen. Complex dynamics and renormalization. Annals of Math. Studies, v. 135, Princeton University Press, 1994.

[McM3] C. McMullen. Renormalization and 3-manifolds which fiber over the circle. Preprint 1994.

[MSS] R. Mañé, P. Sad & D. Sullivan. On the dynamics of rational maps, Ann. scient. Ec. Norm. Sup., v. 16 (1983), 193-217.

[MvS] W. de Melo & S. van Strien. One dimensional dynamics. Springer-Verlag, 1993.

[Sw] G. Swiatek. Hyperbolicity is dense in the real quadratic family. Preprint IMS at Stony Brook 1992/10.

[S1] D. Sullivan. Quasiconformal homeomorphisms and dynamics, topology and geometry. Proc. ICM-86, Berkeley, v. II, 1216-1228.

[S2] D. Sullivan. Bounds, quadratic differentials, and renormalization conjectures. AMS Centennial Publications. **2**: Mathematics into Twenty-first Century (1992).

[S3] D. Sullivan. In: "Topological Methods in Modern Math., A Symposium in Honor of John Milnor's 60th Birthday", Publish or Perish, 1993.

[ST] D. Sullivan & W. Thurston. Extending holomorphic motions. Acta Math., v.157 (1986), 243-257.

[SN] S. van Strien & T. Nowicki. Polynomial maps with a Julia set of positive measure: Fibonacci maps, Preprint IMS at Stony Brook 1994/3.

[Sch] D. Schleicher. The structure of the Mandelbrot set. Preprint 1995.

[Sh] K.Shibayama. Fibonacci sequence of stable periodic orbits for one-parameter families of C^1-unimodal mappings. Preprint 1991.

[Th] W. Thurston. On the geometry and dynamics of iterated rational maps. Preprint, 1984.

[TV] F. Tangerman & P. Veerman. Scaling in circle maps, I. Comm. Math. Phys., v. 134 (1990), 89-107.

[VSK] E.B. Vul, Ya.G. Sinai & K.M. Khanin. Feigenbaum universality and thermodynamic formalism. Russian Math. Surveys, v.39:3 (1984), 1-40.

[W] L. Wenstrom. Parameter scaling for the Fibonacci point. Manuscript in preparation, 1995.

Algebraic K-theory and traces

Dedicated to Wu Chung Hsiang
on the occasion of his 60'th birthday

Ib Madsen
AARHUS UNIVERSITY
AARHUS, DK-8000, DENMARK

In the invitation to speak at the seminar, S.-T. Yau stated the intent of the lectures and the accompanying publication to be that a graduate student, having heard the lecture and read the manuscript, should be able to start research of his or her own on the subject. He added that it was desirable if the manuscript contained some new original results as well. I do not know if this is possible to achieve in a single paper, but it is a noble goal. The present blend between a traditional expository article and a detailed exposition of the subject is in any case my attempt at this goal.

Contents

1 Introduction

The present paper is an attempt to give an overview of topological cyclic homology and its relation to algebraic K-theory. In the 'classical' setting, algebraic K-theory associates to a ring A a space $K(A)$. The homotopy groups of $K(A)$ are Quillen's higher K-groups. They have proved very difficult to calculate, and are, for example, to this day not even known for the ring of rational integers.

The homology of (a component of) $K(A)$ is the group homology of the group $\mathrm{GL}_\infty(A)$ of invertible matrices of the ring. This was early on used by Quillen and Borel to evaluate K-theory of finite fields and the torsion free part of K-theory of algebraic integers, respectively. Later Suslin evaluated the homotopy groups with finite coefficients of K-theory of algebraically closed fields, or what amounts to the same thing, the profinite completion $K(F)^\wedge$. In particular he showed that $K(\mathbb{C})^\wedge$ is equivalent to the profinite completion of the space which classifies complex vector bundles. Bott periodicity then calculates $\pi_i K(\mathbb{C})^\wedge$ to be a copy of the profinite integers when i is even and zero for i odd.

This development inspired another calculational approach to the K-groups, namely via étale K-theory, introduced by Friedlander.

Given a Galois extension $F \subset E$ with group G, $K(F) \sim K(E)^G$. This is no big calculational help, but if one replaces the actual fixed sets by the socalled homotopy fixed sets, a construction introduced by Sullivan in the sixties in connection with his solution of the Adams conjecture, then calculations become possible. The homotopy fixed set is the function space (spectrum)

$$K(E)^{hG} = \mathrm{Map}_G(EG, K(E))$$

where EG is the contractible free G space, and where Map_G denotes the space of G-mappings. The filtration of EG by its skeleta induces a spectral sequence

$$H^*(G, K_*(E)) \Rightarrow \pi_* K(E)^{hG}$$

which in favorable situations can be determined. The étale K-theory of a field F is, very roughly speaking, the homotopy fixed set of $K(\hat{F})^{hG}$ where $\hat{F}$ is the closure of F. In the characteristic zero situation $K(\hat{F})^\wedge \sim K(\mathbb{C})^\wedge$ by Suslin, so the calculation of étale K-theory of fields is intimately tied to Galois cohomology. There has been a lot of efforts by many people to evaluate étale K-theory, and in particular by W. Dwyer, E. Friedlander, S. Mitchell and Bob Thomasson. But the basic question remains: how close is

$$K(F) \to K^{\mathrm{et}}(F)$$

to be a (profinite) homotopy equivalence? In one formulation, the Lichtenbaum-Quillen conjecture asserts this to be the case (above dimension 1) for number fields.

For small values of i, K_i(fields) have been extensively calculated in work of Merkurjev and Suslin. The reader is referred to Suslin's address at the ICM 1990.

In another direction, Waldhausen generalized Quillen's K-theory of rings to include certain 'rings up to homotopy', such as $\Omega^\infty S^\infty(\Omega X_+)$. The resulting functor $A(X)$ is intimately related to the space of automorphisms (homeomorphisms or diffeomorphisms) of X when it is a (high dimensional) manifold.

The approach to K-theory (of rings or spaces) in this paper is to study a certain trace type invariant

$$\mathrm{trc}\colon K(A) \to \mathrm{TC}(A).$$

The target is a topological version of Connes' cyclic homology; We call it topological cyclic homology but maybe trace homology was a better word.

From a superficial viewpoint the cyclotomic trace records the traces of all powers of matrices, so could also be called the characteristic polynomial invariant. It works equally well for Waldhausen's K-theory of spaces, and was introduced in joint work with M. Bökstedt and W.-C. Hsiang [BHM] in order to solve the K-theory analogue of Novikov's conjecture about homotopy invariance of the higher signatures of manifolds. The construction was inspired by ideas of T. Goodwillie. Here however, I shall be mostly concerned with the situation for rings.

There is a map from $\mathrm{TC}(A)$ to another functor denoted $\mathrm{THH}(A)$, the topological Hochschild homology of R, and trc is a lifting of this topological Dennis trace.

Let me briefly sketch the construction of $\mathrm{THH}(A)$. Consider the simplicial abelian group

$$Z_\bullet(A)\colon \quad \cdots A\otimes A\otimes A \overset{\rightarrow}{\underset{\rightarrow}{\rightarrow}} A\otimes A \rightrightarrows A$$

where the face operators sends $a_1\otimes a_2\otimes a_3$ into $a_1a_2\otimes a_3$, $a_1\otimes a_2a_3$, $a_3a_1\otimes a_2$ etc. The homotopy groups of $Z_\bullet(A)$, or what is the same thing, the homology groups of the associated chain complex, are the Hochschild homology groups $\mathrm{HH}_*(A)$.

The basic idea, suggested by T. Goodwillie, is to replace A by the Eilenberg-MacLane spectrum it generates, and $\otimes$ by smash product of spectra. This was carried out by Bökstedt, and leads to a simplicial space $\mathrm{THH}(A)$. The extra structure in $Z_\bullet(A)$ which comes from the cyclic rotation of the tensor factors is also present after the indicated substitutions, and via Connes' theory of cyclic sets, it implies a circle action on $\mathrm{THH}(A)$.

Connes initially defined cyclic homology by replacing $Z_*(A)$ by the complex $C_*(A)$ whose n'th term is $Z_n(A)/C_{n+1}$, the quotient group by the cyclic rotation of factors. It is crude construction to divide out a non-free group action—usually one gets a better theory by instead taking the Borel quotient. This was done in papers of Loday-Quillen and Feigin-Tsygan who replaced the quotients $A^{\otimes n}/C_n$ by $W^{(n)}\otimes_{C_n} A^{\otimes n}$ where $W^{(n)}$ is the standard free $\mathbb{Z}[C_n]$ resolution of $\mathbb{Z}$. In the topological situation of $\mathrm{THH}(A)$ it is better to take fixed sets $\mathrm{THH}(A)^C$ for the various subgroups of the circle. Had the circle action on $\mathrm{THH}(A)$ been free, the fixed sets would have been the Borel orbits $\mathrm{THH}(A)_{hC} = \mathrm{THH}(A)\wedge_C EC_+$. This is not the case, and the fixed sets $\mathrm{THH}(R)^C$ is a mixture of Borel quotients, one for each strata of the action. In our topological situation it turns out that there is a certain map

$$R\colon \mathrm{THH}(A)^{C_n} \to \mathrm{THH}(A)^{C_m}$$

whenever m divides n, which one does not see in the linear setting. This map mixes the stata. We also have the inclusion of fixed sets

$$F\colon \mathrm{THH}(A)^{C_n} \to \mathrm{THH}(A)^{C_m}.$$

The topological cyclic homology $\mathrm{TC}(A)$ is defined to be the homotopy theoretical limit of $\mathrm{THH}(A)^{C_n}$ over the maps R and F as C_n varies over all cyclic subgroups. The basic theory of $\mathrm{THH}(A)$ and $\mathrm{TC}(A)$ is described in chap. 2 below, where we also recall the construction of

$$\mathrm{trc}\colon K(A) \to \mathrm{TC}(A). \tag{1.1}$$

Actually, both $K(A)$, $\mathrm{TC}(A)$ and $\mathrm{THH}(A)$ are spectra in the sense that there are sequences of spaces $K(A)_{\mathbf{R}^n}$ etc. so that $K(A)$ is equivalent to the n'th based loop space of $K(A)_{\mathbf{R}^n}$ etc., and trc preserves this structure. I write $\mathrm{TH}(A)$ for the spectrum $\{\mathrm{THH}(A)_{\mathbf{R}^n}\}$ but do not introduce special notation for the spectra $K(A)$ and $\mathrm{TC}(A)$.

The following chap. 3 presents results of Dundas, Goodwillie and McCarthy. The following theorem is proved in sect. 3.4.

Theorem 1.2 (McCarthy). *For a surjection of rings $f\colon A \to \bar{A}$ with nilpotent kernel,*

$$\begin{array}{ccc} K(A) & \xrightarrow{\mathrm{trc}} & \mathrm{TC}(A) \\ \downarrow & & \downarrow \\ K(\bar{A}) & \xrightarrow{\mathrm{trc}} & \mathrm{TC}(\bar{A}) \end{array}$$

becomes a homotopy Cartesian diagram after profinite completion.

In particular, the relative homotopy groups with finite coefficients of the two vertical maps agree. Earlier results of this nature have appeared in [G4], [G5], [BCCGHM]. The proof is based upon Goodwillie's "calculus of functors"; it is very indirect, and does not in any way produce an explicit inverse from $\mathrm{TC}(A \to \bar{A})$ to $K(A \to \bar{A})$.

The trace (1.1) cannot in general induce an isomorphism of (profinite) homotopy groups. Here is one reason: $\mathrm{TC}(A)$ is constructed out of Eilenberg-MacLane spectra $H(A)$. Now $H(A)^\wedge \sim H(\hat{A})$ since $H(A)$ is characterized by its homotopy groups. This persists to TC,

$$\mathrm{TC}(A)^\wedge \sim \mathrm{TC}(A \otimes \hat{\mathbb{Z}})$$

at least if A is finite over $\mathbb{Z}$. However, it is well known that K-theory does not have this property. Thus (1.1) has little chance of inducing isomorphism on profinite homotopy unless one restricts attention to complete rings.

There is an extension due to B. Dundas of theorem 1.2 to the setting of Waldhausen's functor $A(X)$, namely the following result which is outlined in sect. 3.5.

Theorem 1.3. (Dundas). For any space X the diagram

$$\begin{array}{ccc} A(X) & \xrightarrow{\text{trc}} & \mathrm{TC}(X) \\ \downarrow & & \downarrow \\ K(\mathbb{Z}[\pi_1 X]) & \xrightarrow{\text{trc}} & \mathrm{TC}(\mathbb{Z}[\pi_1 X]) \end{array}$$

becomes homotopy Cartesian after profinite completion.

Let k be a finite field of characteristic $p \neq 0$, and let $W(k)$ be its ring of Witt vectors. Chap. 4 outlines the proof of the following joint result with L. Hesselholt

Theorem 1.4 ([HM]) *For finitely generated $W(k)$-algebras,*

$$\text{trc}\colon K(A)_p^\wedge \to \mathrm{TC}(A)_p^\wedge$$

is a homotopy equivalence (in positive dimensions).

Chapter 4 also gives a new (simpler) proof of one of the main results from [BHM], namely that the assembly map

$$K(\mathbb{Z}) \wedge B\Gamma_+ \to K(\mathbb{Z}\Gamma)$$

is a rational equivalence for a large class of big groups, e.g. for the groups Γ which have finitely generated Eilenberg-MacLane cohomology in each dimension. The simplification of the original proof is made possible by theorem 1.3. Chapter 4 further calculates $\mathrm{TC}(X)^\wedge$ in terms of more traditional functors in algebraic topology; these involve the free loop space of X.

The functor $\mathrm{TC}(A)$ is not very easy to calculate, but it does lent itself to analysis by classical methods of algebraic topology. The basic approach, so far, has been to use the following diagram of (co)fibrations (in the category of spectra)

$$\begin{array}{ccccc} \mathrm{TH}(A)_{hC_{p^n}} & \longrightarrow & \mathrm{TH}(A)^{C_{p^n}} & \xrightarrow{R} & \mathrm{TH}(A)^{C_{p^{n-1}}} \\ \| & & \downarrow{\Gamma} & & \downarrow{\hat{\Gamma}} \\ \mathrm{TH}(A)_{hC_{p^n}} & \xrightarrow{N} & \mathrm{TH}(A)^{hC_{p^n}} & \longrightarrow & \hat{\mathbb{H}}(C_{p^n}, \mathrm{TH}(A)) \end{array}$$

The lower cofibration is usually called the norm cofibration. In order to define it one uses that the spectrum $\mathrm{TH}(A)$ can be extended to an S^1-equivariant spectrum $T(A)$. Roughly speaking this means that there are spaces $T(A)_V$, one for each finite dimensional representation V of S^1 such that $\mathrm{TH}(A) \sim_{S^1} \mathrm{Map}(S^V, T(A)_V)$. Here S^V denotes the one-point compactification of V with its induced S^1-action. The construction of the norm cofibration is due to J. Greenlees and J. P. May.

The point of the diagram is firstly that there are spectral sequences which approximates the terms in the bottom sequence, e.g.

$$\hat{H}^s(C_{p^n}; \pi_t \mathrm{TH}(A)) \Rightarrow \pi_{s+t}\hat{\mathbb{H}}(C_{p^n}\mathrm{TH}(A))$$

where $\hat{H}^s$ denotes Tate cohomology. Secondly, it turns out that the maps Γ and $\hat{\Gamma}$ in many situations are homotopy equivalences in non-negative degrees. This is reminisant of the Segal conjecture (which corresponds to $\mathrm{TH}(\Omega^\infty S^\infty)$) where Γ and $\hat{\Gamma}$ are actual homotopy equivalences. In particular one expects Γ, $\hat{\Gamma}$ to give equivalence (in positive degrees) for integers in local number fields with non-zero residue characteristic.

This has been verified in the unramified situation, $A = W(\mathbb{F}_{p^s})$, where the calculation of TC has been carried through. In order to describe the result, let $\mathrm{im}\, J_p$ be the homotopy fiber in

$$\mathrm{im}\, J_p \to (BU \times \mathbb{Z})_p \xrightarrow{\psi^k - 1} BU_p$$

where ψ^k is the Adams operation for an integer k which generates the units in $\mathbb{Z}/p^2\mathbb{Z}$, i.e. a topological generator of the units of the p-adic integers $\mathbb{Z}_p$. The bottom homotopy group of $\mathrm{im}\, J_p$ is a copy of $\mathbb{Z}_p$, and

$$\pi_{2n-1}(\mathrm{im}\, J_p) = \begin{cases} \mathbb{Z}/p^{v_p(n+1)} & \text{if } n \equiv 0 \ (p-1) \\ 0 & \text{if not} \end{cases}$$

while $\pi_{2n}(\mathrm{im}\, J_p) = 0$ for $n > 0$. ($v_p(-)$ denotes the p-adic valuation). Let $B\,\mathrm{im}\, J_p$ denote the delooping of $\mathrm{im}\, J_p$ with $\pi_i(B\,\mathrm{im}\, J_p) = \pi_{i-1}(\mathrm{im}\, J_p)$. Then one has:

Theorem 1.5. ([BHM2]). *Let $\mathbb{F}_{p^s}$ be the finite field with p^s elements and $A_s = W(\mathbb{F}_{p^s})$ its Witt-vectors. Then for p odd,*

$$\mathrm{TC}(A_s)^\wedge_p \sim \mathrm{im}\, J_p \times B\,\mathrm{im}\, J_p \times SU^\wedge_p \times U^\wedge_p \times \cdots \times U^\wedge_p \quad (s-1 \textit{ copies of } U)$$

where SU is the special unitary group, and U the unitary group.

The proof of theorem 1.5 is a long and complex calculation which requires a thorough knowledge of homotopy theory. It was in fact the first calculation made of the TC functor applied to rings. The general calculational scheme developed in this case was later exploited in a number of less complicated situations. Theorem 1.5 in conjunction with the Dwyer-Mitchell calculation of $K^{\mathrm{et}}(A_s)$, [DM], verifies the conjecture of Lichtenbaum and Quillen for these rings.

The first three sections of chapter 5 give other examples of TC-calculations in situations where theorem 1.4 applies. Sect. 5.1 studies K-theory of group rings of finite groups. In terms of concrete calculations the main result is:

Theorem 1.6. *Let k be a finite field of characteristic $p > 0$, and let C be a cyclic group of p-power order. Then the p-primary part of K-theory is given by*

$$K_{2n-1}(k[C])_{(p)} \cong K_1(k[C])_{(p)}^{\oplus n}$$

and $K_{2n}(k[C])_{(p)} = 0$ for $n > 0$.

The next two sections 5.2 and 5.3 outline joint work with L. Hesselholt. The main result is the following

Theorem 1.7. *Let k be a perfect field of characteristic $p > 0$. Then*

$$K_{2m-1}\big(k[x]/(x^n)\big)_{(p)} = \mathbf{W}_{nm-1}(k)/V_n\mathbf{W}_{m-1}(k)$$

and $K_{2m}(k[x]/(x^n))_{(p)} = 0$ for $n > 0$.

Here $\mathbf{W}(k)$ denotes the big Witt-vectors, that is, $\mathbf{W}(k) = (1 + k[[t]])^{\times}$, the multiplicative group of power series beginning with 1, $\mathbf{W}_r(k)$ is the corresponding truncated version

$$\mathbf{W}_r(k) = (1 + k[[t]])^{\times}/(1 + t^{r+1}k[[t]])^{\times},$$

and V_s is the Verschiebung map which takes a power series $f(t)$ to $f(t^s)$. Sect. 5.3 is just an example; it evaluates the groups $\mathrm{Nil}_*(A)$ for the rings of theorem 1.7.

Finally it is in order to point out that $\mathrm{TC}(A)$ only contains information about K-theory at the residue characteristic. The l-primary part of $K(A)$ for $l \neq p$ is however, for the rings under consideration, already known by theorems of Gabber and Suslin et. al.: one may divide out the radical, cf. [Su], which also contains a thorough account on low dimensional calculations.

It is a pleasure to acknowledge the help I have had from M. Bökstedt, B. Dundas and L. Hesselholt in preparing this paper.

2 Topological cyclic homology

This chapter sets the notation to be used in the rest of the paper, reviews the definition of the functors to be discussed and gives the basic constructions.

2.1 Cyclic constructions.

Let G be a topological monoid and E a two sided G space. For technical reasons we assume that the unit $1 \in G$ is a "good" base point, i.e. $\{1\} \subset G$ be a cofibration. The cyclic bar construction of G with coefficients in E is the simplicial space $N_r^{\mathrm{cy}}(G; E)$ with r-simplices

$$N_r^{\mathrm{cy}}(G; E) = E \times G^r \tag{2.1.1}$$

and simplicial structure maps

$$d_i(e, g_1, \dots, g_r) = \begin{cases} (eg_1, g_2, \dots, g_r), & i = 0 \\ (e, g_1, \dots, g_i g_{i+1}, \dots, g_r), & 0 < i < r \\ (g_r e, g_1, \dots, g_{r-1}), & i = r \end{cases}$$
$$s_i(e, g_1, \dots, g_r) = (e, g_1, \dots, g_{i-1}, 1, g_i, \dots, g_r), \quad 0 \le i \le r.$$

Two special cases have particular interest for us, namely $E = *$ and $E = G$ (with its natural two sided G-structure). In these cases we shorten the notation to

$$N_\bullet G = N_\bullet^{\mathrm{cy}}(G; *), \quad N_\bullet^{\mathrm{cy}} G = N_\bullet^{\mathrm{cy}}(G; G).$$

The simplicial space $N_\bullet^{\mathrm{cy}} G$ has extra structure; it is a *cyclic set* in the sense of Connes: one has the cyclic permutation

$$t_r \colon N_r^{\mathrm{cy}} G \to N_r^{\mathrm{cy}} G, \quad t_r(g_0, \dots, g_r) = (g_r, g_0, \dots, g_{r-1})$$

with the following extra relations, as the reader can easily check,

$$\begin{aligned} d_i t_r &= t_{r-1} d_{i-1}, \quad 1 \le i \le r \\ d_0 t_r &= d_r \\ s_i t_r &= t_{r+1} s_{i-1}, \quad 1 \le i \le r \\ s_0 t_r &= t_{r+1}^2 s_r \\ t_r^{r+1} &= 1. \end{aligned} \tag{2.1.2}$$

Let Δ be the usual simplicial category with objects $[r] = \{0, \dots, r\}$ and order preserving maps, so that a simplicial space is a functor from Δ^{op} to $\{\text{spaces}\}$. It is contained in a category Λ with the same objects but with

$$\Lambda([r], [s]) = \Delta([r], [s]) \times C_{r+1}$$

where C_{r+1} is the cyclic group of order $r+1$. A cyclic space is just a functor from Λ^{op} to {spaces}, see e.g. [J] for further information.

For a simplicial space $X_\bullet$ we let $|X_\bullet|$ denote the usual topological realization,

$$|X_\bullet| = \coprod_{r=0}^{\infty} \Delta^r \times X_r/\sim; \quad (d^i u, x) \sim (u, d_i x), \quad (s^i u, x) \sim (u, s_i x),$$

where $d^i\colon \Delta^{r-1} \to \Delta^r$, $s^i\colon \Delta^r \to \Delta^{r-1}$ are the face and degeneracy operators of the standard simplex. The realization of the cyclic r-simplex $\Lambda[r]_\bullet = \Lambda([\bullet],[r])$ can be calculated to be

$$\Lambda^r \cong \mathbb{R}/\mathbb{Z} \times \Delta^r = S^1 \times \Delta^r. \tag{2.1.3}$$

It is a cocyclic space, that is a functor from Λ into {spaces}. There are two good choices of the homeomorphism in (2.1.3). One can either choose it so that

(i) $t^r(\theta, u_0, \dots, u_r) = (\theta - u_0, u_1, \dots, u_r, u_0)$ or so that

(ii) $t^r(\theta, u_0, \dots, u_r) = \big(\theta - 1/(r+1), u_1, \dots, u_r, u_0\big)$.

In case (i), the cosimplicial maps d^i, s^i are $\mathrm{Id}_{S^1} \times d^i$, $\mathrm{Id}_{S^1} \times s^i$ with d^i, s^i being the usual cosimplicial maps on $\Delta^\bullet$; in case (ii) d^i and s^i depends on the circle coordinate. The realization of cyclic spaces comes equipped with a natural action of S^1. Indeed it is easy to see for a cyclic space $Z_\bullet$ that

$$|Z_\bullet| \cong \coprod_{r=0}^{\infty} \Lambda^r \times Z_r/\approx \tag{2.1.4}$$

where the identifications $\approx$ are

$$(d^i u, z) \approx (u, d_i z), \quad (s^i u, z) \approx (u, s_i z), \quad (t^r u, z) \approx (u, t_r z).$$

The S^1-action on the circle factor of Λ^r descents to the claimed S^1-action on $|Z_\bullet|$. For further information on cyclic spaces we refer the reader to [C], [J], [DHK].

The homotopy theory of spaces X equipped with an action of a group G is governed by the homotopy theory of its fixed sets X^H, $H \subseteq G$ (H closed if G is Lie). In particular a G-map $f\colon X \to Y$ is a weak homotopy equivalence if and only if its induced map on H-fixed set is for all (closed) $H \subseteq G$. Thus it is important to be able to calculate fixed set $|Z_\bullet|^C$ for the realization of a cyclic set, where C is finite cyclic or $C = S^1$. It is not hard to see from (2.1.4) that

$$|Z_\bullet|^{S^1} = \{z \in Z_0 \mid s_0 z = t_1 s_0 z\}$$

but it is harder to use (2.1.4) to get information about $|Z_\bullet|^C$ when $C \subset S^1$ is finite.

There is however a simple devise, *edgewise subdivision*, which can be used to effectively calculate $|Z_\bullet|^C$. Let $X_\bullet$ be any simplicial space, and C a cyclic group of order c. We consider $X_\bullet \colon \Delta^{\mathrm{op}} \to \{\text{spaces}\}$ and define

$$\begin{aligned} &sd_C \colon \Delta \to \Delta \\ &sd_C[r] = [r] \amalg \cdots \amalg [r], \quad c = \text{summands} \\ &sd_C \phi = \phi \amalg \cdots \amalg \phi, \quad \phi \in \Delta([r],[s]). \end{aligned} \tag{2.1.5}$$

The composition $X_\bullet \circ sd_C \colon \Delta^{\mathrm{op}} \to \{\text{spaces}\}$ is the subdivided simplicial space, denoted $sd_C X_\bullet$. Its space of r-simplices is equal to $X_{c(r+1)-1}$.

The diagonal inclusion of Δ^r into the c fold join $\Delta^r * \cdots * \Delta^r$ induces a (non-simplicial) map D from the realization of $sd_C X_\bullet$ into the realization of $X_\bullet$. If $X_\bullet$ is a cyclic space then $|sd_C X_\bullet|$ has a natural $\mathbb{R}/c\mathbb{Z}$ action, which restricts to a simplicial $\mathbb{Z}/c\mathbb{Z}$ action. Indeed $t_{c(r+1)-1}^{r+1}$ acts simplicially on the r-simplices of $sd_C X_\bullet$. From [BHM], sect. 1 we have:

Lemma 2.1.6. *The map $D \colon |sd_C X_\bullet| \to |X_\bullet|$ is a homeomorphism. Moreover, if $X_\bullet$ is cyclic then D is S^1-equivariant when $\mathbb{R}/c\mathbb{Z}$ is identified with the circle in the usual way.* □

For a cyclic space $Z_\bullet$, the action of the $(r+1)$'st power $t_{c(r+1)-1}^{r+1}$ is a simplicial map of $sd_C Z_\bullet$ of order c, so induces a simplicial C-action on $|sd_C Z_\bullet|$, and hence via D an action of C on $|Z_\bullet|$. For example it is not hard to see that

$$sd_C N_\bullet^{\mathrm{cy}}(G) \cong N_\bullet^{\mathrm{cy}}(E, G^c)$$

where $E = G^c$ (c fold Cartesian product) with its componentwise left G^c-action, and right G^c-action given by

$$(g_1, \dots, g_c)(e_1, \dots, e_c) = (g_c e_1, g_1 e_2, \dots, g_{c-1} e_c).$$

The action of C on $sd_C N^{\mathrm{cy}}(G)$ corresponds under the above identification to the cyclic permutation action on E and G^c, so there is a homeomorphism

$$\Delta_C \colon N_\bullet^{\mathrm{cy}}(G) \xrightarrow{\cong} sd_C N_\bullet^{\mathrm{cy}}(G)^C \tag{2.1.7}$$

with Δ_C induced from the diagonal map $G \to G^c$.

We now suppose that our topological monoid is group-like, that is, $\pi_0 G$ is a group. In this case

$$BG = |N_\bullet G|$$

and the canonical map $G \to \Omega BG$ is a weak homotopy equivalence, in sign: $\Omega BG \sim G$. Moreover,

$$B^{\mathrm{cy}}G = |N_\bullet^{\mathrm{cy}} G|$$

is equivalent to the free loop space ΛBG of BG. Indeed, the projection

$$N_\bullet^{\mathrm{cy}}(G;G) \to N_\bullet^{\mathrm{cy}}(G;*)$$

induces a map from $B^{\mathrm{cy}}G$ to BG and the adjoint of the map

$$S^1 \times B^{\mathrm{cy}}G \to B^{\mathrm{cy}}G \to BG$$

defines the equivalence (cf. [G1],[BF])

$$q\colon B^{\mathrm{cy}}G \xrightarrow{\sim} \Lambda BG. \tag{2.1.8}$$

This is not a (weakly homotopy) equivalence of S^1-spaces since the S^1-fixed sets do not agree. However, for each finite subgroup $C \subset S^1$,

$$q^C\colon (B^{\mathrm{cy}}G)^C \xrightarrow{\sim} (\Lambda BG)^C \tag{2.1.9}$$

is an equivalence. This follows easily upon using (2.1.7) and the obvious homeomorphism

$$\Delta_c\colon \Lambda BG \xrightarrow{\cong} (\Lambda BG)^C, \quad \Delta(\lambda)(z) = \lambda(z^c).$$

Indeed, q^C identifies with q under the identifications induced from Δ_C and Δ_c (cf. [BHM], proposition 2.6). Let me give the proof of (2.1.8) when G is a group, and refer to [BF], [G1] for the group-like case. One starts with a rewriting of $N^{\mathrm{cy}}(G)$, namely via the bijection

$$f\colon N_\bullet^{\mathrm{cy}}(G) \xrightarrow{\cong} \mathrm{Ad}G \times_G E_\bullet G,$$

where $\mathrm{Ad}G$ denotes G with conjugation action, and $E_\bullet G$ is the left acyclic bar construction whose k-simplices are $g[g_1|\cdots|g_k]$; the map is defined as

$$f(g_0, \ldots g_k) = g_k \cdots g_1 g_0 [g_1|\cdots|g_{k-1}].$$

The topological realization of $E_\bullet G$ is the free contractible G-space, and $\mathrm{Ad}G \times_G EG \sim \Lambda BG$. Indeed a loop $\lambda(t) \in \Lambda BG$ is mapped into $(g_\lambda, \tilde\lambda(1))$ where $\tilde\lambda(t)$ is a lift of $\lambda\colon [0,1] \to BG$ and g_λ is the holonomy: $g_\lambda \cdot \tilde\lambda(0) = \tilde\lambda(1)$. When G is compact Lie one needs a connection in $EG \to BG$, and $\tilde\lambda$ will be a parallel curve in EG.

The above have generalizations to the nerve and cyclic nerve of a category $\mathcal{C}$. The nerve $N_\bullet \mathcal{C}$ is the simplicial set with

$$N_r\mathcal{C} = \{c_r \xrightarrow{f_r} c_{r-1} \longrightarrow \cdots \xrightarrow{f_1} c_0\},$$

the set of r composable maps. Similarly $N_\bullet^{\mathrm{cy}}(\mathcal{C})$ has r-simplices

$$N_r^{\mathrm{cy}}\mathcal{C} = \{c_r \xrightarrow{f_r} c_{r-1} \longrightarrow \cdots \xrightarrow{f_1} c_0 \xrightarrow{f_0} c_r\}$$

and boundary maps similar to (2.1.1). For categories with only one object, monoids, this agrees with the above constructions. If we restrict the morphisms of $\mathcal{C}$ to be isomorphisms we obtain a subcategory $i\mathcal{C}$, and (2.1.8) generalizes to

$$|N_\bullet^{\mathrm{cy}}(i\mathcal{C})| \sim \Lambda|N_\bullet(i\mathcal{C})|.$$

These more general concepts will be used in the next chapter.

We close this section with a rewriting of $N_\bullet G$, due to Waldhausen, [W2], in the special case where G is a semi-direct product. Let $(V,+)$ be an abelian monoid equipped with a two-sided action of the monoid Γ. Denote by $G = V \rtimes \Gamma$ the semi-direct product with multiplication

$$(v_1, g_1)(v_2, g_2) = (v_1 g_2 + g_1 v_2, g_1 g_2).$$

Let $N_\bullet V$ be the bar construction of $(V,+)$. It inherits a simplicial two sided action of Γ, and we can form the bisimplicial set

$$[r],[s] \mapsto N_r^{\mathrm{cy}}(\Gamma; N_s V).$$

Its diagonal simplicial set with r-simplices $N_r^{\mathrm{cy}}(\Gamma; N_r V)$ is denoted $\delta N_\bullet^{\mathrm{cy}}(\Gamma; N_\bullet V)$. Consider the simplicial map

$$u\colon \delta N_\bullet^{\mathrm{cy}}(\Gamma, N_\bullet V) \to N_\bullet(V \rtimes \Gamma)$$

given on r-simplices by

$$\begin{aligned}&u(v_1, \dots, v_r, \gamma_1, \dots, \gamma_r) = \\ &\big((\gamma_1 \cdots \gamma_r v_1 \gamma_1, \gamma_1), (\gamma_2 \cdots \gamma_r v_2 \gamma_1 \gamma_2, \gamma_2), \dots, (\gamma_r v_r \gamma_1 \cdots \gamma_r, \gamma_r)\big).\end{aligned}$$

The map u can be understood as the composition of two maps: one starts with a rewriting of the left hand side, similar to the above f, and then rearranges factors upon using the semi-direct product. When Γ is a group u is a bijection. In general one has from [W2], lemma 2.3.1:

Lemma 2.1.10. *If Γ is a group-like monoid then u induces a weak homotopy equivalence of topological realizations.* □

The map u will be used in the next chapter for $\Gamma = \mathrm{GL}_k(R)$, $V = \mathrm{M}_k(V)$, where R is a unital ring and V is an R-bimodule. In this case the semi-direct product ring $V \rtimes R$ has

$$\mathrm{GL}_k(V \rtimes R) = \mathrm{M}_k(V) \rtimes \mathrm{GL}_k(R).$$

2.2 Simplicial spaces.

We have already in sect. 2.1 used simplicial sets and spaces. This will continue even more extensively in later sections, and it is in order to collect some of the relevant properties of simplicial sets and spaces.

Let us first point out that we use the word *space* to mean a based compactly generated Hausdorff space, and that all constructions are to be taken in this category.

A map $f\colon X \to Y$ is called k-connected if it induces an isomorphism on homotopy groups in degrees less than k and an epimorphism in degree k, i.e. if the homotopy fiber is $(k-1)$-connected. The convention is that every space is (-2)-connected, non-empty spaces are (-1)-connected and path connected spaces are 0-connected. It is an equivalence if it is k-connected for all k, and in general two spaces X and Y are called equivalent ($X \sim Y$) if they can be connected by a string of equivalences. In almost all cases to be considered, our spaces will have the homotopy type of CW complexes, and in this case $X \sim Y$ iff they are homotopy equivalent in the ordinary sense. The homotopy groups of a simplicial space (or set) $X_\bullet$ will mean the homotopy groups of the topological realization $\|X_\bullet\|$ below.

This is the bigger realization, sometimes called the fat realization, which only depends on the face operators in $X_\bullet$, i.e. on the functor

$$X_\bullet\colon \Delta_m^{\mathrm{op}} \to \{\text{spaces}\}$$

where $\Delta_m \subset \Delta$ is the subcategory of injective maps in Δ. Such a functor is called a Δ-space, [RS], and a presimplicial space in [DM2]. Its realization

$$\|X_\bullet\| = \coprod_{r=0}^{\infty} \Delta^r \times X_r/(d^i u, x) \sim (u, d_i x) \tag{2.2.1}$$

has $|X_\bullet|$ as a quotient when $X_\bullet$ is simplicial.

For simplicial sets,

$$\|X_\bullet\| \to |X_\bullet|$$

is an equivalence, but this is not always true for simplicial spaces.

A simplicial space $X_\bullet\colon \Delta^{\mathrm{op}} \to \{\text{spaces}\}$ is called "good" (or "proper"), [Se1] (or [May1]) if the inclusion of its degenerate simplices

$$\bigcup_{i=0}^{r-1} s_i(X_{r-1}) \subset X_r$$

is a cofibration (an NDR-pair). For such, the two realizations $|X_\bullet|$ and $\|X_\bullet\|$ are equivalent, cf. [Se1], appendix.

Any bisimplicial set (functor from $\Delta^{\mathrm{op}} \times \Delta^{\mathrm{op}}$ into sets) give rise to two "good" simplicial spaces

$$[r] \to |X_{r,\bullet}|, \quad [s] \to |X_{\bullet,s}|.$$

Their realizations are each homeomorphic to the realization of the diagonal simplicial set $\delta X_{\bullet,\bullet}$, and similarly for multisimplicial sets.

The homotopy fiber of a map $f\colon X \to Y$ of spaces with respect to $* \in Y$ is the space

$$\mathrm{hF}(f) = \{(x,\lambda) \in X \times Y^I \mid f(x) = \lambda(0), * = \lambda(1)\}$$

and there is a long exact sequence of homotopy groups

$$\cdots \to \pi_i X \to \pi_i Y \to \pi_{i-1}(\mathrm{hF}(f)) \to \pi_{i-1} X \to \cdots$$

so f is k-connected precisely if $\mathrm{hF}(f)$ is $(k-1)$-connected (for each choice of $*$).

Given a map of simplicial spaces, $f_\bullet\colon X_\bullet \to Y_\bullet$ and a base point $*_\bullet \in Y_\bullet$ there is a natural map

$$\big|[r] \to \mathrm{hF}(X_r \to Y_r)\big| \to \mathrm{hF}\big(|X_\bullet| \to |Y_\bullet|\big). \tag{2.2.2}$$

This is an equivalence if each Y_r is 0-connected, provided $X_\bullet$ and $Y_\bullet$ both are "good". In particular $|f_\bullet|$ is an equivalence when each $f_r\colon X_r \to Y_r$ is an equivalence. The associated fat realizations are equivalent without the goodness assumption.

The homotopy fiber and the dual notion of homotopy cofiber,

$$\mathrm{cof}(f) = (Y \times I) \amalg X / \langle f(x) \sim (y,1), * \sim (y,0)\rangle$$

are special cases of the homotopy limit and the homotopy colimit functor from a small category into spaces, cf. [BK], [G4].

Let us next consider function spaces between pointed spaces X and Y. Denote by $F(X,Y)$ the function space (in the compact open topology) of pointed maps from X to Y.

Suppose X is a pointed CW complex, e.g. the realization of a simplicial set, and that $\dim X \le n$. There is a natural map

$$\phi\colon \big|[r] \to F(X,Y_r)\big| \to F(X,|Y_\bullet|). \tag{2.2.3}$$

If $Y_\bullet$ is a "good" simplicial space and each Y_r is $(\dim X - 1)$-connected, then ϕ is an equivalence. In particular, the loop space of a "good" simplicial space $Y_\bullet$ with each Y_r 0-connected can be computed degreewise:

$$\Omega|Y_\bullet| \sim \big|[r] \to \Omega Y_r\big|$$

cf. [May1], sect. 12.

Given based simplicial sets there is a simplicial version $F_\bullet(X_\bullet, Y_\bullet)$ of the mapping space which we shall occasionally use. Its r-simplices consists of the simplicial maps

$$\delta(\Delta[r]_\bullet \times X_\bullet) \to Y_\bullet$$

which maps $\Delta[r]_\bullet \times *_\bullet$ to the base point of $Y_\bullet$. Here $\Delta[r]_\bullet$ is the simplicial r-simplex with $\Delta[r]_s = \Delta([s],[r])$. More generally, for each based $K_\bullet$,

$$\mathrm{Map}(K_\bullet, F_\bullet(X_\bullet, Y_\bullet)) = \mathrm{Map}(\delta(K_\bullet \wedge X_\bullet), Y_\bullet)$$

where Map denotes the set of based simplicial maps. In particular, we see for $Y_\bullet = \sin_\bullet Y$, the singular complex of Y, that

$$\sin_\bullet F(|X_\bullet|, Y) \cong F_\bullet(X_\bullet, \sin_\bullet Y)$$

(take $K_\bullet = \Delta[r]_\bullet$). Since $Y_\bullet \sim \sin_\bullet |Y_\bullet|$ when $Y_\bullet$ is fibrant (Kan complex) we see in this case that

$$F(|X_\bullet|, |Y_\bullet|) \sim |F_\bullet(X_\bullet, Y_\bullet)|.$$

Let us finally remind the reader that a simplicial group, $X_\bullet \colon \Delta^{\mathrm{op}} \to \{\text{groups}\}$, is always a Kan complex. For simplicial abelian groups $A_\bullet$ and $B_\bullet$, the function complex $\mathrm{s_\bullet Ab}(A_\bullet, B_\bullet)$ has the property that

$$\mathrm{s_\bullet Ab}(A_\bullet, B_\bullet) \cong \mathrm{s_\bullet Ab}\big(\delta A_\bullet \otimes \tilde{\mathbb{Z}}(S^n_\bullet), \delta A_\bullet \otimes \tilde{\mathbb{Z}}(S^n_\bullet)\big).$$

In particular, $\delta A_\bullet \otimes \tilde{\mathbb{Z}}(S^n_\bullet)$ is a deloop of $A_\bullet$. Here $\tilde{\mathbb{Z}}(S^n_\bullet)$ is the free abelian group of the simplicial n-sphere modulo the relations $\lambda *_\bullet = 0$ and $0 \cdot x = 0$. The reader is referred to [Q1] and [May2] for further details on simplicial sets.

Many constructions later in the paper are functors of fixed sets of the topological realizations of cyclic sets and spaces. Examples have already appeared in sect. 2.1. A map of cyclic sets (spaces) $f_\bullet \colon X_\bullet \to Y_\bullet$ induces an S^1-equivariant map. It is an S^1-equivariant homotopy equivalence if $f^C_\bullet \colon X^C_\bullet \to Y^C_\bullet$ induces an equivalence for all closed subgroups of S^1. This includes S^1 itself. But for some purposes of the paper, the S^1 fixed set is exceptional, and only the C-fixed sets for finite C matters. We therefore introduce the notions $X \sim_{C_\infty} Y$ resp. $X \sim_{C_{p^\infty}} Y$ to mean that X and Y can be connected with a sequence of S^1-maps which induce equivalences on all C_n fixed sets, resp. C_{p^n} fixed sets.

In the rest of the paper we shall tacitly assume that our simplicial spaces are "good". This will sometimes have to be verified, but we shall not go into such details below.

2.3 Topological Hochschild homology.

Given a unital ring A and an A-bimodule V we can form its *cyclic construction* $Z_\bullet(A;V)$. It is a simplicial abelian group with r-simplices

$$Z_r(A;V) = V \otimes A^{\otimes r} \tag{2.3.1}$$

and face and degeneracy operators

$$d_i(v \otimes a_1 \otimes \cdots \otimes a_r) = \begin{cases} va_1 \otimes a_2 \otimes \cdots \otimes a_r, & i = 0 \\ v \otimes \cdots \otimes a_i a_{i+1} \otimes \cdots \otimes a_r, & 0 < i < r \\ a_r v \otimes a_1 \otimes \cdots \otimes a_{r-1}, & i = r \end{cases}$$

$$s_i(v \otimes a_1 \otimes \cdots \otimes a_r) = v \otimes \cdots \otimes a_{i-1} \otimes 1 \otimes \cdots \otimes a_r, \quad 0 \le i \le r$$

cf. (2.1.1). When $V = A$ this becomes a cyclic set upon defining

$$t_r(a_0 \otimes \cdots \otimes a_r) = a_r \otimes a_0 \otimes \cdots \otimes a_{r-1}.$$

The topological realization of $Z_\bullet(A;V)$ is denoted $\mathrm{HH}(A;V)$ or when $V = A$ just $\mathrm{HH}(A)$. Its homotopy groups are the Hochschild homology groups,

$$\mathrm{HH}_i(A;V) = \pi_i \mathrm{HH}(A;V).$$

Indeed for any simplicial abelian group $Z_\bullet$ the homotopy groups of $|Z_\bullet|$ can be calculated as the homology of the associated chain complex Z_* with

$$d\colon Z_i \to Z_{i-1}, \quad dz = \sum_{\nu=0}^{i} (-1)^i d_\nu(z)$$

and $Z_*(A;V)$ is the standard Hochschild complex. The space $\mathrm{HH}(A)$ is the topological realization of a cyclic set, so comes equipped with a natural action of S^1, which keeps the base point invariant. Hence it gives a map

$$A\colon S^1_+ \wedge \mathrm{HH}(A) \to \mathrm{HH}(A)$$

which is the identity on the subspace $\mathrm{HH}(A)$. Exterior product with the generator $[S^1] \in \pi_1 S^1$ induces a map from $\mathrm{HH}_r(A)$ to $\mathrm{HH}_{r+1}(A)$. This is Connes' B-operator, cf. [H1].

T. Goodwillie suggested a decade ago to define the topological Hochschild homology analogously by replacing A with the Eilenberg-MacLane spectrum it generates, and the tensor product by smash product of spectra. Some care is needed in order to make these substitutions because smash products of spectra are not easily made strictly associative. M. Bökstedt in [B1] got around this difficulty in a way we now describe; see also [Br], appendix.

Let Top_* denote the category of based spaces and continuous maps, and let $L\colon \mathrm{Top}_* \to \mathrm{Top}_*$ be a continuous functor such that $L(*) = *$, that is, the function

$$F(X,Y) \xrightarrow{L} F(L(X), L(Y))$$

is continuous and maps the constant map to the constant map. Given $X, Y \in \mathrm{Top}_*$, we have maps

$$L(Y) \to L(X \wedge Y)$$

for each $x \in X$, induced from the corresponding inclusion of Y in $X \wedge Y$, so altogether a function

$$\sigma_{X,Y}\colon X \wedge L(Y) \to L(X \wedge Y)$$

and the assumptions on L implies that this is continuous; σ is called the *assembly map.*

Definition 2.3.2. (Bökstedt). A functor with smash product (FSP) is a functor L with an assembly map together with natural transformations

$$1_X\colon X \to L(X)$$
$$\mu_{X,Y}\colon L(X) \wedge L(Y) \to L(X \wedge Y)$$

such that

(i) $\mu_{X,Y} \circ (1_X \wedge \mathrm{id}_{L(Y)}) = \sigma_{X,Y}$

$\mu_{X,Y} \circ (\mathrm{id}_{L(X)} \wedge 1_Y) = L(\pi) \circ \sigma_{Y,X} \circ \pi$

(ii) $\mu_{X \wedge Y, Z} \circ (\mu_{X,Y} \wedge \mathrm{id}_{L(Z)}) = \mu_{X, Y \wedge Z} \circ (\mathrm{id}_{L(X)} \wedge \mu_{Y,Z})$

where π switches factors.

The FSP is called 0-connected if it maps n-connected spaces into n-connected spaces and if

$$\sigma_X\colon S^1 \wedge L(X) \to L(S^1) \wedge L(X) \to L(S^1 \wedge X)$$

is $2n - c$ connected whenever X is n-connected (c independent of n).

Any unital ring A induces a 0-connected FSP which we denote $\tilde{A}$. It takes a based space X into the Dold-Thom construction: the configuration space of particles in X with labels in A:

$$\tilde{A}(X) = \{\Sigma a_i x_i \mid x_i \in X, a_i \in A\}/\langle a \cdot * = 0, 0 \cdot x = *\rangle. \tag{2.3.3}$$

It is a 0-connected FSP and $\tilde{A}(S^n)$ is the Eilenberg-MacLane space of type (A, n) as

$$\pi_i \tilde{A}(X) = \tilde{H}_i(X; A).$$

A topological monoid G induces a 0-connected FSP $\tilde{G}$, namely

$$\tilde{G}(X) = X \wedge G_+ \tag{2.3.4}$$

with the obvious 1_X and $\mu_{X,Y}$.

Definition 2.3.5. A functor with smash product is called commutative if

$$\mu_{X,Y} \circ \pi = L(\pi) \circ \mu_{Y,X}$$

where π switches factors.

The FSP's $\tilde{A}$ and $\tilde{G}$ above are commutative when A and G are commutative. Let I denote the category of finite sets and injective maps. Its objects are the sets $\mathbf{n} = \{1, \dots, n\}$ with $\mathbf{0} = \emptyset$. A morphism $f \in I(\mathbf{n}, \mathbf{m})$ can be written as $\sigma \circ i$ where $\sigma \in \Sigma_m$ and i is the standard inclusion. The Cartesian products I^{r+1} form a cyclic category in that there are structure maps

$$d_i \colon I^{r+1} \to I^r, \quad s_i \colon I^r \to I^{r+1}, \quad t_r \colon I^r \to I^r$$

given by

$$d_i(x_0, \dots, x_r) = \begin{cases} (x_0, \dots, x_i \amalg x_{i+1}, \dots, x_r) & 0 \le i < n \\ (x_n \amalg x_0, x_1, \dots, x_r) & i = n \end{cases}$$
$$s_i(x_0, \dots, x_r) = (x_0, \dots, x_{i-1}, 0, x_i, \dots, x_r)$$
$$t_i(x_0, \dots, x_r) = (x_r, x_0, \dots, x_{r-1}).$$

For $x \in I$ we let S^x be the one point compactification of $\mathbb{R}^x$. For a based space X, consider the functor

$$G_r^X(L) \colon I^{r+1} \to \{\text{spaces}\}$$

given by

$$G_r^X(L, x_0, \dots, x_r) = F(S^{x_0} \wedge \dots \wedge S^{x_r}, L(S^{x_0}) \wedge \dots \wedge L(S^{x_r}) \wedge X),$$

where F denotes the pointed function space. Using the properties of L we find maps

$$d_i \colon G_r^X(L, x_0, \dots, x_r) \to G_{r-1}^X(L, d_i(x_0, \dots, x_r))$$
$$s_i \colon G_{r-1}^X(L, x_0, \dots, x_{r-1}) \to G_r^X(L, s_i(x_0, \dots, x_{r-1}))$$
$$t_i \colon G_r^X(L, x_0, \dots, x_r) \to G_r^X(L, t_i(x_0, \dots, x_r))$$

similar to the maps of sect. 2.1, and we can define a cyclic space $\mathrm{THH}^X_\bullet(L)$ by setting

$$\mathrm{THH}^X_r(L) = \underset{I^{r+1}}{\underrightarrow{\mathrm{holim}}} G^X_r(L). \tag{2.3.6}$$

The realization of $\mathrm{THH}^X_\bullet(L)$ is denoted $\mathrm{THH}^X(L)$; if $X = S^0$ we just write $\mathrm{THH}(L)$.

Lemma 2.3.7. ([B1]) *For a 0-connected* FSP L *and given integer* i,

$$\pi_i \mathrm{THH}^X_r(L) = \pi_i G^X_r(L, x_0, \dots, x_r)$$

provided $x_0, \dots, x_r$ *are sufficiently large.*

Proof. Here is Bökstedt's argument. The category I of finite sets and injective maps has the following structure:

an associative product $\mu\colon I \times I \to I$ (1)

natural transformations between μ and the two projections (2)

a decreasing filtration $F_i I$ with $\mu(F_i I, F_j I) \subset F_{i+j} I$ (3)

Indeed, $\mu(n, m) = n + m$ and $F_i I = \{n \mid n \geq i\}$. Such a category is called a good limit category. These are preserved under Cartesian product, so I^{r+1} is also a good index category.

For $\mathbf{x} = (x_0, \dots, x_r)$, write $G_r(\mathbf{x})$ instead of $G^X_r(L, \mathbf{x})$. To each $\lambda \geq 0$ there exists an i so that $G_r(\mathbf{x}) \to G_r(\mathbf{y})$ is λ connected for each $\mathbf{x} \to \mathbf{y}$ in $F_i = F_i I^{r+1}$. Now it suffices to prove that the following two maps are λ-connected:

$$\underset{\mathbf{x} \in F_i}{\underrightarrow{\mathrm{holim}}}\, G_r(\mathbf{x}) \to \underset{\mathbf{x} \in I^{r+1}}{\underrightarrow{\mathrm{holim}}}\, G_r(\mathbf{x}) \tag{a}$$

$$G_r(\mathbf{y}) \to \underset{\mathbf{x} \in F_i}{\underrightarrow{\mathrm{holim}}}\, G_r(\mathbf{x}), \qquad \mathbf{y} \in F_i \tag{b}$$

The map in (a) is an equivalence; an inverse is induced from $\mu(\mathbf{y}, -)\colon I^{r+1} \to F_i$ for some fixed $\mathbf{y} \in F_i$. This uses property (2) above. To show that (b) is a λ-equivalence, one first argues that the space BF_i (=realization of the nerve) is contractible. Indeed, μ induces a product on BF_i, and by (2) it has a homotopy unit. Condition (2) also yields that $\pi_0 BF_i = 0$. But a connected H-space has a homotopy antipode:

$$-\mathrm{Id}\colon BF_i \to BF_i,$$

with

$$BF_i \xrightarrow{\Delta} BF_i \times BF_i \xrightarrow{-\mathrm{Id}\times\mathrm{Id}} BF_i \times BF_i \xrightarrow{\mu} BF_i$$

homotopy to a constant. Since by (2),

$$\mu \sim pr_2 \colon BF_i \times BF_i \to BF_i$$

we conclude that BF_i is contractible. Finally the projection

$$p \colon \underset{\mathbf{x}\in F_i}{\underrightarrow{\mathrm{holim}}}\, G_r(\mathbf{x}) \to BF_i$$

is a λ-quasifibration in the sense that $G_r(\mathbf{y}) = p^{-1}(\mathbf{y})$ is λ-equivalent to the homotopy fiber. This follows from the last lemma in sect. 1 of [Q2] upon passing to the λ-coskeleton of $G_r(\mathbf{x})$. □

An FSP L induces a ring (pre)spectrum L^S whose n'th term is $L(S^n)$. We note that $\mathrm{THH}^X(L)$ only depends on L^S in the sense that if $L_1 \to L_2$ is a map of FSP's so that $L_1^S \to L_2^S$ is a homotopy equivalence of (pre)spectra then $\mathrm{THH}^X(L_1) \sim \mathrm{THH}^X(L_2)$.

A 0-connected FSP L gives rise to a ring $\pi_0 L$ by linearization, namely

$$\pi_0 L = \underset{n}{\underrightarrow{\lim}}\, \pi_n(L(S^n))$$

($\pi_0 \tilde{A} = A$), and the map

$$G_r(L, x_0, \dots, x_r) \to \pi_0 G_r(L, x_0, \dots, x_r)$$

induces a map $\mathrm{THH}(L) \to \mathrm{HH}(\pi_0 L)$ since

$$\begin{aligned}\pi_{x_0+\cdots+x_r} G(L; x_0, \dots, x_r) &= \tilde{H}_{x_0+\cdots+x_r}(L(S^{x_0}\wedge) \cdots \wedge L(S^{x_r})) \\ &= \pi_0(L) \otimes \cdots \otimes \pi_0(L).\end{aligned}$$

As $\mathrm{THH}^X_\bullet(L)$ is a cyclic space, the realization $\mathrm{THH}^X(L)$ inherits a continuous action of S^1, sect. 2.1, which will be of fundamental importance later in the paper.

There are a number of variations of the construction. First, we may define $\mathrm{THH}(L, M)$ when M is an L-bimodule. This is a functor from pointed spaces to itself with an assembly map

$$\sigma_{X,Y} \colon X \wedge M(Y) \to M(X \wedge Y)$$

and structure maps

$$\begin{aligned} l_{X,Y} &\colon L(X) \wedge M(Y) \to M(X \wedge Y) \\ r_{X,Y} &\colon M(X) \wedge L(Y) \to M(X \wedge Y),\end{aligned}$$

satisfying the obvious compatibility relations which we leave for the reader to explicate. One defines

$$\mathrm{THH}_r^X(L;M) = \underset{\overrightarrow{I^{r+1}}}{\mathrm{holim}}\, F\big(S^{x_0}\wedge\cdots\wedge S^{x_r}, M(S^{x_0})\wedge L(S^{x_1})\wedge\cdots\wedge L(S^{x_r})\wedge X\big)$$

and gets a simplicial space with realization $\mathrm{THH}^X(L;M)$ and a linearization map

$$\mathrm{THH}(L;M) \to \mathrm{HH}(\pi_0 L, \pi_0 M). \tag{2.3.8}$$

This is a rational equivalence when $L = \tilde{A}$, $M = \tilde{V}$.

Second, we may vary the concept of FSP to the simplicial setting and consider simplicial endo-functors of based simplicial sets

$$L_\bullet\colon \mathrm{s_\bullet sets_*} \to \mathrm{s_\bullet sets_*}$$

with properties analogous to the ones given in definition 2.3.2. In this setting the FSP $\tilde{A}$ associated to a ring A is simply

$$\tilde{A}(X_\bullet) = A[X_\bullet]/A\cdot *_\bullet = 0$$

where $A[X_\bullet]$ denotes the simplicial abelian group whose k-simplices is the free A-module with basis X_k. One defines $\mathrm{THH}_\bullet(L_\bullet)$ by using the simplicial function space, assuming $L_\bullet(S^n_\bullet)$ be fibrant, or one can follow $L_\bullet$ by realization, and use the above construction.

Third, there is a variation of $\mathrm{THH}(L)$ which defines $\mathrm{THH}(\mathcal{C})$ of an additive category, cf. [DM2]. The definition is as follows. Consider the functor

$$\tilde{\mathcal{C}}_\bullet\colon \mathrm{s_\bullet sets_*} \to (\mathrm{s_\bullet Ab})^{\mathcal{C}^{\mathrm{op}}\times\mathcal{C}}$$

which to a simplicial set $X_\bullet$ associates the functor from $\mathcal{C}^{\mathrm{op}}\times\mathcal{C}$ to $\mathrm{s_\bullet Ab}$

$$\tilde{\mathcal{C}}_\bullet(X_\bullet)(c_1,c_2) = \mathrm{Hom}_{\mathcal{C}}(c_1,c_2)\otimes\tilde{\mathbb{Z}}(X_\bullet)$$

Write $\mathbf{x} = (x_0,\dots,x_r)$,

$$V_\bullet(\mathcal{C},\mathbf{x}) = \bigvee_{(c_0,\dots,c_r)\in\mathcal{C}^{r+1}} \tilde{\mathcal{C}}_\bullet(S^{x_0}_\bullet)(c_0,c_r)\wedge\cdots\wedge\tilde{\mathcal{C}}_\bullet(S^{x_r})(c_r,c_{r-1}),$$

and

$$G_r(\mathcal{C},\mathbf{x}) = F_\bullet\big(S^{x_0}_\bullet\wedge\cdots\wedge S^{x_r}_\bullet, V_\bullet(\mathcal{C},\mathbf{x})\big).$$

Here $S^{x_0}_\bullet$ is, say the x_0-fold smash product of the simplicial circle $S^1_\bullet = \Delta_\bullet[1]/\partial\Delta_\bullet[1]$, and $F_\bullet$ is the simplicial mapping space. Then, as before,

$$\mathrm{THH}_\bullet(\mathcal{C}) = \underset{\overrightarrow{I^{\bullet+1}}}{\mathrm{holim}}\, G_\bullet(\mathcal{C};\mathbf{x}) \tag{2.3.9}$$

with realization THH($\mathcal{C}$).

If $\mathcal{C}$ is the category of free A-modules of a given rank n, then THH($\mathcal{C}$) is obviously equal to THH($\mathrm{M}_n(A)$). By Morita invariance, cf. proposition 2.6.5 below, this is equivalent to THH(A). If $\mathcal{C} = \mathcal{F}_A$, the category of free A-modules, then the k-simplices of THH($\mathcal{C}$) consists of matrices of varying sizes. By adding zeros in a suitable way to get them to have equal size, one does not change anything up to homotopy, so THH($\mathcal{C}$) $\sim$ THH(A) also in this case. (The reader can easily supply the argument by constructing the required homotopies, step by step). More generally, if $\mathcal{C} = \mathcal{P}_A$ is the category of finitely generated projective modules, then by adding complements to modules and zero homomorphisms, one gets (as pointed out in [DM2]):

Lemma 2.3.10. *For the category of finitely generated projective modules* THH($\mathcal{P}_A$) $\sim$ THH(A). □

The construction $\mathrm{THH}_\bullet(\mathcal{C})$ is clearly a cyclic set, so THH($\mathcal{C}$) has an S^1-action. The equivalence in the above lemma is actually a C_∞ equivalence. This can be seen upon using subdivisions and lemmas 3.10–12 of [BHM].

In general one may associate to $\mathcal{C}$ the simplicial FSP:

$$L_{\mathcal{C}}(X) = \prod_{c_1 \in \mathcal{C}} \bigvee_{c_2 \in \mathcal{C}} \tilde{\mathcal{C}}_\bullet(X)(c_1, c_2).$$

Then THH($\mathcal{C}$) $\simeq$ THH($L_{\mathcal{C}}$) cf. [DM], lemma 1.6.22, so (2.3.9) is not really a generalization. It is however a *very* convenient formulation, as we shall see in the next chapter, and $L_{\mathcal{C}}(X)$ is not functorial in $\mathcal{C}$.

Remark 2.3.11. The ring (pre)spectrum L^S associated to an FSP is very special: it has a strictly associative multiplication, and for commutative L it is strictly commutative. Most of the (pre)spectra which otherwise appear in algebraic topology do not have such a "strict" structure—they are merely "homotopy everything associative" (A_∞-spectra) or "homotopy everything commutative" (E_∞-spectra). Recently Elmendorf, Kriz, Mandell and May have recast the category of A_∞ and E_∞-spectra into what they call $\mathcal{L}$-rings and commutative $\mathcal{L}$-rings, [EKMM]. Such an animal E has an associative product $\mu_{\mathcal{L}}\colon E \wedge_{\mathcal{L}} E \to E$. There is no (strict) unit for $\mu_{\mathcal{L}}$, but one may still define THH(E) by imitating the algebraic construction $Z_\bullet(A)$ of (2.3.1), forgetting the degeneracy operators, cf. (2.2.1). ([EKMM] also introduces S-rings and product $\mu_S\colon E \wedge_S E \to E$ with a unit, and show that the two categories are equivalent, so for S-rings one has $\mathrm{THH}_\bullet(E)$ with degeneracies). More importantly for this paper, Jeff Smith has pointed out that each $\mathcal{L}$-ring

E gives rise to an FSP. Thus FSP's are rich in supply also from the point of view of A_∞ and E_∞-spectra.

In the rest of the paper all FSP's will be assumed to be 0-connected.

2.4 Cyclotomic spectra.

This section constructs from THH(L) an equivariant S^1-spectrum with extra structure, a socalled cyclotomic spectrum.

Let G be a compact Lie group. For any finite dimensional G-representation space V we write S^V for its one point compactification, and if X is a G-space, $\Omega^V X$ for the (based) mapping space $F(S^V, X)$ with its conjugate G-action.

Roughly speaking a G-spectrum T is a G-space T with a specific delooping $T(V)$ for each G-representation, so that T and $\Omega^V T(V)$ are G-equivalent (or even G-homeomorphic). However, due to the many G-automorphisms of V, some real care is needed to make consistent definitions. (For example, the signs which show up for spectra when $G = 1$ blow up to become elements in the Burnside ring of G).

We shall here follow the approach to G-spectra given in [LMS], and we give a brief account before introducing the concept of cyclotomic spectra. Let G be a compact Lie group and $\mathcal{U}$ a "complete G-universe", i.e. an infinite dimensional G-vector space with a G-invariant inner product which contains each finite dimensional representation of G.

A G-*prespectrum* indexed on $\mathcal{U}$ is a collection of G-spaces $t(V)$, one for each finite dimensional G-space $V \subset \mathcal{U}$ together with a transitive system of G-maps

$$\sigma\colon t(V) \to \Omega^{W-V} t(W)$$

Here $W - V$ denotes the orthogonal complement of V in W. It is a G-*spectrum* if the structure maps σ are all homeomorphisms. A map $f\colon t \to t'$ of G-prespectra consists of G-maps $f(V)\colon t(V) \to t'(V)$ which commute strictly with the structure maps. The category of G-prespectra indexed on $\mathcal{U}$ is denoted $G\mathcal{PU}$ and $G\mathcal{SU}$ denotes the full subcategory of G-spectra. The forgetful functor $l\colon G\mathcal{SU} \to G\mathcal{PU}$ has a left adjoint $\mathbf{L}$. It is given by the colimit over the structure maps

$$\mathbf{L}t(V) \cong \varinjlim_{W \subset \mathcal{U}} \Omega^{W-V} t(W),$$

provided that each σ is an inclusion, i.e. induces a homeomorphism onto its image. (This can always be arranged by thickening up t, to such a prespectrum t^τ, cf. [HM], appendix A).

Suppose C is a closed subgroup in G with quotient J and $T \in G\mathcal{SU}$. There are two possible notions of an associated fixed point spectrum in $J\mathcal{SU}^C$, in [A], [LMS] denoted T^C and $\Phi^C T$ respectively. Their V'th spaces are

$$T^C(V) = T(V)^C, \quad \Phi^C T(V) = \varinjlim_{W \subset \mathcal{U}} \Omega^{W^C - V} T(W)^C, \quad V = V^C \qquad (2.4.1)$$

and the structure maps are the evident ones. Since $T(V) \cong \Omega^{W-V} T(W)$ when $V \subset W$ the replacement of a C-equivariant map from S^{W-V} to $T(W)$ with its induced map on C-fixed sets induces a map $s_C \colon T^C \to \Phi^C T$. If $T = \mathbf{L}t$, then $\Phi^C T(V) = \lim \Omega^{W^C - V} t(W)^C$, see e.g. [HM], lemma 1.1.

In the case $G = 1$, the concept of prespectra differ from the usual one in that it is indexed on finite dimensional vector spaces, rather than on just the positive integers n (or $\mathbb{R}^n$). But the two categories are equivalent; the relationship is similar to the relation between a category and its skeleton category. The category of spectra is similar to what used to be called Ω-spectra, where one just demanded that σ be a homotopy equivalence. The functor $T \mapsto \mathbf{L}T^t$ brings us from Ω-spectra to spectra.

We need a few further results. It can all be found in [LMS], chap. 1–2, but the reader which is not accustomed with spectra should first consult [A] to get oriented in the subject.

Let $G \subset H$ be a closed subgroup. There is a pair of adjoint functors

$$i^* \colon G\mathcal{SU} \to G\mathcal{SU}^H, \qquad i_* \colon G\mathcal{SU}^H \to G\mathcal{SU}$$

with i^* the obvious restriction, and

$$i_*(T)(W) = \varinjlim \Omega^{V-W}\big(T(V^H) \wedge S^{V-V^H}\big).$$

Here V runs over the finite dimensional G subspaces which contain W. Given a based G-space X, $\Sigma^\infty(X) \in G\mathcal{SU}^G$ denotes its suspension spectrum, i.e. the spectrum associated with the prespectrum $V \mapsto S^V \wedge X$ for $V \subset \mathcal{U}^G$. Then $i_*(\Sigma^\infty X) = \Sigma^\infty_G X$ is the corresponding equivariant spectrum in $G\mathcal{SU}$.

Maybe the most important construction in the category of spectra is the transfer. Given $T \in G\mathcal{SU}$ and a free G-space E, the transfer is a map

$$\tau \colon j^* T \wedge_G E_+ \to j^*(\Sigma^{-\mathrm{Ad}(G)} T \wedge E_+)^G.$$

Here $j \colon \mathcal{U}^G \to \mathcal{U}$, $\mathrm{Ad}(G)$ the adjoint representation and $\Sigma^{-A} T$ is the function spectrum $F(S^A, T)$ or the equivalently internal delooping $(\Sigma^{-A}T)(V) = T(A \oplus V)$. It follows from [LMS], theorem II.7.1 that τ is a homotopy equivalence. Indeed, II.7.1 proves the result when $T = j_* T_0$, $T_0 \in G\mathcal{SU}^G$. The general case follows because the natural G-map

$$j_* j^* T \wedge E_+ \to T \wedge E_+$$

is a non-equivariant homotopy equivalence, hence as E is G-free a G-equivariant one.

The second result we need is that "induction" and "coinduction" agree, cf. [LMS], theorem II.6.2. Let $T \in GSU^H$, and let $L = T_{\{H\}}(G/H)$ be the tangent space at the base point $\{H\}$, with its H-action. There is a G-equivalence

$$\omega\colon F(G_+, \Sigma^L T)^H \xrightarrow{\sim} G/H_+ \wedge T.$$

See also [HM], sect. 7. In our applications $G = S^1$ or is finite, and $H \subset G$ is finite. In this case we get the equivalences

$$\begin{aligned} \tau\colon \Sigma^A j^*T \wedge_G E_+ &\xrightarrow{\sim} j^*(T \wedge E_+)^G, \\ \omega\colon \Sigma^A F(G/H_+, T) &\xrightarrow{\sim} G/H_+ \wedge T \end{aligned} \tag{2.4.2}$$

with $A = \mathbb{R}$ if $G = S^1$, and $A = 0$ if G is finite.

The smash products above are to be taken in the category of spectra: if X is a G-space and t a G-prespectrum then $t \wedge X_+$ is the prespectrum whose V'th term is $t(V) \wedge X_+$. If T is a G-spectrum then $T \wedge X_+ := \mathbf{L}(X_+ \wedge lT)$. It is worth pointing out that

$$\Phi^C(T \wedge X_+) \sim_G \Phi^C T \wedge X_+^C.$$

This follows from the equivalence $\Phi^C T \sim_G \Phi^C t$, mentioned above.

We will now fix G to be the circle group S^1. Write $\mathbb{C}(n)$ for the one-dimensional representation where $z \in S^1$ acts as multiplication with z^n, and take

$$\mathcal{U} = \bigoplus_{n \in \mathbf{Z}, \alpha \in \mathbf{N}} \mathbb{C}(n)_\alpha.$$

If $C \subset S^1$ is cyclic of order c then $\mathcal{U}^C \subset \mathcal{U}$ is precisely the summands with $n \in c\mathbb{Z}$.

Next we consider the homotopy fiber of $s_C\colon T^C \to \Phi^C T$ when C is a cyclic p-group. Let $j\colon \mathcal{U}^G \to \mathcal{U}^C$ be the inclusion of the G-trivial universe and let D be a J-spectrum. We call j^*T with its J-action forgotten for the underlying non-equivariant spectrum D. The following is proved in [HM], sect. 1 or in [BHM]:

Proposition 2.4.3. *Suppose C is a cyclic p-group. For any S^1-spectrum T there is a cofibration sequence of non-equivariant spectra*

$$T_{hC} \xrightarrow{N} T^C \xrightarrow{s_{C_p}^{C/C_p}} (\Phi^{C_p} T)^{C/C_p}.$$

Here $T_{hC} = EC_+ \wedge_C j^*T$ is the homotopy orbit spectrum. □

The cofibration sequence of proposition 2.4.3 is the C fixed point of $T \wedge (EC_+ \to S^0 \to \widetilde{EC})$. One identifies the terms by use of (2.4.2) and the easy fact that $(T \wedge \widetilde{EC})^{C_p} \xrightarrow{\simeq} \Phi^{C_p} T$, cf. lemma 4.1.2 below.

The circle $G = S^1$ has the nice property that any S^1/C-space X can be viewed as an S^1-space by identifying S^1 with S^1/C via the $|C|$'th root map $\rho_C \colon S^1 \xrightarrow{\cong} S^1/C$. We call this S^1-space $\rho_C^*(X)$. We can also use ρ_C to view S^1/C-spectra as S^1-spectra. Indeed, given an S^1/C-spectrum D indexed on $\mathcal{U}^C$ we have the S^1-spectrum $\rho_C^\# D$ indexed on $\rho_C^* \mathcal{U}^C$, with

$$\rho_C^\# D(V) = \rho_C^* D((\rho_C^{-1})^*(V)).$$

In our case

$$\rho_C^* \mathcal{U}^C = \bigoplus_{\alpha \in \mathbf{N}, n \in c\mathbf{Z}} \mathbb{C}(n/c) = \mathcal{U}$$

so $\rho_C^\# D$ becomes an S^1-spectrum, again indexed on $\mathcal{U}$.

Definition 2.4.4. A *cyclotomic spectrum* is an S^1-spectrum indexed on $\mathcal{U}$ together with an S^1-equivalence

$$r_C \colon \rho_C^\# \Phi^C T \to T$$

for every finite $C \subset S^1$, such that for any pair of finite subgroups the diagram

$$\begin{array}{ccc} \rho_{C_r}^\# \Phi^{C_r} \rho_{C_s}^\# \Phi^{C_s} T & = & \rho_{C_{rs}}^\# \Phi^{C_{rs}} T \\ \downarrow{\scriptstyle \rho_{C_r}^\# \Phi^{C_r} r_{C_s}} & & \downarrow{\scriptstyle r_{C_{rs}}} \\ \rho_{C_r}^\# \Phi^{C_r} T & \xrightarrow{r_{C_r}} & T \end{array}$$

commutes.

The cyclotomic condition is analogous to the property of free loop spaces: $(\Lambda X)^C \cong \Lambda X$, and indeed the S^1-suspension spectrum $\Sigma^\infty_{S^1}(\Lambda X)$ is easily seen to be cyclotomic.

More generally, $\mathrm{THH}^X(L)$ induces a cyclotomic spectrum for every FSP. We proceed to explain this. Let us write $\mathrm{THH}(L; V)$ instead of $\mathrm{THH}^{S^V}(L)$. It is the realization of a cyclic space, so gets an S^1-action from this structure. On the other hand, being a functor in V (or S^V) it has a second S^1-action, and altogether an $S^1 \times S^1$-action. We write $t(L)(V) = \mathrm{THH}(L; V)$ equipped with the diagonal S^1-action. This defines an S^1-prespectrum and we let $T(L)$ be the associated equivariant spectrum, $T(L) = \mathbf{L}t(L)$.

Actually, it is not very hard to see that the adjoint of the natural map

$$S^V \wedge \mathrm{THH}(L; W) \to \mathrm{THH}(L; V \oplus W)$$

is a C-equivalence for each finite subgroup of the circle, so that $T(V) \sim_C \mathrm{THH}(L; V)$, cf. [HM], proposition 1.4.

In order to describe the cyclotomic structure maps r_C we use the subdivision operator sd_C introduced in sect. 2.1. For a cyclic space $Z_\bullet$, $sd_C Z_\bullet$ has a simplicial C-action, and its realization $|sd_C Z_\bullet|$ an $\mathbb{R}/c\mathbb{Z}$ action which extends the $C = \mathbb{Z}/c\mathbb{Z}$ action. The homeomorphism

$$D\colon |sd_C Z_\bullet| \to |Z_\bullet|$$

becomes an S^1-map when $\mathbb{R}/c\mathbb{Z}$ is identified with $\mathbb{R}/\mathbb{Z} = S^1$ by division with c.

We now define a simplicial map

$$r'_C\colon sd_C \mathrm{THH}_\bullet(L; V)^C \to \mathrm{THH}_\bullet(L; V^C)$$

for each cyclic subgroup $C \subset S^1$. Let $c = |C|$. With the notation of sect. 2.3, the r-simplices of $sd_C\mathrm{THH}(L; V)$ is the homotopy colimit

$$sd_C \mathrm{THH}_r(L; V) = \underset{\mathbf{x} \in I^{(r+1)c}}{\underrightarrow{\mathrm{holim}}}\; G^V_{(r+1)c-1}(L; \mathbf{x})$$

where

$$G^V_s(L; \mathbf{x}) = F(S^{x_0} \wedge \cdots \wedge S^{x_s}, L(S^{x_0}) \wedge \cdots \wedge L(S^{x_s}) \wedge S^V).$$

The c-fold diagonal $\Delta_c\colon I^{r+1} \to I^{(r+1)c}$ gives a C-equivariant inclusion

$$\underset{I^{r+1}}{\underrightarrow{\mathrm{holim}}}\, G^V_{c(r+1)-1}(L) \circ \Delta_C \to \underset{I^{(r+1)c}}{\underrightarrow{\mathrm{holim}}}\, G^V_{c(r+1)-1}(L)$$

which induces a homeomorphism of C-fixed sets, and for $\mathbf{x} \in I^{r+1}$,

$$\begin{aligned} &G^V_{c(r+1)-1}(L)(\Delta_C(\mathbf{x})) \\ &\quad = F((S^{x_0})^{(c)} \wedge \cdots \wedge (S^{x_r})^{(c)}, L(S^{x_0})^{(c)} \wedge \cdots \wedge L(S^{x_r})^{(c)} \wedge S^V) \end{aligned}$$

where $Y^{(c)}$ is the c-fold smash power, and the action of C is by cyclic permutation of factors; then $(Y^{(c)})^C$ is the diagonal copy of Y.

The above formula is quite similar to the identification of $sd_C N^{\mathrm{cy}}_\bullet(G) = N^{\mathrm{cy}}_\bullet(E; G^c)$ explained in sect. 2.1, but this time there is no diagonal homeomorphism

$$\Delta_C\colon N^{\mathrm{cy}}_\bullet(G) \xrightarrow{\cong} (sd_C N^{\mathrm{cy}}_\bullet(G))^C.$$

Even in the linear case of $Z_\bullet(R)$ we do not have such a map since $\Delta(r) = r \otimes \cdots \otimes r$ is not linear. However there is a map in the other direction. Indeed, given any two pointed C-spaces Y_1 and Y_2 one has the obvious map

$$r'_C\colon F(Y_1, Y_2)^C \to F(Y_1^C, Y_2^C)$$

which restricts a C-equivariant map to the induced map on the C-fixed points. This gives a map

$$r'_C\colon G^V_{(r+1)c-1}(L,\Delta_C(\mathbf{x}))\to G^{V^C}_r(L,\mathbf{x}),\quad \mathbf{x}\in I^{r+1}$$

and induces a simplicial map

$$r'_C\colon sd_C\mathrm{THH}_\bullet(L;V)^C\to \mathrm{THH}_\bullet(L;V^C).$$

Taking realization and composing with the inverse of the homeomorphism D we have obtained

$$r_C\colon \mathrm{THH}(L;V)^C\to \mathrm{THH}(L;V^C).$$

This is S^1-equivariant, when one identifies the S^1/C-action in the domain with the S^1-action via ρ_C, so induces an S^1-map from

$$\rho_C^*\Phi^C t(L)(W)=\varinjlim_{V\subset\mathcal{U}}\Omega^{V^C-W}\mathrm{THH}(L;V)^C$$

into

$$T(L)(W)=\varinjlim_{V\in\mathcal{U}}\Omega^{V^C-W}\mathrm{THH}(L;V^C).$$

Since $\Phi^C T(L)=\Phi^C t(L)$, we do get a map

$$r_C\colon \rho_C^{\#}\Phi^C T(L)\to T(L)$$

of S^1-spectra. This is an S^1-equivalence by [HM], proposition 1.5, so we have

Theorem 2.4.5. *For every FSP the S^1-spectrum $T(L)$, induced from the prespectrum* $\mathrm{THH}(L;V)$*, is cyclotomic.* □

The essential point in this and the next chapter is the spectrum $T(L)$, but only considered as a spectrum in the usual sense equipped with an action of S^1. To separate out this, let me introduce the notation $\mathrm{TH}(L)$ for this weakened form,

$$\mathrm{TH}(L)=T(L)\mid\mathcal{U}^{S^1}=j^*T(L),\quad j\colon\mathcal{U}^{S^1}\to\mathcal{U}.$$

The reason for introducing the extra notation is to underline the fact that $\mathrm{TH}(L)\wedge E_+$ and $j^*(T(L)\wedge E_+)$, $j\colon\mathcal{U}^C\to\mathcal{U}$ are quite different. If for example E is S^1-free then $(\mathrm{TH}(L)\wedge E_+)^C\sim 0$ whereas $(T(L)\wedge E_+)^C\sim\mathrm{TH}(L)\wedge_C E_+$ by (2.4.2).

We shall continuously use the following special case of proposition 2.4.3; we call it the *fundamental cofibration sequence*

$$\mathrm{TH}(L)_{hC_{p^n}}\longrightarrow\mathrm{TH}(L)^{C_{p^n}}\xrightarrow{R_p}\mathrm{TH}(L)^{C_{p^{n-1}}}. \tag{2.4.6}$$

2.5 Cyclic homology of cyclotomic spectra.

Given any FSP we saw in the last section that the S^1-spectrum $T(L)$ associated to the prespectrum $\mathrm{THH}(L; S^V)$ comes equipped with an S^1-equivalence

$$r_C\colon \rho_C^{\#}\Phi^C T(L) \xrightarrow{\sim} T(L).$$

We now use this structure to define a new functor $\mathrm{TC}(L)$, the topological cyclic homology of L, initially defined in [BHM].

Let $\mathbb{I}$ be the category where objects are the natural numbers, $\mathrm{ob}\,\mathbb{I} = \{1, 2, 3, \dots\}$, and with two morphisms $R_r, F_r\colon n \to m$, whenever $n = rm$, subject to the relations

$$\begin{aligned} &R_1 = F_1 = \mathrm{id}_n \\ &R_r R_s = R_{rs}, \quad F_r F_s = F_{rs} \\ &R_r F_s = F_s R_r. \end{aligned} \tag{2.5.1}$$

For a prime p, we let $\mathbb{I}_p$ be the full subcategory with $ob\mathbb{I}_p = \{1, p, p^2, \dots\}$. A cyclotomic spectrum T defines a functor from $\mathbb{I}$ to the category of non-equivariant spectra. Indeed when $n = lm$ we have two commuting maps

$$R_l, F_l\colon T^{C_n} \to T^{C_m}.$$

Here T^{C_n} and T^{C_m} are considered as ordinary (non-equivariant) spectra. The map F_l, called the *Frobenius* map, is simply the inclusion of fixed points ($C_m \subset C_n$). The map R_l, called the *restriction* map, is the composite

$$R_l\colon T^{C_n} = (\rho_C^{\#} T^C)^{C_m} \xrightarrow{s_C} (\rho_C^{\#}\Phi^C T)^{C_m} \xrightarrow{r_C} T^{C_m}$$

where $C = C_l$ and $s_C\colon T^C \to \Phi^C T$ is the map from (2.4.10), and where r_C is the cyclotomic structure map.

Definition 2.5.2. If T is a cyclotomic spectrum, then

$$\mathrm{TC}(T;p) = \underset{\mathbb{I}_p}{\underset{\longleftarrow}{\mathrm{holim}}}\, T^{C_{p^s}}, \quad \mathrm{TC}(T) = \underset{\mathbb{I}}{\underset{\longleftarrow}{\mathrm{holim}}}\, T^{C_n}.$$

For a functor with smash product L, we write $\mathrm{TC}(L) = \mathrm{TC}(T(L))$ and similarly for $\mathrm{TC}(L;p)$.

The homotopy limit which defines $\mathrm{TC}(T;p)$ may be formed in two steps. First we can take the homotopy limit over F_p (resp. R_p). Since R_p and F_p commute, R_p (resp. F_p) induces a self-map of this homotopy limit, and we may take the homotopy fixed points. More precisely, let

$$\mathrm{TR}(T;p) = \underset{R_p}{\underset{\longleftarrow}{\mathrm{holim}}}\, T^{C_{p^s}}, \quad \mathrm{TF}(T;p) = \underset{F_p}{\underset{\longleftarrow}{\mathrm{holim}}}\, T^{C_{p^s}}. \tag{2.5.3}$$

Then F_p induces an endomorphism of $\mathrm{TR}(T;p)$ and R_p an endomorphism of $\mathrm{TF}(T;p)$, and

$$\mathrm{TC}(T;p) \cong \mathrm{TR}(T;p)^{h\langle F_p\rangle} \cong \mathrm{TF}(T;p)^{h\langle R_p\rangle}.$$

The homotopy inverse limit of a string of maps $\cdots \to X_n \to X_{n-1} \to \cdots$ is a homotopy equivalent to the categorical limit provided each map is a fibration. Here $\langle F_p\rangle$ is the free monoid on F_p and $X^{h\langle F_p\rangle}$ denotes the $\langle F_p\rangle$-homotopy fixed points of X, or in other words, the homotopy fiber of $\mathrm{id} - F_p$. This was the definition used for $\mathrm{TC}(T;p)$ in [BHM].

There is a similar description of $\mathrm{TC}(L)$. Let

$$\mathrm{TR}(T) = \operatorname*{holim}_{\overleftarrow{R}} T^{C_n}, \quad \mathrm{TF}(T) = \operatorname*{holim}_{\overleftarrow{F}} T^{C_n}, \tag{2.5.4}$$

then

$$\mathrm{TC}(T) = \mathrm{TR}(T)^{hF} = \mathrm{TF}(T)^{hR},$$

where hF denotes the homotopy fixed set of the multiplicative monoid of natural numbers acting of $\mathrm{TR}(T)$ through the maps F_s, $s \geq 1$. The inclusions $\{1\} \subset \mathbb{I}_p \subset \mathbb{I}$ induce maps

$$\mathrm{TC}(T) \to \mathrm{TC}(T,p) \to T.$$

The following theorem, basically due to Goodwillie, cf. [HM], sect. 3, tells us that $\mathrm{TC}(L)$ is not really a stronger functor than the collection $\mathrm{TC}(L,p)$ for all primes p.

Theorem 2.5.5. *The projections* $\mathrm{TC}(T) \to \mathrm{TC}(T;p)$ *induce an equivalence of* $\mathrm{TC}(T)$ *with the fiber product of the* $\mathrm{TC}(T;p)$*'s over* T. *Moreover, the functors agree after* p*-adic completion,* $\mathrm{TC}(T)^{\wedge}_p \simeq \mathrm{TC}(T;p)^{\wedge}_p$. □

Remark 2.5.6. T. Goodwillie has introduced the following alternative definition of $\mathrm{TC}(L)$ which has the advantage of allowing an integral description of Waldhausen's reduced A-theory, cf. [G5].

The fixed set $\mathrm{TH}(L)^{C_n}$ has the natural S^1/C_n-action so each $\rho^{\#}_{C_n}\mathrm{TH}(L)^{C_n}$ is a spectrum with an S^1-action (an S^1-spectrum indexed on $\mathcal{U}^{S^1}$). If $n = rm$ then

$$F_r \colon \rho^{\#}_{C_n}\mathrm{TH}(L)^{C_n} \to \rho^{\#}_{C_m}\mathrm{TH}(L)^{C_m}$$

satisfies

$$F_r(\theta^r x) = \theta F_r(x), \quad \theta \in S^1.$$

Let M be the semi-direct product

$$M = \{(r,\theta) \mid r \in \mathrm{N},\ \theta \in S^1,\ \theta r = r\theta^r\}.$$

It acts on

$$\mathrm{TR}(L) = \underset{R}{\operatorname{holim}}\, \rho^{\#}_{C_n} \mathrm{TH}(L)^{C_n}.$$

Goodwillie defines:

$$TC(L) = \mathrm{TR}(L)^{hM} = (\mathrm{TR}(L)^{hS^1})^{h\mathbb{N}},$$

and shows that its p-adic completion is equivalent to $\mathrm{TC}(L,p)^{\wedge}_p$.

In later chapters we shall be concerned with the calculation of $\mathrm{TC}(L)$ primarily for the FSP $\tilde{A}$ associated to a ring, cf. (2.3.3). In this case we write $T(A)$ and $\mathrm{TC}(A)$ etc. instead of $T(\tilde{A})$ and $\mathrm{TC}(\tilde{A})$.

Since $T(L)$ and its fixed points are (-1)-connected spectra, $\mathrm{TC}(L)$ is always (-2)-connected. In [HM], sect. 2 we calculated the component groups $\pi_0 T(A)^{C_{p^n}}$, and in particular:

Theorem 2.5.7. *For a commutative ring A, there is a natural isomorphism*

$$I\colon W(A,p) \to \pi_0 \mathrm{TR}(A,p)$$

where $W(A,p)$ denotes the p-typical Witt vectors. Moreover, the self map F on $\mathrm{TR}(A,p)$ *corresponds to the Frobenius map of Witt-vectors.*

It follows that we have the exact sequence

$$\mathrm{TC}_0(A,p) \longrightarrow W(A,p) \xrightarrow{1-F} W(A,p) \longrightarrow \mathrm{TC}_{-1}(A,p) \longrightarrow 0$$

for the two lowest dimensional homotopy groups of $\mathrm{TC}(A,p)$. The left hand arrow is often injective, but not always.

Addendum 2.5.8. For finite subgroups $H \subset K$ of the circle, there is a map $\Sigma^\infty_{S^1}(S^1/K_+) \to \Sigma^\infty_{S^1}(S^1/H_+)$, namely the Thom collapse map of an equivariant embedding $G/H \subset G/K \times V$. It induces a map of spectra

$$F(\Sigma^\infty_{S^1}(S^1/H_+), T(A)) \to F(\Sigma^\infty_{S^1}(S^1/K_+), T(A))^{S^1},$$

that is a map

$$V\colon T(A)^H \to T(A)^K,$$

well-defined up to homotopy. In particular we get

$$V\colon \pi_0 T(A)^{C_{p^n}} \to \pi_0 T(A)^{C_{p^{n+1}}}.$$

Theorem 2.5.7 extends to the statement that there is an isomorphism

$$I\colon \pi_0 T(A)^{C_{p^n}} \xrightarrow{\cong} W_n(A,p)$$

into the p-typical Witt vectors of length $n+1$ with $\pi_0 F$, $\pi_0 R$ and $\pi_0 V$ corresponding to Frobenius, Restriction and Verschiebung, cf. [HM], theorem 2.3.

We close with two remarks of homotopy theoretic nature.

Remark 2.5.9. Given an $\mathcal{L}$ or S-ring E (cf. 2.3.11), the direct construction THH(E) from [EKMM] is *not* cyclotomic. The price one pays for making the spectrum multiplication $\mu\colon E \wedge_S E \to E$ associative is that there are no "diagonal fixed points" under the cyclic group action on the S-smash powers, and this prevents the cyclotomic property. Passing to Jeff Smith's associated FSP $\tilde{E}$ is one way around this. There might be other ways.

Remark 2.5.10. For a commutative FSP L, one can iterate the construction TC(L) to obtain $\mathrm{TC}^{(n)}(L)$ for each $n \geq 1$, cf. [HM], sect. 3.6. In view of the calculational results of sect. 4 below it is an interesting challenge in homotopy theory to study $\mathrm{TC}^{(n)}(\mathbb{F}_p)$ and $\mathrm{TC}^{(n)}(\mathbb{Z}_p)$.

2.6 The cyclotomic trace.

We begin by defining the K-theory of an FSP. Given L we can consider the associated infinite loop space

$$QL = \varinjlim \Omega^x L(S^x).$$

The components

$$\pi_0 QL = \varinjlim \pi_x L(S^x)$$

is a ring, and we denote by $\widehat{\mathrm{GL}}_1(L) \subset QL$ the subspace of invertible components. This is by definition a group-like monoid.

We let $\mathrm{M}_n(L)$ be the FSP of $n \times n$ matrices over L defined as

$$\mathrm{M}_n(L)(X) = F(\mathbf{n}_+, \mathbf{n}_+ \wedge L(X)), \qquad \mathbf{n} = \{1, \ldots, n\}$$

and set $\widehat{\mathrm{GL}}_n(L) = \widehat{\mathrm{GL}}_1(\mathrm{M}_n(L))$, again a group-like monoid. Direct sum of matrices give maps

$$\widehat{\mathrm{GL}}_n(L) \times \widehat{\mathrm{GL}}_m(L) \to \widehat{\mathrm{GL}}_{n+m}(L)$$

which induces a monoid structure on the disjoint union of the $B\widehat{\mathrm{GL}}_n(L)$. Its group-completion is $K(L)$, that is:

$$K(L) = \Omega B\Big(\coprod_{n=0}^{\infty} B\widehat{\mathrm{GL}}_n(L)\Big) \sim B\widehat{\mathrm{GL}}_\infty(L)^+ \times \mathbb{Z}, \tag{2.6.1}$$

where the superscript + is Quillen's plus construction.

If A is a unital ring, and $\tilde{A}$ the associated FSP, cf. (2.3.3), then $\pi_0 Q(\tilde{A}) = A$, and the natural map $Q\tilde{A} \to A$ is an equivalence. It follows that

$$B\widehat{\mathrm{GL}}_n(\tilde{A}) \to B\widehat{\mathrm{GL}}_n(A)$$

is an equivalence, and in turn that

$$K(\tilde{A}) \to B\mathrm{GL}(A)^+ \times \mathbb{Z} = K(A)$$

is an equivalence. Thus $K(\tilde{A})$ is just another model for Quillen's $K(A)$-space (the version where $K_0(A) = \mathbb{Z}$, rather than the projective class group).

If L is the FSP $\tilde{G}$ of (2.3.4) associated to a topological group-like monoid G, then $K(\tilde{G})$ is a model for Waldhausen's $A(BG)$, again the version with $\pi_0 A(BG) = \mathbb{Z}$.

The space $K(L)$ is an infinite loop space, that is, it is the zero'th space of a connective spectrum which again will be denoted $K(L)$. The deloopings are not as concrete as the deloopings of $\mathrm{TH}(L)$ and $\mathrm{TC}(L)$ above. One has to use the abstract machinery of Segal's Γ-spaces or the equivalent machinery May's operads, or the original approach of Boardman-Vogt.

The *cyclotomic trace* from [BHM] is a spectrum map

$$\mathrm{trc}\colon K(L) \to \mathrm{TC}(L).$$

It is highly technical to construct, so I shall here only give a rough outline of the ideas involved to the extend it throws light on the definition of $\mathrm{TC}(L)$. The interested reader can consult the original source, and [HM], sect. 1.6 for the equivalence of the abstract Γ-space delooping of $\mathrm{TC}(L)$ and the concrete one above.

I begin by recalling K. Dennis' trace map in the linear situation,

$$\mathrm{Tr}\colon K(A) \to \mathrm{HH}(A) \tag{2.6.2}$$

Remember here that $\mathrm{HH}(A)$ denotes the topological realization of the standard cyclic abelian group $Z_\bullet(A)$. We proceed simplicially, and consider

$$N_\bullet \mathrm{GL}_n(A) \xrightarrow{I_\bullet} N_\bullet^{\mathrm{cy}}(\mathrm{GL}_n(A)) \xrightarrow{S_\bullet} Z_\bullet(\mathrm{M}_n(A)) \tag{2.6.3}$$

with

$$I(g_1, \cdots, g_r) = (g_0, g_1, \dots, g_r), \quad g_0 = (g_1 \dots g_r)^{-1}$$
$$S(g_0, \dots, g_r) = g_0 \otimes \cdots \otimes g_n.$$

We have the simplicial map

$$\begin{aligned} &\mathrm{Tr}_\bullet^{(n)} \colon Z_\bullet(\mathrm{M}_n(A)) \to Z_\bullet(A), \\ &\mathrm{Tr}_r^{(n)}(X_0 \otimes \cdots \otimes X_r) = \sum X_0(i_0, i_1) \otimes \cdots \otimes X_r(i_r, i_0). \end{aligned} \tag{2.6.4}$$

It induces a homotopy equivalence

$$\mathrm{Tr}^{(n)} \colon \mathrm{HH}(\mathrm{M}_n(A)) \xrightarrow{\sim} \mathrm{HH}(A). \quad \text{(Morita invariance)}$$

Indeed, if $i \colon A \to \mathrm{M}_n(A)$ is the inclusion which maps $a \in A$ into the matrix with a on the $(1, 1)$ entry and zero elsewhere, then the simplicial map

$$Z_\bullet(i) \colon Z_\bullet(A) \to Z_\bullet(\mathrm{M}_n(A))$$

induces a map from $\mathrm{HH}(A)$ to $\mathrm{HH}(\mathrm{M}_n(A))$ which is an inverse to $\mathrm{Tr}^{(n)}$. Consider the composition of (2.6.3) and (2.6.4):

$$\begin{aligned} &\mathrm{Tr}_\bullet \colon N_\bullet(\mathrm{GL}_n(A)) \to \mathbb{Z}_\bullet(A) \\ &\mathrm{Tr}_r = \mathrm{Tr}_r^{(n)} \circ S_r \circ I_r - s_0^{(r-1)}(n) \end{aligned}$$

where s_0 is the degeneracy operator in $Z_\bullet(A)$. It is easy to check that

$$\begin{array}{ccc} N_\bullet(\mathrm{GL}_n(A)) & \longrightarrow & N_\bullet(\mathrm{GL}_{n+1}(A)) \\ & {\scriptstyle \mathrm{Tr}_\bullet} \searrow \quad \swarrow {\scriptstyle \mathrm{Tr}_\bullet} & \\ & Z_\bullet(A) & \end{array}$$

is commutative, so the topological realization of $\mathrm{Tr}_\bullet$ induces the map in (2.6.2).

The above linear trace map can be generalized to give

$$\mathrm{tr} \colon K(L) \to \mathrm{THH}(L)$$

for each FSP, but two issues have to be addressed: $\widehat{\mathrm{GL}}_n(L)$ has no strict inverses and (2.6.3) does note make sense a priori in $\mathrm{THH}_\bullet(L)$.

There is a standard way to get around the lack of strict inverses by group completing the monoid, see below. For now we simply use (2.1.8):

$$|N^{\mathrm{cy}} B\widehat{\mathrm{GL}}_n(L)| \sim \Lambda B\widehat{\mathrm{GL}}_n(L)$$

and replace $|I_\bullet|$ by the inclusion of $B\widehat{\mathrm{GL}}_n(L)$ into the free loop space as the constant loops. The second map

$$S_\bullet\colon N^{\mathrm{cy}}_\bullet\widehat{\mathrm{GL}}_n(L) \to \mathrm{THH}_\bullet(L)$$

maps a string $(g_0, \dots, g_r)$ into the smash product $g_0 \wedge \cdots \wedge g_r \in \mathrm{THH}_r(L)$ upon thinking of each g_i as a limit of maps $S^{x_i} \to \mathrm{M}_n(L)(S^{x_i})$. Finally we have Morita invariance:

Recall our convention that two S^1-spaces are called C_∞-equivariant if they are connected by a string of S^1-maps which induce equivalences of C-fixed sets for every finite subgroup of S^1.

Proposition 2.6.5. *For every FSP there is a C_∞-equivalence*

$$\mathrm{THH}^X(\mathrm{M}_n(L)) \sim \mathrm{THH}^X(L)$$

which defines a C_∞-equivalence of the associated equivariant spectra $T(\mathrm{M}_n(L))$ and $T(L)$.

Proof. I briefly sketch a proof, modelled upon the linear case treated above. This approach is different from the one of [BHM]. Details can be found in a forthcoming paper by C. Schlichtkrull, [Sch]. See also [DM2]. We can rewrite

$$\mathrm{M}_n(L)(X) = \prod^n \bigvee^n L(X)$$

and have the subfunctor

$$W_n(L)(X) = \bigvee^n \bigvee^n L(X).$$

It is an "FSP without unit". We can restrict the simplicial space

$$\mathrm{THH}_\bullet(L;V)\colon \Delta^{\mathrm{op}} \to \text{spaces} \quad (\mathrm{THH}_\bullet(L;V) = \mathrm{THH}_\bullet^{S^V}(L))$$

to the subcategory of injective maps in Δ^{op}, i.e. we forget degeneracy operators and consider $\mathrm{THH}_\bullet(L;V)$ only as a Δ-space (presimplicial space) in the sense if [RS]. Then $\mathrm{THH}_\bullet(W_n(L);V)$ is defined, and the inclusion of Δ-sets

$$\mathrm{THH}_\bullet(W_n(L);V) \to \mathrm{THH}_\bullet(\mathrm{M}_n(L);V)$$

induces an equivalence upon applying the realization functor $\|\cdot\|$ of Δ-sets. On the other hand, the projection

$$\|\mathrm{THH}_\bullet(\mathrm{M}_n(L);V)\| \to |\mathrm{THH}_\bullet(\mathrm{M}_n(L);V)|$$

is a C-equivalence.

Second, suitable evaluation defines a map from $W_n(L)$ to L, analogous to the linear situation

$$\text{ev}\colon \text{Hom}_A(A^{\oplus n}, A) \otimes A^{\oplus n} \to A,$$

and we can imitate the map of (2.6.4) to get

$$\text{tr}^{(n)}\colon \text{THH}_\bullet(W_n(L); V) \to \text{THH}_\bullet(L; V),$$

This induces the required equivalence. □

The resulting trace map, valid for any FSP, tr: $K(L) \to \text{THH}(L)$, is Bökstedt's topological version of Dennis' trace map. It is far from obvious, however, that tr is a map of spectra. See the final paragraph of this section.

It is time to explain how to lift the topological Dennis trace into the fixed sets $\text{TH}(L)^C$ of the finite subgroups $C \subset S^1$. Suppose first that G is a (topological) group.

The simplicial map (cf. 2.1.7)

$$\delta_C\colon N_\bullet(G) \xrightarrow{I_\bullet} N_\bullet^{\text{cy}}(G) \xrightarrow{\Delta_C} sd_C N_\bullet^{\text{cy}}(G)^C$$

has topological realization homotopic to

$$\delta_c\colon BG \xrightarrow{I} \Lambda BG \xrightarrow{\Delta_c} (\Lambda BG)^C, \quad \Delta_c(\lambda(\theta)) = \lambda(\theta^c)$$

where $c = |C|$ and I is the inclusion into the constant loops.

For a subgroup $C_0 \subset C$, the composition of δ_c with the inclusion of $(\Lambda BG)^C$ into $(\Lambda BG)^{C_0}$ is equal to δ_{c_0} since Δ_c leaves constant loops invariant. On the simplicial level it is therefore not surprising that there is a natural homotopy between δ_{C_0} and the composition

$$|N_\bullet(G)| \xrightarrow{\delta_C} |sd_C N_\bullet^{\text{cy}}(G)|^C \xrightarrow{\text{incl}} |sd_C N_\bullet^{\text{cy}}(G)|^{C_0} \xrightarrow{D} |sd_{C_0} N_\bullet^{\text{cy}}(G)|^{C_0}$$

where D is the subdivision homeomorphism of lemma 2.1.6. Thus if we write

$$F_{C/C_0}\colon |sd_C N_\bullet^{\text{cy}}(G)|^C \longrightarrow |sd_{C_0} N_\bullet^{\text{cy}}(G)|^{C_0}$$
$$R_{C/C_0}\colon |sd_C N_\bullet^{\text{cy}}(G)|^C \xrightarrow{\cong} |sd_{C_0} N_\bullet^{\text{cy}}(G)|^{C_0}, \quad R_{C/C_0} = \Delta_{C/C_0}^{-1}$$

we have

$$F_{C/C_0} \circ \delta_C \sim \delta_{C_0}, \quad R_{C/C_0} \circ \delta_C = \delta_{C_0} \tag{2.6.6}$$

with a specified homotopy in the first relation. There results a diagram

$$\begin{array}{ccc} & & \varprojlim_R |sd_C N^{cy}_\bullet(G)|^C \\ & \overset{\delta_C}{\nearrow} & \\ |N_\bullet(G)| & & \downarrow F \\ & \underset{\delta_C}{\searrow} & \\ & & \varprojlim_R |sd_C N^{cy}_\bullet(G)|^C \end{array}$$

which is homotopy commutative via a specified homotopy equivalence and thus a map

$$\delta\colon |N_\bullet(G)| \to \left(\varprojlim_R |sd_C N^{cy}_\bullet(G)|^C\right)^{hF} \tag{2.6.7}$$

into the homotopy fiber of $F - \mathrm{id}$.

We want to apply (2.6.7) to $G = \widehat{\mathrm{GL}}_k(L)$, so must generalize to group-like topological monoids where $I_\bullet$ a priori does not exist.

The standard way to overcome the lack of strict inverses is to group complete the topological monoid: there are functors $G \mapsto G^\vee$ and $G \mapsto G^\wedge$, and natural transformations $G \leftarrow G^\vee \to G^\wedge$ which induces equivalences of the constructions $N_\bullet(\ )$ and $N^{cy}_\bullet(\ )$ when G is group-like. Here $G^\vee$ is a free monoid and $G^\wedge$ is a topological group, cf. [BF, p. 331] or [G2], sect. I.1.8. Consider the homotopy pull-back

$$\begin{array}{ccc} B'G & \longrightarrow & \underset{R}{\operatorname{holim}} |sd_C N^{cy}_\bullet(G^\vee)|^C \\ \downarrow N & & \downarrow \sim \\ |N_\bullet(G^\vee)| & \xrightarrow{\delta^h} & \underset{R}{\operatorname{holim}} |sd_C N^{cy}_\bullet(G^\wedge)|^C \end{array}$$

where δ^h is the composition of δ with the inclusion of $\varprojlim_R$ into $\underset{R}{\operatorname{holim}}$.

When $G = \widehat{\mathrm{GL}}_k(L)$, the simplicial map

$$S_\bullet\colon N^{cy}_\bullet(\widehat{\mathrm{GL}}_k(L)) \to \mathrm{THH}_\bullet(\mathrm{M}_k(L))$$

is cyclic in the sense of Connes, and the induced maps on C-fixed sets commute with the F and R-maps. One gets a map

$$\underset{R}{\operatorname{holim}} \left|sd_C N^{cy}_\bullet(\widehat{\mathrm{GL}}_k(L))^C\right|^{hF} \to \underset{R}{\operatorname{holim}} \left|sd_C \mathrm{THH}_\bullet(\mathrm{M}_k(L))^C\right|^{hF}.$$

The target is $\mathrm{TC}(\mathrm{M}_k(L))$. It is by (2.6.5) and (2.4.6) equivalent to $\mathrm{TC}(L)$. Thus we have for each k a string of maps

$$\mathrm{trc}\colon B\widehat{\mathrm{GL}}_k \xleftarrow{\sim} B'\widehat{\mathrm{GL}}_k(L) \longrightarrow \mathrm{TC}(\mathrm{M}_k(L)) \xrightarrow{\sim} \mathrm{TC}(L)$$

which in turn induces a map from $K(L)$ to $\mathrm{TC}(L)$, the cyclotomic trace.

In order to see that trc is in fact a map of spectra, one uses e.g. Segal's Γ-structure on $\amalg B\widehat{\mathrm{GL}}_k(L)$ and a corresponding structure on $\amalg\mathrm{THH}(\mathrm{M}_k(L))$, cf. [BHM], sect. 4. Finally, the associated abstract delooping of $\mathrm{TC}(L)$ and the concrete one from sect. 2.5 agree by [HM], sect. 1.6. I return to a different solution to this in the next chapter, but it is in order to mention that the Γ-space approach is based upon the following

Proposition 2.6.8 ([BHM]). *For a product of* FSP*'s there is a* C_∞-*equivalence*

$$\mathrm{TH}(L_1 \times L_2) \sim \mathrm{TH}(L_1) \times \mathrm{TH}(L_2). \qquad \square$$

3 The relative theorems

The end result of this chapter is a proof of the following conjecture from [G5]: Let $f\colon L_1 \to L_2$ be a map of FSP's such that $\pi_0 L_1 \to \pi_0 L_2$ is a surjection of rings with nilpotent kernel. Then

$$\begin{array}{ccc} K(L_1) & \longrightarrow & \mathrm{TC}(L_1) \\ \downarrow & & \downarrow \\ K(L_2) & \longrightarrow & \mathrm{TC}(L_2) \end{array}$$

becomes homotopy Cartesian after profinite completion.

The proof proceeds in three steps, due to Dundas-McCarthy [DM1], McCarthy [Mc] and Dundas [D], respectively, and uses Goodwillie's black magic: *calculus of functors*, [G3], [G4]. The exposition is based on these papers and on [DM2]. I have had invaluable help from B. Dundas with some of the details below.

3.1 Calculus of functors.

Calculus of functors is a general procedure, devised by Goodwillie, for proving relative theorems as above. The reader is referred to [G3], [G4] for more details.

We shall consider certain functors

$$F\colon \mathrm{s}_\bullet\mathrm{sets} \to \{\mathrm{prespectra}\}$$

from the category of simplicial sets (or spaces) to the category of prespectra. I here use prespectra indexed only on $\mathbb{R}^n$, not the coordinate free ones of May.

The functors we consider are supposed to satisfy the following two axioms:

(i) A homotopy $f_t\colon X_1 \to X_2$ induces a natural homotopy $F(f_t)\colon F(X_1) \to F(X_2)$.

(ii) For each $X \in s_\bullet\text{sets}$ and each prime p, the mod p homotopy groups satisfy

$$\pi_i(F(X); \mathbb{F}_p) = \varinjlim \pi_i F(X^{(\alpha)}; \mathbb{F}_p)$$

where $X^{(\alpha)}$ runs over the finite subcomplexes of X.

Condition (i) implies that F is a homotopy functor; (ii) is called the p-limit axiom.

Given such an F and a fixed $(X,x) \in s_\bullet\text{sets}_*$ there is a new functor on $s_\bullet\text{sets}_*$, namely

$$\Phi(Y) = \text{fib}(F(X \vee Y) \to F(X)).$$

Consider the commutative diagram

$$\begin{array}{ccc} \Phi(Y) & \longrightarrow & \Phi(C_+Y) \\ \downarrow & & \downarrow \\ \Phi(C_-Y) & \longrightarrow & \Phi(S^1 \wedge Y) \end{array}$$

where $C_\pm Y$ are the two cones in the reduced suspension $S^1 \wedge Y$. The standard retractions of the cones induce retractions of the two off diagonal terms, and in turn a map

$$\Phi(Y) \to \Omega\Phi(S^1 \wedge Y). \qquad (3.1.1)$$

The homotopy colimit of these maps is called the *differential* of F at (X,x). More importantly for our purpose we have

Definition 3.1.2. The derivative of F at (X,x) is the prespectrum whose n'th term is

$$\partial_x F(X)(\mathbb{R}^n) = \Phi(S^n)$$

and with structure maps

$$S^1 \wedge \partial_x F(X)(\mathbb{R}^n) \to \partial_x F(X)(\mathbb{R}^{n+1})$$

being the adjoints of (3.1.1).

For example the derivative of the functor

$$F(X) = \Sigma^\infty(X_+^n)$$

of the suspension spectrum of the n-fold Cartesian power of X is

$$\partial_x F(X) = \bigvee^n \Sigma^\infty(X_+^{n-1})$$

We next define Goodwillie's concept of analytic functors. The simplest ones are the linear functors. They are the homotopy functors which map a homotopy coCartesian square

$$\begin{array}{ccc} Y_\emptyset & \longrightarrow & Y_{\{2\}} \\ \downarrow & & \downarrow \\ Y_{\{1\}} & \longrightarrow & Y_{\{1,2\}} \end{array}$$

into a homotopy Cartesian square

$$\begin{array}{ccc} L(Y_\emptyset) & \longrightarrow & L(Y_{\{2\}}) \\ \downarrow & & \downarrow \\ L(Y_{\{1\}}) & \longrightarrow & L(Y_{\{1,2\}}), \end{array}$$

and has $F(*) \sim *$.

Here homotopy Cartesian and homotopy coCartesian means that the canonical maps

$$Y_\emptyset \xrightarrow{a} \underleftarrow{\operatorname{holim}}(Y_{\{1\}} \to Y_{\{1,2\}} \leftarrow Y_{\{2\}})$$

$$Y_{\{1,2\}} \xleftarrow{b} \underrightarrow{\operatorname{holim}}(Y_{\{1\}} \leftarrow Y_\emptyset \to Y_{\{2\}})$$

are equivalences.

To define the concept of *analytic functors*, one needs to consider n-dimensional cubes of spaces and spectra, i.e. functors

$$\mathfrak{X}\colon \mathcal{P}(S) \to \mathcal{C}, \quad \mathcal{C} = s_\bullet\text{sets}, \{\text{spectra}\}$$

from the category of posets of the finite set S. If $S = \mathbf{n}$, then $\mathfrak{X}$ is called an n-cube. Generalizing the above, $\mathfrak{X}$ is called k-Cartesian or k-coCartesian if

$$\mathfrak{X}(\emptyset) \xrightarrow{a} \underset{\mathcal{P}_0(S)}{\underleftarrow{\operatorname{holim}}}\, \mathfrak{X}, \quad \mathcal{P}_0 = \mathcal{P}(S) - \{\emptyset\}$$

$$\mathfrak{X}(S) \xleftarrow{b} \underset{\mathcal{P}_1(S)}{\underrightarrow{\operatorname{holim}}}\, \mathfrak{X}, \quad \mathcal{P}_1 = \mathcal{P}(S) - \{S\}$$

are k-equivalences.

Given $U \subset T \subset S$ the face $\partial_U^T \mathfrak{X}$ is the $T - U$ cube given by

$$\partial_U^T \mathfrak{X}(V)\mathfrak{X}(V \cup U).$$

We shall consider ***strongly coCartesian*** cubes, that is, cubes $\mathfrak{X}$ where each 2-dimensional face $\partial_U^T \mathfrak{X}$ is k-coCartesian for all k. This implies in particular that the total cube is homotopy coCartesian.

Definition 3.1.3. A functor $F\colon \mathrm{s_\bullet sets} \to \{\text{spectra}\}$ is called stably n-excisive if the following statement $E_n(c, \kappa)$ is true for some numbers c and κ:

$E_n(c, \kappa)$: Given any strongly coCartesian $(n+1)$-cube $\mathfrak{X}$ with $\mathfrak{X}(\emptyset) \to \mathfrak{X}(\{s\})$ k_s-connected and $k_s \geq \kappa$, then the $(n+1)$-cube $F(\mathfrak{X})$ is $(-c + \Sigma k_s)$-Cartesian.

Definition 3.1.4. A homotopy functor F is called ρ-analytic if for some q, independent of n, F satisfies $E_n(n\rho - q, \rho + 1)$ for all n.

Let (A, P) be a pair of a unitary ring and an A bimodule P. For each based simplicial set $Y_\bullet \in \mathrm{s_\bullet sets_*}$ we have the simplicial ring

$$(A \ltimes P)(Y_\bullet) = A \oplus \tilde{P}(Y_\bullet), \quad \tilde{P}(Y_\bullet) = P[Y_\bullet]/P[*_\bullet]$$

with multiplication

$$(a_1, p_1)(a_2, p_2) = (a_1 a_2, a_1 p_2 + p_1 a_2).$$

We shall see in sect. 3.3 below that the realization of the simplicial functors

$$\begin{aligned} [r] &\to \mathrm{hF}\big(K(A \oplus \tilde{P}(Y_r)) \to K(A)\big) \\ [r] &\to \mathrm{hF}\big(\mathrm{TC}(A \oplus \tilde{P}(Y_r)) \to \mathrm{TC}(A)\big) \end{aligned} \tag{3.1.5}$$

both satisfies $E_n(-2 - n, 0)$ for all n; thus they are (-1)-analytic; hF = homotopy fiber.

The main theorem of Goodwillie's about analytic functors is the following

Theorem 3.1.6. *Suppose $\theta\colon F \to G$ is a natural transformation between ρ-analytic functors such that $\partial_x \theta(X)\colon \partial_x F(X) \to \partial_x G(X)$ is an equivalence of prespectra. Then for every $(\rho + 1)$-connected map $Y \to K$ in $\mathrm{s_\bullet Sets_*}$, the diagram*

$$\begin{array}{ccc} F(Y) & \longrightarrow & G(Y) \\ \big\downarrow & & \big\downarrow \\ F(K) & \longrightarrow & G(K) \end{array}$$

is homotopy Cartesian.

The cyclotomic trace of sect. 2.6 defines a natural transformation between the two functors in (3.1.5), which turns out to satisfy the conditions of the above theorem after profinite completion, cf. sect. 3.2 and sect. 3.3 below, so theorem 3.1.6 implies that

$$\begin{array}{ccc} K(A \oplus \tilde{P}(Y_\bullet))^\wedge & \longrightarrow & \mathrm{TC}(A \oplus \tilde{P}_\bullet(Y))^\wedge \\ \downarrow & & \downarrow \\ K(A)^\wedge & \longrightarrow & \mathrm{TC}(A)^\wedge \end{array} \tag{3.1.7}$$

is homotopy Cartesian, where the upper horizontal line is calculated degree-wise. Indeed, the homotopy fibers of the vertical arrows are the relative theories of (3.1.5), and they vanish for $Y_\bullet = *_\bullet$, so agree by the theorem.

3.2 K- and THH of additive split exact categories.

In this section $\mathcal{C}$ is an additive split exact category, e.g. the category of projective modules $\mathcal{P}_A$ over a ring, or its subcategory $\mathcal{F}_A$ of free modules.

Recall that Waldhausen in [W3] associated to $\mathcal{C}$ a simplicial set (in fact a simplicial category) $S_\bullet\mathcal{C}$. The r-simplices of objects in $S_r\mathcal{C}$ is the set of diagrams

$$\begin{array}{ccccccc} C_1 \rightarrowtail & C_2 \rightarrowtail & C_3 & \rightarrowtail \cdots \rightarrowtail & C_r \\ & \downarrow & \downarrow & & \downarrow \\ & C_{12} \rightarrowtail & C_{13} & \rightarrowtail \cdots \rightarrowtail & C_{1r} \\ & & \downarrow & & \downarrow \\ & & C_{23} & \rightarrowtail \cdots \rightarrowtail & C_{2r} \\ & & & & \downarrow \\ & & & & \vdots \\ & & & & \downarrow \\ & & & & C_{r-1,r} \end{array} \tag{3.2.1}$$

with

$$0 \to C_{ij} \to C_{ik} \to C_{jk} \to 0$$

a (split) exact sequence. Thus $S_r\mathcal{C} = 0$ for $r = 0$, $S_1\mathcal{C} = \mathcal{C}$, and in general $S_r\mathcal{C}$ is the category of flags involving r objects with choice of quotients.

The objects of $S_\bullet\mathcal{C}$ form a simplicial set where d_0 forgets the first row (divides out C_1) and where d_i contracts the flag by forgetting C_i and the row $C_{i,\bullet}$. The degeneracy operator s_i inserts an extra C_i, so for example s_0 and s_1 from $S_1\mathcal{C} = \mathcal{C}$ to $S_2\mathcal{C}$ sends C to $0 \rightarrowtail C$ and $C \rightarrowtail C$, respectively.

The nerve of the isomorphism category $iS_\bullet(\mathcal{C})$ of flags defines a bisimplicial space

$$[s],[r] \to N_s(iS_r\mathcal{C}).$$

The loop space of its realization is Waldhausen's definition of $K(\mathcal{C})$,

$$K(\mathcal{C}) = \Omega|N_\bullet(iS_\bullet\mathcal{C})| \tag{3.2.2}$$

(of course, Waldhausen's definition applies to much more general situations). In order to relate this to the previous definition of K-theory, recall from sect. 2.2 that we can realize the double complex in two steps. Let us first realize the r-direction. There is an obvious map

$$\Delta^1 \times N_s(iS_1(\mathcal{C})) \to |N_s(iS_\bullet\mathcal{C})|$$

(the inclusion of the 1-skeleton), and since $S_0\mathcal{C} = \{0\}$, it factors over

$$\sigma\colon S^1 \wedge N_s(iS_1\mathcal{C}) \to |N_s(iS_\bullet\mathcal{C})|.$$

Realizing the s-direction and adjoining σ we get a map

$$|N_\bullet(\mathcal{C})| \xrightarrow{\mathrm{ad}(\sigma)} \Omega|N_\bullet(iS_\bullet\mathcal{C})| \tag{3.2.3}$$

which turns out to be a *group completion*, cf. [W3], sect. 1.6. When $\mathcal{C} = \mathcal{F}_A$ then

$$|N_\bullet(\mathcal{F}_A)| = \coprod_{n=0}^{\infty} B\mathrm{GL}_n(A)$$

so the above definition of K-theory agrees with the earlier one from sect. 2.6 in this case.

The iterated degeneracy operator in the s-direction defines a map

$$s\colon N_0(iS_\bullet\mathcal{C}) \to N_s(iS_\bullet\mathcal{C})$$

with a one-sided inverse d_0^s, and gives a map

$$s\colon |N_0(iS_\bullet\mathcal{C})| \xrightarrow{\sim} |N_\bullet(iS_\bullet\mathcal{C})|. \tag{3.2.4}$$

Corollary 1.4.1 of [W3] states that (3.2.4) is an equivalence. Thus one can recast (3.2.2) as

$$K(\mathcal{C}) \cong \Omega|\operatorname{ob} S_\bullet\mathcal{C}| = \Omega\big|[r] \to S_r\mathcal{C}\big|. \tag{3.2.5}$$

When $\mathcal{C} = \mathcal{P}_A$, the projective modules, then (3.2.4), and (3.2.6) below, implies that

$$K_0(\mathcal{P}_A) \mathrel{\overset{.}{=}} \pi_1|\operatorname{ob} S_\bullet\mathcal{P}_A| \cong H_1(\operatorname{ob} S_\bullet\mathcal{P}_A) = K_0(A),$$

the projective class group of the ring A.

The $S_\bullet$ construction can be iterated, and defines a (-1)-connected spectrum whose $(n-1)$'st term is $\Omega|\operatorname{ob} S_\bullet^{(n)}\mathcal{C}|$. The natural maps

$$|\operatorname{ob} S_\bullet^{(n-1)}\mathcal{C}| \to \Omega|\operatorname{ob} S_\bullet^{(n)}\mathcal{C}| \tag{3.2.6}$$

are equivalences for $n > 1$, cf. [W3], proposition 1.5.3, so the $S_\bullet$-construction deloops $K(\mathcal{C})$ beyond the first step

$$K(\mathcal{C}) \simeq \Omega^n|\operatorname{ob} S_\bullet^{(n)}\mathcal{C}|, \quad n \geq 1.$$

We now turn to $\mathrm{THH}(\mathcal{C})$, following [DM2]. We have already presented the definition in (2.3.9), and can try to imitate the two key results above, (3.2.3) and (3.2.4), for $N_\bullet(-)$ replaced by $\mathrm{THH}_\bullet(-)$. In fact, since $\mathrm{THH}(-)$ is already a spectrum, one expects that (3.2.3) be an equivalence, and this indeed happens. Here are some details.

We can think of (3.2.2) as

$$K(\mathcal{C}) = \Omega\big|[r] \to |N_\bullet(iS_r\mathcal{C})|\big|$$

and can similarly consider the simplicial space

$$[r] \to \mathrm{THH}(S_r\mathcal{C}) = |\mathrm{THH}_\bullet(S_r\mathcal{C})|$$

which we denote for short $\mathrm{THH}(S_\bullet\mathcal{C})$. There are maps

$$\sigma\colon S^1 \wedge \mathrm{THH}(\mathcal{C}) \to \mathrm{THH}(S_\bullet\mathcal{C})$$
$$s\colon \mathrm{THH}_0(S_r\mathcal{C}) \to \mathrm{THH}_\bullet(S_r\mathcal{C})$$

defined as above.

Theorem 3.2.7. ([DM2]) *The maps σ and s define equivalences*

(i) $\mathrm{THH}(\mathcal{C}) \xrightarrow{\sim} \Omega|\mathrm{THH}(S_\bullet\mathcal{C})|$

(ii) $\varinjlim \Omega^n|\mathrm{THH}_0(S_\bullet^{(n)}\mathcal{C})| \xrightarrow{\sim} \varinjlim \Omega^n|\mathrm{THH}(S_\bullet^{(n)}\mathcal{C})|$

Proof of (i) (sketch). The proof is modelled upon [W3], proposition 1.53. Consider the functor $S_n\mathcal{C} \to \mathcal{C}^n$ which to the flag (3.2.1) associates the n-tuple $(C_1, C_{12}, \dots, C_{n-1,n})$. It induces an equivalence

$$\mathrm{THH}(S_n\mathcal{C}) \xrightarrow{\sim} \mathrm{THH}(\mathcal{C})^n.$$

This is an application of Morita invariance and (2.6.8): the trace of a triangular matrix only depends on the diagonal entries.

Now recall for any simplicial space $X_\bullet$ the simplicial path space construction $P_\bullet X_\bullet$. It has n-simplices $P_nX_\bullet = X_{n+1}$ and face and degeneracy operators are shifted up by 1. The extra degeneracy $s_0\colon X_n \to X_{n+1}$ not used in $P_\bullet X_\bullet$ gives an equivalence $|P_\bullet X_\bullet| \sim X_0$, so $|P_\bullet X_\bullet|$ is contractible when X_0 consists of a single point. Moreover we have a sequence

$$X_1 \xrightarrow{s_0^\bullet} P_\bullet X_\bullet \xrightarrow{d_0} X_\bullet$$

of simplicial sets upon considering X_1 the constant simplicial space. We now have for each r the diagram

$$\begin{array}{ccccc} \mathrm{THH}(\mathcal{C}) & \longrightarrow & \mathrm{THH}(P_r S_\bullet \mathcal{C}) & \longrightarrow & \mathrm{THH}(S_r \mathcal{C}) \\ \| & & \downarrow\sim & & \downarrow\sim \\ \mathrm{THH}(\mathcal{C}) & \longrightarrow & \mathrm{THH}(\mathcal{C})^{r+1} & \longrightarrow & \mathrm{THH}(\mathcal{C})^r \end{array}$$

so the sequence

$$\mathrm{THH}(\mathcal{C}) \to \mathrm{THH}(P_\bullet S_\bullet \mathcal{C}) \to \mathrm{THH}(S_\bullet \mathcal{C})$$

is a degreewise homotopy fibration, and hence becomes a homotopy fibration after realization, since THH$(-)$ is equivalent to an abelian group complex, see (3.2.8) below. Finally $|\mathrm{THH}(P_\bullet S_\bullet \mathcal{C})| \sim *$. □

The proof of (ii) is more delicate and requires some rewritings of THH$(\mathcal{C})$ which we now present. We have for each number x the simplicial abelian group

$$\mathcal{C}^x(c_0, c_1) = \mathrm{Hom}_{\mathcal{C}}(c_0, c_1) \otimes \tilde{\mathbb{Z}}(S^x_\bullet) = \mathcal{C}(x_0, x_1) \otimes \tilde{\mathbb{Z}}(S^x_\bullet)$$

and associated simplicial sets, one for each r,

$$V_{r,\bullet}(\mathcal{C}, \mathbf{x}) = \bigvee_{c_0, \dots, c_r \in \mathcal{C}} \delta\big(\mathcal{C}^{x_0}(c_0, c_r) \wedge \mathcal{C}^{x_1}(c_1, c_0) \wedge \cdots \wedge \mathcal{C}^{x_r}(c_r, c_{r-1})\big)$$

where δ denotes the diagonal in the stated multisimplicial set. There are simplicial maps

$$V_{r,\bullet}(\mathcal{C}, \mathbf{x}) \xrightarrow{d_i} V_{r-1,\bullet}(\mathcal{C}, d_i\mathbf{x})$$
$$V_{r,\bullet}(\mathcal{C}, \mathbf{x}) \xrightarrow{s_i} V_{r+1,\bullet}(\mathcal{C}, s_i\mathbf{x})$$

and we let

$$\mathrm{THH}_{r,\bullet}(\mathcal{C}) = \underset{\mathbf{x} \in I^{r+1}}{\underrightarrow{\mathrm{holim}}}\; s_\bullet\mathcal{C}\big(\delta S^{x_0}_\bullet \wedge \cdots \wedge S^{x_r}_\bullet, V_{r,\bullet}(\mathcal{C}, \mathbf{x})\big).$$

Here $s_\bullet\mathcal{C}$ is the simplicial mapping space. This gives a bisimplicial set $\mathrm{THH}_{r,s}(\mathcal{C})$ whose realization is the THH$(\mathcal{C})$ defined in sect. 2.3.

We now vary the definition by replacing $V_{r,\bullet}(\mathcal{C},\mathbf{x})$ by

$$\begin{aligned} &V^{\oplus}_{r,\bullet}(\mathcal{C},\mathbf{x}) \\ &= \bigoplus_{c_0,\dots,c_r\in\mathcal{C}} \mathcal{C}^{x_0}(c_0,c_r)\otimes\tilde{\mathbb{Z}}(\mathcal{C}^{x_1}(c_1,c_0)\wedge\cdots\wedge\mathcal{C}^{x_r}(c_r,c_{r-1})) \\ &= \bigoplus \mathcal{C}^{x_0}(c_0,c_r)\otimes\tilde{\mathbb{Z}}(\mathcal{C}^{x_1}(c_1,c_0))\otimes\cdots\otimes\tilde{\mathbb{Z}}(\mathcal{C}^{c_r}(c_r,c_{r-1}))\end{aligned} \tag{3.2.8}$$

and write $\mathrm{THH}^{\oplus}_{\bullet,\bullet}(\mathcal{C})$ for the corresponding bisimplicial abelian group. The inclusion of $V_{\bullet,\bullet}(\mathcal{C},\mathbf{x})$ in $V^{\oplus}_{\bullet,\bullet}(\mathcal{C},\mathbf{x})$ induces a simplicial map

$$\theta_\bullet\colon \mathrm{THH}_\bullet(\mathcal{C})\to\mathrm{THH}^{\oplus}_\bullet(\mathcal{C}) \tag{3.2.9}$$

which is an equivalence. This follows from lemma 2.3.7, the well-known isomorphisms

$$\pi_i\tilde{M}(Y)\cong H_i(Y;M)\cong\pi_{i+x}(Y\wedge\tilde{M}(S^x)),\quad i<x,$$

and because the inclusion of the wedge in the product (direct sum) is $2\Sigma x_\nu-1$ connected.

Recall that $S_2\mathcal{C}$ is the category of (split) exact sequences in $\mathcal{C}$. The morphisms are commutative diagrams

$$\begin{array}{ccccccccc} 0 & \longrightarrow & C_0 & \longrightarrow & C_1 & \longrightarrow & C_2 & \longrightarrow & 0 \\ & & \downarrow{\scriptstyle f_0} & & \downarrow{\scriptstyle f_1} & & \downarrow{\scriptstyle f_2} & & \\ 0 & \longrightarrow & C_0' & \longrightarrow & C_1' & \longrightarrow & C_2' & \longrightarrow & 0 \end{array}$$

We use the notation (f_0,f_1,f_2) for this morphism. The simplicial functors

$$d_0,d_1,d_2\colon S_2\mathcal{C}\to\mathcal{C}=S_1\mathcal{C}$$

induce simplicial maps

$$\bar{d}_0,\bar{d}_1,\bar{d}_2\colon \mathrm{THH}^{\oplus}_\bullet(S_2\mathcal{C})\to\mathrm{THH}^{\oplus}_\bullet(\mathcal{C})$$

and we have (in preparation for the proof of theorem 3.2.7 (ii))

Lemma 3.2.10. *For each r, there are natural transformations*

$$T^{(\nu)}_r\colon \mathrm{THH}^{\oplus}_r(\mathcal{C})\to\mathrm{THH}^{\oplus}_r(S_2\mathcal{C}),\quad \nu=1,2$$

such that

$$\bar{d}_0T^{(1)}_r=\mathrm{id},\quad \bar{d}_2T^{(1)}_r=0=\bar{d}_0T^{(2)}_r,\quad \bar{d}_1T^{(1)}_r=\bar{d}_1T^{(2)}_r,\quad \bar{d}_2T^{(2)}_r=s^r_0\circ d^r_0.$$

Proof. Given objects $\mathbf{c} = (c_0, \dots, c_r) \in \mathcal{C}^{r+1}$ and morphisms $\alpha_0 \in \mathcal{C}(c_0, c_r)$, $\alpha_k \in \mathcal{C}(c_k, c_{k-1})$ for $k = 1, \dots, r$ we define objects $\Delta_k^{(\nu)} = \Delta_k^{(\nu)}(\mathbf{c}, \boldsymbol{\alpha})$ of $S_2\mathcal{C}$

$$\Delta_k^{(1)} : 0 \longrightarrow C_r \xrightarrow{i_1} C_r \oplus C_k \xrightarrow{\pi_2} C_k \longrightarrow 0$$

$$\Delta_k^{(2)} : 0 \longrightarrow C_r \xrightarrow{\binom{1}{\beta_k}} C_r \oplus C_k \xrightarrow{(\beta_k, -1)} C_k \longrightarrow 0$$

where $\beta_k = \alpha_{k+1} \cdots \alpha_r$ for $0 \le k < r$ and $\beta_r = 1$.

With these notions we define $t_\bullet^{(\nu)}$

$$t_r^{(1)} : s_\bullet\mathcal{C}(S_\bullet^{\mathbf{x}}, V_{r,\bullet}^{\oplus}(\mathcal{C}, \mathbf{x})) \to s_\bullet\mathcal{C}(S_\bullet^{\mathbf{x}}, V_{r,\bullet}^{\oplus}(S_2\mathcal{C}; \mathbf{x}))$$

$$t_r^{(2)} : s_\bullet\mathcal{C}(S_\bullet^{\mathbf{x}}, V_{r,\bullet}^{\oplus}(\mathcal{C}; \mathbf{x})) \to s_\bullet\mathcal{C}(S_\bullet^{\mathbf{y}}, V_{r,\bullet}^{\oplus}(S_2\mathcal{C}; \mathbf{y}))$$

where $\mathbf{y} = (x_0 + \cdots + x_r, x_1 + \cdots + x_r, \dots, x_r)$ and $S_\bullet^{\mathbf{x}} = S_\bullet^{x_0} \wedge \cdots \wedge S_\bullet^{x_r}$. If $(x_0, \dots, x_r) = (0, \dots, 0)$ the formulas are:

$$t_r^{(1)}(\alpha_0 \otimes \cdots \otimes \alpha_r)$$
$$= \left(1, \left(\begin{smallmatrix} 0 & 0 \\ \alpha_0 & \alpha_0 \end{smallmatrix}\right), \alpha_0\right) \otimes \left(1, \left(\begin{smallmatrix} 1 & 0 \\ 0 & \alpha_1 \end{smallmatrix}\right), \alpha_1\right) \otimes \cdots \otimes \left(1, \left(\begin{smallmatrix} 1 & 0 \\ 0 & \alpha_r \end{smallmatrix}\right), \alpha_r\right)$$

$$t_r^{(2)}(\alpha_0 \otimes \cdots \otimes \alpha_r)$$
$$= \left(\alpha_0 \cdots \alpha_r, \left(\begin{smallmatrix} 0 & 0 \\ \alpha_0 & \alpha_0 \end{smallmatrix}\right), 0\right) \otimes \left(1, \left(\begin{smallmatrix} 1 & 0 \\ 0 & \alpha_1 \end{smallmatrix}\right), \alpha_1\right) \otimes \cdots \otimes \left(1, \left(\begin{smallmatrix} 1 & 0 \\ 0 & \alpha_r \end{smallmatrix}\right), \alpha_r\right).$$

For general $\mathbf{x}$, one needs to replace $\mathcal{C}^x(c, d)$ by the equivalent $s_\bullet\mathcal{C}(c, d \otimes \tilde{\mathbb{Z}}(S^x))$ and one must use suitable suspension maps

$$s_\bullet\mathcal{C}(c, d) \xrightarrow{\sim} s_\bullet\mathcal{C}(c \otimes \tilde{\mathbb{Z}}(S^y), d \otimes \tilde{\mathbb{Z}}(S^y))$$

in order to define both $\Delta_k^{(2)}(\mathbf{c}, \boldsymbol{\alpha})$ and $t_r^{(2)}$. Details are left for the reader to carry out, who can also consult [DM2]. We set

$$T_r^{(\nu)} = \underset{\overrightarrow{I^{r+1}}}{\operatorname{holim}}\, t_r^{(\nu)} : \mathrm{THH}_r^{\oplus}(\mathcal{C}) \to \mathrm{THH}_r^{\oplus}(S_2\mathcal{C}).$$

These are the required maps, and the required relations are obvious to check. □

We assumed $\mathcal{C}$ to be an additive split exact category, so $S_2\mathcal{C}$ is equivalent to $\mathcal{C} \times \mathcal{C}$: there are functors both ways whose composites are naturally isomorphic to the identity. Indeed,

$$S_2\mathcal{C} \xrightarrow{(d_0, d_2)} \mathcal{C} \times \mathcal{C}, \quad \mathcal{C} \times \mathcal{C} \xrightarrow{s_0 \oplus s_1} S_2\mathcal{C}$$

are the two functors. One composite is the identity; the other sends each object to an isomorphic object, and one may easily construct the required *natural* isomorphism.

Functors such as $\mathrm{THH}_r^{\oplus}(\mathcal{C})$ does not map equivalent categories into homotopy equivalent spaces (check e.g. $r = 0$). However the composite functor $\mathrm{THH}_r^{\oplus}(S_\bullet\mathcal{C})$ does have this property.

Lemma 3.2.11. *Let $g_0, g_1 : \mathcal{C} \to \mathcal{D}$ be naturally isomorphic functors between exact categories. Then there is a simplicial functor*

$$G\colon \Delta[1]_\bullet \times S_\bullet\mathcal{C} \to S_\bullet\mathcal{D}$$

which restricts to $S_\bullet f_0$ and $S_\bullet f_1$ at the two ends. Here $\Delta[1]_\bullet$ is the simplicial 1-simplex considered as a discrete category. □

The lemma is proved in [W3], although only stated on objects. Since simplicial homotopies are preserved by functors, the induced maps

$$\mathrm{THH}_r^{\oplus}(S_\bullet\mathcal{C}) \rightrightarrows \mathrm{THH}_r^{\oplus}(S_\bullet\mathcal{D})$$

are homotopic. In particular

$$\mathrm{THH}_r^{\oplus}(S_\bullet S_2\mathcal{C}) \sim \mathrm{THH}_r^{\oplus}(S_\bullet\mathcal{C} \times S_\bullet\mathcal{C}).$$

Consider a functor X from additive exact categories to simplicial or topological groups with $X(0) = 0$ and $X(\mathcal{C} \times \mathcal{D}) \xrightarrow{\sim} X(\mathcal{C}) \times X(\mathcal{D})$. Let $Y(\mathcal{C}) = Y(S_\bullet\mathcal{C})$, a bisimplicial abelian group. Then

$$\bar{d}_1 \sim \bar{d}_0 + \bar{d}_2 = Y(S_2\mathcal{C}) \to Y(\mathcal{C}) \tag{3.2.12}$$

where $\bar{d}_i = Y(d_i)$. This follows from the homotopy commutative diagram

$$\begin{array}{ccccc}
Y(S_2\mathcal{C}) & \xrightarrow{Y(d_0 \times d_1)} & Y(\mathcal{C} \times \mathcal{C}) & \xrightarrow{Y(s_0 \oplus s_2)} & Y(S_2\mathcal{C}) \\
& {\scriptstyle \bar{d}_0 \times \bar{d}_2}\searrow & \uparrow{\scriptstyle \bar{\imath}_1 + \bar{\imath}_2} & \nearrow{\scriptstyle \bar{s}_0 + \bar{s}_1} & \\
& & Y(\mathcal{C}) \times Y(\mathcal{C}) & &
\end{array}$$

The right hand triangle homotopy commutes because it does so after composing with the equivalence $\bar{d}_0 \times \bar{d}_2$, $d_2 s_0 = 0 = d_0 s_1$. The left hand triangle commutes because $\overline{pr}_1 \times \overline{pr}_2$ is a homotopy inverse to $\bar{\imath}_1 + \bar{\imath}_2$. Finally the horizontal composite is homotopic to the identity by lemma 3.2.11, and $\bar{d}_1(\bar{s}_0 + \bar{s}_1)$ is equal to addition.

The functor $\mathrm{THH}_r^{\oplus}(\mathcal{C})$ does not preserve product, but the functor

$$X_r(\mathcal{C}) = \varinjlim \Omega^k \mathrm{THH}_r^{\oplus}(S_\bullet^{(k)}\mathcal{C}) \tag{3.2.13}$$

does. This is formal and true for any functor Z with $Z(0) = 0$ as the map of multisimplicial sets

$$Z(S_\bullet^{(k)}\mathcal{C} \times S_\bullet^{(k)}\mathcal{C}) \to Z(S_\bullet^{(k)}\mathcal{C}) \times Z(S_\bullet^{(k)}\mathcal{C})$$

is an isomorphism when the sum of the $2k$ simplicial degrees is less than $2k$ (because $S_0\mathcal{C} = 0$). In particular the map is $2k$-connected.

Proof of theorem 3.2.7 (ii). With the notation from (3.2.13) have from (3.2.12)

$$\overline{d}_1 \sim \overline{d}_0 + \overline{d}_2 \colon X_r(S_\bullet S_2\mathcal{C}) \to X_r(S_\bullet\mathcal{C}).$$

Lemma 3.2.10 can be applied to X_r as well as to $\mathrm{THH}_r^\oplus$, and shows that the composition

$$X_r(\mathcal{C}) \xrightarrow{d_0^r} X_0(\mathcal{C}) \xrightarrow{s_0^r} X_r(\mathcal{C})$$

is homotopic to the identity. Indeed

$$\mathrm{id} = d_0\mathrm{Tr}^{(1)} + \overline{d}_2\mathrm{Tr}^{(1)} = \overline{d}_1\mathrm{Tr}^{(2)} \sim \overline{d}_0\mathrm{Tr}^{(2)} + \overline{d}_2\mathrm{Tr}^{(2)} = d_0^r s_0^r.$$

The other composition is obviously the identity.

Thus $X_\bullet(\mathcal{C})$ is a simplicial space in which the simplicial structure maps are all homotopy equivalences; for such $X_0(\mathcal{C}) \sim |X_\bullet(\mathcal{C})|$ □

Theorem 3.2.7 allows a slick definition of the topological Dennis trace

$$\mathrm{tr}\colon K(\mathcal{C}) \to \mathrm{THH}(\mathcal{C}),$$

namely as the composite

$$\Omega|S_\bullet\mathcal{C}| \to \Omega|\mathrm{THH}_0(S_\bullet\mathcal{C})| \to \Omega|\mathrm{THH}(S_\bullet\mathcal{C})| \sim \mathrm{THH}(\mathcal{C}) \tag{3.2.14}$$

where the first map is induced from sending an object $\mathbf{C} \in S_r\mathcal{C}$ into $\mathrm{id}_{\mathbf{C}} \in \mathrm{Hom}_{S_\bullet\mathcal{C}}(\mathbf{C},\mathbf{C})$.

We can introduce the spectrum $\mathrm{TH}(\mathcal{C})$ either by iterating the $S_\bullet$-construction or by introducing a dummy variable similar to what we did in the case of $\mathrm{THH}(L)$. The corresponding deloops (spectra) are equivalent by the standard argument which makes use of both deloops:

$$\mathrm{THH}(S_\bullet^{(n)}\mathcal{C}) \sim \Omega^n\mathrm{THH}^{S^n}(S_\bullet^{(n)}\mathcal{C}) \sim \mathrm{THH}^{S^n}(\mathcal{C})$$

(cf. [BM] sect. 1).

If we use the iteration of the $S_\bullet$-construction to define the spectrum, then it is obvious that the map in (3.2.14) is a map of spectra.

Later in the chapter we shall consider $\mathrm{THH}^{\oplus}(\mathcal{C};M)$ where $M\colon \mathcal{C}^0\times\mathcal{C}\to$ Ab. It is defined by replacing $V^{\oplus}_{r,\bullet}(\mathcal{C},\mathbf{x})$ by

$$V^{\oplus}_{r,\bullet}(\mathcal{C};M,\mathbf{x})=\bigoplus M^{x_0}(c_0,c_r)\otimes\tilde{\mathbb{Z}}\mathcal{C}^{x_1}(c_1,c_0)\otimes\cdots\otimes\tilde{\mathbb{Z}}\mathcal{C}^{x_r}(c_r,c_{r-1}).\quad(3.2.15)$$

If $\mathcal{C}$ is the catetegory of projective or free modules and M is an A-bimodule then

$$M^{x_0}(c_0,c_r)=\mathrm{Hom}_A(c_0,c_r\otimes_A M)$$

extends to a functor on $S_\bullet\mathcal{C}$, and the proof of theorem 3.2.7 extends word for word to give

$$\mathrm{THH}^{\oplus}(A,M)\sim\varinjlim_p \Omega^p\big(|\mathrm{THH}^{\oplus}_0(S^{(p)}_\bullet\mathcal{P}_A,M)|\big).$$

Moreover, in this linear situation, one can omit the homotopy colimit over x_0 in the definition of $\mathrm{THH}^{\oplus}_0$. Indeed, for any number x

$$\begin{aligned}\mathrm{Hom}_A(a,b)&\xrightarrow{\sim} s_\bullet\mathcal{P}_A\big(a\otimes\tilde{\mathbb{Z}}(S^x_\bullet),b\otimes\tilde{\mathbb{Z}}(S^x_\bullet)\big)\\&\xrightarrow{\cong} s_\bullet\mathrm{Sets}_*\big(S^x_\bullet,s_\bullet\mathcal{P}_A(a,b\otimes\tilde{\mathbb{Z}}(S^x_\bullet))\big)\end{aligned}$$

where $S^x_\bullet=\Delta[x]_\bullet/\partial$ is the simplicial x-sphere, and $\mathrm{Hom}_A(a,b)$ is considered the constant simplicial group, cf. [Q1]. We have proved

Corollary 3.2.16. *For an A-bimodule M,*

$$\mathrm{THH}(A,M)\sim\varinjlim_p \Omega^p\left|\bigoplus_{\mathbf{c}\in S^{(p)}_\bullet\mathcal{P}_A}\mathrm{Hom}(\mathbf{c},\mathbf{c}\otimes M)\right|.\qquad\square$$

Remark 3.2.17. If we let $\mathbf{x}=\mathbf{0}$ in (3.2.15) we obtain a bisimplicial abelian group $V^{\oplus}_{r,s}(\mathcal{C},M,\mathbf{0})$ which is constant in the s-direction. Following [DM1] we write

$$F_r(\mathcal{C},M)=V_{r,0}(\mathcal{C},M,\mathbf{0})\cong\bigoplus_{c_r\to\cdots\to c_0\in N_r\mathcal{C}}M(c_0,c_r).$$

The homotopy groups of $|F_\bullet(\mathcal{C},M)|$, or equivalently the homology groups of the associated chain complex $F_*(\mathcal{C},M)$, is usually denoted $H_*(\mathcal{C};M)$ and is called the (non-additive) homology of $\mathcal{C}$ with coefficients in M. Dundas and McCarthy proves theorem 3.2.7 for this functor by an argument almost identical to the above. The diagram

$$\begin{array}{ccc}\Omega^\infty|F_\bullet(S^{(\infty)}_\bullet\mathcal{P}_A;M)| & \longrightarrow & \Omega^\infty|\mathrm{THH}^{\oplus}(S^{(\infty)}_\bullet\mathcal{P}_A;M)|\\ \uparrow\sim & & \uparrow\sim\\ \Omega^\infty|F_0(S^{\infty}_\bullet\mathcal{P}_A;M)| & \xrightarrow{\sim} & \Omega^\infty|\mathrm{THH}^{\oplus}_0(S^{(\infty)}_\bullet\mathcal{P}_A;M)|\end{array}$$

then shows that $\pi_*\mathrm{THH}(A;M) \cong H_*(\mathcal{P}_A, M)$. This is a special case of a theorem due to Pirashvili and Waldhausen, [PW].

3.3 Stable K- and TC-theory.

Let A be a ring, V an A-bimodule and $A \ltimes V$ the semiproduct ring. We may replace V by the $(n-1)$-connected simplicial A-bimodule $\tilde{V}(S^n_\bullet)$ and consider the simplicial ring $A \ltimes \tilde{V}(S^n_\bullet)$. This can be thought of as a small deformation of A. We want to measure the difference between $K(A)$ and $K(A \ltimes \tilde{V}(S^n_\bullet))$.

Recall from [W1] that K-theory of a simplicial ring $R_\bullet$ is defined as

$$K(R_\bullet) = \Omega B\left(\coprod_n B\widehat{\mathrm{GL}}_n(R_\bullet)\right) = \mathbb{Z} \times B\widehat{\mathrm{GL}}_\infty(R_\bullet)^+ \tag{3.3.1}$$

where $\widehat{\mathrm{GL}}_n(R_\bullet) \subset \mathrm{M}_n(R_\bullet)$ is the group like simplicial monoid of matrices which map to invertible matrices in $\mathrm{M}_n(\pi_0 R_\bullet)$, and $B\widehat{\mathrm{GL}}_n(R_\bullet) = |N_\bullet(\widehat{\mathrm{GL}}_n(R_\bullet))|$. Alternatively we can use (2.6.1) for the FSP

$$\tilde{R}_\bullet(X) = \big|[p] \to \tilde{R}_p(X)\big|$$

Indeed $K(R_\bullet) \simeq K(\tilde{R}_\bullet)$. There is another, more straightforward possibility, namely to consider the simplicial monoid $\mathrm{GL}(R_\bullet)$ with p-simplices $\mathrm{GL}(R_p)$. This leads to degreewise K-theory, $\big|[p] \to K(R_p)\big|$, which however is not a homotopy invariant of $R_\bullet$, and does not agree with (3.3.1) in general.

For a map of (simplicial) rings $R_\bullet \to S_\bullet$ we write

$$K(R_\bullet \to S_\bullet) = \mathrm{hF}\big(K(R_\bullet) \to K(S_\bullet)\big).$$

Lemma 3.3.2 ([G2]). *Let $R_\bullet$ be a simplicial ring and $I_\bullet \subset R_\bullet$ a (degreewise) square zero ideal. Then*

$$K(R_\bullet \to R_\bullet/I_\bullet) \sim \big|[p] \to K\big(R_p \to R_p/I_p\big)\big|.$$

(There is a little gap in the argument from [G2], lemma I.2.2 where it was used without proof that the diagram

$$\begin{array}{ccc} B\widehat{\mathrm{GL}}(R_\bullet) & \longrightarrow & B\mathrm{GL}(R_\bullet) \\ \downarrow & & \downarrow \\ B\widehat{\mathrm{GL}}(R_\bullet)^+ & \longrightarrow & B\mathrm{GL}(R_\bullet)^+ \quad \text{(Quillen's plus)} \end{array}$$

is homotopy Cartesian. This was repaired in [FOV]).

Definition 3.3.3 ([W2]). The stable K-theory $K^s(A;V)$ is the functor

$$K^s(A;V) = \varinjlim_n \Omega^{n+1} K\big(A \ltimes \tilde{V}(S^n_\bullet) \to A\big)$$

The limit system in the definition, i.e. the maps from $K\big(A \oplus \tilde{V}(S^n_\bullet) \to A\big)$ to $\Omega K\big(A \oplus \tilde{V}(S^{n+1}_\bullet) \to A\big)$, are the ones given in (3.1.1). $K^s(A;V)$ is a spectrum whose k'th space may be given by replacing the $(n+1)$'st loop space in the definition by the $(n+1-k)$'th loop space.

The lemma above shows that we might as well have defined the stable K-theory degreewise as

$$K^s(A,V) = \varinjlim_n \Omega^{n+1} \big|[r] \to K\big(A \ltimes \tilde{V}(S^n_r) \to A\big)\big| \tag{3.3.4}$$

which is the point of view to be used below.

The reader can note the resemblance of K^s with the algebraic "tangent space" of K-theory:

$$TK(A,V) = K(A \ltimes V \to A).$$

In $K^s(A,V)$ one has further made V "small" by passing to the simplicial setting, where one can make V "close to the 0-module" upon replacing it with $\tilde{V}(S^n_\bullet)$, which "approaches 0" in the homotopy sense as $n \to \infty$. Further details on stable K-theory can be found in [K].

The above can be generalized to the setting of FSP's. Indeed, let L be an FSP and M a module over L as in sect. 2.3. One defines

$$(L \ltimes M[n])(X) = L(X) \vee (S^n \wedge M(X))$$

(one could also use $L(X) \vee M(S^n \wedge X)$ as the two definitions give stably equivalent FSP's).

$$\begin{aligned} K^s(L;M) &= \varinjlim \Omega^{n+1} K\big(L \ltimes M[n] \to L\big) \\ \mathrm{TC}^s(L;M) &= \varinjlim \Omega^{n+1} \mathrm{TC}\big(L \ltimes M[n] \to L\big). \end{aligned} \tag{3.3.5}$$

The topological Dennis trace

$$\mathrm{tr}\colon K(L,M) \to \mathrm{TH}(L,M)$$

factors over $K^s(L,M)$ and long ago, Waldhausen conjectured that the resulting map

$$K^s(L;M) \xrightarrow{\mathrm{tr}} \mathrm{TH}(L;M) \tag{3.3.6}$$

is an equivalence.

The rest of the section is a presentation of the Dundas-McCarthy proof of (3.3.6) in the linear situation, corresponding to $L = \tilde{A}$, $M = \tilde{V}$, the FSP's associated with a ring and a bimodule, and of Hesselholt's corresponding result for TC.

Consider the category $\mathcal{P}(A, V)$ of pairs (P, α) of a projective A-module P and an A-linear homomorphism $\alpha\colon P \to P \otimes_A V$. The morphisms from (P, α) to (P', α') are maps $f\colon P \to P'$ such that

$$\begin{array}{ccc} P & \xrightarrow{\alpha} & P \otimes_A V \\ \downarrow{\scriptstyle f} & & \downarrow{\scriptstyle f\otimes 1} \\ P' & \xrightarrow{\alpha'} & P' \otimes_A V \end{array}$$

are commutative.

The K-theory of $\mathcal{P}(A, V)$ will be denoted $K^{\mathrm{cy}}(A; V)$; in the simplicial setting we make the following

Definition 3.3.7. For a simplicial A-bimodule $V_\bullet$,

$$K^{\mathrm{cy}}(A; V_\bullet) = \big|[r] \to K(\mathcal{P}(A; V_r))\big|.$$

Clearly, $K^{\mathrm{cy}}(A; 0) = K(A)$ and we set $\tilde{K}^{\mathrm{cy}}(A; V_\bullet) = \mathrm{hF}\big(K(A; V_\bullet) \to K(A)\big)$. **Lemma 3.3.8.** *There are homotopy equivalences*

(i) $K(A \ltimes V_\bullet \to A) \sim \tilde{K}^{\mathrm{cy}}(A; \delta\tilde{V}_\bullet(S^1_\bullet))$

(ii) $K^s(A; V) \sim \underset{\overrightarrow{n}}{\mathrm{holim}}\,\Omega^{n+1}\tilde{K}^{\mathrm{cy}}(A, \tilde{V}(S^{n+1}_\bullet))$.

Proof. The second statement follows from the first since $N_\bullet\tilde{V}(X_\bullet) = \tilde{V}(S^1_\bullet \wedge X_\bullet)$, so we have left to prove (i).

Since we are considering the relative groups, we may replace $\mathcal{P}(A, V)$ by $\mathcal{F}(A, V)$ in the definitions. But

$$N_\bullet(i\mathcal{F}_V) \sim \coprod_{k=1}^{\infty} N_\bullet(im_k\mathcal{F}_V) \tag{3.3.9}$$

where $m_k\mathcal{F}_V$ is the full subcategories of pairs (A^k, α), $\alpha \in \mathrm{M}_k(V)$ and where i indicates that we are only considering isomorphisms. An r-simplex of

$N_\bullet(im_k\mathcal{F}(A,V))$ is determined by a string $(\alpha_0; f_1, \dots, f_k)$ with $f_i \in \mathrm{GL}_k(A)$ and $\alpha_0 \in \mathrm{M}_k(V)$. Thus

$$N_\bullet(im_k\mathcal{F}(A,V)) \cong N_\bullet^{\mathrm{cy}}(\mathrm{GL}_k(A); \mathrm{M}_k(V)) \tag{3.3.10}$$

upon sending $(\alpha_0; f_1, \dots, f_k)$ into $((f_1 \cdots f_k)^{-1}; f_1, \dots, f_k)$, cf. (2.1.1). From (2.1.10) we have

$$\delta N_\bullet^{\mathrm{cy}}(\mathrm{GL}_k(A), N_\bullet \mathrm{M}_k(V)) \cong N_\bullet(\mathrm{GL}_k(A \ltimes V)) \tag{3.3.11}$$

so, extending (degreewise) to simplicial modules $V_\bullet$, and taking group completions, the result follows. □

Theorem 3.3.12 ([DM1]). *For any A-bimodule M, the trace defines an equivalence*

$$K^s(A,M) \xrightarrow{\sim} \mathrm{THH}(A,M).$$

Proof. We use the model $\mathrm{THH}^{\oplus}$ for THH. Indeed corollary 3.2.16 gives

$$\mathrm{THH}(A,M) \sim \varinjlim \Omega^p \left| \left(\bigoplus_{C \in S_\bullet^{(p)} \mathcal{P}_A} \mathrm{Hom}_{S_\bullet^{(p)}}(\mathbf{C}, \mathbf{C} \otimes_A M) \right) \right| .$$

By definition

$$K^{\mathrm{cy}}(A;M) = \Omega^p |K(S_\bullet^{(p)}\mathcal{P}(A,M))| = \Omega^p \left| \left(\coprod_{C \in S_\bullet^{(p)} \mathcal{P}} \mathrm{Hom}_{S_\bullet^{(p)}}(\mathbf{C}, \mathbf{C} \otimes_A M) \right) \right| ,$$

with $\mathcal{P} = \mathcal{P}_A$. We shall compare these definitions when M is replaced by the simplicial bimodule $W_\bullet = \tilde{M}(S_\bullet^n)$, M applied to the simplicial n-sphere. Both functors are defined degreewise

$$\begin{aligned} \mathrm{THH}(A, W_\bullet) &= |[r] \to \mathrm{THH}(A; W_r)| \\ K^{\mathrm{cy}}(A, W_\bullet) &= |[r] \to K^{\mathrm{cy}}(A; W_r)| . \end{aligned}$$

Actually, we are interested in the relative functor $\tilde{K}^{\mathrm{cy}}(A, W_\bullet)$. Consider the coCartesian diagram

$$\begin{array}{ccc} \left| \coprod_{C \in S_\bullet^{(p)} \mathcal{P}} \mathrm{Hom}_{S_\bullet^{(p)}}(\mathbf{C}, \mathbf{C} \otimes_A W_\bullet) \right| & \longrightarrow & |S_\bullet^{(p)}\mathcal{P}| \\ \Big\downarrow {\scriptstyle -/|S_\bullet^{(p)}\mathcal{P}|} & & \Big\downarrow {\scriptstyle -/|S_\bullet^{(p)}\mathcal{P}|} \\ \left| \bigvee_{C \in S_\bullet^{(p)} \mathcal{P}} \mathrm{Hom}_{S_\bullet^{(p)}}(\mathbf{C}, \mathbf{C} \otimes_A W_\bullet) \right| & \longrightarrow & * \end{array}$$

Each of the spaces are at least $(p-1)$-connected, since the $S_\bullet$-construction applied to any category adds one to the connectivity. It follows the vertical homotopy fiber is $(2p-2)$-equivalent to the space $|S_\bullet^{(p)}\mathcal{P}|$ which was divided out, and hence that the vertical homotopy fibers agree in the same range. Since $p \gg 2n$, it follows that

$$\tilde{K}^{\mathrm{cy}}(A; W_\bullet) \sim_{2n} \Omega^p \left| \bigvee_{\mathbf{C}\in S_\bullet^{(p)}\mathcal{P}} \mathrm{Hom}_{S_\bullet^{(p)}}(\mathbf{C}, \mathbf{C}\otimes_A W_\bullet) \right|.$$

It is clear from the definition of trace given in (3.2.11) that it (under the equivalences above) corresponds to the natural inclusion

$$\bigvee_{\mathbf{C}\in S_\bullet^{(p)}\mathcal{P}} \mathrm{Hom}_{S_\bullet^{(p)}}(\mathbf{C}, \mathbf{C}\otimes_A W_\bullet) \to \bigoplus_{\mathbf{C}\in S_\bullet^{(p)}\mathcal{P}} \mathrm{Hom}_{S_\bullet^{(p)}}(\mathbf{C}, \mathbf{C}\otimes_A W_\bullet).$$

This map is $(p+2n)$-connected. Indeed, the inclusion of a wedge of n-connected spaces into the product is $2n$-connected, so the corresponding map indexed over $S_r^{(p)}\mathcal{P}$ is $(2n+1)$-connected. Thus the homotopy fiber of the map in question is a bisimplicial set $F_{\bullet,\bullet}$ with $|F_{r,\bullet}|$ $(2n+1)$-connected for $r \geq p$ and $|X_{r,\bullet}|$ contractible for $r < p$ (since $|S_\bullet^{(p)}|$ is $(p-1)$-connected. The standard spectral sequence

$$H_r(H_s(F_{\bullet,\bullet})) \Rightarrow H_{r+s}(\delta F_{\bullet,\bullet})$$

is zero for $r < p$ and $s \leq 2n+1$, so gives the connectivity conclusion. The theorem now follows from the equivalences

$$K^s(A; M) = \underrightarrow{\mathrm{holim}}\, \Omega^{n+1} K^{\mathrm{cy}}(A; \tilde{M}(S_\bullet^{n+1}))$$
$$\mathrm{TH}(A; M) = \underrightarrow{\mathrm{holim}}\, \Omega^{n+1} \mathrm{TH}(A; \tilde{M}(S_\bullet^{n+1})). \qquad \square$$

We remark that the above proof also contains a proof of

Addendum 3.3.13. *For a simplicial A bimodule $V_\bullet$,*

(i) $$\tilde{K}^{\mathrm{cy}}(A; V_\bullet) = \varinjlim_p \Omega^p \left| \bigvee_{\mathbf{C}\in S_\bullet^{(p)}\mathcal{P}} \mathrm{Hom}_{S_\bullet^{(p)}}(\mathbf{C}, \mathbf{C}\otimes_A V_\bullet) \right|$$

(ii) $$\tilde{K}^{\mathrm{cy}}(A; \tilde{V}(X_\bullet)) = \varinjlim_p \Omega^p \left| \bigvee_{\mathbf{C}\in S_\bullet^{(p)}\mathcal{P}} \mathrm{Hom}_{S_\bullet^{(p)}}(\mathbf{C}, \mathbf{C}\otimes_A V_\bullet)^{\sim}(X_\bullet) \right|.$$

Theorem 3.3.14 ([H1]). *For any* FSP *L and L-bimodule M, the profinite completions of* $\mathrm{TC}^s(L;M)$ *and* $\mathrm{TH}(L;M)$ *are equivalent.*

Proof (sketch). Recall from (3.3.5) that

$$\mathrm{TC}^s(L;M) = \varinjlim \Omega^{n+1}\mathrm{TC}\,(L \ltimes M[n] \to L)$$

where $L \ltimes M[n]$ is the FSP

$$(L \ltimes M[n])(X) = L(X) \vee (S^n \wedge M(X)).$$

We may decompose

$$(L(S^{x_0}) \vee M[n](S^{x_0})) \wedge \cdots \wedge (L(S^{x_r}) \vee M[n](S^{x_r}))$$

into a wedge, and collect the factors which contain a given number of copies of $M[n](S^x)$. This gives a decomposition of cyclic spaces

$$\mathrm{TH}(L \ltimes M[n]) = \bigvee_{a=0}^{\infty} T_a(L;M[n])$$

with $T_0(L;M[n]) = \mathrm{TH}(L)$. Moreover $T_1(L;M[n])$ is a simplicial spectrum whose k-simplices has exactly one copy of $M[n]$, but sitting at any of the $(k+1)$ positions available, i.e.

$$T_1(L;M[n])_k = C_{k+1+} \wedge \mathrm{TH}(L;M[n]).$$

The realization of this cyclic space is $S^1_+ \wedge \mathrm{TH}(L;M[n])$ with its natural action of S^1 (in the first factor), so

$$T_1(L;M[n]) = S^1_+ \wedge \mathrm{TH}(L;M[n]).$$

The cyclotomic structure map R_p maps

$$R_p\colon T_a(L;M[n])^{C_{p^r}} \to T_{a/p}(L;M[n])^{C_{p^{r-1}}}$$

if $p|a$ and trivially otherwise. By (2.4.6) this map is $(na-1)$-connected. Hence if $(k,p)=1$

$$T_{p^s k}(L;M[n])^{C_{p^r}} \sim_{kpn-1} T_k(L;M[n])^{C_{p^{r-s}}},$$

and again by (2.4.6), $T_k(L;M[n])^{C_{p^{r-s}}} \sim T_k(L;M[n])_{hC_{p^{r-s}}}$, which is $(kn-1)$-connected (as T_k contains k copies of the $(n-1)$-connected $M[n]$).

We are only interested in the range $< 2n$, so $T_{kp^s}(L;M[n])^{C_{p^r}}$ can be disregarded when $k>1$. Thus by theorem 2.5.5,

$$\mathrm{TC}\,(L \ltimes M[n] \to L)^{\wedge}_p \sim_{2n-1} \left(\underset{\mathrm{I}_p}{\operatorname{holim}} \left(\bigvee_{s=0}^{\infty} T_{p^s}(L;M[n])\right)^{C_{p^r}}\right)^{\wedge}_p.$$

Moreover,

$$R_p^{(s)}\colon T_{p^s}(L;M[n])^{C_{p^r}} \sim_{2n-1} T_1(L;M[n])^{C_p^{r-s}} \quad \text{for } r\geq s$$

and $T_{p^s}(L;M[n])^{C_{p^r}} \sim_{2n-1} 0$ if $r<s$. Hence

$$\left(\bigvee_{s=0}^{\infty} T_{p^s}(L;M[n])\right)^{C_{p^r}} \sim_{2n-1} \bigvee_{t=0}^{r} T_1(L;M[n])^{C_{p^t}} = \prod_{t=0}^{r} T_1(L;M[n])^{C_{p^t}}$$

(as we work with spectra, there is no difference between finite wedges and finite products). The R_p-map corresponds to projection on the first r factors, so

$$\underset{\overleftarrow{R_p}}{\operatorname{holim}}\left(\bigvee_{s=0}^{\infty} T_{p^s}(L;M[n])\right)^{C_{p^r}} = \prod_{t=0}^{\infty} T_1(L;M[n])^{C_{p^t}}$$

and by (2.5.4) one concludes that

$$\mathrm{TC}(L\ltimes M[n]\to L)^{\wedge}_p \sim \left(\underset{\overleftarrow{F_p}}{\operatorname{holim}}\, T_1(L;M[n])^{C_{p^t}}\right)^{\wedge}_p.$$

The action of S^1 (and hence C_{p^t}) on

$$T_1(L;M[n]) = S^1_+\wedge \mathrm{TH}(L;M[n])$$

is free, and in this case the action can be divided out, so

$$\underset{\overleftarrow{F_p}}{\operatorname{holim}}\, T_1(L;M[n])^{C_{p^t}} \sim \underset{\longleftarrow}{\operatorname{holim}}\left(S^1/C_{p^t+}\wedge \mathrm{TH}(L;M[n])\right)$$

where the limit on the right is via transfers (in the suspension spectrum $\Sigma^\infty(S/C_{p^t+})$). If we identify $S^1/C_{p^t}=S^1$ then we obtain a (co)fibration of limit systems:

$$\begin{array}{ccccc}
\mathrm{TH}(L;M[n]) & \longrightarrow & S^1_+\wedge \mathrm{TH}(L;M[n]) & \longrightarrow & S^1\wedge \mathrm{TH}(L;M[n]) \\
\downarrow p & & \downarrow F_p & & \downarrow \mathrm{id} \\
\mathrm{TH}(L;M[n]) & \longrightarrow & S^1_+\wedge \mathrm{TH}(L;M[n]) & \longrightarrow & S^1\wedge \mathrm{TH}(L;M[n])
\end{array}$$

This implies a cofibration in the limit. Since

$$\underset{\longleftarrow}{\operatorname{holim}}(\mathrm{TH}(L;M[n]),p)^{\wedge}_p \sim 0$$

we are finished. □

3.4 McCarthy's theorem.

The presentation in this section is my writeup of lectures given by McCarthy in Aarhus, July 1994.

Theorem 3.4.1 (McCarthy). *Let $R \to S$ be a surjection of rings with nilpotent kernel. Then the diagram*

$$\begin{array}{ccc} K(R)^\wedge & \longrightarrow & \mathrm{TC}(R)^\wedge \\ \downarrow & & \downarrow \\ K(S)^\wedge & \longrightarrow & \mathrm{TC}(S)^\wedge \end{array}$$

of profinitely completed spectra is homotopy Cartesian. In particular

$$K(R \to S)^\wedge \sim \mathrm{TC}(R \to S)^\wedge.$$

The obvious induction shows that it suffices to prove the theorem when the kernel is a square zero ideal; this will be assumed in the rest of the section.

Associated to a simplicial ring $R_\bullet$ we have the FSP

$$\tilde{R}_\bullet(X) = \left|[s] \to \tilde{R}_s(X)\right|.$$

We write $\mathrm{TC}(R_\bullet)$ instead of $\mathrm{TC}(\tilde{R}_\bullet)$. If $R_\bullet \to R'_\bullet$ is a simplicial equivalence (i.e. $|R_\bullet| \to |R'_\bullet|$ a homotopy equivalence) then the induced map of FSP's $\tilde{R}_\bullet \to \tilde{R}'_\bullet$ is a stable equivalence in the sense that

$$\varinjlim_n \Omega^n(\tilde{R}_\bullet(S^n)) \to \varinjlim_n \Omega^n(\tilde{R}'_\bullet(S^n))$$

is an equivalence, and in this case

$$\mathrm{TC}(R_\bullet) \xrightarrow{\sim} \mathrm{TC}(R'_\bullet)$$

cf. sect. 2.6, so $\mathrm{TC}(R_\bullet)$ only depends on the homotopy type of $R_\bullet$. On the other hand, we have the possibility of calculating TC degreewise. In contrast to K-theory where the two definitions do not agree in general we have

Proposition 3.4.2. $\mathrm{TC}(R_\bullet) \sim |[s] \to \mathrm{TC}(R_s)|$.

Proof. Since

$$\Omega^{n_0+\cdots+n_k}\left|[s] \to \tilde{R}_s(S^{n_0}) \wedge \cdots \wedge \tilde{R}_s(S^{n_k})\right| \sim$$
$$\left|[s] \to \Omega^{n_0+\cdots+n_k}(\tilde{R}_s(S^{n_0}) \wedge \cdots \wedge \tilde{R}_s(S^{n_k}))\right|$$

we see that the topological Hochschild spectrum $\mathrm{TH}(\tilde{R}_\bullet)$ can be calculated degreewise:

$$\mathrm{TH}(\tilde{R}_\bullet) \sim \left|[s] \to \mathrm{TH}(\tilde{R}_s)\right|.$$

The fundamental cofibration sequence of proposition 2.4.3 then shows that the same assertion is true for fixed sets

$$\mathrm{TH}(\tilde{R}_\bullet)^{C_{p^n}} \sim \left|[s] \to \mathrm{TH}(\tilde{R}_s)^{C_{p^n}}\right|$$

and upon taking inverse limit

$$\mathrm{TF}(\tilde{R}_\bullet, p) \sim \left|[s] \to \mathrm{TF}(\tilde{R}_s, p)\right|$$

cf. (2.5.3) for notation. There is a salient point here: realization does not in general commute with homotopy inverse limits; however in the above situation it does as $\mathrm{TH}(\tilde{R}_s) \sim \Omega\mathrm{THH}(\tilde{R}_s; S^1)$, so $\mathrm{TH}(\tilde{R}_s)$ is equivalent to a Kan simplicial set. For such, realization do commute with homotopy inverse limits.

Finally the homotopy fibrations

$$\mathrm{TC}(\tilde{R}_\bullet, p) \longrightarrow \mathrm{TF}(\tilde{R}_\bullet, p) \xrightarrow{R_p - \mathrm{id}} \mathrm{TF}(\tilde{R}_\bullet, p)$$

$$\mathrm{TC}(\tilde{R}_s, p) \longrightarrow \mathrm{TF}(\tilde{R}_s, p) \xrightarrow{R_p - \mathrm{id}} \mathrm{TF}(\tilde{R}_s, p)$$

show that $\mathrm{TC}(\tilde{R}_\bullet, p)$ can be calculated degreewise. Now apply theorem 2.5.5 to obtain the result for $\mathrm{TC}(\tilde{R}_\bullet)$. □

Lemma 3.4.3 ([G2]). *If the theorem is true in the special case where R is a semi-direct product ring $R = A \ltimes M$ and $S = A$ then it is true in general.*

Proof. Goodwillie associates to S a simplicial ring $\Phi_\bullet(S)$ with a simplicial map $\Phi_\bullet(S) \to S$ (when S is regarded as the constant simplicial ring) such that

(i) $\Phi_r(S)$ is free associative for each r

(ii) $|\Phi_\bullet(S)| \xrightarrow{\phi} S$ is an equivalence.

Indeed, $\Phi_\bullet(S)$ is the simplicial ring with $\Phi_r(S) = (FG)^{r+1}(S)$ where G is the forgetful functor from rings to sets and F its left adjoint free functor: $\Phi_\bullet(S)$ is the "bar-construction", cf. [G2], sect. I.1.6. Write $A_\bullet = \Phi_\bullet(S)$ and consider the (degreewise) pull-back

$$\begin{array}{ccc} B_\bullet & \longrightarrow & A_\bullet \\ \downarrow{\scriptstyle \bar{\phi}} & & \downarrow{\scriptstyle \phi} \\ R & \longrightarrow & S \end{array}$$

Then $M = \ker(B_\bullet \to A_\bullet)$ is the constant ideal $M = \ker(R \to S)$. Since ϕ, and hence $\bar{\phi}$, is an equivalence

$$K(R \to S) \sim K(B_\bullet \to A_\bullet).$$

The latter can be calculated degreewise by lemma 3.3.2,

$$K(B_\bullet \to A_\bullet) \sim |[r] \to K(B_r \to A_r)|\,.$$

Now A_r is free, so $B_r \to A_r$ is a split surjection, and hence $B_r = M_r \ltimes A_r$. With the assumption,

$$K(B_r \to A_r)^\wedge \sim \mathrm{TC}(B_r \to A_r)^\wedge$$

so in conclusion

$$K(B \to A)^\wedge \sim |[r] \to \mathrm{TC}(B_r \to A_r)^\wedge| \sim \mathrm{TC}(B \to A)^\wedge$$

by the previous proposition. □

The idea behind the proof of theorem 3.4.1 is to use calculus of functors on the cyclotomic trace

$$\mathrm{trc}\colon K\left(A \ltimes \tilde{M}(X_\bullet) \to A\right) \to \mathrm{TC}\left(A \ltimes \tilde{M}(X_\bullet) \to A\right)$$

cf. sect. 3.1. First we need:

Proposition 3.4.4. *For any ring A and bimodule M,*

$$\begin{array}{ll} \text{(i)} & X_\bullet \to K(A \ltimes \tilde{M}(X_\bullet)) \\ \text{(ii)} & X_\bullet \to \mathrm{TC}(A \ltimes \tilde{M}(X_\bullet)) \end{array}$$

are (-1)-analytic as functors from based simplicial sets to spectra.

Proof. For K-theory we can use the equivalence of lemma 3.3.8(i),

$$K\left(A \ltimes \tilde{M}(X_\bullet) \to A\right) \sim \tilde{K}^{\mathrm{cy}}(A, \tilde{M}(X_\bullet \wedge S^1_\bullet))$$

and the general fact that a functor

$$F\colon \mathrm{s_\bullet sets_*} \to \{\mathrm{spectra}\}$$

is ρ-analytic if (and only if) $F((-)\wedge S^1_\bullet)$ is $(\rho-1)$-analytic. The latter follows directly from the definition of analyticity. Indeed if F is say 0-analytic, and $\mathfrak{X}$ is a strictly coCartesian $(n+1)$-cube with $\mathfrak{X}(\emptyset) \to \mathfrak{X}(s)$ k_s-connected ($k_s \geq 0$)

then the suspended cube has $\mathfrak{X}(\emptyset) \wedge S^1 \to \mathfrak{X}(s) \wedge S^1$ $(k_s + 1)$-connected, so by assumption

$$a\colon F(\mathfrak{X}(\emptyset) \wedge S^1) \to \underset{S \neq \emptyset}{\operatorname{holim}_{\leftarrow}} F(\mathfrak{X}(S) \wedge S^1)$$

is $(q + \Sigma(k_s + 1))$-connected. Hence if F satisfies the condition $E_n(-q, 1)$ then $F((-) \wedge S^1)$ satisfies $E_n(-n - q - 1, 0)$, so is (-1)-analytic.

To see that $\tilde{K}^{cy}(A, \tilde{M}(X_\bullet))$ is 0-analytic we use the description of addendum 3.3.13:

$$\tilde{K}^{cy}(A, \tilde{M}(X_\bullet)) \sim \operatorname{holim} \Omega^p \left(\bigvee_{\mathbf{C} \in S_\bullet^{(p)} \mathcal{P}} \operatorname{Hom}_{S_\bullet^{(p)}}(\mathbf{C}, \mathbf{C} \otimes_A M)^\sim (X_\bullet) \right).$$

Given a strongly coCartesian $(n + 1)$-cube $\mathfrak{X}$. For given $\mathbf{C} \in S_\bullet^{(p)}$, the cube

$$\operatorname{Hom}_{S_\bullet^{(p)}}(\mathbf{C}, \mathbf{C} \otimes_A M)^\sim (\mathfrak{X})$$

is homotopy Cartesian for each p: this is true for $\tilde{M}(\mathfrak{X})$ for any abelian M.

It follows from the dual Blaker-Massey theorem, [G4], theorem 2.6 that the above strongly Cartesian cube is also $n + \Sigma k_s$ coCartesian. Taking wedge over $\mathbf{C} \in S_\bullet^{(p)}$ we obtain an $(n + p + \Sigma k_s)$-coCartesian cube. (The extra p appears because $S_\bullet^{(p)}$ is $(p - 1)$-connected, cf. the last part of the proof for theorem 3.3.12). By [G4], theorem 2.5, the cube

$$\bigvee_{\mathbf{C} \in S_\bullet^{(p)} \mathcal{P}_A} \operatorname{Hom}_{S_\bullet^{(p)}}(\mathbf{C}, \mathbf{C} \otimes_A M_\bullet)^\sim (\mathfrak{X})$$

is $(p + \Sigma k_s)$-Cartesian, and looping down p times there results a (Σk_s)-Cartesian cube. This proves (i).

The FSP associated to the simplicial ring $A \ltimes \tilde{M}(X_\bullet)$ is equivalent to the FSP which sends $Y_\bullet$ to $\tilde{A}(Y_\bullet) \vee \tilde{M}(X_\bullet \wedge Y_\bullet)$. Thus we have the decomposition of spectra

$$\mathrm{TH}\left(A \ltimes \tilde{M}(X_\bullet)\right) \sim \bigvee_{n=0}^{\infty} T_n\left(A; \tilde{M}(X_\bullet)\right)$$

also used in the proof of theorem 3.3.14.

One now first shows that the functor $M^{(n)}(X_\bullet) = \tilde{M}(X_\bullet) \wedge \cdots \wedge \tilde{M}(X_\bullet)$ is (-1)-analytic. This is a non-trivial task. The functor $T_n\left(A, \tilde{M}(X_\bullet)\right)$ involves n smash copies of $\tilde{M}(X_\bullet)$ in each degree, and is thus (-1)-analytic as well. Hence $\mathrm{TH}\left(A \ltimes \tilde{M}(X_\bullet)\right)$ is (-1)-analytic. The cofibrations of spectra (2.4.6)

$$\mathrm{TH}(A \ltimes \tilde{M}(X_\bullet))_{hC_{p^n}} \to \mathrm{TH}(A \ltimes \tilde{M}(X_\bullet))^{C_{p^n}} \to \mathrm{TH}(A \ltimes \tilde{M}(X_\bullet))^{C_{p^{n-1}}}$$

then give (inductively) that each of the fixed sets is (-1)-analytic. Taking inverse limit we see that

$$X_\bullet \mapsto \mathrm{TF}(A \ltimes \tilde{M}(X_\bullet), p)$$

is (-1)-analytic, and then that $\mathrm{TC}(A \ltimes \tilde{M}(X_\bullet), p)$ has the same property. Apply theorem 2.5.5 to complete the proof. □

Lemma 3.4.5. *The functors*

$$X_\bullet \to K(A \ltimes \tilde{M}(X_\bullet))$$
$$X_\bullet \to \mathrm{TC}(A \ltimes \tilde{M}(X_\bullet))$$

satisfies the p-limit axiom (ii) *of sect. 3.1 for each prime p.* □

This is well-known for K-theory. The proof for TC follows the scheme of the previous lemma: first do TH and then induct over the fundamental cofibrations, (2.4.6).

We next evaluate the differential $\partial_x F$ of the two functors in question, cf. definition 3.1.2.

Lemma 3.4.6. *The functors $K(A \ltimes \tilde{M}(X_\bullet))$ and $\mathrm{TC}(A \ltimes \tilde{M}(X_\bullet))^\wedge_p$ have as differentials the spectrum $|[p] \to \mathrm{TH}(A \times M(X_p); M)|$ and its p-completion, respectively.*

Proof. This is really a consequence of results in the previous section, namely theorems 3.3.12 and 3.3.14.

$$\partial_x K(A \ltimes \tilde{M}(X_\bullet)) = \varinjlim \Omega^{n+1} K\left(A \ltimes \tilde{M}(X_\bullet \vee S^n_\bullet) \to A \ltimes \tilde{M}(X_\bullet)\right).$$

But $\tilde{M}(X_\bullet \vee S^n_\bullet) = \tilde{M}(X_\bullet) \oplus \tilde{M}(S^n_\bullet)$ and thus

$$A \ltimes \tilde{M}(X_\bullet \vee S^n_\bullet) = (A \ltimes \tilde{M}(X_\bullet)) \ltimes \tilde{M}(S^n_\bullet)$$

where on the right hand side the action is through the projection $A \ltimes \tilde{M}(X_\bullet) \to A$. Write $B_\bullet = A \ltimes \tilde{M}(X_\bullet)$. The analogue of lemma 3.3.2 for bisimplicial rings shows that

$$K\left(B_\bullet \ltimes \tilde{M}(S^n_\bullet) \to B_\bullet\right) \sim \left|[p] \to K\left(B_p \ltimes \tilde{M}(S^n_\bullet) \to B_p\right)\right|$$

and by 3.3.8(i) and 3.3.12

$$\varinjlim \Omega^{n+1} K\left(B_p \ltimes \tilde{M}(S^n_\bullet) \to B_p\right) \sim$$
$$\varinjlim \Omega^{n+1} \tilde{K}^{\mathrm{cy}}(B_p, \tilde{M}(S^{n+1}_\bullet)) \sim \mathrm{TH}(B_p; M).$$

Similarly,

$$\varinjlim \Omega^{n+1}\mathrm{TC}(B_p \ltimes \tilde{M}(S^n_\bullet))^\wedge_p \sim \mathrm{TH}(B_p; M)^\wedge_p$$

by theorem 3.3.14, and proposition 3.4.2 supplies the conclusion. □

Finally, one must check that the p-completion of $\partial_x \mathrm{trc}$ induces the equivalence. This follows from the following homotopy commutative diagram of spectra, where $M_\bullet = \tilde{M}(S^m_\bullet)$ for an B-bimodule M:

$$\begin{array}{ccccc}
 & & \widetilde{\mathrm{TC}}(B\oplus M_\bullet)^\wedge_p & \overset{\sim 2m}{\longleftarrow} & \operatorname{holim}\limits_{\longleftarrow} S^1/C_{p^r+} \wedge \mathrm{TH}(R, M_\bullet)^\wedge_p \\
 & \overset{\mathrm{trc}}{\nearrow} & \downarrow & & \downarrow \\
\tilde{K}(B\oplus M_\bullet)^\wedge_p & \overset{\mathrm{tr}}{\longrightarrow} & \widetilde{\mathrm{TH}}(B\oplus M_\bullet)^\wedge_p & \overset{\sim 2m}{\longleftarrow} & S^1_+ \wedge \mathrm{TH}(R, M_\bullet)^\wedge_p \\
 & \underset{\sim 2m+2}{\overset{\mathrm{tr}}{\searrow}} & & & \downarrow \\
 & & \widetilde{\mathrm{TH}}(B; \tilde{M}_\bullet(S^1))^\wedge_p & \underset{\sim 2m}{\longleftarrow} & S^1 \wedge \mathrm{TH}(R, M)^\wedge_p
\end{array}$$

The two upper vertical maps are the natural ones which map a homotopy inverse limit into its initial term. The right-hand vertical composition is an equivalence (cf. the proof of theorem 3.3.14), and the notation is

$$\tilde{K}(B\oplus M_\bullet) = K\,(B\oplus M_\bullet \to B)$$

etc. This completes McCarthy's proof of theorem 3.4.1, as I have understood his Aarhus lectures.

Addendum 3.4.7. (McCarthy) *Suppose $f_\bullet\colon R_\bullet \to S_\bullet$ is a map of simplicial rings and that $\pi_0(|f_\bullet|)$ is surjective and has nilpotent kernel. Then*

$$\begin{array}{ccc}
K(R_\bullet)^\wedge & \longrightarrow & \mathrm{TC}(R_\bullet)^\wedge \\
\downarrow & & \downarrow \\
K(S_\bullet)^\wedge & \longrightarrow & \mathrm{TC}(S_\bullet)^\wedge
\end{array}$$

is homotopy Cartesian.

The proof is the same with the exception of lemma 3.4.3 where one has to add an extra step, passing from nilpotency on the π_0-level to nilpotency on the simplicial ring level, cf. [G2], lemma I.3.3.

3.5 Dundas' theorem.

This section gives a brief outline of the proof from [D] of Goodwillie's conjecture:

Theorem 3.5.1. (Dundas). *Let* $f\colon L_1 \to L_2$ *be a map of FSP's with* $\pi_0(f)$ *surjective and* $\ker \pi_0(f)$ *nilpotent. Then the diagram*

$$\begin{array}{ccc} K(L_1)^\wedge & \longrightarrow & \mathrm{TC}(L_1)^\wedge \\ \downarrow & & \downarrow \\ K(L_2)^\wedge & \longrightarrow & \mathrm{TC}(L_2)^\wedge \end{array}$$

is homotopy Cartesian.

The general idea is to approximate the FSP's L_i by FSP's coming from simplicial rings, and then use McCarthy's theorem 3.4.7 to derive the conclusion. This is similar in spirit to the cosimplicial resolution of a space (simplicial set) by Eilenberg-MacLane spaces.

Let X be a $(k-1)$-connected space (simplicial set) with $k > 1$. By the Hurewicz theorem, $\pi_k X \xrightarrow{\cong} H_k X$ and $\pi_{k+1} X \to H_{k+1} X$ is surjective. In other words, the linearization map $X \xrightarrow{h} \tilde{\mathbb{Z}}X$ is $(k+1)$-connected. The relative version of this is as follows. Suppose $f\colon X \to Y$ is a $(k+1)$-connected map and X is $(k-1)$-connected. Then the 2-cube

$$\begin{array}{ccc} X & \xrightarrow{f} & Y \\ \downarrow{\scriptstyle h} & & \downarrow{\scriptstyle h} \\ \tilde{\mathbb{Z}}X & \xrightarrow{\tilde{\mathbb{Z}}f} & \tilde{\mathbb{Z}}Y \end{array} \qquad (3.5.2)$$

is $(k+2)$-Cartesian in the sense of sect. 3.1.

Indeed, let C be the (homotopy) cofiber of f, and let F be the homotopy fiber. Then F is k-connected and C is $(k+1)$-connected, and the left hand vertical map is the diagram

$$\begin{array}{ccccc} F & \longrightarrow & X & \longrightarrow & Y \\ \downarrow & & \downarrow & & \downarrow \\ \Omega C & \longrightarrow & * & \longrightarrow & C \end{array}$$

is $(k+2)$-connected. (This follows for example from the Serre spectral sequence of the involved homotopy fibrations). On the other hand, $\tilde{\mathbb{Z}}(-)$ sends a cofibration into a fibration, so $\mathrm{hF}(\tilde{\mathbb{Z}}f) \sim \Omega\tilde{\mathbb{Z}}C$: apply $\tilde{\mathbb{Z}}$ the the right hand coCartesian square above. Since C is $(k+1)$-connected, $\Omega C \to \Omega\tilde{\mathbb{Z}}C$ is $(k+2)$-connected. Thus $F \to \mathrm{hF}(\tilde{\mathbb{Z}}f)$ is $(k+2)$-connected; its homotopy fiber is equal to the homotopy fiber of

$$a\colon X \to \underset{\longleftarrow}{\mathrm{holim}}\left(\tilde{\mathbb{Z}}X \xrightarrow{\tilde{\mathbb{Z}}f} \tilde{\mathbb{Z}}Y \longleftarrow Y\right)$$

so (3.5.2) is $(k+2)$-Cartesian. Roughly the same argument proves

Lemma 3.5.3. ([D]). *Let $\mathfrak{X}$ be an $(n+k)$-Cartesian n-cube, $k>1$ such that each sub m-cube is $(m+k)$-Cartesian. Then the $(n+1)$-cube $\mathfrak{X} \to \tilde{\mathbb{Z}}\mathfrak{X}$ is $(n+1+k)$-Cartesian.* □

Starting now with a $(k-1)$-connected space, one can inductively define n-cubes $\mathfrak{Z}_n(X)$ as follows:

$$\mathfrak{Z}_1(x) = \left\{X \to \tilde{\mathbb{Z}}X\right\}, \qquad \mathfrak{Z}_2(X) = \left\{\begin{array}{ccc} X & \xrightarrow{h_X} & \tilde{\mathbb{Z}}X \\ \downarrow{\scriptstyle h_X} & & {\scriptstyle h_{\tilde{\mathbb{Z}}X}}\downarrow \\ \tilde{\mathbb{Z}}X & \xrightarrow{\tilde{\mathbb{Z}}h_X} & \tilde{\mathbb{Z}}\tilde{\mathbb{Z}}X \end{array}\right\},$$

and in general

$$\mathfrak{Z}_n(X) = \left\{\mathfrak{Z}_{n-1}(X) \to \tilde{\mathbb{Z}}(\mathfrak{Z}_{n-1}(X))\right\}.$$

The lemma tells us that $\mathfrak{Z}_n(X)$ is $(n+k)$-Cartesian. For an FSP L, each vertex $\mathfrak{Z}_n(L(X))_S$ defines a new FSP $\mathfrak{Z}_n(L)_S$ with $L = \mathfrak{Z}_n(L)_\emptyset$, and with

$$a_L(X)\colon L(X) \to \underset{\substack{\longleftarrow \\ S\neq\emptyset}}{\mathrm{holim}}\, \mathfrak{Z}_n(L)_S(X)$$

$(n+k)$-connected when X (hence $L(X)$) is $(k-1)$-connected.

One could similarly start with the functor $\tilde{\mathbb{Z}}^q = \tilde{\mathbb{Z}} \circ \cdots \circ \tilde{\mathbb{Z}}$ instead of $\tilde{\mathbb{Z}}$. It is still true that $X \to \tilde{\mathbb{Z}}^q X$ is $(k+1)$-connected for a $(k-1)$-connected X, and one obtains corresponding cubes $\mathfrak{Z}_n^q(L)$ with $a_L^q(X)$ $(n+k)$-connected.

Proposition 3.5.4. *The map a_L induces a map*

$$\mathrm{TC}(L)_p^\wedge \to \underset{\substack{\longleftarrow \\ S\neq\emptyset}}{\mathrm{holim}}\, \mathrm{TC}(\mathfrak{Z}_n(L)_S)_p^\wedge$$

which is $(n-1)$-connected.

Proof. Here is Dundas' argument. It is enough to show that

$$\mathrm{TH}(L) \to \underset{\substack{\longleftarrow \\ S\neq\emptyset}}{\mathrm{holim}}\, \mathrm{TH}(\mathfrak{Z}_n(L)_S)$$

is n-connected, since inductive use of the fundamental cofibration then gives the same conclusion for all C_{p^n}-fixed sets, hence for $\mathrm{TF}(L,p)$, and finally for $\mathrm{TC}(L,p)$ with n replaced by $n-1$.

Now $\mathrm{TH}(L)$ is the prespectrum $\{|\mathrm{THH}_\bullet(L;S^m)|\}_m$, and it suffices to argue that

$$\mathrm{THH}_r(L;S^m) \to \underset{S\neq\emptyset}{\underleftarrow{\mathrm{holim}}}\,\mathrm{THH}_r(\mathfrak{Z}_n(L)_S;S^m)$$

is $(n+m)$-connected for all r. This is lemma 3.5.2 when $r=0$. In general,

$$\mathrm{THH}_r\left(\mathfrak{Z}_n(L)_S;S^m\right) \sim \mathfrak{Z}_n^{r+1}\left(\mathrm{THH}_r(L;S^m)\right)_S. \tag{2}$$

The map is induced from the natural map

$$\sigma\colon \tilde{\mathbb{Z}}^q L(S^{x_0})\wedge\cdots\wedge\tilde{\mathbb{Z}}^q L(S^{x_r}) \to \tilde{\mathbb{Z}}^{q(r+1)}\left(L(S^{x_0})\wedge\cdots\wedge L(S^{x_r})\right).$$

In turn, σ is constructed from iterated use of the assembly map $X\wedge\tilde{\mathbb{Z}}Y \to \tilde{\mathbb{Z}}(X\wedge Y)$. For example, $\tilde{\mathbb{Z}}X\wedge\tilde{\mathbb{Z}}Y \xrightarrow{\sigma} \tilde{\mathbb{Z}}(X\wedge\tilde{\mathbb{Z}}Y) \xrightarrow{\tilde{\mathbb{Z}}\sigma} \tilde{\mathbb{Z}}\left(\tilde{\mathbb{Z}}(X\wedge Y)\right)$. The equivalence statement (2) amounts to the easy fact that $X\wedge\tilde{\mathbb{Z}}(S^n)\to\tilde{\mathbb{Z}}(X\wedge S^n)$ is $(2n-1)$-connected. To finish the proof one applies (3.5.3) with a_L replaced by a_L^q. □

The next result is of similar complexity but I refrain from giving the proof, and refer the reader to [D].

Proposition 3.5.5. *The map*

$$K(L) \to \underset{S\neq\emptyset}{\underleftarrow{\mathrm{holim}}}\,K(\mathfrak{Z}_n(L)_S)$$

is $(n+1)$-connected. □

For $S\neq\emptyset$, $\mathfrak{Z}_n(L)_S$ is equivalent to an FSP associated to a simplicial ring, namely to a simplicial version of $\underset{k}{\underrightarrow{\lim}}\,\Omega^k\mathfrak{Z}_n(L)_S(S^k)$ and $\pi_0\mathfrak{Z}(L)_S = \pi_0 L$, so theorem 3.4.7 applies to show that

$$K\left(\mathfrak{Z}_n(L_1)_S \to \mathfrak{Z}_n(L_2)_S\right)^\wedge \sim \mathrm{TC}\left(\mathfrak{Z}_n(L_1)_S\to\mathfrak{Z}_n(L_2)_S\right)^\wedge$$

when $S\neq\emptyset$. The two previous propositions combine to give the same for $S=\emptyset$. This completes my outline of theorem 3.5.1.

Let G be a topological (or simplicial) monoid homotopy equivalent to ΩX, and $\tilde{G}$ the corresponding FSP, so that $K(\tilde{G})$ is Waldhausen's $A(X)$.

The theorem applies to $\tilde{G} \to \pi_0\tilde{G} = \widetilde{\pi_1 X}$, and to $\widetilde{\pi_1 X} \to \mathbb{Z}[\pi_1 X]^\sim$, so gives a homotopy Cartesian diagram

$$\begin{array}{ccc} A(X)^\wedge & \longrightarrow & \mathrm{TC}(X)^\wedge \\ \downarrow & & \downarrow \\ K(\mathbb{Z}[\pi_1 X])^\wedge & \longrightarrow & \mathrm{TC}(\mathbb{Z}[\pi_1 X])^\wedge \end{array} \tag{3.5.6}$$

The terms on the right-hand side is examined in the next two chapters, and a lot is known. Thus theorem 3.5.1 to some extend reduces the calculation of $A(X)$ to linear K-theory.

4 The absolute theorems

This chapter outlines the proof of the theorem from [HM] that $K(A)$ and $\mathrm{TC}(A)$ agrees after p-adic completion for a large class of p-complete rings, namely for the rings which are finitely generated modules over Witt vectors of perfect fields k of positive characteristic p. It also calculates TC for the FSP's associated with a group like monoid, and gives the relation to Waldhausen's A-functor.

4.1 General approach to TC calculations.

Since $\mathrm{TC}(L)$ is build out of the fixed sets $\mathrm{TH}(L)^C$ the basic calculational problem is to get a hold of $\pi_*\mathrm{TH}(L)^C$ for the cyclic subgroups of the circle. It suffices by theorem 2.5.5 to let C run over the cyclic p-groups, where we have the fundamental cofibration of sect. 2.4

$$\mathrm{TH}(L)_{hC_{p^n}} \longrightarrow \mathrm{TH}(L)^{C_{p^n}} \xrightarrow{R} \mathrm{TH}(L)^{C_{p^{n-1}}}$$

to ease calculations.

Recall that $\mathrm{TH}(L)$ is the restriction of an S^1-invariant spectrum $T(L)$. In the notation of sect. 2.4, $\mathrm{TH}(L) = j^*T(L)$ where $j\colon \mathcal{U}^{S^1} \to \mathcal{U}$. Moreover, the "geometric fix point" spectrum $\Phi^{C_p}T(L)$ of (2.4.1) is equivalent to $T(L)$ by theorem 2.4.5,

$$\rho^{\#}_{C_p}\Phi^{C_p}T(L) \sim_{S^1} T(L).$$

The general approach to the calculation of $\pi_*T(L)^C$ is to replace $T(L)$ by the function spectrum $F(ES^1_+, T(L))$, and to use spectral sequences for calculating the C_{p^n}-fixed points of the function spectrum. This leaves us then for each FSP L with the problem of how close the natural map

$$\pi_*T(L)^{C_{p^n}} \to \pi_*F(ES^1_+, T(L))^{C_{p^n}}$$

is to be an isomorphism. Here ES^1 is the free contractible S^1-space

$$ES^1 = \bigcup_{n=0}^{\infty} S(\mathbb{C}^{n+1}) = \bigcup_{n=0}^{\infty} S^{2n+1}$$

with its standard S^1-action (orbit space $\mathbb{C}P^\infty$), and $F(ES^1_+, T(L))$ is the equivariant S^1-spectrum whose V'th term is $F(ES^1_+, T(L)(V))$, the space of based maps from $ES^1_+ = ES^1 \cup \{+\}$ into the V'th space of $T(L)$, with S^1 acting by conjugation.

Following [GM] we define for each finite p-group C_{p^n},

$$\hat{\mathbb{H}}(C_{p^n}, T(L)) = \left(F(ES^1_+, T(L)) \wedge \tilde{E}S^1\right)^{C_{p^n}} \tag{4.1.1}$$

and call it the C_{p^n}-Tate spectrum of $T(L)$. It is an S^1/C_{p^n}-equivariant spectrum indexed on $\mathcal{U}^{C_{p^n}}$. The space

$$\widetilde{E}S^1 = \bigcup_{n=0}^{\infty} S(\mathbb{C}^n \oplus \mathbb{R}) = \bigcup_{n=0}^{\infty} S^{2n},$$

with S^1-action induced from complex multiplication in C^n, is contractible but not equivariantly: $(\widetilde{E}S^1)^C = S(\mathbb{R}) = S^0$ for each $C \subseteq S^1$.

Lemma 4.1.2. *For any two based C_{p^n}-spaces X and Y, the restriction to C_p-fixed sets induces a weak C_{p^n}/C_p-homotopy equivalence*

$$F(X, Y \wedge \widetilde{E}S^1)^{C_p} \xrightarrow{\sim} F(X^{C_p}, Y^{C_p}).$$

Proof. We may assume X and Y are C_{p^n}-equivariant CW complexes, e.g. by replacing them with the realization of their singular complexes. The singular set of the C_{p^n} space X is X^{C_p}, so $X - X^{C_p}$ has a free C_{p^n}-action,

$$X = X^{C_p} \cup_\partial (\amalg C_{p^n+} \wedge D^{k_i}).$$

Given ϕ: $X^{C_p} \to Y^{C_p} = (Y \wedge \widetilde{E}S^1)^{C_p}$, one can extend ϕ cell by cell to a C_{p^n}-equivariant map from X to $Y \wedge \widetilde{E}S^1$. Indeed the obstructions to extend lie in

$$\pi_0 F(C_{p^n}{}_+ \wedge S^{k_i}, Y \wedge \widetilde{E}S^1)^{C_{p^n}} = \pi_0 F(S^{k_i}, Y \wedge \widetilde{E}S^1) = 0.$$

This proves that the map is surjective on π_0, and hence on π_n by replacing X by $X \wedge S^n$. Injectivity is similar. □

Recall that the smash product of $T(L) \in S^1\mathcal{SU}$ and a based S^1-space X is the spectrification of the obvious prespectrum, or concretely

$$(T(L) \wedge X)(V) = \varinjlim_{W \supset V} \Omega^{W-V}\left(T(L)(W) \wedge X\right). \tag{4.1.3}$$

It follows from lemma 4.1.2 that

$$\Phi^{C_p} T(L) \sim_{C_{p^n}} \left(T(L) \wedge \widetilde{E}S^1\right)^{C_p}$$

and in particular that

$$\hat{\mathbb{H}}\left(C_{p^n}, T(L)\right) \sim \Phi^{C_p} F\left(ES^1_+, T(L)\right)^{C_{p^n}/C_p} \tag{4.1.4}$$

Let $C \subseteq S^1$ be any subgroup. We have the pair of adjoint functors j_* and j^* of sect. 2.4 where $j \colon \mathcal{U}^C \to \mathcal{U}$ is the inclusion, and the maps from (2.4.2),

$$\begin{aligned} &\tau_C \colon j^*T \wedge_C ES^1_+ \to (T \wedge ES^1_+)^C, \quad C \text{ finite} \\ &\tau_{S^1} \colon \Sigma j^*T \wedge_{S^1} ES^1_+ \to (T \wedge ES^1_+)^{S^1}, \quad C = S^1. \end{aligned} \tag{4.1.5}$$

The maps fit together with the non-equivariant transfer maps

$$\begin{aligned} &\operatorname{trf}_C^D \colon j^*T \wedge_D ES^1_+ \to j^*T \wedge_C ES^1_+, \quad D \supset C \\ &\operatorname{trf}_C^{S^1} \colon \Sigma j^*T \wedge_{S^1} ES^1_+ \to j^*T \wedge_C ES^1_+ \end{aligned}$$

in homotopy commutative diagrams, namely

$$\tau_C \circ \operatorname{trf}_C^D \sim F \circ \tau_D, \quad \tau_C \circ \operatorname{trf}_C^{S^1} \sim F \circ \tau_{S^1} \tag{4.1.6}$$

where F denotes inclusion of fixed sets as usual, cf. [A], [LMS].

Since $\widetilde{E}S^1 = ES^1 * S^0$, the unreduced suspension of ES^1, there is an S^1-equivariant cofibration sequence

$$ES^1_+ \to S^0 \to \widetilde{E}S^1 \to \Sigma(ES^1_+) \to \cdots$$

which induces a cofibration of equivariant spectra upon smashing it with the S^1-equivariant function spectrum $F\left(ES^1_+, T(L)\right)$. We take C_{p^n}-fixed sets and apply (4.1.4) and (4.1.5) to get the norm cofibration of [GM]:

$$\mathrm{TH}(L) \wedge_{C_{p^n}} ES^1_+ \to F(ES^1_+, \mathrm{TH}(L))^{C_{p^n}} \to \hat{\mathbb{H}}(C_{p^n}, T(L)). \tag{4.1.7}$$

By definition it appears that $\hat{\mathbb{H}}(C_{p^n}, T(L))$ depends on the full equivariant structure of $T(L)$, and not only on $\mathrm{TH}(L)$, but this is not really the case. The adjunction $j_*\mathrm{TH}(L) \to T(L)$ induces a map

$$\hat{\mathbb{H}}(C_{p^n}, j_*\mathrm{TH}(L)) \to \hat{\mathbb{H}}(C_{p^n}, T(L))$$

which also fits into the cofibration sequence above; it must be an equivalence by a 5-lemma argument. Thus we shall often write $\hat{\mathbb{H}}(C_{p^n}, \mathrm{TH}(L))$ instead of $\hat{\mathbb{H}}(C_{p^n}, T(L))$. We shall also use the costumary abbreviations

$$\begin{aligned} \mathrm{TH}(L)_{hC_{p^n}} &= \mathrm{TH}(L) \wedge_{C_{p^n}} ES^1_+ \\ \mathrm{TH}(L)^{hC_{p^n}} &= F(ES^1_+, \mathrm{TH}(L))^{C_{p^n}}. \end{aligned}$$

With these notions we have

Proposition 4.1.8. *There is a homotopy commutative diagram of cofibrations (of non-equivariant spectra)*

$$\begin{array}{ccccc} \mathrm{TH}(L)_{hC_{p^n}} & \xrightarrow{N} & \mathrm{TH}(L)^{C_{p^n}} & \xrightarrow{R} & \mathrm{TH}(L)^{C_{p^{n-1}}} \\ \big\downarrow \mathrm{id} & & \big\downarrow \Gamma & & \big\downarrow \hat{\Gamma} \\ \mathrm{TH}(L)_{hC_{p^n}} & \xrightarrow{N^h} & \mathrm{TH}(L)^{hC_{p^n}} & \xrightarrow{R^h} & \hat{\mathbb{H}}(C_{p^n}, \mathrm{TH}(L)) \end{array}$$

Remark 4.1.9. The S^1-fixed set of $\mathrm{TH}(L)$ is contained in $\mathrm{THH}_0(L)$, cf. sect. 2.1, and is of no relevance. In particular the upper horizontal sequence in (4.1.7) has no analogue for S^1 fixed sets. But the lower sequence does have an S^1-version, namely

$$\Sigma \mathrm{TH}(L)_{hS^1} \to \mathrm{TH}(L)^{hS^1} \to \hat{\mathbb{H}}(S^1, \mathrm{TH}(L))$$

with the right-hand term defined by (4.1.1) upon replacing the C_{p^n} fixed set by the S^1 fixed set, cf. [GM].

Example 4.1.10. In the special case of the identity FSP, $L(X) = X$, $T(L)$ is the equivariant sphere spectrum,

$$T(L)(W) \sim_{C_\infty} \varinjlim_{V} \Omega^{V-W} S^W, \quad V \subset \mathcal{U}^C$$

cf. lemma 4.4.4 below. In this case the diagram of proposition 4.1.7 is completely known. Listing only the 0'th terms of the spectra we have

$$\begin{aligned} \mathrm{TH}(\mathrm{Id})^{C_{p^n}} &\sim \Omega^\infty S^\infty(BC_{p^n}{}_+) \times \cdots \times \Omega^\infty S^\infty(BC_{p+}) \times \Omega^\infty S^\infty(S^0) \\ \mathrm{TH}(\mathrm{Id})_{hC_{p^n}} &\sim \Omega^\infty S^\infty(BC_{p^n}{}_+) \end{aligned}$$

where $\Omega^\infty S^\infty(X_+) = \varinjlim \Omega^k(S^k \wedge X_+)$. The map R is the projection onto the last n factors. Moreover, the affirmed Segal conjecture tells us that the profinite completions of Γ_n and $\hat{\Gamma}_n$ are equivalences for all n.

One may get information about the homotopy groups of the terms in the norm cofibration by spectral sequences. Let M be a coefficient group (usually $M = \mathbb{Z}_p$ or $M = \mathbb{F}_p$). To ensure convergence of the spectral sequences I will assume that $\pi_*(T(L); M)$ is a finitely generated $\mathbb{Z}_p$-module in each degree.

The spectral sequences were set up in [BM], sect. 2 and in [GM], sect. 10; in [BM] by using a topological version (due to Greenlees) of the complete resolution in usual Tate cohomology of groups and in [GM] by the dual viewpoint where one uses the equivariant Postnikov tower of the spectrum. In our case, the spectral sequences takes the form

$$\text{(i)}\quad E^2_{s,t}\left(T(L)_{hC_{p^n}}; M\right) = H_s\left(C_{p^n}; \pi_t(T(L); M)\right) \Rightarrow \pi_{s+t}\left(T(L)_{hC_{p^n}}; M\right)$$

$$\text{(ii)}\quad E^2_{-s,t}\left(T(L)^{hC_{p^n}}; M\right) = H^s\left(C_{p^n}; \pi_t(T(L); M)\right) \Rightarrow \pi_{t-s}\left(T(L)^{hC_{p^n}}; M\right)$$

$$\text{(iii)}\quad E^2_{-s,t}\left(\hat{\mathbb{H}}(C_{p^n}, T(L)); M\right) = \hat{H}^s\left(C_{p^n}; \pi_t(T(L); M)\right) \Rightarrow \pi_{t-s}\left(\hat{\mathbb{H}}(C_{p^n}; T(L)); M\right)$$

The spectral sequences are concentrated in the upper half plane, the differentials take $E^r_{s,t}$ to $E^r_{s-r,t+r-1}$, and for commutative L the last two spectral sequences have ring structure (with the differentials being derivations) when M is a p-adic ring with p odd. Since the C_{p^n}-action comes from an S^1-action $\pi_*(T(L); M)$ has trivial C_{p^n}-action. Thus for p odd:

$$E^2\left(T(L)^{hC_{p^n}}; \mathbb{F}_p\right) = E\{u_n\} \otimes S\{t\} \otimes \pi_*(T(L); \mathbb{F}_p) \tag{4.1.11}$$
$$E^2\left(\hat{\mathbb{H}}(C_{p^n}, T(L)); \mathbb{F}_p\right) = E\{u_n\} \otimes S\{t, t^{-1}\} \otimes \pi_*(T(L); \mathbb{F}_p)$$

with $\deg(u_n) = (-1, 0)$, $\deg(t) = (-2, 0)$ and $\pi_t(T(L); \mathbb{F}_p)$ sitting in degree $(0, t)$.

In the above H_s, H^s and $\hat{H}^s$ denotes group homology, group cohomology and group Tate cohomology. They are related by the formulas:

$$\hat{H}^{-s}(G; A) = \begin{cases} H^{-s}(G; A), & s \leq 0 \\ H_{s-1}(G; A), & s > -1 \\ \ker\left(\text{Norm}\colon H_0(G; A) \to H^0(G; A)\right), & s = -1 \\ \operatorname{coker}\left(\text{Norm}\colon H_0(G; A) \to H^0(G; A)\right), & s = 0 \end{cases}$$

When $G = C_{p^n}$ and $pA = 0$ then Norm $= 0$, so we see that

$$E^2_{-s,t}\left(\hat{\mathbb{H}}(C_{p^n}; T(L)); \mathbb{F}_p\right) = \begin{cases} E^2_{-s,t}\left(T(L)^{hC_{p^n}}; \mathbb{F}_p\right), & s \geq 0 \\ E^2_{-s-1,t}\left(T(L)_{hC_{p^n}}; \mathbb{F}_p\right), & s < 0 \end{cases}$$

It is important for calculation of $\pi_*(\mathrm{TC}(L); \mathbb{F}_p)$ to identify the R-map, or in the setting of the norm cofibration to identify $\pi_*(R^h)$. This is connected with the differentials in the spectral sequence for $\hat{\mathbb{H}}(C_{p^n}, T(L))$ which cross over the line $-s = 1/2$ in $E^r_{-s,t}$. Indeed, the maps in

$$T(L)^{hC_{p^n}} \xrightarrow{R^h} \hat{\mathbb{H}}(C_{p^n}, T(L)) \xrightarrow{\partial} \Sigma T(L)_{hC_{p^n}}$$

induce homomorphisms of spectral sequences

$$\begin{aligned} E^r(R^h)\colon E^r_{-s,t}\left(T(L)^{hC_{p^n}}; M\right) &\to E^r_{-s,t}\left(\hat{\mathbb{H}}(C_{p^n}, T(L)); M\right) \\ E^r(\partial)\colon E^r_{-s,t}\left(\hat{\mathbb{H}}(C_{p^n}, T(L)); M\right) &\to E^r_{-s-1,t}\left(T(L)_{hC_{p^n}}; M\right) \end{aligned} \tag{4.1.12}$$

with $E^r(R^h)$ surjective for $s \geq 0$, and $E^r(\partial)$ injective for $s < 0$. In a situation where one can calculate the spectral sequences one will also know $E^\infty(R^h)$ and $E^\infty(\partial)$, and hence since the spectral sequences converge,

$$\begin{aligned} E^0\pi_* R^h\colon E^0\pi_*\left(T(L)^{hC_{p^n}}; M\right) &\to E^0\pi_*\left(\hat{\mathbb{H}}(C_{p^n}, T(L)); M\right) \\ E^0(\pi_*\partial)\colon E^0\pi_*\left(\hat{\mathbb{H}}(C_{p^n}, T(L)); M\right) &\to E^0\pi_{*-1}\left(T(L)_{hC_{p^n}}; M\right) \end{aligned}$$

In general this is of course not sufficient to give, say $\pi_* R^h$; there might be filtration shifts. The following lemma goes a long way to overcome this difficulty.

Lemma 4.1.13. *If $\alpha \in E^0\pi_{s+t}\left(T(L)^{hC_{p^n}}; M\right)$ is in the kernel of $E^0(\pi_* R^h)$ then there exists an element $\beta \in \pi_{s+t}\left(T(L)_{hC_{p^n}}; M\right)$ with $E^0\pi_*(N^h)(\beta) = \alpha$.*

Proof. This is a special case of [BM], theorem 2.15. The argument can be outlined as follows. By assumption $E^\infty(R^h)(\alpha) = 0$. The reason must be that there exists an $r > s$ such that α belongs to the image of

$$d^r\colon E^r_{r-s,t-r+1}\left(\hat{\mathbb{H}}(C_{p^n}, T(L)); M\right) \to E^r_{-s,t}\left(\hat{\mathbb{H}}(C_{p^n}, T(L)); M\right)$$

say $\alpha = d^r(\gamma)$. Now $\gamma = E^r(\partial)(\overline{\beta})$ and $\overline{\beta}$ will be an infinite cycle in $E^r\left(T(L)_{hC_{p^n}}; M\right)$. Thus $\overline{\beta}$ represents an element of $E^\infty\left(T(L)_{hC_{p^n}}; M\right)$, and one can pick a suitable representative. More details can be found in [BM], sect. 2. □

The d^2-differential in the spectral sequences is connected to the action

$$A\colon S^1_+ \wedge \mathrm{TH}(L) \to \mathrm{TH}(L)$$

as follows. The stable homotopy $\pi_1^s(S^1_+) \cong \pi_4(\Sigma^3(S^1_+)) \xrightarrow{\cong} \pi_4(S^4) \oplus \pi_4(S^3)$ is $\mathbb{Z} \oplus \mathbb{Z}/2$, generated by the σ = id and the Hopf map η. Thus we get operators

$$[S^1], \eta\colon \pi_i(\mathrm{TH}(L)) \longrightarrow \pi_{i+1}(S^1_+ \wedge \mathrm{TH}(L)) \xrightarrow{A_*} \pi_{i+1}\mathrm{TH}(L)$$

where the first map is exterior product with σ and η, respectively. There are induced operations

$$\hat{H}^s(C_{p^n}; \pi_t(\mathrm{TH}(L); M)) \to \hat{H}^s(C_{p^n}; \pi_{t+1}(\mathrm{TH}(L); M))$$

which we can compose with the periodicity isomorphism

$$\hat{H}^s(C_{p^n}; \pi_{t+1}(\mathrm{TH}(L); M)) \xrightarrow{\cong} \hat{H}^{s+2}(C_{p^n}; \pi_{t+1}(\mathrm{TH}(L); M))$$

to get maps $[S^1]_\#$, $\eta_\#$.

Proposition 4.1.14. *In the spectral sequence $E^r_{*,*}\left(\hat{\mathbb{H}}(C_{p^n}, \mathrm{TH}(L)); M\right)$, the d^2-differential*

$$d^2\colon \hat{H}^s(C_{p^n}; \pi_t(\mathrm{TH}(L); M)) \to \hat{H}^{s+2}(C_{p^n}; \pi_t(\mathrm{TH}(L); M))$$

is equal to $[S^1]_\#$, provided η acts trivially on $\pi_*(\mathrm{TH}(L); M)$. □

This is proved in [H2] when C_{p^n} is replaced by S^1, and the above can be deduced from this case. The assumption that $\eta_\#$ be zero is satisfied for the linear FSP's $L = \tilde{A}$ associated with a ring because $\mathrm{TH}(A) \sim \mathrm{TH}^\oplus(A)$ is a product of Eilenberg-MacLane spectra.

We have left to consider the ***homotopy limit problem***, i.e. the homotopical behavior of

$$\hat{\Gamma}_n\colon \mathrm{TH}(L)^{C_{p^{n-1}}} \to \hat{\mathbb{H}}(C_{p^n}, \mathrm{TH}(L)).$$

In the special case of L = Id it is a homotopy equivalence, but this is too much to expect in general. The domain is a (-1)-connected spectrum, but this is often false for the right hand side, e.g. when $L = \tilde{\mathbb{F}}_p$ as we shall see in sect. 4.2 below. The best one could hope for would be that $\pi_i(\hat{\Gamma}_n)$, and hence also $\pi_i(\Gamma_n)$, be isomorphisms for $i \geq 0$. This unfortunately is also not true. For the FSP $\tilde{A}$ associated with truncated polynomial algebras $A = k[t]/(t^n)$, the two sides have different homotopy groups in all even dimensions; this is an easy consequence of sect. 5.2. The only completely general theorem is the following result of S. Tsalidis:

Theorem 4.1.15. ([T]) *Suppose*

$$\pi_i(\hat{\Gamma}_1)\colon \pi_i(\mathrm{TH}(L); \mathbb{F}_p) \to \pi_i\left(\hat{\mathbb{H}}(C_p, \mathrm{TH}(L)); \mathbb{F}_p\right)$$

is an isomorphism for $i \geq i_0$. Then the same is true for $\pi_i(\hat{\Gamma}_n)$ for all $n \geq 1$. □

Tsalidis' proof is similar to the induction step from C_p to C_{p^n} in the proof of the affirmed Segal conjecture.

Calculations from [H2] show that if $\pi_i(\hat{\Gamma}_1; \mathbb{F}_p)$ is an isomorphism in non-negative degrees for a ring A then the same is the case for the polynomial algebra $A[t]$ and more generally for any smooth A-algebra. In [BM1] and in sect. 5.4 below the assumption of theorem 4.1.15 is established for $A = W(\mathbb{F}_{p^s})$, with $i_0 = 0$. Optimistically one would hope for

Conjecture 4.1.16. For a regular ring A,

$$\pi_i(\hat{\Gamma}_1; \mathbb{F}_p) \colon \pi_i\left(\mathrm{TH}(A); \mathbb{F}_p\right) \to \pi_i\left(\hat{\mathbb{H}}(C_p, \mathrm{TH}(A)); \mathbb{F}_p\right)$$

is an isomorphism when $i \geq 0$.

Note that the statement is equivalent to the assertion that

$$\hat{\Gamma}_n \colon \mathrm{TH}(A)^{C_{p^n}} \to \hat{\mathbb{H}}\left(C_{p^{n+1}}, \mathrm{TH}(A)\right)[0, \infty)$$

becomes a homotopy equivalence after p-adic completion, with $[0, \infty)$ indicating (-1)-connected cover.

4.2 The spectrum $\mathrm{TC}(\mathbb{F}_p)$.

This section illustrates sect. 4.1 by completely determining the spectra $\mathrm{TH}(\mathbb{F}_p)^{C_{p^n}}$ and $\mathrm{TC}(\mathbb{F}_p)$. The calculation was originally carried out in [M], but [HM], sect. 4.1–3 is a better place to look for additional details.

For any ring, THH(A) is the realization of a simplicial abelian group, cf. sect. 3.2, so its homotopy type is determined by its homotopy groups:

$$\mathrm{TH}(A) \sim \bigvee_{n=0}^{\infty} \Sigma^n H(\pi_n \mathrm{TH}(A)) \sim \prod_{n=0}^{\infty} \Sigma^n H(\pi_n \mathrm{TH}(A)) \tag{4.2.1}$$

where $H(-)$ is the Eilenberg-MacLane spectrum with $\pi_0 H(B) = B$ and $\pi_i H(B) = 0$ for $i \neq 0$, and Σ^n is the suspension functor.

One may filter TH(A) by skeletons, since it is the realization of a simplicial construction. This leads to a spectral sequence,

$$E^2(A) = \mathrm{HH}_*(A_A) \Rightarrow \pi_*(\mathrm{TH}(A); \mathbb{F}_p) \tag{4.2.2}$$

with $\mathcal{A}_A = H_*(H(A); \mathbb{F}_p)$. This spectral sequence was used by Bökstedt to calculate $\mathrm{TH}(\mathbb{F}_p)$. I refer the reader to [B1] or [HM], sect. 4.2 for details. Different calculational methods can be found in [Br] or [FLS].

The 0-skeleton of $\mathrm{TH}(A)$ is the Eilenberg-MacLane spectrum $H(A)$, and one may use the S^1-action to get the map

$$\sigma\colon S^1_+ \wedge HA \to S^1_+ \wedge \mathrm{TH}(A) \to \mathrm{TH}(A). \tag{4.2.3}$$

For $A = \mathbb{F}_p$ we have $\tau_0 \in \pi_1(H\mathbb{F}_p; \mathbb{F}_p)$ and can consider $\sigma_*([S^1] \wedge \tau_0) \in \pi_2(\mathrm{TH}(\mathbb{F}_p); \mathbb{F}_p)$, where $[S^1] \in \pi_1^S(S^1_+)$ was defined in the previous section.

Theorem 4.2.4. ([B1], [Br]). *The reduction*

$$\mathrm{red}_p\colon \pi_2\mathrm{TH}(\mathbb{F}_p) \to \pi_2(\mathrm{TH}(\mathbb{F}_p); \mathbb{F}_p)$$

is an isomorphism, and

$$\pi_*\mathrm{TH}(\mathbb{F}_p) = S_{\mathbb{F}_p}\{\sigma\},$$

the polynomial algebra on σ of degree 2 with $\mathrm{red}_p(\sigma) = \sigma_*([S^1] \wedge \tau_0)$. □

Combined with (4.1.11) we can explicate the E^2-terms of the spectral sequence $\hat{E}^r(C_{p^n}; M) = E^r\left(\hat{\mathbb{H}}(C_{p^n}, T(\mathbb{F}_p)); M\right)$ for $M = \mathbb{F}_p, \mathbb{Z}_p$ to be

$$\hat{E}^2(C_{p^n}; \mathbb{F}_p) = E_{\mathbb{F}_p}\{u_n\} \otimes S_{\mathbb{F}_p}\{t, t^{-1}\} \otimes E_{\mathbb{F}_p}\{e_1\} \otimes S_{\mathbb{F}_p}\{\sigma\}$$
$$\hat{E}^2(C_{p^n}; \mathbb{Z}_p) = E_{\mathbb{F}_p}\{u_n\} \otimes S_{\mathbb{F}_p}\{t, t^{-1}\} \otimes S_{\mathbb{F}_p}\{\sigma\}$$

except if $p = 2$ and $n = 1$ where the first two terms are replaced by $S\{u_1, u_1^{-1}\}$. The mod p Bockstein operator maps $e_1\sigma^l$ to σ^l for $l \geq 0$. For p odd, $\hat{E}^r(C_{p^n}; \mathbb{F}_p)$ is a spectral sequence of algebras. If $p = 2$ there is the usual trouble with products in $\pi_*(T; \mathbb{F}_2)$ but in all cases, $\hat{E}^r(C_{p^n}; \mathbb{F}_p)$ is an algebra over $\hat{E}^r(C_{p^n}; \mathbb{Z}_p)$.

Lemma 4.2.5. *The non-zero differentials in $\hat{E}^r(C_{p^n}; \mathbb{F}_p)$ are generated from $d^2e_1 = t\sigma$ in the module structure over $\hat{E}^r(C_{p^n}; \mathbb{Z}_p)$. In particular*

$$\pi_*\left(\hat{\mathbb{H}}(C_{p^n}, \mathrm{TH}(\mathbb{F}_p)); \mathbb{F}_p\right) \cong E_{\mathbb{F}_p}\{u_n\} \otimes S_{\mathbb{F}_p}\{t, t^{-1}\}, \quad p \text{ odd or } n > 1$$
$$\pi_*\left(\hat{\mathbb{H}}(C_2, \mathrm{TH}(\mathbb{F}_2)); \mathbb{F}_2\right) \cong S_{\mathbb{F}_2}\{u_1, u_1^{-1}\}$$

with $\deg(t) = -2$, $\deg u_n = -1$.

Proof. Since $e_1 = \sigma_*(\tau_0)$, $\tau_0 \in \pi_1(H\mathbb{F}_p; \mathbb{F}_p)$ and $\mathrm{red}_p(\sigma) = \sigma_*([S^1] \wedge \tau_0)$ we have in the notation of proposition 4.1.14,

$$[S^1]_\#(e_1) = \sigma, \qquad [S^1]_\#(1) = 0,$$

and hence $[S^1]_\#(e_1\sigma^l) = \sigma^{l+1}$. The d^2-differential then follow from (4.1.14), and a routine cohomology calculation gives

$$\hat{E}^3(C_{p^n}; \mathbb{F}_p) = E_{\mathbb{F}_p}\{u_n\} \otimes S_{\mathbb{F}_p}\{t, t^{-1}\}$$

(with $u_1^2 = t$ if $p = 2$ and $n = 1$). For degree reasons there can be no further differentials. For p odd (and $p = 2$, $n = 1$) this is a free commutative algebra in the graded sense, and the stated value of the mod p homotopy is immediate. If $p - 2$ and $n > 1$ one uses that the mod p Bockstein on u_n is trivial. □

For $n = 1$, the mod p Bockstein relation $\beta(u_1) = t$ gives that

$$\pi_*\hat{\mathbb{H}}(C_p, \mathrm{TH}(\mathbb{F}_p)) = S_{\mathbb{F}_p}\{t, t^{-1}\}$$

(with $t = u_1^2$ if $p = 2$). We next check the assumption of theorem 4.1.15.

Lemma 4.2.6. *The homomorphism*

$$\pi_i(\hat{\Gamma}_1; \mathbb{F}_p)\colon \pi_i(\mathrm{TH}(\mathbb{F}_p); \mathbb{F}_p) \to \pi_i\left(\hat{\mathbb{H}}(C_p, \mathrm{TH}(\mathbb{F}_p)); \mathbb{F}_p\right)$$

is an isomorphism when $i \geq 0$.

Proof. Since $\hat{\Gamma}_1\colon \mathrm{TH}(\mathbb{F}_p) \to \hat{\mathbb{H}}(C_p, \mathrm{TH}(\mathbb{F}_p))$ is multiplicative, it suffices to see that $\pi_2(\hat{\Gamma}_1; \mathbb{F}_p)$ is an isomorphism.

Continuing the cofibration diagram of (4.1.8), $n = 1$, to the right, gives a homotopy commutative square of S^1-spectra

$$\begin{array}{ccccc}
\mathrm{TH}(\mathbb{F}_p) & \xrightarrow{\partial} & \Sigma\rho^\#_{C_p}(\mathrm{TH}(\mathbb{F}_p)_{hC_p}) & \xrightarrow{\Sigma N} & \Sigma\rho^\#_{C_p}\mathrm{TH}(\mathbb{F}_p)^{C_p} \\
\downarrow{\scriptstyle\hat{\Gamma}_1} & & \downarrow{\scriptstyle\mathrm{id}} & & \downarrow{\scriptstyle\Gamma_1} \\
\rho^\#_{C_p}\hat{\mathbb{H}}(C_p, \mathrm{TH}(\mathbb{F}_p)) & \xrightarrow{\partial^h} & \Sigma\rho^\#_{C_p}(\mathrm{TH}(\mathbb{F}_p)_{hC_p}) & \xrightarrow{\Sigma N^h} & \Sigma\rho^\#_{C_p}\mathrm{TH}(\mathbb{F}_p)^{hC_p}
\end{array}$$

Here as usual $\rho^\#_{C_p}$ indicates that the S^1/C_p-spectra are to be considered as S^1-spectra under the p'th root isomorphism $S^1 \to S^1/C_p$.

Now $\sigma = [S^1]_\#(\tau_0)$, so we are done if we can show that $e_0 = \pi_i(\partial; \mathbb{F}_p)(\tau_0)$ is non-zero in $\pi_0\left(\mathrm{TH}(\mathbb{F}_p)_{hC_p}; \mathbb{F}_p\right)$, and $[S^1]_\#(e_0) \neq 0$.

The spectral sequence $E^r(\mathrm{TH}(\mathbb{F}_p);\mathbb{Z}_p)$ gives $\pi_0\mathrm{TH}(\mathbb{F}_p) = \mathbb{Z}/p$, and by (2.5.8) $\pi_0\mathrm{TH}(\mathbb{F}_p)^{C_p} = \mathbb{Z}/p^2$. The fundamental cofibration thus induces the exact non-split sequence

$$0 \longrightarrow \pi_0\mathrm{TH}(\mathbb{F}_p)_{hC_p} \xrightarrow{N} \pi_0\mathrm{TH}(\mathbb{F}_p)^{C_p} \xrightarrow{R} \pi_0\mathrm{TH}(\mathbb{F}_p) \longrightarrow 0$$

so $\pi_0(N;\mathbb{F}_p) = 0$, and $\pi_1(\partial;\mathbb{F}_p)$ must be surjective. Finally, the inclusion

$$T(\mathbb{F}_p) \wedge_{C_p} S^1_+ \to T(\mathbb{F}_p) \wedge_{C_p} ES^1_+$$

coming from $S^1 \subset ES^1$ induces a monomorphism on $\pi_i(-;\mathbb{F}_p)$ for $i = 0, 1$. The homeomorphism

$$\rho^{\#}_{C_p}\left(\mathrm{TH}(\mathbb{F}_p) \wedge_{C_p} S^1_+\right) \to \mathrm{TH}(\mathbb{F}_p) \wedge S^1_+, \quad (x,\theta) \mapsto (\theta^{-1}x, \theta)$$

map the diagonal S^1-structure in the domain to the extended S^1-structure in the range. Hence

$$[S^1]_\# : \pi_0\left(\rho^{\#}_{C_p}\mathrm{TH}(\mathbb{F}_p) \wedge_{C_p} S^1_+;\mathbb{F}_p\right) \to \pi_1\left(\rho^{\#}_{C_p}\mathrm{TH}(\mathbb{F}_p) \wedge_{C_p} S^1_+;\mathbb{F}_p\right)$$

must be injective. □

The spectrum $\mathrm{TH}(\mathbb{F}_p)$ is p-complete, and inductive use of the fundamental cofibration (2.4.6) implies the same for $\mathrm{TH}(\mathbb{F}_p)^{C_{p^n}}$ for each n. Thus

$$\pi_*\mathrm{TH}(\mathbb{F}_p)^{C_{p^n}} = \pi_*\left(\mathrm{TH}(\mathbb{F}_p)^{C_{p^n}};\mathbb{Z}_p\right).$$

Proposition 4.2.7. *For $n \geq 1$,*

$$\pi_*T(\mathbb{F}_p)^{C_{p^n}} = S_{\mathbb{Z}/p^{n+1}}\{\sigma_n\}$$

with $\deg\sigma_n = 2$. *Moreover,* $F(\sigma_n) = \sigma_{n-1}$ *and* $R(\sigma_n) = \lambda_n p\sigma_{n-1}$ *with* $\lambda_n \in \mathbb{Z}/p^n$ *a unit.*

Proof. Theorem 4.1.15 shows that

$$\hat{\Gamma}: \pi_*\mathrm{TH}(\mathbb{F}_p)^{C_{p^n}} \to \pi_*\hat{\mathbb{H}}\left(C_{p^n}, \mathrm{TH}(\mathbb{F}_p)\right)$$

is an isomorphism in non-negative degrees. For the target, the integral spectral sequence $\hat{E}^r(C_{p^n};\mathbb{Z}_p)$ has

$$\hat{E}^2 = E_{\mathbb{F}_p}\{u_n\} \otimes S_{\mathbb{F}_p}\{t, t^{-1}\} \otimes S_{\mathbb{F}_p}\{\sigma\}.$$

The elements t and σ are infinite cycles. Indeed the inclusion of S^1 fixed sets into C_{p^n} fixed sets gives a map

$$\hat{\mathbb{H}}\left(S^1, \mathrm{TH}(\mathbb{F}_p)\right) \to \hat{\mathbb{H}}\left(C_{p^n}, \mathrm{TH}(\mathbb{F}_p)\right)$$

cf. (4.1.9), and an induced map of spectral sequence. The E^2-term of the range is

$$E^2\left(\hat{\mathbb{H}}(S^1, \mathrm{TH}(\mathbb{F}_p)); \mathbb{Z}_p\right) = S_{\mathbb{F}_p}\{t, t^{-1}\} \otimes S_{\mathbb{F}_p}\{\sigma\},$$

so is concentrated in even total degrees. Thus $E^2 = E^\infty$. On the other hand it injects into the $\hat{E}^2$ above. Thus $t^k\sigma^l$ are all infinite cycles.

We claim that u_n survives to $\hat{E}^{2n+1}(C_{p^n}; \mathbb{Z}_p)$ and that $d^{2n+1}(u_n) = t^{n+1}\sigma^n$. Indeed, the first non-trivial differential on u_n must be of the form

$$d^{2r+1}(u_n) = t^{r+1}\sigma^r$$

for some r. Given this it is easy to solve the spectral sequence. In particular

$$E^0\pi_0\hat{\mathbb{H}}\left(C_{p^n}; \mathrm{TH}(\mathbb{F}_p)\right) = \mathbb{F}_p^{\oplus r}$$

generated by $1, t\sigma, \dots, (t\sigma)^{r-1}$, $(d^{2r+1}(u_n t^{-1}) = (t\sigma)^r)$. Since $\pi_0\hat{\mathbb{H}}\left(C_{p^n}, \mathrm{TH}(\mathbb{F}_p)\right) \cong \pi_0\mathrm{TH}(\mathbb{F}_p)^{C_{p^{n-1}}}$ is $\mathbb{Z}/p^n$ by (2.5.8), we conclude that $r = n$. Moreover,

$$E^0\pi_{2k}\hat{\mathbb{H}}\left(C_{p^n}, \mathrm{TH}(\mathbb{F}_p)\right) = \mathbb{F}_p^{\oplus n}$$

generated by $\sigma^k, \sigma^{k+1}t, \dots, \sigma^{k+n}t^n$, and $\pi_{2k+1}\hat{\mathbb{H}}\left(C_{p^n}, \mathrm{TH}(\mathbb{F}_p)\right) = 0$. Since in addition $\pi_{2k}\left(\hat{\mathbb{H}}(C_{p^n}, \mathrm{TH}(\mathbb{F}_p)); \mathbb{F}_p\right)$ is a single copy of $\mathbb{F}_p$ we must have

$$\pi_{2k}\hat{\mathbb{H}}\left(C_{p^n}; \mathrm{TH}(\mathbb{F}_p)\right) = \mathbb{Z}/p^n$$

for all $k \geq 0$. One more application of theorem 4.1.15 gives the stated homotopy groups. The inclusion F corresponds under $\hat{\Gamma}$ to the inclusion

$$\hat{\mathbb{H}}\left(C_{p^{n+1}}, \mathrm{TH}(\mathbb{F}_p)\right) \xrightarrow{F^h} \hat{\mathbb{H}}\left(C_{p^n}, \mathrm{TH}(\mathbb{F}_p)\right)$$

so $\pi_{2k}(F^h)$ must be surjective, and we can pick the generator to satisfy $F(\sigma_n) = \sigma_{n-1}$.

Finally the exact sequence

$$\pi_2 T(\mathbb{F}_p)^{C_{p^n}} \xrightarrow{R} \pi_2 T(\mathbb{F}_p)^{C_{p^{n-1}}} \xrightarrow{\partial_*} \pi_1 T(\mathbb{F}_p)_{hC_{p^n}} \xrightarrow{N} \pi_1 T(\mathbb{F}_p)^{C_{p^n}},$$

with $\pi_1 T(\mathbb{F}_p)^{C_{p^n}} = 0$ and $\pi_1 T(\mathbb{F}_p)_{hC_{p^n}} = \mathbb{Z}/p$, yields the stated value of R. □

Corollary 4.2.8. $\mathrm{TC}(\mathbb{F}_p) \simeq H\mathbb{Z}_p \vee \Sigma^{-1}H\mathbb{Z}_p$

Proof. We use the cofibration sequence of sect. 2.5,

$$\mathrm{TC}(\mathbb{F}_p,p)\longrightarrow \mathrm{TR}(\mathbb{F}_p,p)\xrightarrow{F-1}\mathrm{TR}(\mathbb{F}_p,p).$$

The previous proposition yields

$$\pi_k\mathrm{TR}(\mathbb{F}_p,p)=\varprojlim_R \pi_k\mathrm{TH}(\mathbb{F}_p)^{C_{p^n}}=\begin{cases}0 & \text{for } k>0\\ \mathbb{Z}_p & \text{for } k=0\end{cases}$$

so that

$$\pi_0\mathrm{TC}(\mathbb{F}_p,p)=\mathbb{Z}_p,\quad \pi_{-1}\mathrm{TC}(\mathbb{F}_p,p)=\mathbb{Z}_p$$

and $\pi_k\mathrm{TC}(\mathbb{F}_p,p)=0$ otherwise. Finally, $\mathrm{TC}(\mathbb{F}_p)$ is p-complete and by theorem 2.5.5 equal to $\mathrm{TC}(\mathbb{F}_p,p)$. □

4.3 The absolute theorem: linear case.

This section sketches the proof of theorem 1.3 of the introduction. It is joint work with L. Hesselholt, and further details can be found in [HM], sect. 4.5, 5.1, 5.2 and [HM], appendix B.

We fix a perfect field k of positive characteristic p, and consider algebras A over the (p-typical) Witt vectors $W(k)$ which are finitely generated as modules; for short: *finite* $W(k)$-algebras. If k is finite the assumption is that A be a finite $\mathbb{Z}_p$-algebra. We use the notation

$$K_i(A;\mathbb{Z}_p)=\pi_i(K(A)_p^\wedge)$$
$$\mathrm{TC}_i(A;\mathbb{Z}_p)=\pi_i(\mathrm{TC}(A)_p^\wedge)$$

and want to prove

Theorem 4.3.1. *For finite $W(k)$-algebras, the cyclotomic trace*

$$\mathrm{trc}\colon K_i(A;\mathbb{Z}_p)\to \mathrm{TC}_i(A;\mathbb{Z}_p)$$

is an isomorphism, for $i\geq 0$.

The ring of Witt vectors $W(k)$ is a P.I.D and is p-adically complete. Since A is finite over $W(k)$,

$$A=\varprojlim A/p^nA,$$

and we can introduce the continuous version of the functors:

$$K^{\mathrm{top}}(A)=\operatorname{holim}_{\longleftarrow} K(A/p^nA),\quad \mathrm{TC}^{\mathrm{top}}(A)=\operatorname{holim}_{\longleftarrow}\mathrm{TC}(A/p^nA).$$

There are exact sequences

$$0 \to \varprojlim{}^{(1)} K_{i+1}(A/p^nA;\mathbb{Z}_p) \to K_i^{\mathrm{top}}(A;\mathbb{Z}_p) \to \varprojlim K_i(A/p^nA;\mathbb{Z}_p) \to 0$$
$$0 \to \varprojlim{}^{(1)} \mathrm{TC}_{i+1}(A/p^nA;\mathbb{Z}_p) \to \mathrm{TC}_i^{\mathrm{top}}(A;\mathbb{Z}_p) \to \varprojlim \mathrm{TC}_i(A/p^nA;\mathbb{Z}_p) \to 0$$

cf. [BK], p. 249 and p. 299.

The proof of theorem 4.3.1 is broken down into three statements to be considered separately below:

$$\text{(i)} \qquad K_i(A/pA;\mathbb{Z}_p) \xrightarrow{\cong} \mathrm{TC}_i(A/pA;\mathbb{Z}_p), \qquad i \geq 0$$
$$\text{(ii)} \qquad \mathrm{TC}_i(A;\mathbb{Z}_p) \xrightarrow{\cong} \mathrm{TC}_i^{\mathrm{top}}(A;\mathbb{Z}_p), \qquad i \geq 0$$
$$\text{(iii)} \qquad K_i(A;\mathbb{Z}_p) \xrightarrow{\cong} K_i^{\mathrm{top}}(A;\mathbb{Z}_p), \qquad i \geq 0$$

Indeed, given (i), McCarthy's theorem 3.4.12 show that

$$\mathrm{trc}\colon K_i(A/p^nA;\mathbb{Z}_p) \to \mathrm{TC}_i(A/p^nA;\mathbb{Z}_p)$$

is an isomorphism for all $i \geq 0$, and hence by the short exact sequences above that

$$\mathrm{trc}\colon K_i^{\mathrm{top}}(A;\mathbb{Z}_p) \xrightarrow{\cong} \mathrm{TC}_i^{\mathrm{top}}(A;Z_p), \quad i \geq 0.$$

Use of (ii) and (iii) completes the proof.

I begin with (i). For $A = W(k)$, $A/pA = k$. If k is finite then $K(k)_p^\wedge \simeq H\mathbb{Z}_p$ by [Q3]. For general perfect fields the same holds by [Kr]. We must therefore first extend Corollary 4.2.8 to general perfect fields. The result we need is

Theorem 4.3.2. *For a perfect field of characteristic $p > 0$, there is a homotopy equivalence* $\mathrm{TR}(k,p) \sim HW(k)$

Given this, we can calculate $\mathrm{TC}(k,p)$ from the cofibration

$$\mathrm{TC}(k,p) \longrightarrow \mathrm{TR}(k,p) \xrightarrow{1-F} \mathrm{TR}(k,p),$$

since by theorem 2.5.7 we know that

$$\pi_0 F\colon \pi_0(\mathrm{TR}(k,p);\mathbb{F}_p) \to \pi_0(\mathrm{TR}(k,p);\mathbb{Z}_p)$$

induces the Frobenius homomorphism of Witt vectors. Moreover

$$\ker(F - \mathrm{id}\colon W(k) \to W(k)) = W(k^F)$$

and $k^F = \mathbb{F}_p$ so $W(k^F) = \mathbb{Z}_p$. Thus theorem 4.3.2 gives

$$\mathrm{TC}_i(k;\mathbb{Z}_p) = \begin{cases} 0, & i \geq 0 \\ \mathbb{Z}_p, & i = 0 \\ \operatorname{cok}(F - \mathrm{id}\colon W(k) \to W(k)), & i = -1 \end{cases} \tag{4.3.3}$$

and hence $K_i(k;\mathbb{Z}_p) \cong \mathrm{TC}_i(k;\mathbb{Z}_p)$ for $i \geq 0$.

Proof of 4.3.2. For a perfect field of positive characteristic the usual Hochschild homology groups $\mathrm{HH}_*(k)$ vanish in higher degrees, and $\mathrm{HH}_0(k) = k$. It then follows from the spectral sequence (4.2.2) that

$$\pi_*(\mathrm{TH}(k)) = k \otimes \pi_*\mathrm{TH}(\mathbb{F}_p).$$

The cofibration sequence

$$\mathrm{TH}(k)_{hC_{p^n}} \xrightarrow{N} \mathrm{TH}(k)^{C_{p^n}} \xrightarrow{R} \mathrm{TH}(k)^{C_{p^{n-1}}}$$

was derived from taking C_{p^n} fixed points, so $\mathrm{TH}(k)^{C_{p^n}}$ acts on it. In particular, the homotopy groups are $\pi_0\mathrm{TH}(k)^{C_{p^n}}$-modules, and by (2.5.8) $W_{n+1}(k)$-modules. The inclusion $\mathbb{F}_p \subset k$ induces $W_{n+1}(k)$-homomorphisms:

(i) $\quad W_{n+1}(k) \otimes \pi_i\mathrm{TH}(\mathbb{F}_p)_{hC_{p^n}} \to \pi_i\mathrm{TH}(k)_{hC_{p^n}}$

(ii) $\quad W_{n+1}(k) \otimes \pi_i\mathrm{TH}(\mathbb{F}_p)^{C_{p^n}} \to \pi_i\mathrm{TH}(k)^{C_{p^n}}$

(iii) $\quad W_{n+1}(k) \otimes \pi_i\mathrm{TH}(\mathbb{F}_p)^{C_{p^{n-1}}} \to \pi_i\mathrm{TH}(k)^{C_{p^{n-1}}}$

Now $\pi_i\mathrm{TH}(\mathbb{F}_p)^{C_{p^{n-1}}} = \mathbb{Z}/p^n$ and $W_{n+1}(k) \otimes \mathbb{Z}/p^n = W_n(k)$, so the domain of (iii) is $W_n(k) \otimes \pi_i\mathrm{TH}(\mathbb{F}_p)^{C_{p^{n-1}}}$. We may inductively assume the third arrow to be an isomorphism. Thus we are done by the 5-lemma, if we can show that (i) is an isomorphism. This follows from the spectral sequence

$$H_*\left(C_{p^n};\pi_*\mathrm{TH}(k)\right) \Rightarrow \pi_*\mathrm{TH}(k)_{hC_{p^n}}.$$

Indeed, it is a spectral sequence of $W_{n+1}(k)$-modules when the $W_{n+1}(k)$-structure on the E^2-term is via $F^n\colon W_{n+1}(k) \to W_1(k) = k$ and

$$W_{n+1}(k) \otimes (F^n)^{\#}\pi_i\mathrm{TH}(\mathbb{F}_p) \cong (F^n)^{\#}\pi_*(\mathrm{TH}(k)).$$

We conclude that the homomorphisms in (i), (ii) and (iii) are isomorphisms. Now (4.2.8) gives

$$\pi_*\mathrm{TH}(k)^{C_{p^n}} = S_{W_{n+1}(k)}\{\sigma_n\} \tag{4.3.4}$$

with $R(\sigma_n) = \lambda_n p \sigma_{n-1}$ with $\lambda_n \in W_n(\mathbb{F}_p) = \mathbb{Z}/p^n$ a unit. In conclusion,

$$\varprojlim \pi_*\mathrm{TH}(k)^{C_{p^n}} = \begin{cases} 0 & \text{for } * > 0 \\ W(k) & \text{for } * = 0 \end{cases}$$

$$\varprojlim{}^{(1)} \pi_*\mathrm{TH}(k)^{C_{p^n}} = 0$$

as the limit system is obviously Mittag-Leffler, cf. [BK], p. 256. □

Theorem 4.3.5. *If A is a semi-simple k-algebra then $K_i(A; Z_p) \cong \mathrm{TC}_i(A; \mathbb{Z}_p)$ for $i \geq 0$.*

Proof. Both functors preserve products so it suffices to do the case of a simple algebra. If $A = \mathrm{M}_n(k)$ then we are done by Morita invariance:

$$K_i(\mathrm{M}_n(k)) = K_i(k), \quad \mathrm{TC}_i(\mathrm{M}_n(k)) = \mathrm{TC}_i(k)$$

and theorem 4.3.2. In general, we only know that

$$A \otimes_k k' \cong \mathrm{M}_n(k')$$

for a Galois extension k' with $|k' : k|$ prime to p. (The existence of such a k' is a consequence of the lack of p-torsion in the Brauer group $\mathrm{Br}(k)$). Finally, the horizontal compositions in the diagram

$$\begin{array}{ccccc} K_i(A; \mathbb{Z}_p) & \xrightarrow{i_*} & K_i(A \otimes_k k'; \mathbb{Z}_p) & \xrightarrow{i^*} & K_i(A; \mathbb{Z}_p) \\ \downarrow & & \downarrow & & \downarrow \\ \mathrm{TC}_i(A; \mathbb{Z}_p) & \xrightarrow{i_*} & \mathrm{TC}_i(A \otimes'_k; \mathbb{Z}_p) & \xrightarrow{i^*} & \mathrm{TC}_i(A; \mathbb{Z}_p) \end{array}$$

are isomorphisms since $|k' : k|$ is a unit of $\mathbb{Z}_p$, and the middle arrow is an isomorphism. (Here i^* is the composition of the functors applied to $A \otimes_k k' \to \mathrm{End}_A(A \otimes_k k')$ and Morita invariance). □

Corollary 4.3.6. *If A satisfies the assumption of theorem 4.3.1, then* $\mathrm{trc}\colon K_i^{\mathrm{top}}(A; \mathbb{Z}_p) \to \mathrm{TC}_i^{\mathrm{top}}(A; \mathbb{Z}_p)$ *is an isomorphism for $i \geq 0$.*

Proof. We are reduced to check that

$$\mathrm{trc}\colon K_i(A/pA; \mathbb{Z}_p) \to \mathrm{TC}_i(A/pA; \mathbb{Z}_p)$$

is an isomorphism. But A/pA is artenian, so its radical J is nilpotent. Thus by theorem 3.4.1 it is enough that the cyclotomic trace induce isomorphism for the algebra $(A/pA)/J$, which is semi-simple. Apply theorem 4.3.5. □

Theorem 4.3.7. *In the situation of theorem 4.3.1, the natural map*

$$\mathrm{TC}_i(A; \mathbb{Z}_p) \to \mathrm{TC}_i^{\mathrm{top}}(A; \mathbb{Z}_p)$$

is an isomorphism.

Proof. It is enough to prove the statement with $\mathbb{F}_p$ coefficients: a map of p-complete spaces is a homotopy equivalence if the induced homomorphism on mod p homotopy groups is an isomorphism.

The functor which to A associates the Eilenberg-MacLane spectrum HA is continuous, $\pi_i HA \cong \underleftarrow{\lim}\, \pi_i HA/p^n A$ when $A = \underleftarrow{\lim}\, A/p^n A$. The same is true for the r fold smash product, $HA^{(r)} = HA \wedge \cdots \wedge HA$,

$$\pi_*(HA^{(r)}; \mathbb{F}_p) \xrightarrow{\cong} \pi_* \left(\underset{n}{\underleftarrow{\operatorname{holim}}}\, H(A/p^n A)^{(r)}; \mathbb{F}_p \right).$$

This is an easy calculation based on the isomorphism

$$\pi_*(HA^{(r)}; \mathbb{F}_p) \cong H_*(HA^{(r-1)}; k) \oplus H_{*-1}(HA^{(r-1)}; k)$$

cf. [HM], lemma 5.1. It implies that the k-simplices

$$\mathrm{THH}_k(A)^\wedge_p \simeq \underleftarrow{\operatorname{holim}}\, \mathrm{THH}_k(A)^\wedge_p.$$

The simplicial group model $\mathrm{THH}^\oplus_\bullet$ for $\mathrm{THH}_\bullet$, cf. sect. 2.4, is a Kan complex, and for such homotopy inverse limits commutes with realizations, so we get

$$\mathrm{THH}(A)^\wedge_p \sim \underleftarrow{\operatorname{holim}}\, \mathrm{THH}(A/p^n A)^\wedge_p.$$

The same relation the holds for the spectra $\mathrm{TH}(A)$ and $\mathrm{TH}(A/p^n A)$.

Finally inductive use of the fundamental cofibration sequence shows that the fixed sets $(\mathrm{TH}(A)^{C_{p^n}})^\wedge_p$ are continuous, and since $\mathrm{TC}(A)^\wedge_p$ is a homotopy inverse limit construction, $\mathrm{TC}(A)^\wedge_p$ must be continuous. □

Theorem 4.3.8. *For the rings in theorem 4.3.1,*

$$K_i(A; \mathbb{Z}_p) \xrightarrow{\cong} K_i^{\mathrm{top}}(A; \mathbb{Z}_p)$$

Proof. Let F be the field of fractions of $W(k)$, and let $E = A \otimes_{W(k)} F$ with radical $J(E)$. Then $J = A \cap J(E)$ is a nilpotent ideal of A and it suffices, again the theorem 3.4.1, to show the theorem for A/J. But

$$A/J \otimes_{W(k)} F = E/J(E)$$

is semi-simple, and for such algebras results of Gabber, Suslin and Suslin-Yufryakov give the result, cf. [HM], appendix B for more details. □

Theorem 4.3.1 is probably the optimal result for K-theory calculations by traces. One would have liked to have a similar isomorphism for other rings, and in particular for the ring of rational integers. But

$$\mathrm{TC}_i(A; \mathbb{Z}_p) \xrightarrow{\cong} \mathrm{TC}_i(\underleftarrow{\lim}\, A/p^n A; \mathbb{Z}_p)$$

at least when A is finite over $\mathbb{Z}$. Indeed, this holds for the functor $A \mapsto (HA)^{\wedge}_p$ and hence adapting the argument of theorem 4.3.7 also for $\mathrm{TC}(A)^{\wedge}_p$. But K-theory does not have this property. One would also like to drop the finiteness assumption on A, and could wonder what would happen for $A = k[[X]]$. For such a ring the arguments proving theorem 4.3.7 and theorem 4.3.8 break down. In the first case for the simple reason that the r fold tensor power of A is not $k[[X_1, \dots, X_r]]$ - one needs completed tensor products.

4.4 The absolute theorem: group-like case.

This section examines $\mathrm{TC}(L)$ for a certain class of FSP's which include the $\tilde{G}$ of (2.3.4). The results are mostly a reformulation of parts of [BHM].

Definition 4.4.1. An FSP L is called group-like if the associated cyclotomic spectrum $T(L)$ satisfies the following condition:
For each finite cyclic group C there is an equivariant map of spectra

$$\sigma_C \colon \Phi^C T(L) \to T(L)^C,$$

natural with respect to inclusions $C_1 \subset C_2$, such that σ_C splits the natural map $s_C \colon T(L)^C \to \Phi^C T(L)$, $s_C \circ \sigma_C = \mathrm{id}$.

For group-like L, the fundamental cofibration

$$\mathrm{TH}(L)_{hC_{p^n}} \longrightarrow \mathrm{TH}(L)^{C_{p^n}} \xrightarrow{R} \mathrm{TH}(L)^{C_{p^{N-1}}}$$

is split by the map

$$S_{n-1} \colon \mathrm{TH}(L)^{C_{p^{n-1}}} \to \mathrm{TH}(L)^{C_{p^n}}$$

coming from the identification of $\rho^{\#}_{C_p} \Phi^{C_p} T(L)$ with $T(L)$, and

$$RS_{n-1} = \mathrm{id}, \quad FS_{n-1} \sim S_{n-2}F \quad (n \geq 2). \tag{4.4.2}$$

We recall from (4.1.5) that the fiber of R was identified as $\mathrm{TH}(L)_{hC_{p^n}}$ by the transfer map

$$\tau_{C_{p^n}} \colon \mathrm{TH}(L)_{hC_{p^n}} = \mathrm{TH}(L) \wedge_{C_{p^n}} ES^1_+ \xrightarrow{\sim} (T(L) \wedge ES^1_+)^{C_{p^n}}.$$

Naturality of transfers shows that

$$\begin{array}{ccc}
\mathrm{TH}(L)_{hC_{p^n}} & \xrightarrow{\sim} & T(L)^{C_{p^n}} \\
\downarrow{\scriptstyle \tau_n} & & \downarrow{\scriptstyle F} \\
TH(L)_{hC_{p^{n-1}}} & \xrightarrow{\sim} & T(L)^{C_{p^{n-1}}}
\end{array}$$

is homotopy commutative with τ_n being a suitable transfer map.

Proposition 4.4.3. *For a group-like FSP there is a homotopy Cartesian diagram*

$$\begin{array}{ccc} \mathrm{TC}(L)_p^\wedge & \longrightarrow & \left(\underset{\overleftarrow{\tau_n}}{\mathrm{holim}}\,\mathrm{TH}(L)_{hC_{p^n}}\right)_p^\wedge \\ \downarrow & & \downarrow{\scriptstyle \mathrm{proj}} \\ \mathrm{TH}(L)_p^\wedge & \xrightarrow{FS_0-1} & \mathrm{TH}(L)_p^\wedge \end{array}$$

Proof. The splittings of (4.4.2) give equivalences

$$\bigvee_{i=0}^{n} \mathrm{TH}(L)_{hC_{p^i}} \xrightarrow{\sim} \mathrm{TH}(L)^{C_{p^n}}$$

such that on the left hand side R corresponds to projection. Hence

$$\mathrm{TR}(L,p) \sim \prod_{i=0}^{\infty} \mathrm{TH}(L)_{hC_{p^i}}.$$

Under this equivalence $F(x_0, x_1, \ldots) = (Fx_1 + FSx_0, Fx_2, \ldots)$, and the diagram

$$\begin{array}{ccccc} \mathrm{TH}(L) & \longrightarrow & \prod_{i=0}^{\infty} \mathrm{TH}(L)_{hC_{p^i}} & \longrightarrow & \prod_{i=1}^{\infty} \mathrm{TH}(L)_{hC_{p^i}} \\ \downarrow{\scriptstyle FS_0-\mathrm{id}} & & \downarrow{\scriptstyle F-\mathrm{id}} & & \downarrow{\scriptstyle F-\mathrm{id}} \\ \mathrm{TH}(L) & \longrightarrow & \prod_{i=0}^{\infty} \mathrm{TH}(L)_{hC_{p^i}} & \longrightarrow & \prod_{i=1}^{\infty} \mathrm{TH}(L)_{hC_{p^i}} \end{array}$$

gives the cofibration

$$\mathrm{hF}(FS_0 - \mathrm{id})_p^\wedge \to \mathrm{TC}(L,p)_p^\wedge \to \underset{\overleftarrow{\tau_n}}{\mathrm{holim}}\left(\mathrm{TH}(L)_{hC_{p^i}}\right)_p^\wedge$$

upon taking vertical homotopy fibers. Apply theorem 2.5.5. □

Lemma 4.4.4. *For the identity FSP, $T(\mathrm{Id}) \sim_{C_\infty} \Sigma_{S^1}^\infty(S^0)$, where the right-hand side is the equivariant sphere spectrum.*

Proof. Recall from sect. 2.1 the subdivision S^1-homeomorphism

$$|sd_C\mathrm{THH}_\bullet(L;V)| \xrightarrow{D} |\mathrm{THH}_\bullet(L;V)|,$$

where C is a finite cyclic group of order c.

The space of 0-simplices in $sd_C\mathrm{THH}_\bullet(\mathrm{Id};V)$ is equal to $\mathrm{THH}_{c-1}(\mathrm{Id};V)$ and there is a natural C-map

$$i_C\colon \varinjlim \Omega^{m\mathbf{R}C}(S^{m\mathbf{R}C}\wedge S^V) \to |sd_C\mathrm{THH}_\bullet(\mathrm{Id};V)|$$

which is a C-homotopy equivalence onto the space of 0-simplices. The simplicial structure maps are C-homotopy equivalences, so the topological realization is C-homotopy equivalent to the space of 0-simplices, cf. sect. 2.2. Hence i_C is a C-homotopy equivalence. The diagram

$$\begin{array}{ccc} |sd_C\mathrm{THH}_\bullet(L;V)| & \xrightarrow{\ D\ } & |\mathrm{THH}_\bullet(L;V)| \\ & {\scriptstyle i_C}\nwarrow \quad \nearrow{\scriptstyle i} & \\ & S^V & \end{array}$$

is commutative. It follows that the S^1-map

$$\Sigma^\infty_{S^1}(S^V) \to T(\mathrm{Id})(V)$$

induced by i is a C-homotopy equivalence for each finite C. □

For any FSP L and monoid G we may define a new FSP by

$$L[G](X) = L(X)\wedge G_+. \tag{4.4.5}$$

If $L = \mathrm{Id}$ this is precisely $\tilde{G}$ of (2.3.4). If $L = \tilde{A}$ for a commutative ring A, the map $\tilde{A}[G] \to \widetilde{A[G]}$ is a stable equivalence, so there are equivalences

$$K(\tilde{A}[G]) \sim K(A[G]), \quad \mathrm{TC}(\tilde{A}[G]) \sim \mathrm{TC}(A[G])$$

for every discrete group. When G is a group-like topological monoid, the cyclic classifying space $B^{\mathrm{cy}}G = |N^{\mathrm{cy}}_\bullet(G)|$ was identified in sect. 2.1 to be the free loop space ΛBG of the ordinary classifying space BG. Moreover, if δ_C is the composite homeomorphism

$$\delta_C\colon |N^{\mathrm{cy}}_\bullet(G)| \xrightarrow{\Delta_C} |sd_C N^{\mathrm{cy}}_\bullet(G)|^C \xrightarrow{D} |N^{\mathrm{cy}}_\bullet(G)|$$

then there is a commutative diagram ([BHM], proposition 2.5)

$$\begin{array}{ccc} B^{\mathrm{cy}}G & \xrightarrow{\ \delta_C\ } & (B^{\mathrm{cy}}G)^C \\ \downarrow & & \downarrow \\ \Lambda BG & \xrightarrow{\ \Delta_c\ } & (\Lambda BG)^C, \quad \Delta_c(\lambda)(z) = \lambda(z^c) \end{array} \tag{4.4.6}$$

Given any cyclotomic spectrum T and any space X, the spectrum smash product $T \wedge \Lambda X_+$ is again cyclotomic. Indeed, there is a canonical map from right to left:

$$\Phi^C(T \wedge \Lambda X_+) \sim \Phi^C T \wedge (\Lambda X)^C_+$$

which is an S^1/C-equivalence, and

$$r_C \wedge \Delta_c^{-1} \colon \rho_C^\# \Phi^C T \wedge \rho_C^\# \Lambda X_+^C \to T \wedge \Lambda X_+$$

defines the required equivalence, cf. sect. 2.4.

Lemma 4.4.7. *There is an S^1-equivalence of cyclotomic spectra, $T(L[G]) \sim_{S^1} T(L) \wedge \Lambda BG_+$, provided G is group-like.*

Proof. Consider the bi-simplicial space $X_{\bullet,\bullet}(G;V)$ with

$$X_{k,l}(G;V) = \underset{\mathbf{x}\in I^{k+1}}{\operatorname{holim}_{\to}} F\left(S^{x_0} \wedge \cdots \wedge S^{x_k}, F(S^{x_0}) \wedge \cdots \wedge F(S^{x_k}) \wedge G_+^l \wedge S^V\right).$$

Cyclic permutation of factors make it a bi-cyclic space. The map

$$X_{k,l}(1;V) \wedge G_+^l \to X_{k,l}(G;V) \tag{1}$$

becomes highly connected as an equivariant map as V runs through the S^1-universe $\mathcal{U}$ (one needs $\dim V^C \to \infty$ for all $C \subseteq S^1$).

The diagonal complex $\delta X_{\bullet,\bullet}(G;V)$ is precisely $\mathrm{THH}_\bullet(L[G];V)$ with realization $\mathrm{THH}(L[G];V)$. On the other hand, if we instead first realize the l-direction and then the k-direction and use (1), then we get a highly connected S^1-map

$$\mathrm{THH}(L;V) \wedge \Lambda BG_+ \to \mathrm{THH}(L[G];V).$$

Use of subdivision and (4.4.6) shows that the corresponding map on C-fixed sets become highly connected when V runs over $\mathcal{U}$, so the two prespectra are equivalent. Moreover, the corresponding cyclotomic structure maps agree. Apply spectrification. □

Corollary 4.4.8. *The FSP $\tilde{G}$ is group-like if G is.*

Proof. The previous result tells us that $T(\tilde{G}) \sim_{C_\infty} \Sigma^\infty_{S^1}(\Lambda BG_+)$. But the suspension spectrum satisfies the requirement of (4.4.1). This is a consequence of the tom Dieck-Segal splitting, valid for any based S^1-space X:

$$\Sigma^\infty_{S^1}(X)^C \sim_{S^1/C} \bigvee_{H \subseteq C} \Sigma^\infty_{S^1/C} E_{S^1/C}(C/H)_+ \wedge_{C/H} X^H$$

$$\Phi^C(\Sigma^\infty_{S^1}(X)) \sim \Sigma^\infty_{S^1/C}(X^C)$$

Here $E_G(\Gamma)$ is the G-equivariant model of $E\Gamma$. The map s_C is the projection onto the factor $C = H$ and σ_C is the obvious inclusion, cf. [tD], [LMS]. $\square$

The next theorem is similar to lemma 5.15 of [BHM], but avoids the assumption that T has finite p-type. It contradicts the "counter-example" presented in [BHM], p. 498–499, which is wrong. The mistake occurs in the identification of $(t_m^{m-1})_r$ on p. 499. The mistake was pointed out by T. Goodwillie, and the proof below is due to him.

Lemma 4.4.9. *For any equivariant S^1-spectrum T, the S^1-transfer induces an isomorphism*

$$\pi_*(\Sigma T_{hS^1};\mathbb{F}_p) \to \pi_*(\operatorname{holim}_{\leftarrow} T_{hC_{p^n}};\mathbb{F}_p).$$

Proof. The skeletons of ES^1 are the spheres $S^{2k-1} \in \mathbb{C}^k$ with the standard action of S^1. There is the cofibration diagram

$$\begin{array}{ccccc} S^1 \wedge (S_+^{2k-1} \wedge_{S^1} T) & \longrightarrow & S^1 \wedge (S_+^{2k+1} \wedge_{S^1} T) & \longrightarrow & S^1 \wedge (S^{2k+1}/S^{2k-1} \wedge_{S^1} T) \\ \downarrow{\scriptstyle \mathrm{trf}} & & \downarrow{\scriptstyle \mathrm{trf}} & & \downarrow{\scriptstyle \mathrm{trf}} \\ S_+^{2k-1} \wedge_{C_{p^n}} T & \longrightarrow & S_+^{2k+1} \wedge_{C_{p^n}} T & \longrightarrow & S^{2k+1}/S^{2k-1} \wedge_{C_{p^n}} T \end{array} \qquad (1)$$

Now $S^{2k+1}/S^{2k-1} \simeq_{S^1} S_+^1 \wedge S^{2k}$; the S^1-action on the right hand side is the diagonal action with $S^{2k} = \Sigma(S^{2k-1})$. However for any S^1-space or spectrum X,

$$S_+^1 \wedge X \cong_{S^1} S_+^1 \wedge |X|, \quad (z,x) \mapsto (z, z^{-1}x)$$

where the bars indicate X with no S^1-action. In particular,

$$S_+^1 \wedge S^{2k} \wedge T \simeq_{S^1} S_+^1 \wedge |S^{2k} \wedge T|$$

and the upper right hand term in (1) may be identified as

$$S^1 \wedge S^{2k+1}/S^{2k-1} \wedge_{S^1} T \simeq S^1 \wedge |S^{2k} \wedge T|.$$

Moreover, the right-hand vertical map in (1) can be identified as the smash product of the transfer

$$\tau\colon S^1 \wedge \Sigma^\infty(S_+^1/S^1) \to \Sigma^\infty(S_+^1/C_{p^n}) \qquad (2)$$

with $|S^{2k} \wedge T|$. The transfers

$$\tau_n\colon \Sigma^\infty(S_+^1/C_{p^{n-1}}) \to \Sigma^\infty(S_+^1/C_{p^n})$$

of the C_p-covering $S^1/C_{p^{n-1}} \to S^1/C_{p^n}$ are known as follows. If we identify S^1/C_{p^n} with S^1 (via $\rho_{C_{p^n}}$), and use the splitting

$$\Sigma^\infty(S^1_+) = \Sigma^\infty(S^1) \vee \Sigma^\infty(S^0)$$

induced by the projections, then τ_n becomes the matrix

$$\tau_n = \begin{pmatrix} \mathrm{id} & 0 \\ \eta & p \end{pmatrix} \tag{3}$$

with $\eta \in \pi_1(\Sigma^\infty(S^0)) = \mathbb{Z}/2$ the non-trivial element. This can be seen for example by using ω of (2.4.2). Since the transfers in the limit system

$$\mathrm{trf}_n\colon S^{2k+1} \,/\, S^{2k-1} \wedge_{C_{p^{n-1}}} T \to S^{2k+1}/S^{2k-1} \wedge_{C_{p^n}} T$$

can be identified with $\tau_n \wedge |S^{2k} \wedge T|$,

$$\underset{\mathrm{trf}_n}{\underleftarrow{\mathrm{holim}}}\, S^{2k+1}/S^{2k-1} \wedge_{C_{p^{n-1}}} T \simeq \underset{\tau_n}{\underleftarrow{\mathrm{holim}}}\, \Sigma^\infty(S^1_+) \wedge |S^{2k} \wedge T|,$$

and we obtain from (3) a cofibration

$$S^1 \wedge S^{2k} \wedge T \to \underset{\tau_n}{\underleftarrow{\mathrm{holim}}}\, \Sigma^\infty(S^1_+) \wedge |S^{2k} \wedge T| \to \underset{p}{\underleftarrow{\mathrm{holim}}}\, S^{2k} \wedge T.$$

We can calculate the mod p homotopy groups of the right hand term by the exact sequence

$$0 \to \underleftarrow{\lim}^{(1)} \pi_{i-1}(S^{2k} \wedge T; \mathbb{F}_p) \to \pi_i(\underleftarrow{\mathrm{holim}}(S^{2k} \wedge T); \mathbb{F}_p) \to \underleftarrow{\lim}\, \pi_i(S^{2k} \wedge T; \mathbb{F}_p) \to 0.$$

The outer terms vanish, so in conclusion

$$\pi_i(S^1 \wedge S^{2k} \wedge T; \mathbb{F}_p) \cong \pi_i(\underset{\mathrm{trf}_n}{\underleftarrow{\mathrm{holim}}}\, S^{2k+1}/S^{2k-1} \wedge_{C_{p^n}} T; \mathbb{F}_p),$$

and comparing with (2) it follows that the right-hand vertical maps in (1) induces an isomorphism

$$\pi_i(\Sigma(S^{2k+1}/S^{2k-1} \wedge_{S^1} T); \mathbb{F}_p) \xrightarrow{\cong} \pi_i(\underleftarrow{\mathrm{holim}}\, S^{2k+1}/S^{2k-1} \wedge_{C_{p^n}} T; \mathbb{F}_p).$$

We can finally make the obvious induction over k. □

Remark 4.4.10. The lemma can be restated as a homotopy equivalence of p-completed spaces,

$$(S^1 \wedge T_{hS^1})^\wedge_p \simeq (\underleftarrow{\mathrm{holim}}\, T_{hC_{p^n}})^\wedge_p.$$

Corollary 4.4.11. *For a group-like FSP, there is a homotopy Cartesian diagram of (non-equivariant) spectra*

$$\begin{array}{ccc} \mathrm{TC}(L)^\wedge_p & \longrightarrow & (\Sigma\mathrm{TH}(L)_{hS^1})^\wedge_p \\ \downarrow & & \downarrow{\scriptstyle \mathrm{trf}_{S^1}} \\ \mathrm{TH}(L)^\wedge_p & \xrightarrow{FS_0-\mathrm{id}} & \mathrm{TH}(L)^\wedge_p \end{array}$$

Moreover, if $L = \tilde{G}$ *for a group-like monoid, then* $\mathrm{TH}(L) = \Sigma^\infty(\Lambda BG_+)$ *and* $FS_0 = \Sigma^\infty(\Delta_{p+})$ *where* $\Delta_p(\lambda)(z) = \lambda(z^p)$.

Proof. Only the last point need any explanation. It comes from the Segal-tom Dieck splitting used in the proof of corollary 4.4.8:

$$\begin{aligned} \Sigma^\infty_{S^1}(\Lambda BG_+)^{C_p} &\sim \Sigma^\infty(\Lambda BG_{hC_p+}) \vee \Sigma^\infty(\Lambda BG_+^{C_p}) \\ &\cong \Sigma^\infty(\Lambda BG_{hC_p+}) \vee \Sigma^\infty(\Lambda BG_+) \end{aligned}$$

where the last homeomorphism is $\mathrm{id} \vee \Sigma^\infty_+(\Delta_p^{-1})$. The map F becomes the sum of the transfer

$$\Sigma^\infty_+(\Lambda BG_{hC_p}) \to \Sigma^\infty(\Lambda BG_+)$$

and the inclusion

$$\Sigma^\infty(\Lambda BG_+^{C_p}) \to \Sigma^\infty(\Lambda BG_+)$$

and

$$S_0\colon \Sigma^\infty(\Lambda BG_+) \to \Sigma^\infty_+(\Lambda BG_{hC_p}) \vee \Sigma^\infty(\Delta BG_+^{C_p})$$

is the inclusion in the second factor via $\Sigma^\infty_+(\Delta_p)$ □

Recall for an FSP L that we write $\pi_0 L$ for the associated ring $\pi_0 L = \varinjlim \pi_n L(S^n)$.

Theorem 4.4.12. *Suppose* L *is an FSP so that* $\pi_0 L$ *is a finite* $W(k)$*-algebra for some perfect field* k *of characteristic* p. *Then*

$$\mathrm{trc}\colon K(L)^\wedge_p \to \mathrm{TC}(L)^\wedge_p$$

is a homotopy equivalence.

Proof. Dundas' theorem 3.5.1 gives the homotopy Cartesian square

$$\begin{array}{ccc} K(L)^\wedge_p & \longrightarrow & \mathrm{TC}(L)^\wedge_p \\ \downarrow & & \downarrow \\ K(\pi_0 L)^\wedge_p & \longrightarrow & \mathrm{TC}(\pi_0 L)^\wedge_p \end{array}$$

and the bottom arrow is a homotopy equivalence by theorem 4.3.1. □

4.5 The K-theory assembly map.

For a discrete group G and a commutative ring R, $\mathrm{GL}_n(R[G])$ contains $\mathrm{GL}_n(R) \times G$ as a subgroup, namely as the tensor product of $(n \times n)$-matrices over R and elements $g \in G$ considered as (1×1)-matrices over $R[G]$. Taking classifying spaces gives a map

$$B\mathrm{GL}_n(R) \times BG \to B\mathrm{GL}_n(R[G]).$$

This induces a map of spectra

$$a_K \colon K(R) \wedge BG_+ \to K(R[G])$$

usually called the assembly map. Indeed, one may either use Segal's Γ-space definition, May's operad version or Waldhausen's definition of $K(A)$ to do the details, or one can use the device of ring suspensions as in the original source, [L1].

The study of a_K has long been promoted by W. C. Hsiang, who e.g. in [Hs], conjectured that a_K is a rational injection, provided R is regular and BG is a finite complex. The conjecture is often called the K-theory Novikov conjecture. The reason is that there is a similar assembly map in L-theory, initially constructed by F. Quinn,

$$a_L \colon L(R) \wedge BG_+ \to L(R[G])$$

and (rational) injectivity of a_L (for $R = \mathbb{Z}$ and BG a manifold) translates via the surgery exact sequence to Novikov's original conjecture about the homotopy invariance of the higher signatures.

The definition of a_K extends to the case of FSP's to give a map of spectra

$$a_K \colon K(L) \wedge BG_+ \to K(L[G]).$$

(Here G could be any group-like monoid, and thus BG any space. For $L = \mathrm{Id}$ the above becomes Waldhausen's assembly map $A(*) \wedge X_+ \to A(X)$). The study of the assembly map when $L = \mathrm{Id}$ was the main motivation behind [BHM]. We can now present a somewhat easier proof of the main result from [BHM], thanks to Dundas' relative theorem 3.5.1.

There is an obvious assembly map

$$\mathrm{THH}(L; V) \wedge \Lambda BG_+ \to \mathrm{THH}(L[G]; V)$$

(cf. lemma 4.4.7) and hence via the inclusion

$$BG \to \Lambda BG$$

an assembly map

$$\mathrm{THH}(L;V)\wedge BG_{+}\to \mathrm{THH}(L[G];V).$$

This passes to an assembly map of cyclotomic spectra and induces

$$a_{\mathrm{TC}}\colon \mathrm{TC}(L)\wedge BG_{+}\to \mathrm{TC}(L[G])$$

so that the diagram

$$\begin{array}{ccc} K(L)\wedge BG_{+} & \xrightarrow{a_K} & K(L[G]) \\ \big\downarrow{\scriptstyle \mathrm{trc}\wedge\mathrm{id}} & & \big\downarrow{\scriptstyle \mathrm{trc}} \\ \mathrm{TC}(L)\wedge BG_{+} & \xrightarrow{a_{\mathrm{TC}}} & \mathrm{TC}(L[G]) \end{array} \qquad (4.5.1)$$

is commutative.

For each FSP L, we can from its p-adic completion L_p, $L_p(S) = L(S)^{\wedge}_{p}$. (It should be remembered that $X^{\wedge}_{p}\wedge Y^{\wedge}_{p}$ is not p-complete; but this causes no problems because we are always completing the functors on the outside, so there are no unpleasant surprises in $\mathrm{THH}(L_p)^{\wedge}_{p}$ etc.)

Theorem 4.5.2. *For a discrete group G, the assembly map*

$$a_K\colon K(\mathrm{Id}_p)\wedge BG_{+}\to K(\mathrm{Id}_p[G])$$

becomes split injective after p-adic completion.

Proof. We compose with the cyclotomic trace and consider

$$\begin{array}{ccc} (K(\mathrm{Id}_p)\wedge BG_{+})^{\wedge}_{p} & \xrightarrow{a_K} & K(\mathrm{Id}_p[G])^{\wedge}_{p} \\ \big\downarrow & & \big\downarrow \\ (\mathrm{TC}(\mathrm{Id}_p)\wedge BG_{+})^{\wedge}_{p} & \xrightarrow{a_{\mathrm{TC}}} & \mathrm{TC}(\mathrm{Id}_p[G])^{\wedge}_{p} \end{array} \qquad (1)$$

Now corollary 4.4.11 gives the homotopy Cartesian diagram

$$\begin{array}{ccc} \mathrm{TC}(\mathrm{Id}_p[G])^{\wedge}_{p} & \longrightarrow & (\Sigma(\Sigma^{\infty}_{S^1}(\Lambda BG_{+}))_{hS^1})^{\wedge}_{p} \\ \big\downarrow & & \big\downarrow \\ \Sigma^{\infty}(\Lambda BG_{+})^{\wedge}_{p} & \xrightarrow{\Sigma^{\infty}(\Delta_p)-\mathrm{id}} & \Sigma^{\infty}(\Lambda BG_{+})^{\wedge}_{p} \end{array} \qquad (2)$$

upon using the obvious equivalence between $\mathrm{TH}(\mathrm{Id}_p)^{\wedge}_{p}$ and $\mathrm{TH}(\mathrm{Id})^{\wedge}_{p}$ together with lemma 4.4.4 and lemma 4.4.7.

The component group $\pi_0(\Lambda BG)$ is the set of free homotopy classes of maps from the circle into BG, and hence equal to the conjugary classes of elements in G. Let $\Lambda_{[1]}BG$ be the component of the identity element. There are S^1-equivariant maps

$$\Lambda BG_+ \xrightarrow{\text{proj}} \Lambda_{[1]}BG_+ \xleftarrow{\text{incl}} BG_+. \tag{3}$$

The inclusion is a homotopy equivalence, but not an equivariant one. Anyway, the weak statement is enough to ensure that

$$ES^1 \times_{S^1} BG \longrightarrow ES^1 \times_{S^1} \Lambda_{[1]}BG$$

is a homotopy equivalence, and since

$$\Sigma^\infty_{S^1}(\Lambda BG_+)_{hS^1} = \Sigma^\infty(ES^1 \times_{S^1} \Lambda BG_+)$$

diagram (1) projects to the homotopy Cartesian diagram

$$\begin{array}{ccc} (\mathrm{TC}(\mathrm{Id}_p) \wedge BG_+)^\wedge_p & \longrightarrow & \Sigma(\Sigma^\infty BS^1_+ \wedge BG_+)^\wedge_p \\ \downarrow & & \downarrow \\ \Sigma^\infty(BG_+)^\wedge_p & \longrightarrow & \Sigma^\infty(BG_+)^\wedge_p \end{array}$$

Moreover,

$$(\mathrm{TC}(\mathrm{Id}_p) \wedge BG_+)^\wedge_p \xrightarrow{a_{\mathrm{TC}}} \mathrm{TC}(\mathrm{Id}_p[G])^\wedge_p \xrightarrow{\text{proj}} (\mathrm{TC}(\mathrm{Id}_p) \wedge BG_+)^\wedge_p$$

is the identity, and thus a_{TC} is split injective after p-adic completion. Now apply theorem 4.4.12 and diagram (1) to conclude the proof. □

Soulé proved in [Sou] that

$$\pi_{4n+1}(K(\mathbb{Z}); \mathbb{Q}_p) \to \pi_{4n+1}(K(\mathbb{Z}_p); \mathbb{Q}_p) \tag{4.5.3}$$

is an isomorphism provided the p-adic L-function $L_p(1+2n, \omega^{-2n}) \neq 0$ (both groups are equal to $\mathbb{Q}_p$). This is certainly the case for regular primes and maybe always. Soulé proved (4.5.3) by using the étale cohomology invariant. It was reproved in [BHM] by cyclotomic trace considerations. One can use (4.5.3) to translate theorem 4.5.2 into a rational statement, namely

Theorem 4.5.4. ([BHM]). *If G is a discrete group for which each $H_i(BG; \mathbb{Z})$ is finitely generated, then the K-theory assembly map*

$$a_K \colon K(\mathbb{Z}) \wedge BG_+ \longrightarrow K(\mathbb{Z}G)$$

induce an injection on rational homotopy groups.

Proof. The linearization maps

$$K(\mathrm{Id}) \to K(\mathbb{Z}), \quad K(\mathrm{Id}[G]) \to K(\mathbb{Z}G)$$

are rational equivalences, essentially because the homotopy groups of $\Sigma^\infty(S^0)$ are finite in positive degrees, cf. [W1]. Thus it suffices to show the statement for

$$a_K\colon K(\mathrm{Id}) \wedge BG_+ \to K(\mathrm{Id}[G]).$$

We have

$$K(\mathrm{Id}[G]) \to \mathrm{TC}(\mathrm{Id}[G]) \to (\mathrm{TC}(\mathrm{Id}) \wedge BG_p)^\wedge_p$$

and must show

$$\mathrm{trc} \wedge \mathrm{id}_{BG}\colon K(\mathrm{Id}) \wedge BG_+ \to (\mathrm{TC}(\mathrm{Id}) \wedge BG_+)^\wedge_p$$

is rational injective. This is the case because

$$K(\mathrm{Id})^\wedge_p \to K(\mathrm{Id}_p)^\wedge_p \simeq \mathrm{TC}(\mathrm{Id}_p)^\wedge_p$$

is rationally the same as $K(\mathbb{Z})^\wedge_p \to K(\mathbb{Z}_p)^\wedge_p$, and because we can choose p to be a regular prime and apply (4.5.3). □

Remark 4.5.5. It would be nice if the above argument could be extended to L-theory, and thus proving the original Novikov conjecture for the groups with finitely generated Eilenberg-MacLane homology. There is a variant of $\mathrm{TC}(R)$, namely the topological Dihedral homology $\mathrm{TD}(R)$, which imitates the linear construction of [L2]. It is the fixed set of a suitable involution on $\mathrm{TC}(R)$, $\mathrm{TD}(R) = \mathrm{TC}(R)^{\mathbb{Z}/2}$, and there is a map from Hermitian K-theory into $\mathrm{TD}(R)$, at least when $1/2 \in R$. The basic problem with this approach however, is that $\mathrm{TD}(R)^\wedge_p \to \mathrm{TD}(R \otimes \mathbb{Z}_p)^\wedge_p$ is again an equivalence (under suitable finiteness conditions on R). But in contrast to (4.5.3), $L(\mathbb{Z}) \to L(\mathbb{Z}_p)$ is rationally trivial for all primes, so one cannot extend the K-theory proof directly.

There might be a chance of proceding indirectly as follows. Let E be the maximal abelian extension of $\mathbb{Q}_p$, and let A be the integers of E. If one could produce a signature type rationally injective map from $L(\mathbb{Z}[g])$ to $K(A[G])$, or maybe into some completion $\mathcal{A}(G)$ of $A[G]$, like the C^*-algebra associated with $\mathbb{C}[G]$, then one could study the K-theory assembly map on $A[G]$ (or $\mathcal{A}(G)$) using the techniques above.

In this connection one should remember the theorems of Suslin that $K(\hat{E})^\wedge_p \simeq K(\mathbb{C})^\wedge_p$ for the algebraic closure of E and that $K(\mathbb{C})^\wedge_p \simeq BU^\wedge_p$. The latter equivalence comes from the roots of unity: the map $BS^1 \to B\mathrm{GL}_1(\mathbb{C}) \to K(\mathbb{C})$ extends to $\Omega^\infty S^\infty(BS^1) \to K(\mathbb{C})$, and gives via the splitting $\Omega^\infty S^\infty(BS^1) \sim BY \times X$ the required map from BU to $K(\mathbb{C})$, I believe.

The same procedure gives a map from $BU_p^\wedge \to K(A)_p^\wedge$ because $\mu(A) = \mathbb{Q}/\mathbb{Z}$ and $B(\mathbb{Q}/\mathbb{Z})_p^\wedge \sim (BS^1)_p^\wedge$.

This remark represents years of discussions with W. C. Hsiang.

The main interest in the assembly map a_K lies in its relationship to automorphism groups of manifolds. For a group-like monoid, such as $G = \Omega X$, $K(\mathrm{Id}[G])$ is Waldhausen's $A(X)$ and in particular $K(\mathrm{Id}) = A(*)$, so that the assembly map takes the form

$$a_A\colon A(*) \wedge X_+ \to A(X).$$

Waldhausen defined the spectrum $\mathrm{Wh}^{\mathrm{top}}(X)$ to be the cofiber of a_A.

For a manifold M, the space of topological pseudo isotopies $\mathcal{P}^{\mathrm{top}}(M)$ is defined as the space of homeomorphisms of $\mathrm{M}^n \times I$ which is the identity on $\mathrm{M}^n \times 0 \cup \partial M \times I$. A celebrated result of Waldhausen [W4] states that

$$\Omega^2 \mathrm{Wh}^{\mathrm{top}}(M) \sim \underset{\underset{k}{\longrightarrow}}{\mathrm{holim}}\, \mathcal{P}^{\mathrm{top}}(M \times D^k). \tag{4.5.6}$$

Moreover, the stability theorem of K. Igsa, [I] asserts that the map

$$\mathcal{P}^{\mathrm{top}}(M) \to \underset{\longrightarrow}{\mathrm{holim}}\, \mathcal{P}^{\mathrm{top}}(M \times D^k)$$

is $(\dim M - 7)/3$-connected, at least if M is smoothable

Farell and Jones has in [FJ] shown that for a negatively curved manifold M, $\mathrm{Wh}^{\mathrm{top}}(S^1)$ determines $\mathrm{Wh}^{\mathrm{top}}(M)$. Thus it would be of considerable interest to determine $\mathrm{Wh}^{\mathrm{top}}(S^1)$. Theorem 3.5.1, proposition 4.4.3 and corollary 4.4.11 reduces this to the problem of studying the linearization map

$$L^{\langle 1\rangle}\colon \mathrm{TC}^{\langle 1\rangle}(S^1,p)_p^\wedge \longrightarrow \mathrm{TC}^{\langle 1\rangle}(\mathbb{Z}[t,t^{-1}],p)_p^\wedge$$

where $\mathrm{TC}^{\langle 1\rangle}(-)$ is the cofiber of a_{TC}. Indeed, the K-theory assembly map $S^1_+ \wedge K(\mathbb{Z}) \to K(\mathbb{Z}[t,t^{-1}])$ is an equivalence, so the fiber of $L^{\langle 1\rangle}$ is $\mathrm{Wh}^{\mathrm{top}}(S^1)_p^\wedge$. See also remark 5.4.8 below. See [M] for more details.

5 Calculations in K-theory

This chapter evaluates the higher K-groups $K_i(R;\mathbb{Z}_p)$ with p-adic coefficients in a number of cases where the K-groups were not previously known. The rings we consider are all of the type where the absolute theorem of the previous chapter applies, and the functor we actually calculate is $\mathrm{TC}(R)_p^\wedge$.

5.1 On the K-theory of group rings.

Let A be a finite algebra over $W(k)$, the Witt vectors of a finite field k of characteristic p. For a finite group G, the group ring $A[G]$ is again finite, so

$$K_i(A[G];\mathbb{Z}_p) \cong \mathrm{TC}_i(A[G];\mathbb{Z}_p).$$

By general induction theory, cf. [O1]

$$K(A[G])^{\wedge}_p \simeq \operatorname*{holim}_{\longleftarrow} K(A[\Gamma])^{\wedge}_p$$

where Γ runs over the hyper-elementary subgroups of G, that is, the subgroups of the form $\Gamma = C_N \ltimes P$ where P a p-group and $(N,p) = 1$. It follows that $A[\Gamma]$ decomposes into a product of twisted group rings $B^t[P]$ for unramified extensions B/A.

We here study the case of an untwisted group-ring $A[P]$. In terms of explicit values our main result is

Theorem 5.1.1. *For a perfect field k of characteristic $p > 0$, $K_{2n-1}(k[C_{p^N}];\mathbb{Z}_p) = K_1(k[C_{p^N}];\mathbb{Z}_p)^{\oplus n}$ and $K_{2n}(k[C_{p^N}];\mathbb{Z}_p) = 0$ when $n > 0$.*

The K_1-group on the left is the p-part of the units $k[C_{p^N}]^\times$ which is easily calculated, cf. theorem 5.1.16 below. Note also that $k[C_{p^N}]/\mathrm{rad} = k$, so that $K_i(k[C_{p^N}];\mathbb{Z}_l) = K_i(k;\mathbb{Z}_l)$ for $(l,p) = 1$.

Our starting point is lemma 4.4.7,

$$T(A[P]) \sim_{S^1} T(A) \wedge \Lambda BP_+.$$

Let $X(P)$ denote the conjugacy classes of elements in P. Then $\pi_0(\Lambda BP) = X(P)$, and the p'th power map $\Delta\colon X(P) \to X(P)$ has $\Delta^N X(P) = 1$ when P has exponent p^N. Define a filtration of $X(P)$,

$$\{1\} = X_0(P) \subset X_1(P) \subset \cdots \subset X_N(P) = X(P), \quad X_k(P) = \{g | g^{pk} = 1\}$$

and a corresponding filtration of $\Lambda = \Lambda BP$

$$\Lambda_0 \subset \Lambda_1 \subset \cdots \subset \Lambda_N = \Lambda \qquad (5.1.2)$$

where $\Lambda_k = \coprod_{\gamma \in X_k(P)} \Lambda_\gamma BP$ is the set of components corresponding to the listed conjugacy classes. We note (from [BHM], sect. 7) that

$$\Lambda_\gamma BP \sim BC_P(\gamma)$$

the classifying space of the centralizer of γ.

We are interested in the C_{p^n}-action on $T(A[P])$. The p'th power map $\Delta\colon \Lambda \xrightarrow{\cong} \Lambda^{C_p} \subset \Lambda$ maps Λ_k homeomorphically into $\Lambda_{k-1}^{C_p}$, so in (5.1.2), $\Lambda_N - \Lambda_{N-1}$ is the free stratum, and

$$\Delta\colon \Lambda_k - \Lambda_{k-1} \to (\Lambda_{k-1} - \Lambda_{k-2})^{C_p}$$

is a homeomorphism for $1 < k \leq N$. Let $\mathrm{TC}^{\langle 1\rangle}(A[G],p)$ denote the cofiber of the assembly map from sect. 4.5,

$$\mathrm{TC}(A,p) \wedge BP_+ \xrightarrow{\alpha_{\mathrm{TC}}} \mathrm{TC}(A[P],p) \longrightarrow \mathrm{TC}^{\langle 1\rangle}(A[P],p), \tag{5.1.3}$$

and write $T \ltimes X = T \wedge X_+$.

Proposition 5.1.4. *One has*

$$\mathrm{TC}^{\langle 1\rangle}(A[P],p) \simeq \underset{\overleftarrow{F}}{\mathrm{holim}}\, (T(A) \ltimes (\Lambda_1 - \Lambda_0))^{C_{p^n}},$$

where the limit runs over inclusions of fixed sets.

Proof. In the proof we write $B = BP$. The inclusion $i\colon B \to \Lambda$ of B into the constant loops induces a cofibration sequence of cyclotomic spectra

$$T(A) \ltimes B \to T(A) \ltimes \Lambda \to T(A) \wedge \Lambda/B$$

This gives a cofibration sequence of fixed sets, and hence the cofibration sequence

$$\underset{\overleftarrow{F,R}}{\mathrm{holim}}(T(A) \ltimes B)^{C_{p^n}} \to \mathrm{TC}(A[G],p) \to \underset{\overleftarrow{F,R}}{\mathrm{holim}}(T(A) \wedge \Lambda/B)^{C_{p^n}}.$$

Now $\Delta = \mathrm{id}$ on B, and since B has trivial S^1-action,

$$\underset{\overleftarrow{F,R}}{\mathrm{holim}}(T(A) \ltimes B)^{C_{p^n}} = (\underset{\overleftarrow{F,R}}{\mathrm{holim}}\, T(A)^{C_{p^n}}) \ltimes B = \mathrm{TC}(A,p) \ltimes B.$$

It follows that

$$\mathrm{TC}^{\langle 1\rangle}(A[G],p) = \underset{\overleftarrow{F,R}}{\mathrm{holim}}(T(A) \wedge \Lambda/B)^{C_{p^n}}. \tag{1}$$

We examine the right-hand side in two steps. First we evaluate the homotopy limit over R and then we use the cofibration

$$\underset{\overleftarrow{F,R}}{\mathrm{holim}}(T(A)\wedge\Lambda/B)^{C_{p^n}} \longrightarrow \underset{\overleftarrow{R}}{\mathrm{holim}}(T(A)\wedge\Lambda/B)^{C_{p^n}} \xrightarrow{F-1} \underset{\overleftarrow{R}}{\mathrm{holim}}(T(A)\wedge\Lambda/B)^{C_{p^n}}. \tag{2}$$

We use the decomposition

$$\Lambda/B = \Lambda_0/B \vee (\Lambda_1 - \Lambda_0)_+ \vee \cdots \vee (\Lambda_N - \Lambda_{N-1})_+$$

and the corresponding decomposition

$$(T(A) \wedge \Lambda/B)^{C_{p^n}} = (T(A) \wedge \Lambda_0/B)^{C_{p^n}} \vee \bigvee_{k=1}^{N} (T(A) \ltimes (\Lambda_k - \Lambda_{k-1}))^{C_{p^n}}.$$

There are the following easy consequences of the cyclotomic structure on $T(A) \ltimes \Lambda$, cf. lemma 4.4.7:

(3)

(i) $\Lambda_0^{C_p}/B \overset{\Delta}{\cong} \Lambda_1/B = \Lambda_0/B \vee (\Lambda_1 - \Lambda_0)_+$

(ii) $R\colon (T(A) \wedge \Lambda_0/B)^{C_{p^n}} \xrightarrow{\simeq} (T(A) \wedge \Lambda_0/B)^{C_{p^{n-1}}} \vee (T(L) \ltimes (\Lambda_1 - \Lambda_0))^{C_{p^{n-1}}}$

(iii) $R\colon (T(A) \ltimes (\Lambda_k - \Lambda_{k-1}))^{C_{p^n}} \to (T(L) \ltimes (\Lambda_{k+1} - \Lambda_k))^{C_{p^{n-1}}}, \quad 1 \le k < N$

(iv) $R\colon (T(A) \ltimes (\Lambda_N - \Lambda_{N-1}))^{C_{p^n}} \to 0$

The fundamental cofibration applied to $T = T(A) \wedge \Lambda_0/B$ shows that (3,ii) is a homotopy equivalence. Indeed $(T(A) \wedge \Lambda_0/B)_{hC_{p^n}} \sim 0$ since the inclusion of B in Λ_0 is a non-equivariant homotopy equivalence. If we write

$$X_n = \bigvee_{k=2}^{N} (T(A) \ltimes (\Lambda_k - \Lambda_{k-1}))^{C_{p^n}}$$

$$Y_n = (T(A) \wedge \Lambda_0/B)^{C_{p^n}} \vee (T(A) \ltimes (\Lambda_1 - \Lambda_0))^{C_{p^n}}$$

and consider the cofibration sequence of limit systems

$$(X_n, R) \to ((T(A) \wedge \Lambda/B)^{C_{p^n}}, R) \to (Y_n, R)$$

it follows from (3,iii–iv) that $R^{N-1}\colon X_n \to X_{n-N+1}$ is null-homotopic. Hence $\underleftarrow{\mathrm{holim}} X_n \simeq 0$, and

$$\underleftarrow{\mathrm{holim}} (T(A) \wedge \Lambda/B)^{C_{p^n}} \simeq \underleftarrow{\mathrm{holim}}\, Y_n.$$

Inductive use of (3,ii) yields

$$Y_n \simeq \bigvee_{i=0}^{n-1} (T(A) \ltimes (\Lambda_1 - \Lambda_0))^{C_{p^i}}$$

and that $R\colon Y_n \to Y_{n-1}$ corresponds to the obvious projection. Therefore

$$\underset{R}{\underleftarrow{\mathrm{holim}}}\, Y_n \simeq \prod_{i=0}^{\infty} (T(A) \ltimes (\Lambda_1 - \Lambda_0))^{C_{p^i}}$$

Now it is easy to see that

$$F\colon \prod_{i=0}^{\infty}(T(A)\ltimes(\Lambda_1-\Lambda_0))^{C_{p^i}} \to \prod_{i=0}^{\infty}(T(A)\ltimes(\Lambda_1-\Lambda_0))^{C_{p^i}}$$

sends $(t_0, t_1, \dots)$ to $(Ft_1, Ft_2, \dots)$ where on the right-hand side

$$F\colon (T(A)\ltimes(\Lambda_1-\Lambda_0))^{C_{p^k}} \to (T(A)\ltimes(\Lambda_1-\Lambda_0))^{C_{p^{k-1}}}$$

is just inclusion of fixed sets. Thus by (2),

$$\underset{F,R}{\underleftarrow{\operatorname{holim}}}(T(A)\wedge\Lambda/B)^{C_{p^n}} \simeq \underset{F}{\underleftarrow{\operatorname{holim}}}\,(T(A)\ltimes(\Lambda_1-\Lambda_0))^{C_{p^n}} \qquad \square$$

If P has exponent p then $\Lambda_1-\Lambda_0$ is a free C_{p^n} space, so

$$(T(A)\ltimes(\Lambda_1-\Lambda_0))^{C_{p^n}} \sim T(A)\ltimes_{C_{p^n}}(\Lambda_1-\Lambda_0) \sim (T(A)\ltimes(\Lambda_1-\Lambda_0))_{hC_{p^n}}$$

and lemma 4.4.9 gives

$$\mathrm{TC}^{\langle 1\rangle}(A[P])^{\wedge}_{p} \sim (\Sigma(\mathrm{TH}(A)\ltimes(\Lambda_1-\Lambda_0))_{hS^1})^{\wedge}_{p}\,. \tag{5.1.5}$$

For more general P, there is a spectral sequence

$$E^1_{k,l} = \pi_{k+l-1}\left((\mathrm{TH}(A)\ltimes(\Lambda_k-\Lambda_{k-1}))_{hS^1};\mathbb{Z}_p\right) \Rightarrow \pi_*\left(\mathrm{TC}^{\langle 1\rangle}(A[G];\mathbb{Z}_p)\right)$$

which might be of use in some situations. In this connection, I note from [J], theorem B that the homology of the homotopy S^1 orbit is closely related to cyclic homology, namely

$$HC_n(C_*(G)) = H_n(\Lambda BG_{hS^1})$$

where $C_*(G)$ denotes the singular chain complex; for discrete G this is equivalent to the group ring. Thus the E^1-term above is a twisted version of certain subgroups of cyclic homology groups associated with the filtration (5.1.2). If one takes a Postnikov decomposition of $\mathrm{TH}(A)$ one obtains a second spectral sequence which converges to the E^1-term and starts out with cyclic homology.

For $A=\mathbb{Z}_p$ with p odd one can in a range instead use theorem 4.4.11 with $L=\mathrm{Id}_p[P]$. Indeed,

$$\mathrm{TH}(\mathrm{Id}_p[P]) \to \mathrm{TH}(\mathbb{Z}_p[P])$$

is $(2p-3)$-connected. The same is then the case when one replaces $\mathrm{TH}(-)$ by $\mathrm{TR}(-)$, and it follows that

$$\mathrm{TC}(\mathrm{Id}_p[P]) \to \mathrm{TC}(\mathbb{Z}_p[P])$$

is $(2p-4)$-connected. One the other hand for a p-group

$$\Delta_p\colon \Lambda BP/BP \to \Lambda BP/BP$$

is nilpotent, so theorem 4.4.11 yields the homotopy Cartesian square

$$\begin{array}{ccc}
\mathrm{TC}(\mathrm{Id}_p[P]) & \longrightarrow & \Sigma^\infty(\Sigma_+(\Lambda BP_{hS^1})) \\
\Big\Updownarrow & & \Big\downarrow \mathrm{trf}_{S^1} \\
\Sigma^\infty(BP_+) & \xrightarrow{\;0\;} & \Sigma^\infty(BP_+)
\end{array}$$

This gives the exact sequence

$$K_n(\mathbb{Z}_pP;\mathbb{Z}_p)/H_n(P;\mathbb{Z}_p) \to \mathrm{TC}_{n-1}(\mathbb{Z}_pP) \to H_n(P;\mathbb{Z}_p) \to \cdots \tag{5.1.6}$$

exact for $n \le 2p-4$, cf. conjecture 0.1 from [O2]. I leave for the reader to wonder about $p=2$.

I now specialize to $P = C_{p^N}$, the cyclic group of order N, where one can be more explicit.

The components of $\Lambda = \Lambda BC_{p^N}$ are indexed by C_{p^N}, and are denoted Λ_g, $g \in C_{p^N}$. Two elements g_1, g_2 of the same order have S^1-homeomorphic components since there is an automorphism $\phi \in \mathrm{Aut}(C_{p^N})$ with $\phi(g_1) = g_2$ which induces $\phi\colon \Lambda_{g_1} \xrightarrow{\cong} \Lambda_{g_2}$. Moreover, for each component corresponding to a non-generator, one has the S^1-homeomorphism

$$\rho^{\#}_{C_p}\Lambda_g^{C_p} \cong \coprod\{\Lambda_h \mid h^p = g\} \tag{5.1.7}$$

induced by the p'th power map $\Delta\colon \Lambda_h \longrightarrow \Lambda_g^{C_p}$.

Lemma 5.1.8. *For any cyclotomic spectrum T and $k \ge l$ there is a cofibration sequence of spectra*

$$(\rho^{\#}_{C_{p^l}} T^{C_{p^l}} \ltimes \Lambda_g)^{C_{p^{k-l}}} \to (T \ltimes \Lambda_{g^{p^l}})^{C_{p^k}} \to \bigvee^{p-1} \bigvee_{j=1}^{l-1} (T \ltimes \Lambda_{g^{p^j}})^{C_{p^{k-l+j}}}.$$

Proof. The l'th iterate $\Delta^l\colon \Lambda_g \to \Lambda_{g^{p^l}}$ embeds Λ_g into one component of $\Lambda_{g^{p^l}}^{C_{p^l}}$, and $\Delta^l(\Lambda_g)$ is (non-equivariantly) equivalent to the ambient space $\Lambda_{g^{p^l}}$. The cofibration of the lemma is induced from

$$\Delta^l(\Lambda_g)_+ \to \Lambda_{g^{p^l}+} \to \Lambda_{g^{p^l}}/\Delta^l(\Lambda_g)$$

upon applying the functor $\rho^{\#}_{C_{p^k}}(T\wedge(-))^{C_{p^k}}$. Since $\Delta^l(\Lambda_g)$ is fixed under C_{p^l},

$$\rho^{\#}_{C_{p^l}}\left(T\ltimes\Delta^l(\Lambda_g)\right)^{C_{p^l}}\sim_{S^1}\rho^{\#}_{C_{p^l}}T^{C_{p^l}}\ltimes\Lambda_g.$$

We use (2.4.3) to calculate the cofiber. Indeed, $\left(T\wedge\Lambda_{g^{p^l}}/\Delta^l(\Lambda_g)\right)_{hC_{p^k}}\sim 0$ so that

$$\left(T\wedge\Lambda_{g^{p^l}}/\Delta^l(\Lambda_g)\right)^{C_{p^k}}\sim\rho^{\#}_{C_p}\Phi^{C_p}\left(T\wedge\Lambda_{g^{p^l}}/\Delta^l(\Lambda_g)\right)^{C_{p^{k-1}}}$$
$$\sim\left(T\wedge\rho^{\#}_{C_p}\Delta(\Lambda_{g^{p^{l-1}}})/\Delta^l(\Lambda_g)\right)^{C_{p^{k-1}}}\vee\bigvee_{j=1}^{p-1}\left(T\ltimes\Lambda_{g^{p^{l-1}}+jp^{N-1}}\right)^{C_{p^{k-1}}}.$$

Each of the $p-1$ wedge terms are equivalent to $(T\ltimes\Lambda_{g^{p^{l-1}}})^{C_{p^{k-1}}}$, and we can iterate. □

The point of the lemma is that the component $\Lambda_{g^{p^l}}$ has been replaced by the simpler components $\Lambda_g,\dots,\Lambda_{g^{p^{l-1}}}$, simpler w.r.t. the C_{p^k}-action. For example, the action of C_{p^k} on Λ_g is free when g is a generator of C_{p^N}. For every equivariant S^1-spectrum T,

$$\begin{aligned}&\text{(i)}\quad(\Sigma T_{hS^1})^{\wedge}_p\sim(\operatorname{holim}_{\leftarrow}T_{hC_{p^n}})^{\wedge}_p\\&\text{(ii)}\quad(T^{hS^1})^{\wedge}_p\sim(\operatorname{holim}_{\leftarrow}T^{hC_{p^n}})^{\wedge}_p\\&\text{(iii)}\quad\hat{\mathbb{H}}(S^1,T)^{\wedge}_p\sim\operatorname{holim}_{\leftarrow}\hat{\mathbb{H}}(C_{p^n},T)^{\wedge}_p\end{aligned}\tag{5.1.9}$$

The first equivalence is lemma 4.4.9, the second is an easy consequence of the definitions, and is just an equivariant version of the relation $\operatorname{holim}_{\rightarrow}BC_{p^n}\sim(\mathbb{C}P^{\infty})^{\wedge}_p$. The third equivalence follows by comparing the norm fibration for C_{p^n} and S^1, cf. remark 4.1.9. We consider the convergent sequences with

$$E^2(T^{hS^1};\mathbb{Z}_p)=S_{\mathbb{Z}_p}\{t\}\otimes\pi_*(T;\mathbb{Z}_p)$$
$$E^2(\hat{\mathbb{H}}(S^1,T);\mathbb{Z}_p)=S_{\mathbb{Z}_p}\{t,t^{-1}\}\otimes\pi_*(T;\mathbb{Z}_p)$$

cf. [HM1], [GM] for convergence.

Proposition 5.1.10. *If $g\in C_{p^N}$ is a generator, then the Tate spectrum*

$$\hat{\mathbb{H}}(S^1,\rho^{\#}_{C_{p^l}}T(k)^{C_{p^l}}\ltimes\Lambda_g)^{\wedge}_p\sim *$$

Proof. We use $\mathbb{Z}_p$ coefficients and have

$$\hat{E}^2_{*,*}=S_{\mathbb{Z}_p}\{t,t^{-1}\}\otimes S_{W_{l+1}(k)}\{\sigma\}\otimes H_*(\Lambda_g;\mathbb{Z}/p^{l+1})$$

with $t \in \hat{E}^2_{-2,0}$, $\sigma \in \hat{E}_{0,2}$ and $H_*(\Lambda_g; \mathbb{Z}/p^{l+1}) \subset \hat{E}^2_{0,*}$, cf. (4.3.4).

The spectrum $T(k)^{C_{p^l}}$ is a product of Eilenberg MacLane spectra, since it is a module over $\mathrm{TR}(k) \sim HW(k)$, and the $\hat{d}^2$-differential is this given by

$$[S^1]_\#: H_t(\Lambda_g; \mathbb{Z}/p^{l+1}) \to H_{t+1}(\Lambda_g; \mathbb{Z}/p^{l+1})$$

induced from the action

$$S^1 \times \Lambda_g \to \Lambda_g$$

cf. proposition 4.1.14. The evaluation of loops at 1 gives a non-equivariant homotopy equivalence $\Lambda_g \to BC_{p^N}$, so

$$H_*(\Lambda_g; \mathbb{Z}/p^{l+1}) = E_{\mathbb{Z}/p^{l+1}}\{y_1\} \otimes \Gamma_{\mathbb{Z}/p^{l+1}}\{x_2\}$$

with $\deg(y_1) = 1$, $\deg x_2 = 2$ and with $\Gamma\{x_2\}$ being the divided polynomial algebra. We show in lemma 5.1.12 below that $[S^1]_\#$ multiplies by y_1. Hence

$$\hat{d}^2\,(t^s\gamma_n(x_2)\sigma^r) = t^{s+1}\gamma_n(x_2)y_1\sigma^r, \quad s \in \mathbb{Z}, \quad n \geq 0,$$

and $\hat{E}^3$=0. □

Proposition 5.1.11. *For a generator $g \in C_{p^N}$,*

$$\pi_*\left((\rho^\#_{C_{p^l}} T(k)^{C_{p^l}} \ltimes \Lambda_g)^{hS^1}; \mathbb{Z}_p\right) = S_{W_{l+1}(k)}\{\sigma\} \otimes \tilde{H}_*(BC_{p^l}; \mathbb{Z}_p).$$

Proof. The spectral sequence for the homotopy S^1 fixed set has E^2-term

$$E^2_{*,*} = S_{\mathbb{Z}_p}\{t\} \otimes S_{W_{l+1}(k)}\{\sigma\} \otimes H_*(\Lambda_g; \mathbb{Z}/p^{l+1})$$

with differentials as above. This time, however t^{-1} is not present, so there is no differential to kill the classes $\gamma_n(x_2)y_1\sigma^r$. Thus

$$E^3_{*,*} = S_{W_{l+1}(k)}\{\sigma\} \otimes y_1\Gamma_{\mathbb{Z}/p^{l+1}}\{x_2\},$$

all concentrated on one vertical line, and $E^3_{*,*} = E^\infty_{*,*}$. □

Lemma 5.1.12. *If $g \in C_{p^N}$ is a generator, then the action $S^1 \times \Lambda_g \to \Lambda_g$ induces multiplication by $y_1 \in H_1(\Lambda_g; \mathbb{Z}_p)$ on $H_*(\Lambda_g; \mathbb{Z}/p^{l+1})$.*

Proof. Let $\tilde{g}$: $S^1 \to BC_{p^N}$ represent the homotopy class corresponding to $g \in C_{p^N}$. Consider $\tilde{g}$ as an element of ΛBC_{p^N}. Since C_{p^N} is abelian, BC_{p^N} is an abelian topological group. The map f: $BC_{p^N} \to \Lambda BC_{p^N}$ with $f(b)(z) = b\tilde{g}(z)$ lands in Λ_g since we may connect b with a path to $1 \in BC_{p^N}$. Moreover,

f is a homotopy equivalence, since its composition with the evaluation map is homotopic to the identity. The lemma now follows from the homotopy commutative diagram

$$\begin{array}{ccccc} S^1 \times B & \xrightarrow{1\times f} & S^1 \times \Lambda B & \xrightarrow{\text{action}} & \Lambda B \\ \Big\downarrow{\scriptstyle g\times 1} & & & & \Big\downarrow{\scriptstyle \text{ev}} \\ B \times B & & \xrightarrow{\qquad\quad \text{mult} \qquad\quad} & & B \end{array}$$

□

We return to the calculation of the p-adic homotopy groups of $\mathrm{TC}^{\langle 1 \rangle}(kC_{p^N})$. They are by proposition 5.1.4 equivalent to

$$\pi_* \left(\bigvee^{p-1} \operatorname*{holim}_{\overleftarrow{F}} \left(T(k) \ltimes \Lambda_{g^{p^{N-1}}} \right)^{C_{p^n}} ; \mathbb{Z}_p \right)$$
$$= \bigoplus^{p-1} \lim_{\longleftarrow} \pi_* \left((T(k) \ltimes \Lambda_{g^{p^{N-1}}})^{C_{p^n}} ; \mathbb{Z}_p \right),$$

where g generates C_{p^n}. The idea is to use the cofibration sequence of lemma 5.1.8 inductively for $l = 1, \dots, N-1$. One has

$$\operatorname*{holim}_{\overleftarrow{F}} (T(k) \ltimes \Lambda_g)^{C_{p^n}} \sim \operatorname*{holim}_{\overleftarrow{F}} (T(k) \ltimes \Lambda_g)_{hC_{p^n}} \sim (T(k) \ltimes \Lambda_g)^{hS^1}$$

after p-completion. This follows from (5.1.9,i) and proposition 5.1.10. Proposition 5.1.11 shows inductively that all p-adic homotopy is concentrated in odd degrees. In particular we get, for each l, short exact sequences

$$0 \to \pi_* \left(\rho^{\#}_{C_{p^l}} T(k)^{C_{p^l}} \ltimes \Lambda_g \right)^{hS^1} \to \pi_* \operatorname*{holim}_{\overleftarrow{F}} \left(T(k) \ltimes \Lambda_{g^{p^l}} \right)^{C_{p^n}}$$
$$\to \bigoplus^{p-1} \bigoplus_{j=1}^{l-1} \pi_* \operatorname*{holim}_{\overleftarrow{F}} \left(T(k) \ltimes \Lambda_{g^{p^j}} \right)^{C_{p^n}} \to 0$$

of homotopy groups with $\mathbb{Z}_p$ coefficients. These sequences are also split exact. Indeed the left hand term consists of a sum of groups $W_{l+1}(k) = W(k)/p^{l+1}$, so it suffices to check that

$$p^{l+1} \pi_* \left(\operatorname*{holim}_{\overleftarrow{F}} T(k) \ltimes \Lambda_{g^{p^l}} \right)^{C_{p^n}} = 0. \qquad (5.1.13)$$

This on the other hand is a consequence of induction theory, upon using a result of C. Schlichtkrull, [Sch], which I now describe.

Let L be an FSP and consider the functor

$$\mathrm{TF}(L[G],p) = \underset{\overleftarrow{F}}{\mathrm{holim}}\, T(L[G])^{C_{p^n}}.$$

For $\Gamma \subset G$ of finite index we have the map

$$\mathrm{Ind}_G^\Gamma \colon \mathrm{TF}(L[G],p) \to \mathrm{TF}(L[\Gamma],p)$$

given as the composition of the functor applied to $L[G] \to \mathrm{End}_{\mathrm{L}[\Gamma]}(\mathrm{L}[\mathrm{G}])$ with Morita equivalence. Now

$$\mathrm{TF}(L[G],p) \sim \underset{\overleftarrow{F}}{\mathrm{holim}}\, (T(L) \ltimes \Lambda BG)^{C_{p^n}}$$

decomposes into components,

$$\mathrm{TF}(L[G],p) \sim \bigvee_{[g]\in X(G)} \underset{\overleftarrow{F}}{\mathrm{holim}}\, \left(T(L) \ltimes \Lambda_{[g]} BG\right)^{C_{p^n}}$$

with $\Lambda_{[g]}BG = B_{S^1}C_G(g)$, the classifying space of the centralizer with some action of S^1. It follows that Ind_G^Γ decomposes into components,

$$\mathrm{Ind}_G^\Gamma([g],[\gamma]) \colon \underset{\overleftarrow{F}}{\mathrm{holim}}\, \left(T(L) \ltimes \Lambda_{[g]} BG\right)^{C_{p^n}} \to \underset{\overleftarrow{F}}{\mathrm{holim}}\, \left(T(L) \ltimes \Lambda_{[\gamma]} B\Gamma\right)^{C_{p^n}}.$$

Theorem 5.1.14. ([Sch]) (i) $\mathrm{Ind}_G^\Gamma([g],[\gamma]) = 0$ *if* $\gamma \notin [g]$. (ii) *If* $\gamma \in [g]$ *then* Ind_G^Γ *is induced from the* S^1*-equivariant covering* $\Lambda_\gamma B\Gamma \to \Lambda_g BG$. □

(The theorem verifies in particular conjecture 7.14 of [BHM]; it undoubtly generalizes to simplicial groups, and should be of help in the study of transfers in Waldhausen's A-theory).

Corollary 5.1.15. *In the limit over* k*, the cofibration sequences of lemma 5.1.8 become split, for* $T = T(k)$.

Proof. The terms in the limit sequence are modules over $K(k)_p^\wedge = HW(k)$ via the cyclotomic trace, so it suffices to check that the homotopy exact sequence is split. This was above reduced to the statement (5.1.13). We use theorem 5.1.14(i) with $G = C_{p^n}$, $\Gamma = C_{p^{N-l-1}}$ to conclude that

$$\mathrm{Res}^{C_{p^N}}_{C_{p^{N-l-1}}} \circ \mathrm{Ind}^{C_{p^{N-l-1}}}_{C_{p^N}} \colon \mathrm{TF}(k[C_{p^N}],p) \to \mathrm{TF}(k[C_{p^N}],p)$$

is trivial on $\underleftarrow{\mathrm{holim}}\left(T(k) \ltimes \Lambda_{g^{p^l}}\right)^{C_{p^n}}$. On the other hand the composition induces multiplication by the index p^{l+1} on homotopy. □

Theorem 5.1.16. *For a perfect field of characteristic $p > 0$,*

$$\pi_{2n-1}\mathrm{TC}^{\langle 1\rangle}(k[C_{p^N}]) = \pi_1\left(\mathrm{TC}^{\langle 1\rangle}(k[C_{p^N}])\right)^{\oplus n},$$
$$\pi_{2n}\mathrm{TC}^{\langle 1\rangle}(k[C_{p^N}]) = 0,\ n \geq 0$$

Moreover,

$$\pi_1\mathrm{TC}^{\langle 1\rangle}(kC_{p^N}) = \left(W(k)/p^N\right)^{\oplus(p-1)} \oplus \bigoplus_{j=1}^{N-1} \left(W(k)/p^{N-j}\right)^{\oplus(p-1)(p^j-p^{j-1})}.$$

Proof. This follows from corollary 5.1.15 and proposition 5.1.11 upon collecting terms. □

We have left to determine the exact homotopy sequence of

$$\mathrm{TC}(k) \ltimes BC_{p^N} \to \mathrm{TC}(k[C_{p^N}]) \to \mathrm{TC}^{\langle 1\rangle}(k[C_{p^N}]). \tag{5.1.17}$$

From (4.3.3) we have

$$\mathrm{TC}(k) \sim H\mathbb{Z}_p \vee \Sigma^{-1}H(\mathbb{Z}_p). \tag{5.1.18}$$

when k is finite. Thus

$$\pi_i\left(\mathrm{TC}(k) \ltimes BC_{p^N}\right) = H_i(BC_{p^N};\mathbb{Z}_p) \oplus H_{i+1}(BC_{p^N};\mathbb{Z}_p),$$

with one copy of $\mathbb{Z}/p^N$ in each degree.

Lemma 5.1.19. *The homotopy exact sequence of (5.1.17) reduces to the exact sequence*

$$0 \longrightarrow H_{2n-1}(BC_{p^N};\mathbb{Z}_p) \longrightarrow \mathrm{TC}_{2n-1}(k[C_{p^N}])$$
$$\longrightarrow \mathrm{TC}^{\langle 1\rangle}_{2n-1}(k[C_{p^N}]) \xrightarrow{\partial_*} H_{2n-1}(BC_{p^N};\mathbb{Z}_p) \longrightarrow 0.$$

Proof. We must argue that ∂_* is surjective. This is true for $n = 1$ because $\mathrm{TC}^{\langle 1\rangle}_2(k[C_{p^N}]) = 0$ and because the K-theory assembly map is clearly injective in dimension zero.

For $n > 1$ we use that (5.1.17) is a module over $\mathrm{TR}(k) \ltimes BC_{p^N}$, and hence over $\mathrm{TR}(\mathbb{F}_p) \ltimes BC_{p^N}$. Thus

$$\partial\colon \mathrm{TC}^{\langle 1\rangle}(k[C_{p^N}]) \to \Sigma\mathrm{TC}(k) \ltimes BC_{p^N} \to H\mathbb{Z}_p \ltimes BC_{p^N}$$

commutes with the resulting actions

$$\mathrm{TC}_*^{\langle 1\rangle}(k[C_{p^N}]) \otimes H_*(BC_{p^N};\mathbb{Z}/p^N) \to \pi_*\left(\mathrm{TC}^{\langle 1\rangle}(k[C_{p^N}]);\mathbb{Z}/p^N\right)$$
$$H_*(BC_{p^N};\mathbb{Z}_p) \otimes H_*(BC_{p^N};\mathbb{Z}/p^N) \to H_*(BC_{p^N};\mathbb{Z}/p^N)$$

The second map has the property

$$H_1(BC_{p^N};\mathbb{Z}_p) \cdot H_{2n}(BC_{p^N};\mathbb{Z}/p^N) = H_{2n+1}(BC_{p^N};\mathbb{Z}/p^N),$$

and since $H_{2n+1}(BC_{p^N};\mathbb{Z}_p) = H_{2n+1}(BC_{p^N};\mathbb{Z}/p^N)$, surjectivity of ∂_* in dimension 1 gives surjectivity in general. □

Since $\mathrm{TC}_i(k[C_{p^N}]) = K_i(k[C_{p^N}];\mathbb{Z}_p)$ has exponent p^N, lemma 5.1.19 yields the abstract isomorphism

$$\mathrm{TC}_{2n-1}(k[C_{p^N}]) \cong \mathrm{TC}_{2n-1}^{\langle 1\rangle}(k[C_{p^N}]).$$

This proves theorem 5.1.1.

It seems clear that one should be able to calculate $K_*(k[P])$ for more complicated p-groups. It is also natural to attack $K_*(A[P])$ for other base rings, and in particular for $A = \mathbb{Z}_p$ cf. sect. 5.4 below.

I conclude with some remarks about the twisted group ring case, inspired by [O1], ch. 12. Let E be any finite extension of $\mathbb{Q}_p$ and $A \subset E$ the ring of integers. Given a p-group P and any homomorphism $t\colon P \to \mathrm{Gal}(E/\mathbb{Q}_p)$ we have the twisted group ring $A^t[P]$. It contains the untwisted group ring $A[P_0]$, $P_0 = \mathrm{Ker}(t)$. Theorem 12.3 of [O1] states that the inclusion induces an isomorphism

$$K_1(A[P_0])_{P/P_0} \xrightarrow{\cong} K_1(A^t[P]), \qquad (5.1.20)$$

where the left hand side denotes the coinvariants of the action induced from $P/P_0 \to \mathrm{Aut}(A) \times \mathrm{Out}(P_0)$. Olivers argument is based upon the integral p-adic logarithm, close in spirit to $\pi_1(\mathrm{trc})$; one may wonder if (5.1.20) generalizes to the statement

$$\mathrm{TC}(A^t[P]) \sim \mathrm{TC}(A[P_0])_{hP/P_0} \quad ?$$

5.2 K-theory of $k[x]/(x^n)$.

This section outlines joint work with Lars Hesselholt. The main result is Theorem 5.2.8 below. A detailed account can be found in [HM], sect. 6–8, when $n = 2$ and will appear in [HM2] when $n > 2$.

Let $\Pi_n = \{0, 1, x, \dots, x^{n-1}\}$, considered as a pointed monoid with 0 as base point and with $x^i = 0$ for $i \geq n$. We form the cyclic construction $N_\bullet^{\mathrm{cy}}(\Pi_n)$. Its set of k-simplices is the $(k+1)$-fold smash power of Π_n, so consists of $k+1$ tuples $(x^{i_0}, \dots, x^{i_k})$ with $(x^{i_0}, \dots, x^{i_k}) = 0$ if some $i_\nu \geq n$; $N_\bullet^{\mathrm{cy}}(\Pi_n)$ becomes a cyclic set when we give it the structure maps of sect. 2.1.

The argument of lemma 4.4.7 gives for any ring A (or even FSP) the equivalence of equivariant spectra

$$T(A[x]/(X^n)) \sim_{S^1} T(A) \wedge |N_\bullet^{\mathrm{cy}}(\Pi_n)|, \tag{5.2.1}$$

There is an analogue of the component decomposition of $N_\bullet^{\mathrm{cy}}(G) = \Lambda BG$, namely

$$N_\bullet^{\mathrm{cy}}(\Pi_n) = \bigvee_{s=0}^{\infty} N_\bullet^{\mathrm{cy}}(\Pi_n; s)$$

where $N_k^{\mathrm{cy}}(\Pi_n; s)$ consists of simplices $(x^{i_0}, \dots, x^{i_k})$ with $\Sigma i_\nu = s$, and $0 \in N_k^{\mathrm{cy}}(\Pi_n; s)$ for all s. The simplex $(x^{(s)}) = (x, \dots, x) \in N_{s-1}^{\mathrm{cy}}(\Pi_n; s)$ is represented by a cyclic map

$$i_{s,\bullet} \colon \Lambda[s-1]_\bullet \to N_\bullet(\Pi_n; s)$$

of the standard cyclic $(s-1)$-simplex. Its realization becomes a map

$$i \colon S^1 \times \Delta^{s-1} \to |N_\bullet(\Pi_n; s)|,$$

cf. (2.1.3). Since $(x^n, x, \dots, x) \in N_{s-n}(\Pi_n; s)$ is the base point, the composite of $i_{s,\bullet}$ with the iterated face operator $d_{s-n+1} \circ \cdots \circ d_s$, maps the corresponding face $S^1 \times \Delta^{s-n}$ to zero. Moreover, as $(x^{(s)})$ is invariant under cyclic permutations, i_s maps the orbit $S^1 \times C_s \cdot \Delta^{s-1} = C_s \cdot (S^1 \times \Delta^{s-1})$ to zero. All in all we obtain a map

$$\hat{i}_s \colon S^1 \times_{C_s} \Delta^{s-1} / S^1 \times_{C_s} C_s \cdot \Delta^{s-n} \to |N_\bullet^{\mathrm{cy}}(\Pi_n; s)|$$

and it is not hard to prove:

Lemma 5.2.2. *The map $\hat{i}_s$ is an S^1-equivariant homeomorphism.* □

For $n = 2$, the domain of $\hat{i}_s$ is $S^1 \times_{C_s} \Delta^{s-1} / \partial(S^1 \times_{C_s} \Delta^{s-1})$. We consider $\Delta^{s-1} \subset \mathbb{R}C_s$ to be the simplex spanned by the group elements $g^i \in \mathbb{R}C_s$.

It projects homeomorphically to the reduced regular representation $\mathbb{R}C_s - \mathbb{R}$ with $\mathbb{R} \subset \mathbb{R}C_s$ the invariant line through $\sum_{i=0}^{s-1} g^i$. Hence we have:

$$\begin{aligned} S^1 \times_{C_s} \Delta^{s-1}/\partial(S^1 \times_{C_s} \Delta^{s-1}) \\ &\cong S^1 \times_{C_s} D(\mathbb{R}C_s - \mathbb{R})/\partial(S^1 \times_{C_s} D(\mathbb{R}C_s - \mathbb{R})) \\ &\cong S^1_+ \wedge_{C_s} S^{\mathbb{R}C_s - \mathbb{R}}. \end{aligned}$$

If s is odd then $\mathbb{R}C_s - \mathbb{R}$ is a complex representation and

$$S^1_+ \wedge_{C_s} S^{\mathbb{R}C_s - \mathbb{R}} \cong_{S^1} S^1_+/C_s \wedge S^{\mathbb{R}C_s - \mathbb{R}}$$

with diagonal S^1-action on the right hand side. If s is even then $\mathbb{R}C_s - \mathbb{R} = \mathbb{R}_- \oplus V_s$ with V_s complex, and

$$S^1_+ \wedge_{C_s} S^{\mathbb{R}C_s - \mathbb{R}} \cong \operatorname{cof}\left(S^1_+/C_{s/2} \xrightarrow{\Delta} S^1_+/C_s\right) \wedge S^{V_s}$$

with Δ the natural projection.

The above description of $|N^{\text{cy}}_\bullet(\Pi_2; s)|$ has the following generalization when $n > 2$. We use $\mathbb{C}(n)$ to denote the complex S^1-representation where the action of $z \in S^1$ is multiplication with z^n. Suppose $dn < s \leq (d+1)n$, and write

$$V_s = \mathbb{C}(1) \oplus \mathbb{C}(2) \oplus \cdots \oplus \mathbb{C}(d). \tag{5.2.3}$$

It is an S^1-module and hence by restriction to $C_s \subset S^1$ also a C_s-module.

Theorem 5.2.4. ([HM2]). *Suppose $n \geq 2$ and $dn < s \leq (d+1)n$. Then*

$$S^1 \times_{C_s} \Delta^{s-1}/S^1 \times_{C_s} C_s \cdot \Delta^{s-n} \sim_{S^1} \begin{cases} S^1_+/C_s \wedge S^{V_s}, & s < (d+1)n \\ \operatorname{cofib}\left(S^1_+/C_{d+1} \to S^1_+/C_s\right) \wedge S^{V_s}, & s = (d+1)n \end{cases}$$

Proof. (Outline). The proof is based upon the concept of regular cyclic polytopes of D. Gale, [G]. Let $\pi_d(g) = (\xi_s, \xi_s^2, \ldots, \xi_s^d)$, $\xi_s = e^{2\pi i/s}$. The image $P_{s,d} = \pi_d(\Delta^{s-1}) \subset V_s$ is a regular cyclic polytope. Its structure of facets (=codim 1 faces) is completely described in [G]. Using this we prove in [HM2] that

$$\pi_d(\Delta^{s-1}/C_s \cdot \Delta^{s-n}) \sim P_{s,d}/Q_{s,d} \sim S^{V_s}$$

for $dn < s < (d+1)n$ where $Q_{s,d} = \pi_d(C_s \cdot \Delta^{s-n})$. Next, the socalled Buenos Aires formula, [BAG], gives explicit generators of the homology

$$H_*(N^{\text{cy}}_\bullet(\Pi_n); \mathbb{Z}) = \mathrm{HH}_*(\mathbb{Z}[x]/(x^n); \mathbb{Z})$$

in terms of the simplices of $N^{\text{cy}}_\bullet(\Pi_n; s)$. This is used to show that

$$S^1_+ \wedge_{C_s} \Delta^{s-1}/S^1_+ \wedge_{C_s} C_s \cdot \Delta^{s-n} \sim S^1_+ \wedge_{C_s} S^{V_s}$$

when $dn < s < (d+1)n$. The case $s = (d+1)n$ is somewhat more complicated, and will not be outlined here. □

We also need to know the cyclotomic structure of $T\left(A[x]/(x^n)\right)$, similar to lemma 4.4.7, and must calculate the geometric fixed points:

$$\begin{aligned}\rho^{\#}_{C_p}\Phi^{C_p}T\left(A[x]/(x^n)\right) &\sim_{C_{p^\infty}} \rho^{\#}_{C_p}\Phi^{C_p}T(A)\wedge\rho^{\#}_{C_p}|N^{\mathrm{cy}}_{\bullet}(\Pi_n)|^{C_p}\\ &\sim_{C_{p^\infty}} T(A)\wedge\rho^{\#}_{C_p}|sd_{C_p}N^{\mathrm{cy}}_{\bullet}(\Pi_n)|^{C_p}.\end{aligned}$$

Comparing with (2.1.7), the isomorphism

$$\Delta_{C_p}\colon N^{\mathrm{cy}}_{\bullet}(\Pi_n,s)\xrightarrow{\cong} sd_{C_p}N^{\mathrm{cy}}_{\bullet}(\Pi_n;sp)^{C_p}$$

gives an S^1-map

$$\Delta^{-1}\colon \rho^{\#}_{C_p}sd_{C_p}N^{\mathrm{cy}}_{\bullet}(\Pi_n;sp)^{C_p}\xrightarrow{\cong} N^{\mathrm{cy}}_{\bullet}(\Pi_n;s)$$

which when composed with the above gives the required C_{p^∞}-equivalence

$$\rho^{\#}_{C_p}\Phi^{C_p}T\left(A[x]/(x^n)\right)\sim_{C_{p^\infty}} T\left(A[x]/(x^n)\right).$$

It is clear from the definition of V_s that $\rho^{\#}_{C_p}V_{sp}^{C_p}\cong_{S^1}V_s$, and we also have $S^1/C_s\cong_{S^1}\rho^{\#}_{C_p}(S^1/C_{ps})$. This yields a C_{p^∞}-equivalence

$$r_{C_p}(s)\colon \rho^{\#}_{C_p}\Phi^{C_p}\left(T(A)\wedge S^1_+/C_{ps}\wedge S^{V_{ps}}\right)\to T(A)\wedge S^1_+/C_s\wedge S^{V_s}.$$

The proof of theorem 5.2.4 contains the following

Addendum 5.2.5. *The cyclotomic structure of $T(A[x]/(x^n))$ is given by* $\bigvee_{s=0}^{\infty}r_{C_p}(s)$. □

Since we are working in the category of equivariant spectra, $T(A)\wedge S^{V_s}$ is equal to the V_s'th deloop $T(A)(V_s)$ of $T(A)$. With this interpretation $r_{C_p}(s)$ induces

$$R\colon T(A)(V_{ps})^{C_{p^m}}\to T(A)(V_s)^{C_{p^{m-1}}}$$

and we can form the homotopy inverse limit over these maps

Denote by $\widetilde{\mathrm{TC}}(A[x]/(x^n))$ the reduced space, i.e. the homotopy fiber of $\mathrm{TC}(A[x]/(x^n))\to\mathrm{TC}(A)$. Then

$$\mathrm{TC}\left(A[x]/(x^n)\right)\sim\mathrm{TC}(A)\times\widetilde{\mathrm{TC}}\left(A[x]/(x^n)\right).$$

For any S^1-equivariant spectrum $T\in S^1\mathcal{SU}$ and any finite dimensional S^1-module W in the Universe we have the map

$$V_n\colon T(W)^{C_{p^r}}\to T(W)^{C_{np^r}}$$

constructed from equivariant transfers, cf. (2.5.8). If $n = p^{v_p(n)}n'$ with $(p, n') = 1$ we write

$$V_n^{(p)}: T(W)^{C_{p^r}} \to T(W)^{C_{p^{r+v_p(n)}}}$$

instead of $V_{p^{v_p(n)}}$. Then we have the following analogue of [HM], addendum 7.2:

Theorem 5.2.6. *The spectrum $\widetilde{\mathrm{TC}}(A[x]/(x^n))^\wedge_p$ is equivalent to the product of the p-adic completions of*

$$\Pi \left\{ \Sigma \operatorname{holim}_{\overleftarrow{R}} T(A)(V_{p^i l})^{C_{p^i}} \mid (l, p) = 1,\ n' \nmid l \right\}$$

and

$$\Pi \left\{ \operatorname{cof} \left(\Sigma \operatorname{holim}_{\overleftarrow{R}} T(A)(V_{p^i l})^{C_{p^{i-v_p(n)}}} \xrightarrow{V_n^{(p)}} \Sigma \operatorname{holim}_{\overleftarrow{R}} T(A)(V_{p^i l})^{C_{p^i}} \right) \;\middle|\; (l, p) = 1,\ n' \mid l \right\},$$

where in the second factor $T(A)(V_{p^i l})^{C_{p^{i-v_p(n)}}} = 0$ if $i < v_p(n)$.

Proof. We use the description

$$\widetilde{\mathrm{TC}}(-)^\wedge_p \longrightarrow \widetilde{\mathrm{TF}}(-)^\wedge_p \xrightarrow{R-1} \widetilde{\mathrm{TF}}(-),$$

so shall first determine $\widetilde{\mathrm{TF}}(A[x]/(x^n), p)$, the homotopy inverse limit of $\tilde{T}(A[x]/(x^n))^{C_{p^n}}$ under the inclusion of fixed sets. For fixed m,

$$\bigvee_{s=1}^{\infty} (T(A)(V_s) \wedge S^1_+/C_s)^{C_{p^m}} \to \prod_{s=1}^{\infty} (T(A)(V_s) \wedge S^1_+/C_s)^{C_{p^m}} \tag{1}$$

is an equivalence of spectra. Indeed, since we are only interested in p-completions

$$T(A)(V_s) \wedge S^1_+/C_s \sim T(A)(V_s) \wedge S^1_+/C_{p^{v_p(s)}}$$

and

$$\left(T(A)(V_s) \wedge S^1_+/C_{p^{v_p(s)}}\right)^{C_{p^m}} \sim \left(\rho^\#_{C_{p^r}} T(A)(V_s)^{C_{p^r}} \wedge S^1_+\right)^{C_{p^{m-r}}}$$

with $r = \min(v_p(s), m)$. The action of $C_{p^{m-r}}$ on S^1_+ is free, so can be divided out, and when we use the S^1-action on $\rho^\#_{C_{p^r}} T(A)(V_s)^{C_{p^r}}$ to untwist the action, we get

$$(T(A)(V_s) \wedge S^1_+/C_s)^{C_{p^m}} \sim T(A)(V_s)^{C_{p^r}} \wedge S^1_+/C_{p^{m-r}}, \tag{2}$$

with $r = m$ when $v_p(s) \geq m$. The cofibration sequence of proposition 4.1.8 takes the form

$$T(A)(V_s)_{hC_{p^r}} \longrightarrow T(A)(V_s)^{C_{p^r}} \xrightarrow{R} T(A)(V_{s/p})^{C_{p^{r-1}}}$$

and inductive use show that the connectivity of $T(A)(V_s)^{C_{p^r}}$ tends to infinity with s. This proves (1). Next for fixed s,

$$\underset{\overleftarrow{F}}{\text{holim}}\, (T(A)(V_s) \wedge S^1_+/C_s)^{C_{p^m}} \sim \Sigma T(A)(V_s)^{C_{p^{v_p(s)}}} \tag{3}$$

after p-completion. This follows from (2) with $r = v_p(s) \leq m$, since the F-map corresponds to the transfers

$$\Sigma^\infty_+(S^1/C_{p^{m+1-v_p(s)}}) \to \Sigma^\infty_+(S^1/C_{p^{m-v_p(s)}})$$

which fit into a cofibration diagram

$$\begin{array}{ccccc}
\Sigma^\infty(S^1/C_{p^{m+1-r}}) & \longrightarrow & \Sigma^\infty_+(S^1/C_{p^{m+1-r}}) & \longrightarrow & \Sigma^\infty(S^0) \\
\Big\downarrow \sim & & \Big\downarrow & & \Big\downarrow p \\
\Sigma^\infty(S^1/C_{p^{m-r}}) & \longrightarrow & \Sigma^\infty_+(S^1/C_{p^{m-r}}) & \longrightarrow & \Sigma^\infty(S^0)
\end{array}$$

Here $\Sigma^\infty_+(X)$ is the suspension spectrum of X_+. Now smash with $T(A)(V_s)^{C_{p^r}}$ to obtain (3), cf. proof of lemma 4.4.9. The above together with theorem 5.2.4 yields

$$\begin{aligned}
\widetilde{\mathrm{TF}}\,(A[x]/(x^n), p) \sim & \prod_{n' \nmid s} \Sigma T(A)(V_s)^{C_{p^{v_p(s)}}} \\
& \times \prod_{n' \mid s} \operatorname{cof}\left(\Sigma T(A)(V_s)^{C_{p^{v_p(s)-v_p(n)}}} \xrightarrow{V_n^{(p)}} \Sigma T(A)^{C_{p^{v_p(s)}}}\right)
\end{aligned}$$

after p-completion. The homotopy fiber of $R - \text{id}$ corresponds to taking homotopy inverse limit over R. □

So far we have not specified the ground ring A, but to obtain explicit calculation we now restrict A to be a perfect field k of positive characteristic p, where we have the following result from [HM], sect. 8.1.

Proposition 5.2.7. *Let $V \subset \mathcal{U}$ be a complex S^1-module. The non-zero homotopy groups of $T(k)(V)^{C_{p^m}}$ is concentrated in even degrees greater that or equal to $\dim V^{C_p}$. They are explicitly given as*

$$\pi_{2i} T(k)(V)^{C_{p^m}} = W_s(k) \quad \textit{if} \quad \dim V^{C_{p^{m-s+1}}} \leq 2i < \dim V^{C_{p^{m-s}}}$$

for $s = 1, \ldots m$, *and*

$$\pi_{2i} T(k)(V)^{C_{p^m}} = W_{m+1}(k) \quad \textit{if} \quad 2i > \dim V.$$

Proof. The argument is similar to the proof of proposition 4.2.7 and (4.3.4). One first treats the case $k = \mathbb{F}_p$.

We remember that $T(k)(V) \sim T(k) \wedge S^V$. It follows that the inclusion $V^{C_p} \subset V$ induces an equivalence

$$\hat{\mathbb{H}}\left(C_{p^m}, T(k)(V^{C_p})\right) \xrightarrow{\sim} \hat{\mathbb{H}}\left(C_{p^m}, T(k)(V)\right).$$

Indeed, the cofiber is $\hat{\mathbb{H}}(C_{p^m}, T(k) \wedge S^{V - V^{C_p}})$ and $S^{V - V^{C_p}}$ is a free C_{p^m}-space, build up from free C_{p^m}-cells, and the obvious induction over cells reduces us to show that

$$\hat{\mathbb{H}}\left(C_{p^m}, T(k) \wedge (C_{p^m +} \wedge S^i)\right) = 0.$$

This follows e.g. from the spectral sequence of sect. 4.1, since Tate cohomology groups vanish on free modules.

We next use the following analogue of proposition 4.1.8:

$$\begin{array}{ccccc} T(k)(V)_{hC_{p^m}} & \xrightarrow{N} & T(k)(V)^{C_{p^m}} & \xrightarrow{R} & T(k)(V^{C_p})^{C_{p^{m-1}}} \\ \downarrow{\scriptstyle \mathrm{id}} & & \downarrow{\scriptstyle \Gamma_{m,V}} & & \downarrow{\scriptstyle \hat{\Gamma}_{m,V}} \\ T(k)(V)_{hC_{p^m}} & \xrightarrow{N^h} & T(k)(V)^{hC_{p^m}} & \xrightarrow{R^h} & \hat{\mathbb{H}}\left(C_{p^m}, T(k)(V)\right) \end{array} \tag{1}$$

For $m = 1$,

$$\hat{\Gamma}_{1,V} \colon T(k)(V^{C_p}) \to \hat{\mathbb{H}}\left(C_p, T(k)(V^{C_p})\right) \sim \hat{\mathbb{H}}\left(C_p, T(k)(V)\right)$$

induces an equivalence on π_i for $i \geq \dim V^{C_p}$. This is simply lemma 4.2.6 suspended $\dim V^{C_p}$ times. Theorem 4.1.15 implies that $\Gamma_{m,V}$ and $\hat{\Gamma}_{m,V}$ induce isomorphisms on homotopy groups in the same range. The spectral sequence

$$\hat{H}^*\left(C_{p^m}, \pi_* T(\mathbb{F}_p)(V)\right) \Rightarrow \pi_* \hat{\mathbb{H}}\left(C_{p^m}, T(\mathbb{F}_p)(V)\right)$$

is isomorphic to the spectral sequence for $V = 0$ reindexed by shifting bidegrees up by $(0, \dim V)$. Hence the argument of proposition 4.2.7 gives

$$\pi_* \hat{\mathbb{H}}\left(C_{p^m}, T(\mathbb{F}_p)(V)\right) = S_{\mathbb{Z}/p^m}\{\sigma, \sigma^{-1}\}[\dim V] \tag{2}$$

and it follows from (1) that $\pi_i T(\mathbb{F}_p)(V)^{C_{p^m}} = \mathbb{Z}/p^{m+1}$ if i is even and $i \geq \dim V^{C_p}$. Moreover, the upper horizontal sequence in (1) yields that

$$R \colon T(\mathbb{F}_p)(V)^{C_{p^m}} \to T(\mathbb{F}_p)(V^{C_p})^{C_{p^{m-1}}}$$

is $(\dim V - 1)$-connected, so induction on m gives the claimed homotopy groups for $k = \mathbb{F}_p$.

Finally, the argument going from $\mathbb{F}_p$ to the perfect field k is similar to the one presented in sect. 4.3. □

Let $\mathbf{W}(k)$ denote the big Witt vectors of k, i.e. $\mathbf{W}(k) = (1 + Xk[[X]])^\times$, the multiplicative group of power series which begins with 1. Write $\mathbf{W}_m(k)$ for the truncated Witt-vectors

$$\mathbf{W}_m(k) = (1 + Xk[[X]])^\times / (1 + X^{m+1}k[[X]])^\times.$$

Let $V_n \colon \mathbf{W}(k) \to \mathbf{W}(k)$ be the Verschiebung: it sends a polynomium $p(X)$ to $p(X^n)$, and induces an injection

$$V_n \colon \mathbf{W}_{m-1}(k) \to \mathbf{W}_{nm-1}(k).$$

Theorem 5.2.8. ([HM2]). *For a perfect field k of characteristic $p > 0$,*

$$K_{2m-1}\left(k[x]/(x^n); \mathbb{Z}_p\right) = \mathbf{W}_{nm-1}(k)/V_n\mathbf{W}_{m-1}(k)$$

and $K_{2m}(k[x]/(x^n); \mathbb{Z}_p) = 0$ for $m > 0$.

Proof. We are in a situation where $K_*(-; \mathbb{Z}_p)$ and $\mathrm{TC}_*(-; \mathbb{Z}_p)$ agree, and shall calculate the latter. I shall only treat the case $(p, n) \neq 1$; the other case is less complicated.

Suppose first $n' \mid l$ and choose m in the range

$$\dim_{\mathbb{C}} V_{p^{r-1}l} < m \leq \dim_{\mathbb{C}} V_{p^r l} \tag{1}$$

with notation as in theorem 5.2.6. By definition,

$$\dim_{\mathbb{C}} V_{p^{r-1}l} = \begin{cases} \left[\frac{p^r l}{n}\right] & \text{if } r < v_p(n) \\ \left[\frac{p^r l}{n}\right] - 1 & \text{if } r \geq v_p(n) \end{cases}$$

so the above condition is equivalent to

$$\begin{aligned} &p^{r-1}l - n < mn \leq p^r l - n \quad \text{if } r > v_p(n) \\ &p^{r-1}l < mn \leq p^r l - n \quad \text{if } r = v_p(n) \\ &p^{r-1}l < mn \leq p^r l \quad \text{if } r < v_p(n) \end{aligned} \tag{2}$$

Now

$$\begin{aligned}\pi_{2m-1}\left(\Sigma \operatorname*{holim}_{\overleftarrow{R}} T(k)(V_{p^i l})^{C_{p^i}}\right) &\cong \pi_{2m-2} \operatorname*{holim}_{\overleftarrow{R}} T(k)(V_{p^i l})^{C_{p^i}}\\ &\cong \pi_{2m-2} T(k)(V_{p^{r-1}l})^{C_{p^{r-1}}}\\ &\cong W_r(k)\end{aligned}$$

by theorem 5.2.7, and similarly

$$\pi_{2m-1}\left(\Sigma \operatorname*{holim}_{\overleftarrow{R}} T(k)(V_{p^i l})^{C_{p^{i-v_p(n)}}}\right) \cong W_{r-v_p(n)}(k).$$

Thus the second factor in theorem 5.2.6 (where $n' \mid l$) contributes

$$\bigoplus_{(l,p)=1,\ n'\mid l} \operatorname{cok}\left(W_{r(m,l)-v_p(n)}(k) \to W_{r(m,l)}(k)\right)$$

to $\mathrm{TC}_{2m-1}(k[x]/(x^n);\mathbb{Z}_p)$, when $r = r(m,l)$ denotes the unique number which satisfies (2). In other words, the contribution is

$$\begin{aligned}&\bigoplus\{W_{v_p(n)}(k) \mid (l,p)=1,\ n' \mid l,\ l < pmn'\}\oplus\\ &\bigoplus\{W_{r(m,l)}(k) \mid (l,p)=1,\ n' \mid l,\ l > pmn'\}.\end{aligned} \tag{3}$$

Similar considerations show that the first factor in theorem 4.2.6 contributes

$$\bigoplus\{W_{r(m,l)}(k) \mid (p,l)=1,\ n' \nmid l\}, \tag{4}$$

where this time $r = r(m,l)$ is the unique number with $p^{r-1}l < mn \leq p^r l$. Finally, it is easy to see that the direct sum of (3) and (4) is isomorphic to $\mathbf{W}_{mn-1}(k)/V_n(\mathbf{W}_{m-1}(k))$. □

Remark 5.2.9. The above argument shows that

$$\pi_{2m-1}\left(\prod_{(l,p)=1} \operatorname*{holim}_{\overleftarrow{R}} T(k)(V_{p^i l})^{C_{p^i}}\right) = \mathbf{W}_{mn-1}(k)$$

and more generally that

$$\pi_{2m-1}\left(\prod_{(l,p)=1} \operatorname*{holim}_{\overleftarrow{R}} T(k)(V_{p^i l})^{C_{p^{i-\nu}}}\right) = p^{\nu}\mathbf{W}_{mn-1}(k).$$

The difference between the two cases $(p,n) = 1$ and $(p,n) \neq 1$ in theorem 5.2.8 is just that V_n in the first case gives an isomorphism on the subproduct with $n \mid l$; in the second case there is a cokernel whose size depends on the p-adic valuation of n.

I should point out that the low dimensional groups $K_i(k[x]/(x^n);\mathbb{Z}_p)$, $i \leq 3$, were determined previously, and that Thomas Geisser asked us to use the present techniques to work out the groups for general i; he even conjectured the correct answer.

5.3 Nil calculations.

McCarthy's relative theorem makes it possible to calculate the socalled Nil-groups of rings A which contain a nilpotent ideal I for which A/I is a regular ring. In this situation, we have the cofibration sequence

$$\mathrm{NK}(A)^\wedge \to \mathrm{TC}\,(A[t] \to A/I[t])^\wedge \to \mathrm{TC}\,(A \to A/I)^\wedge \tag{5.3.1}$$

with $\Omega\mathrm{NK}(A)^\wedge \sim \mathrm{Nil}(A)^\wedge$. I illustrate the situation with an explicit calculation for the rings $A_n = k[x]/(x^n)$ of the previous section. Further details and examples are to appear in [HM2].

Lemma 5.2.2 and (5.2.1) shows that

$$T(A[t]) \sim_{C_\infty} T(A) \wedge N^{\mathrm{cy}}(\Pi_\infty) \sim_{C_\infty} T(A) \wedge \bigvee_{s=0}^{\infty} S^1_+/C_s$$

and we can apply theorem 5.2.6 for the ring

$$A_n[t] = k[t,x]/(x^n).$$

This expresses $\mathrm{TC}(A_n[t])$ in terms of $\Sigma \underleftarrow{\mathrm{holim}} T(k[t])(V_{p^i l})^{C_{p^i}}$ with $(p,l)=1$. Write $\tilde{T}(k[t]) = T\,(k[t] \to k)$, and let $\sim_p$ denote equivalence after p-adic completions. Then

$$\begin{aligned}\tilde{T}(k[t])(V_{p^i l})^{C_{p^i}} &\sim \left(T(k)(V_{p^i l})^{C_{p^i}} \wedge \bigvee_{s=1}^{\infty} S^1_+/C_s\right)^{C_{p^i}} \\ &\sim_p \bigvee_{(\nu,p)=1} \bigvee_{j=0}^{\infty} \left(T(k)(V_{p^i l})^{C_{p^i}} \wedge S^1_+/C_{p^j}\right),\end{aligned}$$

since $S^1/C_{p^j \nu} \sim_p S^1/C_{p^j}$ when $(\nu,p)=1$, and one has as in sect. 5.2:

$$\begin{aligned}\left(T(k)(V_{p^i l}) \wedge S^1_+/C_{p^j}\right)^{C_{p^i}} &\sim T(k)(V_{p^i l})^{C_{p^{\min(i,j)}}} \wedge S^1_+/C_{p^{\max(i,j)}} \\ &\sim \Sigma T(k)(V_{p^i l})^{C_{p^{\min(i,j)}}} \vee T(k)(V_{p^i l})^{C_{p^{\min(i,j)}}}.\end{aligned}$$

For fixed k,

$$\pi_k \operatorname*{holim}_{\overleftarrow{R}} T(k[t])(V_{p^i l})^{C_{p^i}} = \pi_k \Sigma T(k[t])(V_{p^i l})^{C_{p^i}}$$

if i is sufficiently large; the precise value of i is given in the proof of theorem 5.2.8. It follows from remark 5.2.9 and the above that

$$\begin{aligned} \pi_{2m-1} \prod_{(p,l)=1} \Sigma \operatorname*{holim}_{\overleftarrow{R}} T(k[t])(V_{p^i l})^{C_{p^i}} &\cong \pi_{2m} \left(\prod_{(p,l)=1} \operatorname*{holim}_{\overleftarrow{R}} T(k[t])(V_{p^i l}) \right)^{C_{p^i}} \\ &\cong \bigoplus_{(\nu,p)=1} \left(\bigoplus_{j=1}^{r} p^j \mathbf{W}_{mn-1}(k) \oplus \bigoplus^{\infty} \mathbf{W}_{mn-1}(k) \right), \end{aligned}$$

where p^r is the exponent of $\mathbf{W}_{mn-1}(k)$. This can also be written as

$$\bigoplus_{(\nu,p)=1} \mathbf{W}_{mn-1}[y]/\mathbf{W}_{mn-1}[py].$$

We divide out the image of the Verschiebung $V_n \colon \mathbf{W}_{m-1}(k) \to \mathbf{W}_{mn-1}(k)$ to get

Theorem 5.3.1. *The groups* $\mathrm{NK}_{2m}(k[x]/(x^n))$ *and* $\mathrm{NK}_{2m-1}(k[x]/(x^n))$ *are isomorphic and are given as an infinite sum of* $\Lambda_n[y]/\Lambda_n[py]$ *with* $\Lambda_n = \mathbf{W}_{mn-1}(k)/V_n\mathbf{W}_{m-1}(k)$. □

There are more canonical ways to present the result, e.g. by using the deRham-Witt complex of Deligne and Illusie. I refer the reader to [HM2].

5.4 On the K-theory of local class fields.

It is natural to attempt to generalize the calculations of the previous sections to rings of integers in local class fields, $A = \mathrm{int}(E)$ with $E/\mathbb{Q}_p$ abelian (or even to local number fields). Such fields appear as centers in group rings $\mathbb{Q}_p[G]$, and their integers are centers in the corresponding maximal orders $\mathcal{M}_p(G)$,

$$\mathbb{Z}_p G \subset \mathcal{M}_p(G) \subset \mathbb{Q}_p G.$$

If $E/\mathbb{Q}_p$ is unramified, then $A \cong W(\mathbb{F}_{p^s})$ is a factor in $\mathbb{Z}_p[C_f]$, $(p, f) = 1$, and one can use lemma 4.4.7,

$$T(\mathbb{Z}_p[C_f])^{\wedge}_p \sim (T(\mathbb{Z}_p) \wedge \Lambda BC_{f+})^{\wedge}_p \sim (T(\mathbb{Z}_p) \wedge C_{f+})^{\wedge}_p$$

to get the cofibration sequence

$$\mathrm{TC}(W(\mathbb{F}_{p^s}))^\wedge_p \longrightarrow \mathrm{TF}(\mathbb{Z}_p)^\wedge_p \xrightarrow{R^s-1} \mathrm{TF}(\mathbb{Z}_p)^\wedge_p,$$

cf. [BM2]. Thus the unramified case is of the same complexity as $A = \mathbb{Z}_p$, where one has the calculational methods of sect. 4.1. In outline the calculation of $\mathrm{TC}(\mathbb{Z}_p)$ is similar to the calculation of $\mathrm{TC}(\mathbb{F}_p)$, but the details are of a different magnitude of difficulties.

The first problem is to verify Conjecture 4.1.16 for the rings in question, i.e. to show that

$$\hat{\Gamma}\colon \mathrm{TH}(A)^\wedge_p \to \hat{\mathbb{H}}(C_p, \mathrm{TH}(A)) \tag{5.4.1}$$

induces isomorphisms on homotopy groups in non-negative degrees. This was done in [BM1], sect. 5 for $A = \mathbb{Z}_p$, p odd, and in [R] for $p = 2$. I will go through the p odd case below; it was not so well presented in [BM1]. First recall from [B2]:

Theorem 5.4.2. *The* mod p *homotopy groups of* $\mathrm{TH}(\mathbb{Z}_p)$ *are*

$$\pi_*(\mathrm{TH}(\mathbb{Z}_p); \mathbb{F}_p) \cong E\{e_{2p-1}\} \otimes S\{f_{2p}\}$$

where the subscripts indicate degrees. Moreover, the Bockstein operator on f_{2p} *is* e_{2p-1}, *and the reduction map from* $\mathrm{TH}(\mathbb{Z}_p)$ *to* $\mathrm{TH}(\mathbb{F}_p)$ *maps* f_{2p} *non-trivially.* □

The reader with no access to [B2] may consult [HM] for an outline. Recall from lemma 4.4.4 the S^1-map

$$\iota\colon \Sigma^\infty_{S^1}(S^0)^\wedge_p \to \mathrm{TH}(\mathrm{Id}_p)$$

which induces equivalence on all C_{p^n} fixed sets. The inclusion

$$\Sigma^\infty_{S^1}(S^0)^{S^1} \to \Sigma^\infty(S^0)$$

is split, and the splitting induces a map

$$f\colon \Sigma^\infty(S^0) \to \Sigma^\infty_{S^1}(S^0)^{S^1} \to \Sigma^\infty_{S^1}(S^0)^{hS^1} \to \mathrm{TH}(\mathrm{Id}_p)^{hS^1}.$$

The homotopy ring $\pi_*(\Sigma^\infty(S^0); \mathbb{Z}_p)$ is of course unknown, but it contains the direct summand

$$E\{a_{2p-3}\} \otimes S\{b_{2p-2}\} \subset \pi_*(\Sigma^\infty(S^0); \mathbb{F}_p), \quad p \text{ odd}. \tag{5.4.3}$$

The first element outside the direct summand lies in degree $2p^2 - 2p - 2$. There is a similar statement for $p = 2$. We compose f with the map

$$g\colon \mathrm{TH}(\mathrm{Id}_p)^{hS^1} \xrightarrow{L} \mathrm{TH}(\mathbb{Z}_p)^{hS^1} \longrightarrow F(S^3_+, \mathrm{TH}(\mathbb{Z}_p))^{S^1}$$

where L comes from linearization $\mathrm{Id}_p \to \tilde{\mathbb{Z}}_p$, and the second map is restriction to the skeleton $S^3_+ \subset ES^1_+$. The cofibration sequence $S^1_+ \to S^3_+ \to S^3/S^1$ and the S^1-equivalence $S^3/S^1 \sim S^1_+ \wedge S^2$ yield the fibration sequence

$$\Omega^2\mathrm{TH}(\mathbb{Z}_p) \xrightarrow{i} F(S^3_+, \mathrm{TH}(\mathbb{Z}_p))^{S^1} \longrightarrow \mathrm{TH}(\mathbb{Z}_p).$$

The composition $g \circ f$ maps the homotopy fiber of $\Sigma^\infty(S^0) \to H\mathbb{Z}_p$ into the fiber $\Omega^2\mathrm{TH}(\mathbb{Z}_p)$,

$$l\colon hF(\Sigma^\infty(S^0) \to H\mathbb{Z}_p) \to \Omega^2\mathrm{TH}(\mathbb{Z}_p), \quad i \circ l = g \circ f.$$

On homotopy groups one has

$$l_*(a_{2p-3}) = \Omega^2 e_{2p-1}, \qquad l_*(b_{2p-2}) = \Omega^2 f_{2p}. \tag{5.4.4}$$

This is a consequence of the statement that the composition

$$S^1_+ \wedge H\mathbb{Z}_p \xrightarrow{\iota} \mathrm{TH}(\mathbb{Z}_p) \xrightarrow{\mathrm{proj}} \Sigma^{2p-1}H\mathbb{Z}/p,$$

with ι the inclusion of the cyclic 0-skeleton, represents the suspension of the first Steenrod operation P^1, cf. [BM1], lemma 5.3 for details. We next consider the diagram of spectral sequences

$$\begin{array}{ccc} E^r_{-p,q}\left(\mathrm{TH}(\mathrm{Id}_p)^{hS^1};\mathbb{F}_p\right) & \longrightarrow & E^r_{-p,q}\left(\mathrm{TH}(\mathbb{Z}_p)^{hS^1};\mathbb{F}_p\right) \\ \downarrow & & \downarrow \\ E^r_{-p,q}\left(\mathrm{TH}(\mathrm{Id}_p)^{hC_p};\mathbb{F}_p\right) & \longrightarrow & E^r_{-p,q}\left(\mathrm{TH}(\mathbb{Z}_p)^{hC_p};\mathbb{F}_p\right) \end{array} \tag{5.4.5}$$

In fiber degree $q \le 2p^2 - 2p - 2$, the E^2-terms are:

$$\begin{aligned} E^2\left(\mathrm{TH}(\mathrm{Id}_p)^{hC_p};\mathbb{F}_p\right) &= E\{u_1\} \otimes S\{t\} \otimes E\{a_{2p-3}\} \otimes S\{b_{2p-2}\} \\ E^2\left(\mathrm{TH}(\mathbb{Z}_p)^{hC_p};\mathbb{F}_p\right) &= E\{u_1\} \otimes S\{t\} \otimes E\{e_{2p-1}\} \otimes S\{f_{2p}\} \\ E^2\left(\mathrm{TH}(\mathrm{Id}_p)^{hS^1};\mathbb{F}_p\right) &= S\{t\} \otimes E\{a_{2p-3}\} \otimes S\{b_{2p-2}\} \\ E^2\left(\mathrm{TH}(\mathbb{Z}_p)^{hS^1};\mathbb{F}_p\right) &= S\{t\} \otimes E\{e_{2p-1}\} \otimes S\{f_{2p}\}. \end{aligned}$$

The vertical maps in (5.4.5) are the inclusions. It is well-known to homotopy theorists (see e.g. [BM1], sect. 3) that

$$E^2(\mathrm{TH}(\mathrm{Id}_p);\mathbb{F}_p) \cong E^{2(p-2)}(\mathrm{TH}(\mathrm{Id}_p);\mathbb{F}_p)$$

and that

$$d^{2p-2}(t) = t^p a_{2p-3}, \quad d^{2p-2}(u_1) = 0, \quad d^{2p-1}(u_1) = t^p b_{2p-2}. \tag{5.4.6}$$

The horizontal maps in (5.4.5) are zero (at least in fiber degrees $\leq 2p^2-2p-2$) but this is due to the filtration shift indicated by (5.4.4).

Proposition 5.4.7. *Let p be an odd prime. In the spectral sequence $E^r(\mathrm{TH}(\mathbb{Z}_p)^{hC_p};\mathbb{F}_p)$, the elements te_{2p-1} and tf_{2p} are infinite cycles. Moreover $E^2 = E^{2p}$ and*

$$d^{2p}(t) = t^{p+1}e_{2p-1}, \quad d^{2p}(u_1) = 0, \quad \text{and } d^{2p+1}(u_1) = t^{p+1}f.$$

Proof. Let $T = \mathrm{TH}(\mathbb{Z}_p)$ or $T = \mathrm{TH}(\mathrm{Id}_p)$. Consider the Postnikov tower

$$T[0,0] \leftarrow T[0,1] \leftarrow \cdots \leftarrow T[0,q] \leftarrow \cdots$$

with inverse limit T. Here $T[0,q]$ has homotopy groups precisely in degree t for $0 \leq t \leq q$, and in this range they are equal to the homotopy groups of T. The Postnikov tower can be taken to be functorial (e.g. by using J. Moore's simplicial construction of it), so each term has an S^1-action.

The homotopy groups of the Postnikov tower defines an exact couple, which gives the spectral sequence we are looking at. It has

$$E^2_{-p,q} = \pi_{q-p}F(EC_{p+},T[q,q])^{C_p} \cong \pi_{q-p}F(BC_{p+},T[q,q]) \cong H^p(BC_p,\pi_q T)$$

and the differentials d^{r+1} are induced from the additive relations

$$\pi_{q-p}F(EC_{p+},T[q,q])^{C_p} \longleftarrow \pi_{q-p}F(EC_{p+},T[q,q+r-1])^{C_p} \xrightarrow{\partial_*} \pi_{q-p-1}F(EC_{p+},T[q+r,q+r])^{C_p}.$$

Here ∂_* is the connecting homomorphism in the homotopy exact sequence of the fibration

$$T[q,q+r-1] \leftarrow T[q,q+r] \leftarrow T[q+r,q+r].$$

We shall now compare the situation for $\mathrm{TH}(\mathrm{Id}_p)$ and $\mathrm{TH}(\mathbb{Z}_p)$. To shorten notation, write

$$F[s,t] = F(EC_{p+},\mathrm{TH}(\mathrm{Id}_p)[s,t])^{C_p}$$
$$F_{\mathbb{Z}}[s,t] = F(EC_{p+},\mathrm{TH}(\mathbb{Z}_p)[s,t])^{C_p}$$

and let $\pi_*(-) = \pi_*(-;\mathbb{Z}_p)$. Then (5.4.4) translates as follows: the additive relations

$$\pi_{2p-3}F[2p-3,2p-3] \twoheadleftarrow \pi_{2p-3}F[2p-3,2p-1] \xrightarrow{i_*} \pi_{2p-3}F_{\mathbb{Z}}[2p-3,2p-1] \xleftarrow{\cong} \pi_{2p-3}F_{\mathbb{Z}}[2p-1,2p-1]$$
$$\pi_{2p-2}F[2p-2,2p-2] \xleftarrow{\cong} \pi_{2p-2}F[2p-2,2p] \xrightarrow{i_*} \pi_{2p-2}F_{\mathbb{Z}}[2p-2,2p] \hookleftarrow \pi_{2p-2}F_{\mathbb{Z}}[2p,2p] \qquad (1)$$

give well-defined maps (from left to right) which take a_{2p-3} and b_{2p-2} into the generators te_{2p-1} and tf_{2p} of

$$\pi_{2p-3}F_{\mathbf{Z}}[2p-1,2p-1] = H^2(BC_p;\pi_{2p-1}\mathrm{TH}(\mathbb{Z}_p))$$
$$\pi_{2p-2}F_{\mathbf{Z}}[2p,2p] = H^2(BC_p;\pi_{2p}\mathrm{TH}(\mathbb{Z}_p))$$

For example, the first additive relation is well-defined because l_* annihilates the generator $u_1b_{2p-2} \in H^1(BC_p;\pi_{2p-2}\mathrm{TH}(\mathrm{Id}_p))$: it maps to an element of filtration degree 3, according to (5.4.4).

The elements a_{2p-3} and b_{2p-2} are infinite cycles in the spectral sequence for $\mathrm{TH}(\mathrm{Id}_p)^{hC_p}$, being in the image of f_*. This means that they lift to elements of $\pi_{2p-3}F[2p-3,\infty]$ and $\pi_{2p-2}F[2p-2,\infty]$. It follows that te_{2p-1} and tf_{2p} lift to $\pi_{2p-3}F_{\mathbf{Z}}[2p-1,\infty]$ and $\pi_{2p-2}F_{\mathbf{Z}}[2p,\infty]$, so are infinite cycles.

Let us prove that $d^{2p}(u_1) = 0$ and $d^{2p+1}(u_1) = t^{p+1}f_{2p}$ and leave the easier differential $d^{2p}(t) = t^{p+1}e_{2p-1}$ to the reader. The additive relation defining $d^{2p}(u_1)$ is

$$\pi_{-1}F[0,0] \xleftarrow{\cong} \pi_{-1}F[0,2p-4] \leftarrowtail \pi_{-1}F[0,2p-3] \xrightarrow{\partial_*} \pi_{-2}F[2p-2,2p-2]. \quad (2)$$

Indeed u_1 lies in the subgroup $\pi_{-1}F[0,2p-3]$ because $d^{2p-2}(u_1) = 0$ (and not equal to $t^{p-1}u_1a_{2p-3}$). Because of the filtration shift represented by (5.4.4), it is better to consider the additive relations

$$\begin{array}{ccccc} A\colon \pi_{-1}F[0,0] & \longleftarrow & \pi_{-1}F[0,2p-4] & \xrightarrow{\partial'_*} & \pi_{-2}F[2p-3,2p] \\ \Big\downarrow h_* & & \Big\downarrow l_* & & \Big\downarrow h_* \\ A\mathbb{Z}\colon \pi_{-1}F_{\mathbf{Z}}[0,0] & \longleftarrow & \pi_{-1}F_{\mathbf{Z}}[0,2p-4] & \xrightarrow{\partial'_*} & \pi_{-2}F_{\mathbf{Z}}[2p-3,2p] \end{array} \quad (3)$$

where ∂'_* is the connecting homomorphism in the homotopy exact sequence of

$$F[0,2p-4] \leftarrow F[0,2p] \leftarrow F[2p-3,2p].$$

Theorem 5.4.2 and (5.4.3) gives

$$\pi_{-2}F[2p-3,2p] \cong \pi_{-2}F[2p-3,2p-2]$$
$$\pi_{-2}F[2p-3,2p] \cong \pi_{-2}F[2p-1,2p]$$

and hence exact sequences

$$\begin{array}{ccccccc} 0 \to \pi_{-2}F[2p-2,2p] & \xrightarrow{i} & \pi_{-2}F[2p-3,2p] & \xrightarrow{j} & \pi_{-2}F[2p-3,2p-3] \to 0 \\ \Big\downarrow & & \Big\downarrow l_* & & \\ 0 \to \pi_{-2}F[2p,2p] & \xrightarrow{i_{\mathbf{Z}}} & \pi_{-2}F_{\mathbf{Z}}[2p-3,2p] & \xrightarrow{j_{\mathbf{Z}}} & \pi_{-2}F[2p-3,2p-1] \to 0 \end{array}$$
(4)

One has the following values of the groups involved:

$$\begin{aligned}
\pi_{-2}F[2p-2,2p] &\cong H^{2p}(BC_p;\pi_{2p-2}\mathrm{TH}(\mathrm{Id}_p)) &&= \mathbb{F}_p\langle t^p b_{2p-2}\rangle\\
\pi_{-2}F[2p-3,2p-3] &\cong H^{2p-1}(BC_p;\pi_{2p-3}\mathrm{TH}(\mathrm{Id}_p)) &&= \mathbb{F}_p\langle t^{p-1}u_1a_{2p-3}\rangle\\
\pi_{-2}F[2p,2p] &\cong H^{2p+2}(BC_p;\pi_{2p}\mathrm{TH}(\mathbb{Z}_p)) &&= \mathbb{F}_p\langle t^{p+1}f_{2p}\rangle\\
\pi_{-2}F[2p-3,2p-1] &\cong H^{2p+1}(BC_p;\pi_{2p-1}\mathrm{TH}(\mathbb{Z}_p)) &&= \mathbb{F}_p\langle t^p u_1 e_{2p-1}\rangle
\end{aligned}$$

I claim that $j_{\mathbf{Z}}\circ l_*\circ i=0$, giving the left hand vertical arrow in (4). Indeed the generator of $\pi_{-2}F[2p-2,2p]$ is t^pb_{2p-2}, hence in the image of the product map

$$\pi_{-2p}F[0,0]\otimes\pi_{2p-2}F[2p-2,2p-1]\to\pi_{-2}F[2p-2,2p-1].$$

The homomorphism

$$\pi_{2p-2}F[2p-2,2p-1]\to\pi_{2p-2}F_{\mathbf{Z}}[2p-2,2p-1]$$

is zero by (5.4.4), and the claim follows by using the product

$$\pi_{-2p}F_{\mathbf{Z}}[0,0]\otimes\pi_{2p-2}F_{\mathbf{Z}}[2p-3,2p-1]\to\pi_{-2}F_{\mathbf{Z}}[2p-3,2p-1].$$

On the other hand

$$d^{2p-2}(u_1)=j\circ A(u_1),\qquad d^{2p}_{\mathbf{Z}}(u_1)=j_{\mathbf{Z}}\circ A_{\mathbf{Z}}(u_1)\tag{5}$$

and by (3), $j_{\mathbf{Z}}A_{\mathbf{Z}}(u_1)=j_{\mathbf{Z}}l_*A(u_1)$. Since $d^{2p-2}(u_1)=0$, one concludes that $d^{2p}_{\mathbf{Z}}(u_1)=0$. Finally the differential $d^{2p-1}(u_1)=t^pb$ shows that $l_*A(u_1)\neq0$ and that it belongs to the image of i in diagram (4). But the left hand vertical arrow in (4) is an isomorphism; use of products as above and (5.4.4) completes the proof. □

Corollary 5.4.8. *For p odd,*

$$\pi_*\left(\hat{\mathbb{H}}(C_p,\mathrm{TH}(\mathbb{Z}_p));\mathbb{F}_p\right)=E\{e_{2p-1}\}\otimes S\{t^p,t^{-p}\}.$$

Moreover,

$$\hat{\Gamma}_*:\pi_*(\mathrm{TH}(\mathbb{Z}_p);\mathbb{F}_p)\to\pi_*\left(\hat{\mathbb{H}}(C_p,T(\mathbb{Z}_p));\mathbb{F}_p\right)$$

is an isomorphism in non-negative degrees.

Proof. The non-zero differentials in $E^r\left(\hat{\mathbb{H}}(C_p,\mathrm{TH}(\mathbb{Z}_p));\mathbb{F}_p\right)$ are by (5.4.7):

$$d^{2p}(t^i)=it^{i+p}e_{2p-1},\quad d^{2p+1}(u_1)=t^{p+1}f,\quad(i\in\mathbb{Z})$$

and a routine calculation gives $E^{2p-2}_{*,*} \cong E\{u_1\} \otimes S\{t^p, t^{-p}\}$. For degree reasons $E^{2p+2}_{*,*} = E^\infty_{*,*}$. This proves the first statement. The commutative diagram

$$\begin{array}{ccc} T(\mathbb{Z}_p) & \longrightarrow & T(\mathbb{F}_p) \\ \downarrow{\scriptstyle \hat{\Gamma}} & & \downarrow{\scriptstyle \hat{\Gamma}} \\ \hat{\mathbb{H}}(C_p, T(\mathbb{Z}_p)) & \longrightarrow & \hat{\mathbb{H}}(C_p, T(\mathbb{F}_p)) \end{array},$$

together with lemma 4.2.4 and theorem 5.4.2 tells us that $\hat{\Gamma}_*(f_{2p}) = t^{-2p}$ and $\hat{\Gamma}_*(e_{2p-1}) = e_{2p-1}$. □

The corollary implies that

$$\mathrm{TH}(\mathbb{Z}_p) \xrightarrow{\hat{\Gamma}} \hat{\mathbb{H}}(C_p, \mathrm{TH}(\mathbb{Z}_p))[0,\infty)$$

is a p-adic equivalence, and theorem 4.1.15 then gives

$$\mathrm{TF}(\mathbb{Z}_p, p)^\wedge_p \sim \left(\operatorname{holim}_{\overleftarrow{F}} T(\mathbb{Z}_p)^{hC_{p^n}} \right)^\wedge_p \sim (T(\mathbb{Z}_p)^{hS^1})^\wedge_p.$$

The homotopy groups $\pi_*(T(\mathbb{Z}_p)^{hS^1}; \mathbb{F}_p)$ and $\pi_*\left(\hat{\mathbb{H}}(S^1; T(\mathbb{Z}_p)); \mathbb{F}_p\right)$ were calculated in [BM2] by solving the involved spectral sequences, and

$$R^s_* : \pi_*(\mathrm{TF}(\mathbb{Z}_p, p); \mathbb{F}_p) \to \pi_*(\mathrm{TF}(\mathbb{Z}_p, p); \mathbb{F}_p)$$

was determined. This was enough to give the values of $\mathrm{TC}_*(W(\mathbb{F}_{p^s}); \mathbb{F}_p)$. The groups turn out to be v_1-periodic, i.e.

$$v_1 : \mathrm{TC}_*(W(\mathbb{F}_{p^s}); \mathbb{F}_p) \xrightarrow{\cong} \mathrm{TC}_{*+2p-2}(W(\mathbb{F}_{p^s}); \mathbb{F}_p)$$

($v_1 = b_{2p-2}$), and this together with other tricks leads to the proof of theorem 1.5 of the introduction. I refer to [BM2] for the details.

For odd primes p, theorem 1.5 states that

$$\mathrm{TC}(\mathbb{Z}_p)^\wedge_p[0,\infty) \sim \operatorname{im} J_p \times B \operatorname{im} J_p \times SU^\wedge_p. \tag{5.4.9}$$

This is true as (-1)-connected spectra when one use the deloopings arising from Bott periodicity on the right hand side. For $p = 2$ there are added complications. For example, mod 2 homotopy groups of a ring spectrum does not in general form a ring. At the time of writing $\mathrm{TC}_*(\mathbb{Z}_2)$ has not been completely determined, but preliminary calculations of J. Rognes suggests that

$$\pi_*(\mathrm{TC}(\mathbb{Z}_2), \mathbb{F}_2) \cong \pi_*(\operatorname{im} J_2 \times B\operatorname{im} J_2 \times SU^\wedge_2; \mathbb{F}_2).$$

One expects that a twisted version of (5.4.9) is true for $p = 2$, cf. [BM2], sect. 6. I stress that $\operatorname{im} J_2$ is the *complex* J space at 2, i.e. the homotopy fiber of $\psi^k - 1\colon (BU \times \mathbb{Z})_2^\wedge \to BU_2^\wedge$.

For geometric reasons it is important to study the relative K-theory $K(\mathrm{Id} \to \mathbb{Z})$, by theorem 3.5.1 equal to $\mathrm{TC}(\mathrm{Id} \to \mathbb{Z})$. Indeed, a celebrated theorem due to F. Waldhausen states that

$$K(\mathrm{Id}) \sim \Omega^\infty S^\infty \times \mathrm{Wh}^{\mathrm{Diff}}(*)$$

where $\Omega^2 \mathrm{Wh}^{\mathrm{Diff}}(*) \sim \underrightarrow{\mathrm{holim}} \mathrm{Diff}(D^n, D^{n-1}_-)$ with $D^{n-1}_- \subset \partial D^n$ the lower hemisphere, and where $\Omega^\infty S^\infty$ is the zero'th term in the sphere spectrum.

Conjecture 5.4.10. For each odd p we have to split fibration

$$\mathrm{cok}J_p \to (\Omega^\infty S^\infty)_p^\wedge \overset{e}{\underset{\longleftarrow}{\longrightarrow}} \operatorname{im} J_p. \tag{1}$$

There is a similar split fibration

$$X_p \to \Omega^\infty S^\infty (S^1 \wedge \mathbb{C}P^\infty)_p^\wedge \overset{e'}{\underset{\longleftarrow}{\longrightarrow}} SU_p^\wedge. \tag{2}$$

Here the map from $S^1 \wedge \mathbb{C}P^\infty \to SU$ is adjoint to the map which classifies the reduced canonical line bundle, and e' is its 'universal' extension. The S^1-transfer

$$\tau\colon \Omega^\infty S^\infty (S^1 \wedge \mathbb{C}P^\infty) \to \Omega^\infty S^\infty$$

induces a map $\tau_p'\colon X_p \to \mathrm{cok}J_p$ and a map from $SU_p^\wedge$ to $\operatorname{im} J_p$ with fiber $SU_p^\wedge$. Let $\operatorname{im} \tilde{J}_p$ be the 0-connected cover of $\operatorname{im} J_p$. I conjecture that

$$\mathrm{TC}(\mathrm{Id} \to \mathbb{Z})_p^\wedge \sim \mathrm{cok}J_p \times B\mathrm{cok}J_p \times B \operatorname{im} \tilde{J}_p \times hF(\tau_p'). \tag{3}$$

The difficult part is to prove that the restriction of $\mathrm{TC}(\mathrm{Id}_p) \to \mathrm{TC}(\mathbb{Z}_p)$ to the $SU_p^\wedge$ factor of (2) is the deloop of $\psi^k - 1\colon BU_p^\wedge \to BU_p^\wedge$; this gives the factor $B \operatorname{im} \tilde{J}_p$ in (3).

The outstanding problem which remains is to determine $\mathrm{TC}(A)_p^\wedge$ in ramified situations. There are at least two approaches. One can attempt to use that A appears as the center in a maximal order $\mathcal{M}_p(G) \subset \mathbb{Q}_p[G]$, and use the ideas of sect. 5.1 to calculate $\mathrm{TC}(\mathbb{Z}_p[G])_p^\wedge$. But this leaves one with following problem, interesting in its own right:

Problem 5.4.11. Give a calculable trace description of $K(\mathbb{Z}_p[G] \to \mathcal{M}_p(G))$.

One knows by the localization theorem in K-theory a categorical description of $K(\mathbb{Z}_p[G] \to \mathbb{Q}_p[G])$, and hence of $K(\mathbb{Z}_p[G] \to \mathcal{M}_p(G))$, namely as the K-theory of cohomological trivial modules. But despite a lot of efforts by Bökstedt and the author, (5.4.11) remains unsolved (even for $G = C_p$).

A second approach is to follow sect. 4.1, starting with a calculation of $\mathrm{TH}(A)$. Recently, A. Lindenstrass has determined $\mathrm{TH}(A)$ for quadratic ramified extensions of $\mathbb{Z}_2$. In general one should have

Conjecture 5.4.12. *Let A be totally ramified and let $\pi \in A$ be the prime element ($A/\pi A = \mathbb{F}_p$). Then*

$$\pi_* \mathrm{TH}(A, A/\pi) \cong E_{\mathbb{F}_p}\{a_1\} \otimes S_{\mathbb{F}_p}\{a_2\}$$

with $\deg a_i = i$. □

Conjecture 5.1.12 yields $\pi_*(\mathrm{TH}(A); \mathbb{F}_p)$ as well, but I do not know if (5.4.1) is an equivalence in this case.

Finally of course there is the deep problem of determining the relative K-theory $K\left(\mathbb{Z}_{(p)} \to \mathbb{Z}_p\right)$ but this is a different story altogether.

Bibliography

[A] J. F. Adams, *Prerequisites (on equivariant theory) to Carlsson's lecture,* Algebraic topology, Aarhus 1982, LNM **1051**, 483–532, Springer.

[B1] M. Bökstedt, *Topological Hochschild homology,* preprint, Bielefeld.

[B2] M. Bökstedt, *Topological Hochschild homology of* $\mathbb{Z}$ *and* $\mathbb{Z}/p$, preprint, Bielefeld.

[BAG] The Buenos Aires cyclic homology group, *Cyclic homology of algebras with one generator,* K-theory **5** (1991), 51–68.

[BF] D. Burghelea, Z. Fiedorowicz, *Cyclic homology and algebraic* K*-theory of spaces II,* Topology **25** (1986).

[BHM] M. Bökstedt, W. C. Hsiang, I. Madsen, *The cyclotomic trace and algebraic* K*-theory of spaces,* Invent. Math. **111** (1993), 865–940.

[BK] A. K. Bousfield, D. M. Kan, *Homotopy limits, completions and localizations,* LNM **304** (1972), Springer-Verlag.

[BM1] M. Bökstedt, I. Madsen, *Topological cyclic homology of the integers,* Astérisque **226** (1994), 57–145.

[BM2] M. Bökstedt, I. Madsen, *Algebraic* K*-theory of local number fields: the unramified case,* preprint no 20, Aarhus university (1994).

[Br] L. Breen, *Extensions du groupe additif,* Publ. Math. I.H.E.S. **48** (1978), 39–123.

[BV] J. M. Boardman, R. Vogt, *Homotopy invariant algebraic structures on topological spaces,* LNM **347** (1973), Springer-Verlag.

[C] A. Connes, *Cohomologie cyclic et functors* Ext^n, C. R. Acad. Sci. Paris **296** (1983), 953–958.

[D] B. Dundas, *Relative K-theory and topological cyclic homology,* preprint, Aarhus University (1995).

[tD1] T. tom Dieck, *Transformation groups and representations theory,*LNM **766** (1979).

[tD2] T. tom Dieck, *Transformation groups,* De Gruyler studies in mathematics **8** (1987).

[DHK] W. Dwyer, M. Hopkins, D. Kan, *The homotopy theory of cyclic sets,* Trans. Amer. Math. Soc. **291** (1985), 281–289.

[DM] W. Dwyer, S. Mitchell, *On the K-theory spectrum of a ring of algebraic integer,*preprint 1994.

[DM1] B. Dundas, R. McCarthy, *Stable K-theory and topological Hochschild homology,* Ann. Math. **140** (1994), 685–701.

[DM2] B. Dundas, R. McCarthy, *Topological Hochschild homology of ring functors and exact categories,* preprint.

[EKMM] A. D. Elmendorf, I. Kriz, M. A. Mandell, J. P. May, *Rings, Modules, and algebras in stable homotopy theory.*

[FJ] T. Farrell, L. Jones, *Rigidity in geometry and topology,* Proc. ICM Kyoto (1990).

[FLS] V. Franjou, J. Lannes, L. Schwartz: *Autour de la cohomologie de MacLane des corps finis,* Invent. Math. **115** (1994), 513–538.

[FOV] Z. Fiedorowicz, C. Ogle, R. Vogt, *Volodin K-theory of A_∞ ring spaces,* Topology **32** (1993), 329–352.

[G] D. Gale, *Neighborly and cyclic polytopes,* Proc. Symp. Pure Math. **7** (1963), 255–232.

[G1] T. Goodwillie, *Cyclic homology, derivations and the free loop space,* Topology **24** (1985), 187–215.

[G2] T. Goodwillie, *Relative algebraic K-theory and cyclic homology of spaces,* Ann. of Math **124** (1986), 347–402.

[G3] T. Goodwillie, *Calculus I: The first derivative,* K-theory **4** (1990), 1–27.

[G4] T. Goodwillie, *Calculus II: Analytic functors,* K-theory **5** (1992), 295–332.

[G5] T. Goodwillie, Proc. ICM Kyoto (1990).

[G6] T. Goodwillie, *Notes on the cyclotomic trace,* MSRI preprint.

[GM] J. Greenlees, J. P. May, *Generalized Tate cohomology,* Mem AMS vol **113**, No 543 (1995).

[H1] L. Hesselholt, *Stable topological cyclic homology is topological Hochschild homology,* Asterisque **226** (1994), 178–192.

[H2] L. Hesselholt, *One the p-typical curves in Quillen's K-theory,* Preprint 1995.

[Hs] W.-C. Hsiang, *Geometric applications of algebraic K-theory,* Proc. ICM Warszawa (1983), 99–118.

[HM] L. Hesselholt, I. Madsen, *On the K-theory of finite algebras over Witt vectors of perfect fields,* preprint no 2., Aarhus University (1995).

[HM1] L. Hesselholt, I. Madsen, *The S^1-Tate spectrum for J,* Bol. Soc. Math. Mex. vol **37** (1992), 215–240.

[HM2] L. Hesselholt, I. Madsen, *Cyclic polytopes and the K-theory of truncated polynomial algebras* (in preparation).

[I] K. Igusa, *The stability theorem for smooth pseudoisotopies,* K-theory **2** (1988), 1–355.

[J] J. D. S. Jones, *Cyclic homology and equivariant homology,* Invent. Math. **87** (1987), 403–423.

[K] C. Kassel, *La K-theorie stable,* Bull. Soc. Math. France **110** (1982), 381–416.

[Kr] C. Kratzor, *λ-structure on K-theorie algebrique,* Comment Math. Helv. **55** (1980), 233–254.

[L1] J.-L. Loday, *K-théorie algebrique et représentations de groupes,* Ann. Sci. Ec. Norm. Sup. IV, ser **9** (1976), 309–377.

[L2] J.-L. Loday, *Homologies diedrale et quaternionique,* Advances in Math. **66** (1987), 119–148.

[LMS] G. Lewis, J. P. May, M. Steinberger, *Stable equivariant homotopy theory,* LNM **1213**, Springer-Verlag.

[M] I. Madsen, *The cyclotomic trace in algebraic K-theory,* Proceedings ECM, Paris 1992, Progress in Math. vol. **120**.

[May1] J. P. May, *The geometry of infinite loop spaces,* LNM **271** (1992), Springer-Verlag.

[McC] R. McCarthy, *Two lectures in Aarhus,* July 1994.

[O1] R. Oliver, *Whitehead groups of finite groups,* London Math. Soc. Lecture Notes **132** (1988), Cambridge Univ. Press.

[O2] R. Oliver, K_2 *of p-adic group rings of abelian p-groups,* Math Z. **195** (1987), 505–558.

[PW] T. Parashvili, F. Waldhausen, *MacLane homology and topological Hochschild homology,* J. Pure Appl. Alg **82** (1992), 81–99.

[Q1] D. Quillen, *Homotopical algebra,* LNM **43** (1967), Springer-Verlag.

[Q2] D. Quillen, *Higher algebraic K-theory I,* LNM **341**, 85–147 (1973), Springer-Verlag.

[Q3] D. Quillen, *On the cohomology and K-theory of general linear groups over finite fields,* Annals of Math **96** (1972), 552–586.

[R] R. Rognes, *The homotopy limit problem for* $\mathrm{TH}(\mathbb{Z})$ *at 2,* preprint, Oslo University, 1994

[RS] C. Rourke, B. Sanderson, Δ*-sets I, II,* Q. J. Math. Oxf. **22** (1971), 321–338, 465–485.

[Sch] C. Schlichtkrull, *On the induction map in topological cyclic homology,* (in preparation).

[Se1] G. Segal, *Categories and cohomology theories,* Topology **13** (1974), 293–312.

[Se2] G. Segal, *Equivariant stable homotopy theory,* ICM, Nice (1976), 59–63.

[Sou] C. Soulé, *On the higher p-adic regulators,* LNM **854** (1980), 472–501.

[Su] A. Suslin, Algebraic K-theory of fields, Proc. ICM, Berkeley (1986), 222–243.

[T] S. Tsalidis, Topological Hochschild homology and the homotopy limit problem, preprint 1995.

[W1] F. Waldhausen, *Algebraic K-theory of topological spaces I,* Proc. Symp. Pure Math **32**, part I, AMS (1978), 35-60.

[W2] F. Waldhausen, *Algebraic K-theory of topological spaces II,* LNM **763**, 356–394, Springer-Verlag.

[W3] F. Waldhausen, *Algebraic K-theory of spaces,* LNM **1126**, 318–419 (1985), Springer-Verlag.

[W4] F. Waldhausen, *Algebraic K-theory of spaces, a manifold approach,* Canadian Math. Soc. Conf. Proc. CMS-AMS vol **2** (1982), 141–184.

The classification of conformal dynamical systems

Curtis T. McMullen *
UNIVERSITY OF CALIFORNIA,
BERKELEY CA 94720

Contents

1 Introduction

Consider the group generated by reflections in a finite collection of disjoint circles in the plane. Under reflection through their partners, the circles nest down to a Cantor set on which every orbit is dense. Poincaré christened

*Research partially supported by the NSF.

such conformal dynamical systems *Kleinian groups* and noted they extend to isometries of hyperbolic 3-space.

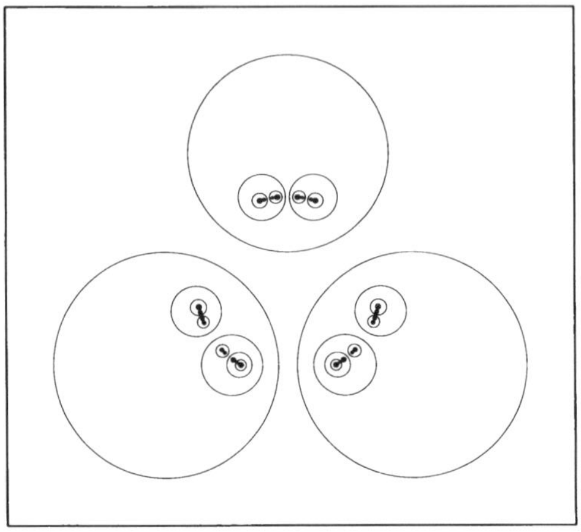

Figure 1. A reflection group.

The dynamics of a single conformal endomorphism of the sphere can exhibit similar structure, and these *iterated rational maps* were studied by Fatou and Julia in the 1920s.

Interest in iterated rational maps was rekindled in the 1980s, when Douady and Hubbard revealed the complexity of the family of quadratic polynomials. Contemporaneously, Thurston's 3-dimensional insights revolutionized the theory of Kleinian groups.

Sullivan then discovered a dictionary between these two classical subjects. Introducing quasiconformal methods to the setting of rational maps, he translated Ahlfors' finiteness theorem into a solution of the long-outstanding problem of wandering domains.

In this paper we illustrate the dictionary by discussing the classification problem for conformal dynamical systems.

We begin with the Teichmüller space of a dynamical system, which classifies deformations and leads to the finiteness theorems of Ahlfors and Sullivan. A more detailed picture is available for expanding dynamical systems, and we examine the conjecture that such systems are dense. Then we present the program for a topological classification of Kleinian groups via 3-manifolds, along with hints of a parallel theory for rational maps. The Hausdorff dimension, ergodic theory and local connectivity of limit sets and Julia sets are also discussed. Finally we describe the role of renormalization in recent progress on the classification of quadratic polynomials.

Notes and references are collected at the end.

2 Finiteness theorems and Teichmüller space

Definitions. A *Kleinian group* Γ is a discrete subgroup of the conformal automorphism group $\mathrm{Aut}(\widehat{\mathbb{C}})$ of the Riemann sphere. We make the essential assumption that Γ is finitely generated. For convenience, we also assume Γ is torsion-free, orientation-preserving, and every abelian subgroup has infinite index. Elements of Γ can be represented concretely as Möbius transformations $\gamma(z) = (az + b)/(cz + d)$.

According to the dynamics of Γ, the sphere is partitioned into the *limit set* Λ and the *domain of discontinuity* Ω. The limit set is the locus of chaotic behavior. It is a compact, perfect set which can be defined as:

- the smallest closed Γ-invariant set with $|\Lambda| > 2$;
- the set of accumulation points of any orbit $\Gamma x \subset \widehat{\mathbb{C}}$;
- the closure of the set of repelling fixed points of $\gamma \in \Gamma$; or
- the set of points near which Γ does not form a normal family.[1]

On the other hand, Γ acts freely and conformally on $\Omega = \widehat{\mathbb{C}} - \Lambda$, preserving the *Poincaré metric* of constant curvature -1 provided by the uniformization theorem. Thus $X = \Omega/\Gamma$ is a disjoint union of hyperbolic Riemann surfaces.

A classical result in Kleinian groups is:

Theorem 2.1 (Ahlfors' finiteness theorem) *The quotient surface X is isomorphic to a finite union of compact Riemann surfaces with a finite number of points removed.*

In particular, every component of Ω has a nontrivial stabilizer; there is no *wandering component.*

Now let $f : \widehat{\mathbb{C}} \to \widehat{\mathbb{C}}$ be a *rational map*, whose iterates $f^n = f \circ f \circ \ldots \circ f$ are to be studied. We will assume the topological degree d of f is at least two, so $f(z) = P(z)/Q(z)$ for two relatively prime polynomials with $d = \max(\deg P, \deg Q) \geq 2$.

We again obtain a partition of the sphere into the *Julia set* J and the *Fatou set* Ω according to the dynamics of f. The Julia set is a closed, perfect set, defined as

- the smallest closed set with $f^{-1}(J) = J$ and $|J| > 2$;
- the closure of the set of repelling periodic points of f; or
- the set of points near which $\langle f^n \rangle$ does not form a normal family.

[1] A family of maps $\gamma_\alpha : U \to \widehat{\mathbb{C}}$ is *normal* if every sequence has a subsequence converging uniformly on compact subsets of U.

As f is proper, it maps components of Ω to components. Perceiving the analogy with Ahlfors' theorem, in the early 1980s Sullivan resolved a problem left open by Fatou and Julia in the 1920s by establishing:

Theorem 2.2 (No wandering domains) *Every component of the Fatou set eventually cycles, and there are only finitely many periodic components.*

One then has a fairly complete topological picture of the dynamics of f, since a component Ω_0 of the Fatou set of period p is of exactly one of the following types:

1. *An attractive basin:* there is an $x \in \Omega_0$, fixed by f^p, with
$$0 < |(f^p)'(x)| < 1,$$
attracting all points of Ω_0 under iteration of f^p. (This is the generic case).

2. *A superattractive basin:* as above, but x is a critical point of f^p, so $(f^p)'(x) = 0$.

3. *A parabolic basin:* there is a point x in $\partial\Omega_0$ with $(f^p)'(x) = 1$, attracting all points of Ω_0.

4. *A Siegel disk:* Ω_0 is conformally isomorphic to the unit disk, and f^p acts by an irrational rotation.

5. *A Herman ring:* Ω_0 is isomorphic to an annulus, and f^p acts by an irrational rotation.

Teichmüller theory. Underlying these finiteness theorems is a description of the Teichmüller space of a conformal dynamical system.

It is a remarkable fact that a *measurable* field of tangent ellipses of bounded eccentricity determines a complex structure on the sphere. This ellipse field is recorded by a $(-1,1)$-form $\mu(z)d\overline{z}/dz$ with $\|\mu\|_\infty < 1$. Since the sphere admits a unique complex structure, there is a homeomorphism $\phi : \widehat{\mathbb{C}} \to \widehat{\mathbb{C}}$ such that μ is the pullback of the standard structure. The conditions on μ imply that ϕ is a *quasiconformal map*: it has derivatives in L^2, and $\overline{\partial}\phi = \mu\partial\phi$.

The *Teichmüller space* Teich(Γ) classifies Kleinian groups Γ' equipped with quasiconformal conjugacies ϕ to Γ. Points in this space are determined by Γ-invariant complex structures μ on $\widehat{\mathbb{C}}$. The discussion is simplified by Sullivan's:

Theorem 2.3 (No invariant line fields) *The limit set of a finitely generated Kleinian group carries no measurable Γ-invariant field of tangent lines.*

Thus any Γ-invariant μ vanishes on the limit set (otherwise the major axes of the corresponding ellipse field would give invariant lines). On the other hand, $\mu|\Omega$ descends to a complex structure on the quotient Riemann surface X, yielding:

Corollary 2.4 *There is an isomorphism between the Teichmüller space of* Γ *and that of its quotient Riemann surface* $X = \Omega/\Gamma$ *:*

$$\mathrm{Teich}(\Gamma) \cong \mathrm{Teich}(X).$$

Given a group G, let

$$\mathcal{V}(G) = \mathrm{Hom}(G, \mathrm{Aut}(\widehat{\mathbb{C}}))/\mathrm{conjugation}$$

denote the *representation variety* of G, and let $AH(G) \subset \mathcal{V}(G)$ be the subspace of discrete, faithful representations.

A conjugacy ϕ from Γ to Γ' determines an isomorphism of groups, so we have a map

$$\delta : \mathrm{Teich}(\Gamma) \to AH(\Gamma) \subset \mathcal{V}(\Gamma).$$

Since Γ is finitely generated, $\mathcal{V}(\Gamma)$ is finite-dimensional; the fibers of δ are discrete, and therefore the Teichmüller space of X is also finite-dimensional. In other words, each component of X has finitely many moduli, so it is isomorphic to a compact Riemann surface with a finite number of points removed.

This completes the proof of Ahlfors' finiteness theorem, except for the fact that infinitely many components of X might be triply-punctured spheres, which have no moduli; this possibility was ruled out by Greenberg.

The rigidity of the triply-punctured sphere is equivalent to the vanishing of any holomorphic quadratic differential with at worst simple poles at the punctures. By considering differentials of higher order, one obtains:

Theorem 2.5 (Bers) *If* Γ *is generated by* N *elements, then the hyperbolic area of* Ω/Γ *is bounded by* $4\pi(N-1)$.

The line fields problem for rational maps. The absence of invariant line fields is equivalent to the assertion that a finitely generated Kleinian group is quasiconformally *rigid* on its limit set. In particular, if $\Lambda = \widehat{\mathbb{C}}$ then Γ has a trivial Teichmüller space — it admits no deformations.

A literal translation of this result is false for rational maps, due to the Lattès examples.

Let $T = \mathbb{C}/L$ be a complex torus, and let $F : T \to T$ be the map $F(x) = nx$ for $n > 1$. The quotient of T by the equivalence relation $x \sim -x$ is the Riemann sphere; the quotient map $\wp : X \to \widehat{\mathbb{C}}$ can be given by the Weierstrass $\wp$-function, which presents T as a twofold cover of the sphere branched over four points. Since $F(-x) = -F(x)$, the dynamical system F descends to a rational map f such that the diagram

$$\begin{array}{ccc} T & \xrightarrow{F} & T \\ \wp\downarrow & & \wp\downarrow \\ \widehat{\mathbb{C}} & \xrightarrow{f} & \widehat{\mathbb{C}} \end{array}$$

commutes.

Repelling periodic points for F are dense on the torus, so the Julia set of f is the whole sphere. The map $x \mapsto nx$ preserves the family of horizontal lines in the plane, so F has an invariant line field on T (tangent to the foliation by parallel closed geodesics.) This line field descends to the Riemann sphere (with singularities at the four branch points) to give an invariant line field for f.

The existence of this line field reflects the fact that, although $J(f) = \widehat{\mathbb{C}}$, f is not ***rigid***: it can be deformed by changing the shape of the torus T.

A fundamental problem is to show these are the only examples:

Conjecture 2.6 (No invariant line fields for rational maps) *The Julia set of a rational map f carries no invariant line field, except if f is a Lattès example.*

We will see later (Corollary 3.7) that this conjecture implies the density of expanding dynamical systems.

The Teichmüller space of f. The space Teich(f) classifies rational maps g equipped with quasiconformal conjugacies ϕ to f. It maps with discrete fibers to the moduli space

$$\mathcal{V}_d = \mathrm{Rat}_d\,/\mathrm{conjugation}$$

of all rational maps of degree d, so it is finite-dimensional.

To describe it, we form an analogue of the quotient surface for a Kleinian group. Let $\widehat{J} \supset J$ denote the closure of the grand orbits of all critical points and periodic points of f.[2] Let $\widehat{\Omega} = \widehat{\mathbb{C}} - \widehat{J}$; then $f : \widehat{\Omega} \to \widehat{\Omega}$ is a covering map, hence a ***local isometry*** for the Poincaré metric. Decompose $\widehat{\Omega}$ as $\Omega^{\mathrm{dis}} \sqcup \Omega^{\mathrm{fol}}$ depending on whether or not the grand orbits are locally discrete.

The space $X = \Omega^{\mathrm{dis}}/f$ is a union of Riemann surfaces; each attracting or parabolic basin contributes a torus or cylinder with punctures to X.

The space Ω^{fol} consists of points tending to Siegel disks, Herman rings or superattracting cycles. Each component of Ω^{fol} is a dynamically foliated punctured disk or annulus.

Since we do not know if the Julia set of f carries an invariant line field or not, we need to include a factor $M_1(J, f)$ consisting of all f-invariant μ on J with $\|\mu\| < 1$. Then we have:

[2]Two points x, y belong to the same ***grand orbit*** if $f^n(x) = f^m(y)$ for some $n, m > 0$.

Theorem 2.7 *The Teichmüller space of a rational map is canonically the product of the finite-dimensional Teichmüller space of the quotient surface $\Omega^{\rm dis}/f$ and a pair of finite polydisks:*

$$\mathrm{Teich}(f) \cong \mathrm{Teich}(\Omega^{\rm dis}/f) \times \mathrm{Teich}(\Omega^{\rm fol}, f) \times M_1(J, f).$$

Proof of no wandering domains. We can now complete the proof of Sullivan's finiteness theorem. First, a wandering domain would give rise to an open disk (possibly with finitely many punctures) as a component of $\Omega^{\rm dis}/f$; since the Teichmüller space of a disk is infinite-dimensional, this is a contradiction. Each attracting basin or Herman ring contributes at least one complex modulus to f, so the total number of periodic regions of these two types is bounded. The remaining types of regions are bounded by counting critical points and indifferent cycles.

Using quasiconformal surgery, one has the sharper estimate:

Theorem 2.8 (Shishikura) *Let N be the total number of cycles of periodic components in the Fatou set, H the number of cycles of Herman rings, and I the number of nonparabolic indifferent cycles in the Julia set. Then*

$$N + H + I \;\leq\; 2\deg(f) - 2.$$

Examples. Let X be a hyperbolic Riemann surface of finite area. Then X can be presented as the quotient of the unit disk Δ by the action of a *Fuchsian group* $\Gamma \subset \mathrm{Aut}(\Delta)$. By Schwarz reflection Γ acts on the sphere as a Kleinian group, with limit set $\Lambda = S^1$. Letting $\overline{X}$ denote the complex conjugate of X, we have $\Omega/\Gamma = X \sqcup \overline{X}$ and thus $\mathrm{Teich}(\Gamma) = \mathrm{Teich}(X) \times \mathrm{Teich}(\overline{X})$.

As one moves away from the Fuchsian locus while remaining in $\mathrm{Teich}(\Gamma)$, the limit set becomes a fractal curve of dimension strictly greater than one.[3] This curve can be geometrically complicated but it remains a *quasicircle*: that is, it is a Jordan curve such that for any $x, y \in \Lambda$, the smaller subarc $\Lambda(x, y)$ joining x to y satisfies

$$\operatorname{diam} \Lambda(x, y) \leq K|x - y|.$$

Finally when one reaches the boundary of $\mathrm{Teich}(\Gamma)$ in $AH(\Gamma)$, some points in Λ can become identified and the topology of the limit set abruptly changes.

The case where X is a once-punctured torus is particularly easy to compute with, because $\pi_1(X) \cong \mathbb{Z} * \mathbb{Z}$ is a free group; see Figure 2. The simplest type of point on the boundary of Teichmüller space is depicted in the final frame: a simple curve on the torus has been pinched to a *cusp*, splitting Ω into countably many components.

A proper holomorphic map or *Blaschke product* $f : \Delta \to \Delta$ is like a Fuchsian group; by Schwarz reflection again, it extends to a rational map on the sphere. If f has a fixed point in the disk then this point is attracting or superattracting and the Julia set $J = S^1$.

[3]Poincaré writes "Ces deux domaines sont séparés par une ligne L, si l'on peut appeler cela une ligne."

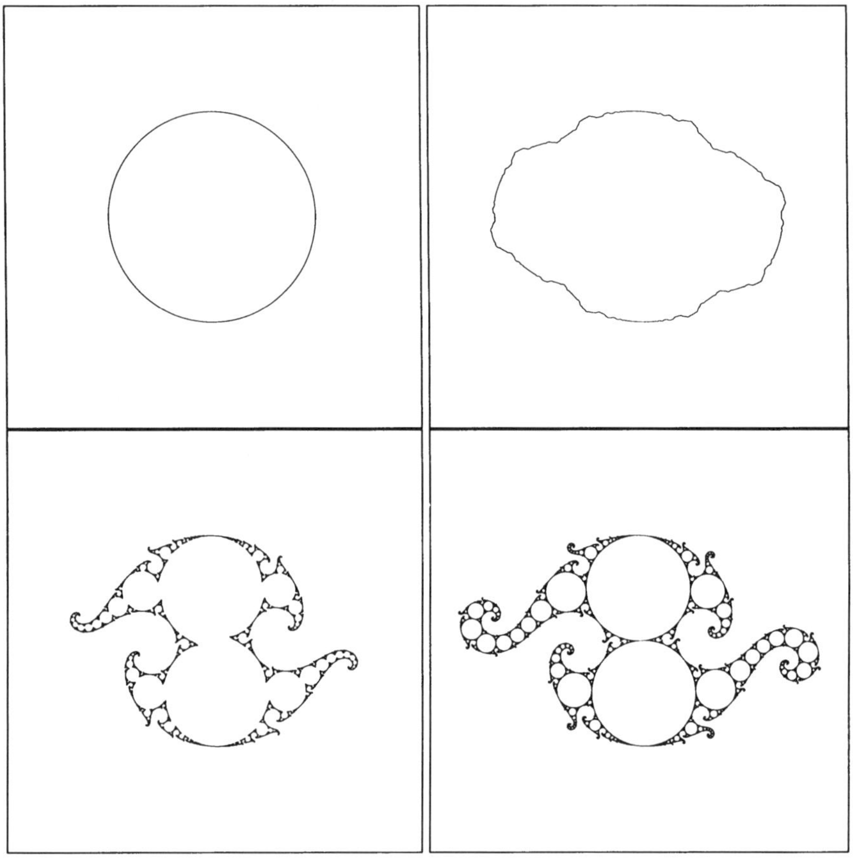

Figure 2. Limit sets.

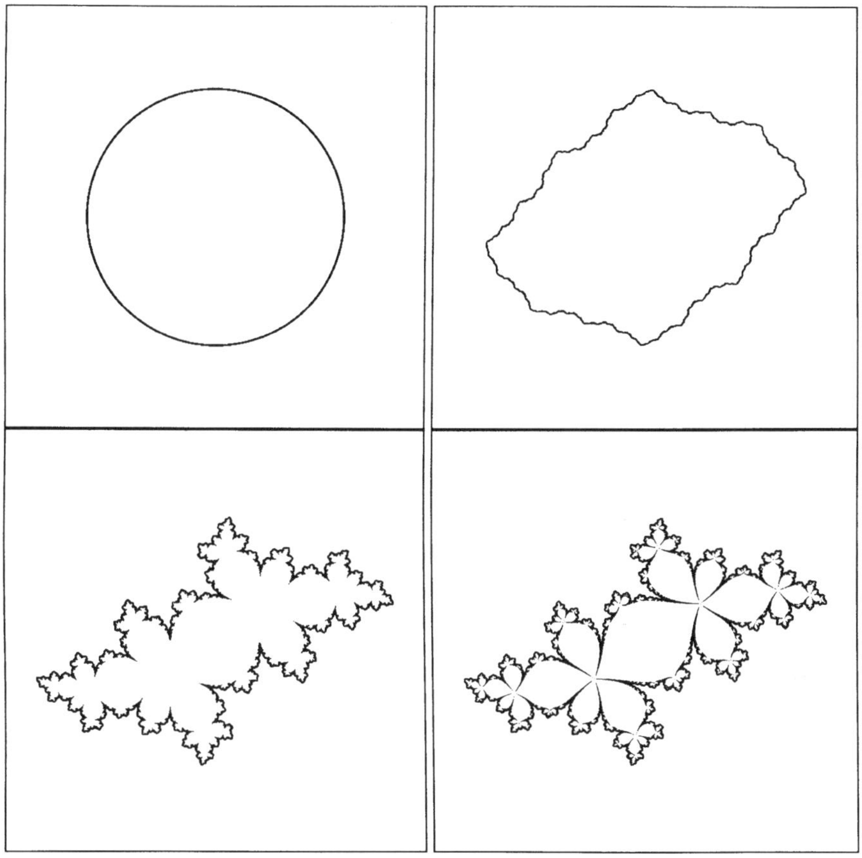

Figure 3. Julia sets.

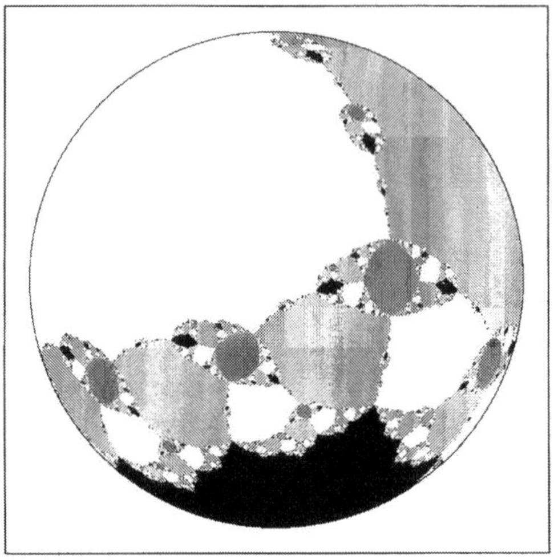

Figure 4. Newton's method for $4z^3 - 2z + 1$

The simplest Blaschke product is $f(z) = z^2$. For $|\lambda| < 1$ the maps $f_\lambda(z) = \lambda z + z^2$ are all quasiconformally conformally conjugate on their Julia sets, which are again fractal quasicircles. For $0 < |\lambda| < 1$, these maps belong to a single Teichmüller space; $z = 0$ is an attracting fixed point yielding a once-punctured torus as the quotient Riemann surface. Figure 3 depicts Julia sets for λ approaching $\exp(2\pi i 3/5)$; in the first three frames, $|\lambda| < 1$, while in in the final frame the attracting fixed point has become parabolic, splitting the Fatou set into countably many components, five of which meet at the origin. The pictures are centered at the critical point $z = -\lambda/2$.

Figure 4 depicts the dynamics of a more complicated rational map, namely Newton's method for a cubic polynomial. This map has superattracting cycles at the three roots of the polynomial, and another superattracting cycle of period two. (Initial guesses in the basin of this last cycle do not converge to a root.) Regions in the figure are shaded to indicate which of the four attractors they tend to.

3 Expanding dynamics

Crucial to the modern theory of Kleinian groups is the insight that the action of Γ extends isometrically to ***hyperbolic space*** $\mathbb{H}^3$. Thus Γ determines a ***Kleinian manifold***

$$\overline{M} = M \cup \partial M = (\mathbb{H}^3 \cup \Omega)/\Gamma$$

with a complete metric of constant negative curvature on its interior and a conformal structure on its boundary.

Conversely, let M be a 3-manifold with a metric of constant curvature -1; pursuant to our standing assumptions about Γ, we will always assume M is oriented and $\pi_1(M)$ is finitely generated. Then M gives rise to a Kleinian group, by letting $\Gamma = \pi_1(M)$ acting on the compactification of $\widetilde{M} \cong \mathbb{H}^3$ by the Riemann sphere.

Perhaps the most prominent gap in the dictionary is summed up by:

Question 3.1 *Is there a 3-dimensional geometric object naturally associated to a rational map f?*

The best behaved dynamical systems have *expanding* properties. For Kleinian groups, expansion is reflected in the geometry of M.

To describe this relation, let $\mathrm{hull}(\Lambda) \subset \mathbb{H}^3$ denote the *convex hull* of the limit set, i.e. the smallest convex set containing all geodesics with both endpoints in Λ. Let $K = \mathrm{hull}(\Lambda)/\Gamma \subset M$ be the *convex core* of M; it is the smallest convex set containing all the closed geodesics in M. We say M is *geometrically finite* if a unit neighborhood of K has finite volume. A *cusp* of M is a component of the thin part with parabolic fundamental group.

The following are equivalent:

- The convex core of M is compact.
- The Kleinian manifold $\overline{M}$ is compact.
- M is geometrically finite without cusps.
- For all $x \in \Lambda$ there is a $\gamma \in \Gamma$ such that $\|\gamma'(x)\| > 1$ with respect to the spherical metric.

In this case we say Γ is *expanding*, or *convex cocompact.*

The examples in Figure 2 are geometrically finite, but not expanding, because they have cusps. Limit sets of expanding groups appear in Figures 5 and 6 below.

For rational maps, the nonexpanding behavior is closely tied to the critical points of f. Let

$$P = \overline{\bigcup_{n>0,\, f'(c)=0} f^n(c)}$$

denote the *postcritical set* of f, i.e. the closure of the forward orbits of the critical points. Then the following are equivalent:

- $P \cap J = \emptyset$.
- There are no critical points or parabolic cycles in the Julia set.
- All critical points converge to attracting or superattracting cycles under iteration.
- For all $x \in J$ there is an $n > 0$ such that $\|(f^n)'(x)\| > 1$.

In this case we say f is *expanding.*[4] The Newton's method example in Figure 4 and the first three frames in Figure 3 are expanding.

Expanding dynamical systems have many robust features.

1. *Openness.* The space of expanding maps is open in $\mathcal{V}_d$, and the space of expanding Kleinian groups is open in $\mathcal{V}(G)$.

2. *Structural stability.* Any dynamical system close enough to an expanding one is quasiconformally conjugate to it on its limit set or Julia set.

3. *Finiteness.* An expanding dynamical system can be described by a finite amount of data. For a Kleinian group the topology of a convex submanifold suffices. For a rational map one may find a pair of finitely connected neighborhoods $U \subset V$ of the Julia set such that $f : U \to V$ is a covering map.

4. *Quasi-self-similarity.* Because of expansion, any small ball meeting the Julia set or limit set can be blown up to definite size with bounded distortion. Thus all features of J or Λ are replicated everywhere and at all scales.

5. *Fractal geometry.* Apart from the case where $\Lambda = \widehat{\mathbb{C}}$, the Julia set and limit set have Hausdorff dimension δ with $0 < \delta < 2$, and finite, positive Hausdorff measure in this dimension. The Hausdorff measure can be constructed dynamically as a weak limit of probability measures

$$\lim_{s \searrow \delta} \frac{1}{C_s} \sum_{\gamma(y)=x} \|\gamma'(y)\|^{-s} \delta_y$$

for Kleinian groups, and

$$\lim_{s \searrow \delta} \frac{1}{C_s} \sum_{f^n(y)=x} \|(f^n)'(y)\|^{-s} \delta_y$$

for rational maps; here x is any point with an infinite orbit outside the limit set or Julia set. (The constants C_s normalize the total mass to 1.)

If a rational map f is expanding, then there is a finite set A (the union of the attracting and superattracting cycles) such that $f^n(x)$ converges to this finite attractor for all x in the Fatou set (an open, dense full measure subset of the sphere.) Thus we have a simple picture of the asymptotic behavior of a typical point.

Experimentally, expanding groups and rational maps are ubiquitous. We can now state two central conjectures.

[4] We avoid using the more common terminology *hyperbolic*, to minimize confusion with hyperbolic manifolds.

Conjecture 3.2 *Let G be a torsion-free group which does not contain $\mathbb{Z} \oplus \mathbb{Z}$. Then convex compact groups are dense in $AH(G)$.*

Conjecture 3.3 *For any degree d, the expanding rational maps are dense in the space of all rational maps of degree d.*

The theory of *holomorphic motions* is a basic tool in the construction of quasiconformal conjugacies, which leads to the following pair of tantalizing results:

Theorem 3.4 (Mañé-Sad-Sullivan) *Structural stability is dense in* Rat_d.

Theorem 3.5 (Sullivan) *A structurally stable Kleinian group is expanding.*

To prove the density of convex cocompact groups, it would suffice to translate the first result to the setting of Kleinian groups. The problem is that for rational maps, one can work in the full moduli space of rational maps $\mathcal{V}_d$, while for Kleinian groups one must restrict attention to the subspace $AH(G)$ of discrete faithful representations in $\mathcal{V}(G)$. The topology of $AH(G)$ is not well-understood; indeed conjecture 3.2 is equivalent to the assertion that $AH(G)$ is the closure of its interior.

Generalizing the second result to the setting of rational maps would prove the density of expanding dynamics. Here one knows:

Theorem 3.6 *There is an open dense set $U \subset \mathcal{V}_d$ such that any f, g in the same component U_0 of U are quasiconformally conjugate on the whole sphere. Thus U_0 is uniformized by the Teichmüller space of f.*

By counting dimensions, one can show that a structurally stable rational map f with no invariant line fields on its Julia set must be expanding. This yields:

Corollary 3.7 *The no invariant line fields conjecture implies the density of expanding rational maps.*

4 Topology of hyperbolic 3-manifolds

We now turn to the *classification problem* for rational maps and Kleinian groups. Ultimately, one would like to

1. attach a *combinatorial invariant* to each dynamical system,
2. describe all invariants which arise, and
3. parameterize all systems with the same invariants.

The last goal is fulfilled by the Teichmüller theory discussed in §2, so it is enough to find a combinatorial invariant which determines the *quasiconformal conjugacy class* of a rational map or Kleinian group.

This section presents the conjectural classification of Kleinian groups. The measurable and topological dynamics on the sphere are also briefly discussed.

3-manifolds. The classification of a Kleinian group Γ up to quasiconformal conjugacy is the same as the classification of $M = \mathbb{H}^3/\Gamma$ up to quasi-isometry. A combinatorial invariant for M can be presented in terms of 3-dimensional topology.

To begin with, suppose Γ is expanding. Then the desired combinatorial invariant is simply the compact Kleinian manifold $N = \overline{M}$, regarded as a topological space. It is not hard to see that the topology of N determines the quasi-isometry type of M. Indeed, given an end of M diffeomorphic to $S \times [0, \infty)$, choose a hyperbolic metric $\rho(x)dx^2$ on S; then the model metric $e^t\rho(x)dx^2 + dt^2$ is quasi-isometrically correct on this end.

Now suppose M is geometrically finite. Some additional information must be included to specify the cusps of M. In this case the combinatorial invariant is a *pared manifold* (N, P), where N is compact and P is a submanifold of ∂N. The pair (N, P) is determined by the conditions (a) M is diffeomorphic to the interior of N and (b) $\overline{M}$ is diffeomorphic to $N - P$. The *parabolic locus* P contains every torus component of ∂N; the rest of P consists of a finite number of annuli on ∂N. We then have:

Theorem 4.1 (Marden) *The topology of the pared manifold (N, P) determines M up to quasi-isometry.*

Finally assume only that M is an orientable hyperbolic 3-manifold with $\pi_1(M)$ finitely generated. In this case the conjectural invariant is a triple (N, P, ϵ), where ϵ is the *ending lamination.* To even define this invariant, we need to first assume:

Conjecture 4.2 (Tameness) *A hyperbolic 3-manifold M with finitely generated fundamental group is* topologically tame*: that is, M is diffeomorphic to the interior of a compact manifold N.*

It is known that M is homotopy equivalent to a compact manifold, but examples like the Whitehead continuum show the ends of a 3-manifold with finitely generated fundamental group need not be products.

Assuming M is diffeomorphic to int(N), we again can locate a parabolic locus $P \subset N$ specifying the finite set of cusps. A cusp can be thought of as a simple curve on ∂N whose length in M has shrunk to zero. The ending lamination records a generalization of this phenomenon, in which an "irrational" simple curve has been pinched.

Given a surface S, endowed with a hyperbolic metric for convenience, a *lamination* $\lambda \subset S$ is a closed subset which is a union of disjoint simple geodesics. To construct the ending lamination, fix a diffeomorphism between $\operatorname{int} N$ and M, so that a neighborhood of P maps into the cusps of M. Consider any sequence of simple closed curves γ_n on a component S of $\partial N - P$ whose geodesic representatives in M converge to S. Then a subsequence converges, in a natural sense, to a lamination λ on S.[5]

The ending lamination ϵ is the union of all λ which arise this way.

Conjecture 4.3 (The ending lamination conjecture) *The pared manifold (N, P) and the ending lamination ϵ determine M up to quasi-isometry.*

This conjecture is open even for $N = S \times [0,1]$ where S is a compact surface. Recently (March 1995) Minsky has announced:

Theorem 4.4 *The ending lamination conjecture is true for*

$$(N, P) = (S \times I, \partial S \times I),$$

where S is a torus with a disk removed.

The space of laminations on a punctured torus is a circle (since a lamination is determined by a line in $H_1(S, \partial S, \mathbb{R})$). Bers defined a compactification of Teichmüller space using limits of quasifuchsian groups; by Minsky's result, the compactifying points for a punctured torus are determined by their laminations, and we have:

Corollary 4.5 *The Bers' boundary of the Teichmüller space of a punctured torus is a Jordan curve.*

Here are two more general results bearing on the tameness and ending lamination conjectures.

A map $f : S \to N$ of a surface into a 3-manifold is *incompressible* if $f_* : \pi_1(S, p) \to \pi_1(N)$ is injective for all $p \in S$. A compact 3-manifold has *incompressible boundary* if $\partial N \subset N$ is incompressible. An open 3-manifold M has *incompressible ends* if it contains an embedded compact submanifold with incompressible boundary, such that $N \hookrightarrow M$ is a homotopy equivalence.

Theorem 4.6 (Thurston, Bonahon) *Any hyperbolic 3-manifold with incompressible ends is topologically tame.*

Theorem 4.7 (Minsky) *Any two hyperbolic manifolds M_1 and M_2 with incompressible ends, injectivity radii bounded below and the same combinatorial invariant (N, ϵ) are quasi-isometric.*

[5] The precise definition of convergence is that there is a projective measured lamination $[\lambda_0]$ with support λ such that $[\gamma_n] \to [\lambda_0]$.

All of these results are based on a study of incompressible surfaces $f : S \to M$. The map f can always be deformed to a *pleated surface*, which induces a hyperbolic metric on S. Since the hyperbolic geometry of S is controlled (for example, the total area of S is $2\pi|\chi(S)|$, by Gauss-Bonnet), one obtains control on the geometry of M near $f(S)$. By constructing families of pleated surfaces exiting the ends of M, one obtains global estimates for its 3-dimensional geometry and topology.

Constructing hyperbolic manifolds. To have a more complete picture, we need to describe which combinatorial invariants can arise. To begin with, the data already discussed satisfies a few topological constraints.

- Since the universal cover of M is $\mathbb{H}^3$, N is *irreducible* — every embedded S^2 bounds a ball.
- By the classification of cusps, N is *atoroidal* — every $\mathbb{Z} \oplus \mathbb{Z}$ subgroup of $\pi_1(N)$ comes from a torus boundary component.
- The parabolic locus P is incompressible, it contains every torus component of ∂N, and any incompressible cylinder $f : (S^1 \times I, S^1 \times \partial I) \to (N, P)$ is homotopic into P.

Let us define a general pared manifold to be a pair (N, P) with the properties above. A 3-manifold is *Haken* if it can be built up from 3-balls inductively, by gluing along incompressible submanifolds of the boundary.

We can now state a remarkable existence theorem:

Theorem 4.8 (Thurston) *Every Haken pared manifold (N, P) can be realized by a geometrically finite hyperbolic manifold.*

Any irreducible 3-manifold with nonempty boundary is Haken. Thus one has a complete characterization of expanding Kleinian groups whose limit set is not the whole sphere:

Corollary 4.9 *A 3-manifold N is homeomorphic to the Kleinian manifold $\overline{M}$ of a convex cocompact group with $\Lambda \neq \widehat{\mathbb{C}}$ if and only if:*

- *N is compact, irreducible, orientable,*
- *$\partial N \neq \emptyset$ and*
- *$\pi_1(N)$ does not contain $\mathbb{Z} \oplus \mathbb{Z}$.*

Corollary 4.10 *For any Kleinian group Γ, $AH(\Gamma)$ contains a geometrically finite group.*

Proof. If $M = \mathbb{H}^3/\Gamma$ is closed, then it is geometrically finite; otherwise M has the homotopy type of an irreducible atoroidal manifold with boundary, so Thurston's theorem applies.

■

Unfortunately a characterization of *closed* hyperbolic manifolds is not yet known. Thurston's geometrization program for 3-manifolds includes as a special case:

Conjecture 4.11 *A closed irreducible orientable 3-manifold is hyperbolic if and only if its fundamental group is infinite and does not contain* $\mathbb{Z} \oplus \mathbb{Z}$.

By Mostow rigidity these remaining cases account for only countably many Kleinian groups up to conjugacy.

To conclude we mention what is known about the ending lamination ϵ. The triple (N, P, ϵ) has some topological constraints which we will not make precise; we simply call a triple *admissible* if it satisfies them. Building on work of Thurston, Ohshika established:

Theorem 4.12 *Let* (N, P) *be a pared manifold with nonempty, incompressible boundary. Then every admissible triple* (N, P, ϵ) *can be realized by a hyperbolic manifold* M.

When $P = \emptyset$ the proof is by approximating M by convex cocompact manifolds. This yields:

Corollary 4.13 *Let* N *be a compact, orientable irreducible 3-manifold with incompressible boundary, such that* $\pi_1(N)$ *does not contain* $\mathbb{Z} \oplus \mathbb{Z}$. *Then the ending lamination conjecture implies the density of convex cocompact groups in* $AH(\pi_1(N))$.

Ergodic theory of Kleinian groups. Since Γ preserves the Lebesgue measure class on the sphere, it determines a measurable dynamical system. Here are two conjectured properties of the dynamics on the limit set.

Conjecture 4.14 (Ahlfors' measure zero problem) *If the limit set* $\Lambda \neq \widehat{\mathbb{C}}$, *then it has measure zero.*

Conjecture 4.15 (Ergodicity of the geodesic flow) *If* $\Lambda = \widehat{\mathbb{C}}$, *then* Γ *acts ergodically on* $\widehat{\mathbb{C}} \times \widehat{\mathbb{C}}$. *Equivalently, the geodesic flow on the unit tangent bundle* $T_1(M)$ *is ergodic.*

These conjectures are stronger than the No Invariant Line Fields Theorem 2.3. Indeed, suppose Λ carries an invariant line field; then for $(x, y) \in \Lambda \times \Lambda$ we can use parallel transport along the hyperbolic geodesic from x to y to measure the angle $\theta(x, y) \pmod{\pi}$ between the lines at x and y. The function $\theta(x, y)$ is Γ-invariant and essentially nonconstant, so Γ does not act ergodically on $\Lambda \times \Lambda$.

A triumph of the three-dimensional point has been its contributions to these originally two-dimensional problems. Due to work of Thurston, Bonahon and Canary, one now knows:

Theorem 4.16 *If $M = \mathbb{H}^3/\Gamma$ is topologically tame, then the conjectures above are true: either Λ has measure zero, or Γ acts ergodically on $\widehat{\mathbb{C}} \times \widehat{\mathbb{C}}$.*

Because of Theorem 4.6, the hypothesis can be verified in many cases: for example, the theorem applies to any Kleinian group Γ which is not isomorphic to a free product $A * B$. But it is still unknown if Ahlfors' conjecture is true for a Kleinian group Γ isomorphic to $\mathbb{Z} * \mathbb{Z}$.

On the other hand, by work of Sullivan, Tukia and Bishop-Jones one now knows:

Theorem 4.17 *The limit set of a finitely generated Kleinian group Γ has Hausdorff dimension 2 if and only if Γ is geometrically infinite.*

Local connectivity of limit sets. It is interesting to try to reconstruct the *topological dynamics* of a Kleinian group acting on the sphere from its combinatorial invariant (N, P, ϵ). This is possible when the limit set is locally connected — that is, when it has a basis of connected open sets.

The following result is well-known to experts:

Theorem 4.18 *If Γ is geometrically finite, then every component of Λ is locally connected.*

Proof. The group Γ is built up from groups with connected limit sets by the combination theorems, as can be seen by decomposing $\overline{M}$ along compressing disks. Thus one can reduce to the case where Λ is connected. The diameters of the components of Ω tend to zero, by a theorem of Maskit, so it suffices to show the boundary of every component Ω_0 of Ω is locally connected. We can assume Γ stabilizes Ω_0. If Ω has just two components, the limit set is a quasicircle by a theorem of Maskit and we are done. Otherwise we have a geometrically finite surface group with accidental parabolics; the limit set is also locally connected in this case, by a result of Abikoff.

∎

Peano curves. By work of Cannon-Thurston and Minsky, one also knows local connectivity of Λ and a topological model for the action when $\Gamma \cong \pi_1(S)$ is a surface group with injectivity radius bounded below. Sphere-filling Peano curves arise as special cases: that is, when $\Lambda = \widehat{\mathbb{C}}$ there is a surjective continuous map $f : S^1 \to \widehat{\mathbb{C}}$, conjugating a Fuchsian action of $\pi_1(S)$ to the action of Γ.

5 Topology of rational maps

For rational maps, even a conjectural combinatorial classification has yet to be formulated. In this section we discuss two cases where a rich theory is developing: expanding maps and polynomials.

Expanding components. Let $\mathrm{Exp}_d \subset \mathcal{V}_d$ denote the open set of expanding rational maps of degree d. An *expanding component* U is a connected component of Exp_d.

Any two maps in the same expanding component are quasiconformally conjugate on their Julia sets. An expanding component is partitioned into strata, each parameterized by a Teichmüller space. There is a unique top-dimensional stratum, characterized as those $f \in U$ such that the critical points have disjoint, infinite forward orbits.

One would like an invariant to classify expanding components. Such an invariant can be formulated when the Julia set is connected.

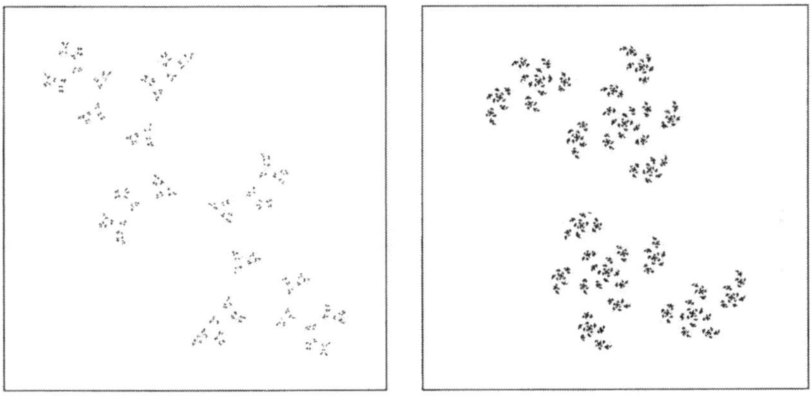

Figure 5. Cantor sets

Incompressible dynamics. Properties of a compact Kleinian manifold $\overline{M}$ are often reflected in the topology of its limit set. For example, $\overline{M}$ has incompressible boundary if and only if its limit set is *connected.* (Any essential loop in Ω represents an element of $\pi_1(\partial M)$ which is trivial in M.)

Figure 5 depicts the limit set for a *Schottky group* on the left, where $\overline{M}$ is a handlebody of genus two; this manifold has highly compressible boundary, and the limit set is a Cantor set. On the right is the Julia set for an quadratic polynomial whose finite critical point escapes to infinity; J is also a Cantor set, and it seems reasonable to regard this rational map as maximally compressible.

The classification of expanding components U is best understood in the case of *connected* Julia sets — that is, the incompressible case.

Indeed, if J is connected then there is a unique $f_0 \in U$, the *center* of U, such that all critical points eventually map to periodic critical points. The map f_0 also forms the unique zero-dimensional Teichmüller stratum in U. Then a finite combinatorial invariant for U is the simply *isotopy class* $[f_0]$ of the branched covering

$$f_0 : (S^2, P(f_0)) \to (S^2, P(f_0)).$$

(Here $P(f_0) = \bigcup_{n>0, f_0'(c)=0} f_0^n(c)$ is the postcritical set.)

Thurston's combinatorial characterization of rational maps

- shows $[f_0]$ determines the rational map f_0, and hence the expanding component U of which it is the center, uniquely; and
- describes all branched coverings which can arise.

To state this theorem, let $g : S^2 \to S^2$ be a topological branched covering of degree $d > 1$. The map g is *critically finite* if $|P(g)| < \infty$, where $P(g)$ is the union of the forward orbits of the branch points.

Consider a finite system Γ of disjoint simple closed curves γ on $S^2 - P(g)$, no two parallel, and none contractible or peripheral (i.e. enclosing a single point of $P(g)$). The system Γ is *g-invariant* if every component α of $g^{-1}(\gamma)$ is contractible, peripheral or isotopic to some $\delta \in \Gamma$. Define $A(\Gamma) : \mathbb{R}^\Gamma \to \mathbb{R}^\Gamma$ by the formula

$$A_{\delta\gamma} = \sum_{\alpha} \frac{1}{\deg(g : \alpha \to \gamma)}$$

where the sum is taken over components α of $g^{-1}(\gamma)$ which are isotopic to δ. Let $\lambda(\Gamma)$ denote the spectral radius of $A_{\delta\gamma}$. Then we have:

Theorem 5.1 (Thurston) *A critically finite branched cover $g : S^2 \to S^2$ is realized by an expanding rational map f if and only if:*

- *every branch point of g eventually maps to a periodic branch point, and*
- *$\lambda(\Gamma) < 1$ for every g-invariant curve system Γ.*

The rational map f is unique up to conformal conjugacy.

A slightly more general statement characterizes *all* critically finite (not necessarily expanding) rational maps.

The fact that $\lambda(\Gamma) < 1$ for a rational map is already an interesting observation. This can be seen by thickening the curves in Γ to annuli C_γ, and observing that the geometry of $P(f)$ gives an upper bound for the vector of conformal moduli $[m_\gamma] = [\text{mod}(C_\gamma)]$. The vector $[A_{\delta\gamma} m_\gamma]$ is a strict lower bound for the moduli of annuli which can be manufactured from the preimages of C_γ under f, so the Perron-Frobeníus eigenvalue of $A_{\delta\gamma}$ must be contracting.

Thurston's criterion is rather implicit. Even fixing the degree d of g and the size $p = |P(g)|$ of the postcritical set, there are usually infinitely many isotopy classes of branched coverings g, only finitely many of which represent rational maps. For a more satisfactory classification, one needs to solve the following combinatorial problem:

Problem 5.2 *Find an efficient way to organize and enumerate all critically finite rational maps of a given degree.*

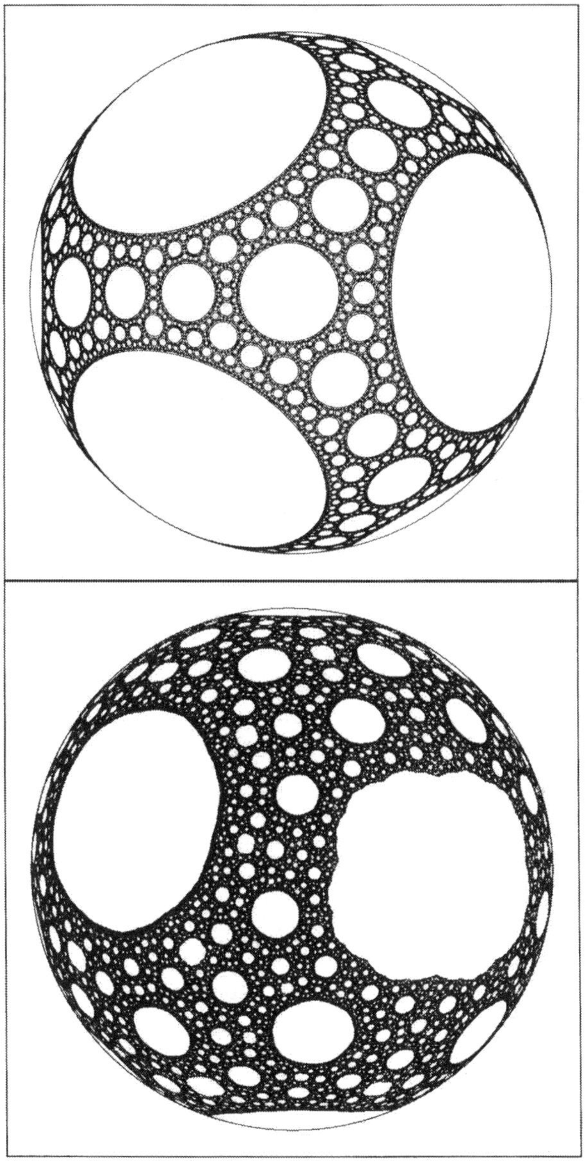

Figure 6. Sierpiński curves.

Acylindrical dynamics. A 3-manifold N with incompressible boundary is *acylindrical* if any map $f : (S^1 \times I, \partial(S^1 \times I)) \to (N, \partial N)$, injective on π_1, is homotopic into ∂N. A compact Kleinian manifold $\overline{M}$ is acylindrical if and only if its limit set Λ is a *Sierpiński curve.* This means Λ is obtained from the sphere by removing a countable dense set of open disks, bounded by disjoint Jordan curves whose diameters tend to zero.

Explicit examples of expanding rational maps with Sierpiński Julia sets have been given by Milnor-Tan and Pilgrim, and the notion of a "cylinder" for rational maps is under development.

Thurston showed that degeneration of Kleinian groups is related to cylinders, and that $AH(\pi_1(\overline{M}))$ is *compact* when $\overline{M}$ is acylindrical.

Question 5.3 *Let U be an expanding component with center f_0. If the Julia set of f_0 is a Sierpiński curve, is $\overline{U} \subset \mathcal{V}_d$ compact?*

At present, not one expanding component is known to have compact closure.

Figure 6 depicts the limit set of an acylindrical Kleinian group above, and the Julia set of an acylindrical expanding rational map below. By a result of Whyburn, these two fractal sets are homeomorphic.

A sequence of rational maps can degenerate by collapsing towards a map of lower degree. Morgan and Shalen have used $\mathbb{R}$-trees to study degenerations of hyperbolic manifolds, and Shishikura applied simplicial trees to some special degenerations of rational maps, so perhaps trees can help resolve the question above.

The general case. A compact 3-manifold with *compressible* boundary can be cut along a finite number of compressing disks to obtain incompressible pieces. Inverting this procedure, combination theorems of Klein and Maskit allow one to build up a Kleinian group with disconnected limit set from a finite number of subgroups with connected limit sets.

It should be possible to treat expanding rational maps with disconnected Julia sets by an analogous procedure.

Problem 5.4 *Develop decomposition and combination theorems for rational maps.*

One would like a systematic theory including existing constructions like *mating* and *tuning,* as well a theory for disconnected Julia sets — just as the Haken theory discusses cutting along general incompressible subsurfaces, not just disks. For example, some degree two rational maps can be expressed as matings in more than one way, just as a 3-manifold can contain several embedded incompressible surfaces.

Problem 5.5 *Give a combinatorial invariant for a general rational map f.*

One expects that every Kleinian group is a limit of geometrically finite groups of *fixed* topological type. In contrast to this, a rational map with every cycle repelling cannot belong to the closure of any expanding component $U \subset \mathcal{V}_d$. Thus it is challenging to predict even the ***topology*** of a sequence of expanding maps f_n converging to f.

A combinatorial invariant for rational maps of degree two near a Lattès example has been developed by Bernard.

Polynomials. Using laminations, one can formulate a combinatorial invariant for many polynomials.

Let $f(z) = z^d + a_2 z^{d-2} + \ldots + a_d$ be a monic centered polynomial of degree $d > 1$ with connected Julia set. The set of all such polynomials forms the *connectedness locus* $\mathcal{C}_d$. The coefficients $(a_2, \ldots, a_d)$ of f determine an embedding of $\mathcal{C}_d$ as a compact, connected subset of $\mathbb{C}^{d-1}$. Any polynomial with connected Julia set is conformally conjugate to one in $\mathcal{C}_d$.

The *filled Julia set* $K(f) \subset \mathbb{C}$ is the union of $J(f)$ and the bounded components of the Fatou set; it is compact set bounded by $J(f)$. The complement $\widehat{\mathbb{C}} - K(f)$ is the basin of attraction of the point at infinity. By the Riemann mapping theorem, there is a unique conformal map

$$\phi_f : (\mathbb{C} - \overline{\Delta}) \to (\mathbb{C} - K(f))$$

such that $\phi_f(z)/z \to 1$ as $z \to \infty$. The map ϕ_f conjugates z^d to f; that is, that is, $f(\phi_f(z)) = \phi_f(z^d)$.

This conjugacy facilitates the encoding of the combinatorics of f by a lamination. As can be seen in Figure 3, as f varies in $\mathcal{C}_d$ the topology of the Julia set $J(f)$ can change from a circle to a circle with many points identified. The lamination records these identifications.

Let $S^1 = \mathbb{R}/\mathbb{Z}$, and identify S^1 with the boundary of the unit disk Δ via the map $t \mapsto \exp(2\pi i t)$. For $t \in S^1$, let

$$R_t = \{\phi_f(r \exp(2\pi i t)) \ : \ 1 < r < \infty\}$$

be the *external ray* at angle t. Clearly $f(R_t) = R_{dt}$. The ray R_t *lands* at $p \in J(f)$ if

$$\lim_{r \searrow 1} \phi_f(r \exp(2\pi i t)) = p.$$

It is known that R_t lands for every rational angle t, and if t is periodic under $t \mapsto dt$, then the landing point p is a parabolic or repelling periodic point for f.

A *lamination* $\lambda \subset S^1 \times S^1$ is an equivalence relation such that the convex hulls of distinct equivalence classes are disjoint. The ***rational lamination*** $\lambda_{\mathbb{Q}}(f)$ is defined by $t \sim t'$ if $t = t'$, or if t and t' are rational and the external rays R_t and $R_{t'}$ land at the same point in the Julia set $J(f)$. (See Figure 7 for an example where $\{1/7, 2/7, 4/17\}$ is a single equivalence class under $\lambda_{\mathbb{Q}}(f)$.)

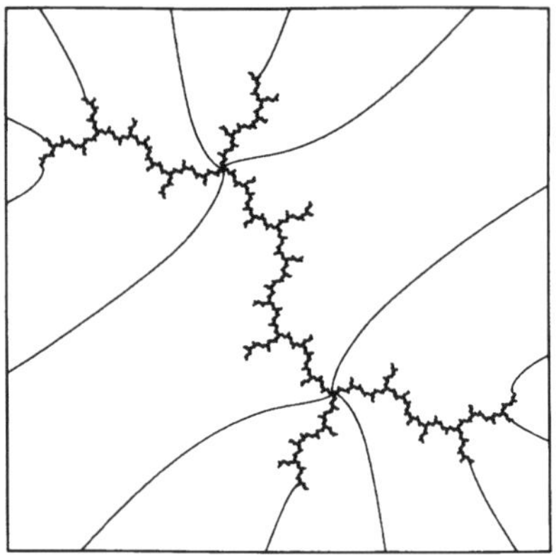

Figure 7. External rays for $f(z) = z^2 + i$.

A map $f \in C_d$ with no indifferent cycles is *combinatorially rigid* if it is determined up to quasiconformal conjugacy on its Julia set by its rational lamination. More precisely, for any $g \in C_d$ with no indifferent cycles such that $\lambda_{\mathbb{Q}}(g) = \lambda_{\mathbb{Q}}(f)$, the conformal map

$$\phi_g \circ \phi_f^{-1} : (\mathbb{C} - K(f)) \to (\mathbb{C} - K(g))$$

should extend to a quasiconformal map on the whole Riemann sphere.

We can now state:

Conjecture 5.6 (Combinatorial rigidity) *Every polynomial with connected Julia set and no indifferent cycle is combinatorially rigid.*

One also conjectures a relative version of the density of expanding rational maps.

Conjecture 5.7 *Expanding polynomials are dense in the connectedness locus C_d and in the space of all polynomials of degree d.*

Measure of Julia sets. A direct translation the Ahlfors measure zero conjecture to the setting of rational maps appeared plausible for many years. Recently however the following result has been announced:

Theorem 5.8 (Nowicki-van Strien) *Let $d > 0$ be an even integer, and let $f(z) = z^d + c$ where $c \in \mathbb{R}$. Suppose the closest returns of $f^n(0)$ to the origin occur along the Fibonacci sequence $n = 1, 2, 3, 5, 8, \ldots$. Then for d large enough, the Julia set of f has positive measure.*

A pair of Fibonacci maps can also be combined to give a polynomial which is not ergodic on its Julia set. On the other hand, Ahlfors' conjecture was motivated by the problem of invariant complex structures on the limit set, and these Fibonacci maps carry no invariant line fields.

Question 5.9 *When $J(f)$ has positive measure, does it have finitely many ergodic components? Is their number bounded by the number of critical points?*

Though one expects the expanding maps to be open and dense in $\mathcal{V}_d$, they do not fill out the space of rational maps in the sense of measure. In fact a random rational map has a definite chance of being ergodic:

Theorem 5.10 (Rees) *The set of rational maps such that $J(f) = \widehat{\mathbb{C}}$ and f acts ergodically on the sphere has positive measure in the space of all rational maps of degree d.*

Local connectivity of Julia sets. If $f \in \mathcal{C}_d$ and its Julia set is locally connected, then $J(f)$ is homeomorphic to S^1/λ for some lamination λ. However, it is known that there are Julia sets which are *not* locally connected. A concrete example is provided by $f(z) = e^{2\pi i t}z + z^2$, where θ is an irrational number which is very well approximated by rationals — e.g. $t = \sum 1/a_n$ where $\{a_1, a_2, a_3, \dots\} = \{2, 2^2, 2^{2^2}, \dots\}$.

Expanding rational maps are better behaved. A rational map f is *geometrically finite* if every critical point either has a finite forward orbit, or converges to an attracting, superattracting or parabolic cycle. Any expanding map is geometrically finite, and we have:

Theorem 5.11 (Tan-Yin) *Every preperiodic component of the Julia set of a geometrically finite rational map is locally connected.*

6 Renormalization

Much progress towards the density of expanding dynamics and combinatorial rigidity has been recently achieved in the setting of quadratic polynomials. The methods also shed light on questions of local connectivity and measure. The very formulation of the results involves the idea of *renormalization*.

For quadratic polynomials $P_c(z) = z^2 + c$, the connectedness locus $\mathcal{C}_2$ is the same as the *Mandelbrot set* $M \subset \mathbb{C}$. The Mandelbrot set can also be described as the set of c such that $P_c^n(0)$ does not tend to infinity. By work of Douady, Hubbard and Yoccoz we have:

Theorem 6.1 *Let c belong to the Mandelbrot set M. Then either*

- *P_c has an indifferent cycle and M is locally connected at c; or*
- *P_c has no indifferent cycle, and M is locally connected at c if and only if P_c is combinatorially rigid.*

Corollary 6.2 (Douady-Hubbard) *Local connectivity of the Mandelbrot set implies the density of expanding maps in the space of quadratic polynomials.*

Proof. If M is locally connected then we have combinatorial rigidity in degree two.

Density of expanding dynamics can only fail if there is a component Q of $\operatorname{int}(M)$ consisting of a single quasiconformal conjugacy class, such that $f_0 \in Q$ has all finite cycles repelling and an invariant line field on its Julia set. The points $B_n \subset \partial Q$ with an indifferent cycle of period n are nowhere dense, so there is is a $g \in \partial Q$ with no indifferent cycle. The rational lamination changes only at parabolic points, so $\lambda_{\mathbb{Q}}(g) = \lambda_{\mathbb{Q}}(f_0)$.

Combinatorial rigidity would then imply g and f_0 are quasiconformally conjugate on their Julia sets. But then $J(g)$ would carry an invariant line field, contradicting the fact that $g \in \partial M$.

■

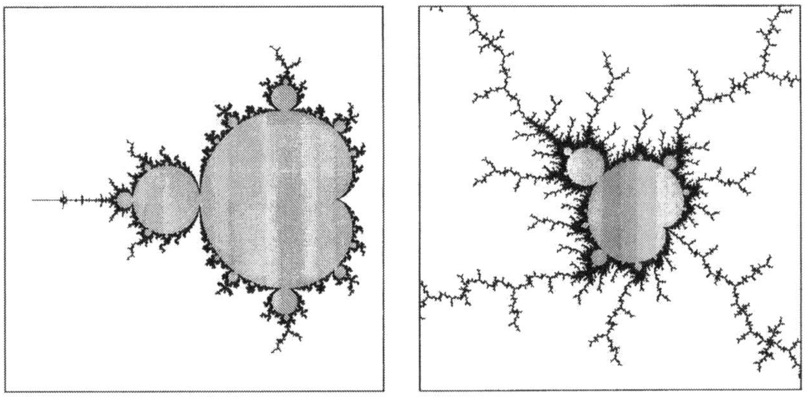

Figure 8. Small copies of M in M.

Quadratic-like maps. A remarkable feature of the Mandelbrot set is that it contains countably many small copies of itself (Figure 8). To explain these and other reappearances of M, Douady and Hubbard introduced the theory of ***polynomial-like maps.*** A polynomial-like map $f : U \to V$ is a proper holomorphic map between disks in the plane, such that $\overline{U}$ a compact subset of V. The filled Julia set of f is defined by $K(f) = \bigcap_1^\infty f^{-n}(V)$, and the Julia set $J(f)$ is defined as its boundary. The filled Julia set consists of the points which never escape from the domain of f under iteration. A basic theorem asserts that $f|K(f)$ is quasiconformally conjugate to $p|K(p)$ for some polynomial p.

Consider now a quadratic polynomial $f(z) = z^2 + c$ with connected Julia set. A *quadratic-like* map is a polynomial-like map of degree two. We say an iterate f^n is *renormalizable* if there are disks U_n and V_n, containing the critical point $z = 0$, such that $f^n : U_n \to V_n$ is a quadratic-like map with connected Julia set. (Note that f^n itself has degree 2^n.) If there are infinitely many such n, then f is *infinitely renormalizable.*

A point $s \in M$ is *superstable* of period p if $z = 0$ has period p under $P_s(z)$. Douady and Hubbard construct for each such s a *tuning homeomorphism* $M \to M_s \subset M$, written $c \mapsto s * c$, such that if $c \neq 1/4$, then $P^p_{s*c}(z)$ is renormalizable and conjugate on its filled Julia set to $P_c(z)$. Thus renormalization explains the small copies of M inside of M. Any infinitely renormalizable map lies inside a countable nest of small Mandelbrot sets.

Example. The two left frames in Figure 9 depict the filled Julia sets of $f_i(z) = z^2 + s_i$, $i = 1, 2$, where s_1 and s_2 are superstable of period two and three respectively; the latter is known as Douady's rabbit. The upper right frame depicts the filled Julia set for $f(z) = z^2 + s_1 * s_2$; here the critical point has period six, and f^2 is renormalizable. The final frame shows the filled Julia set of a renormalization $f^2 : U_2 \to V_2$; it is homeomorphic to the rabbit.

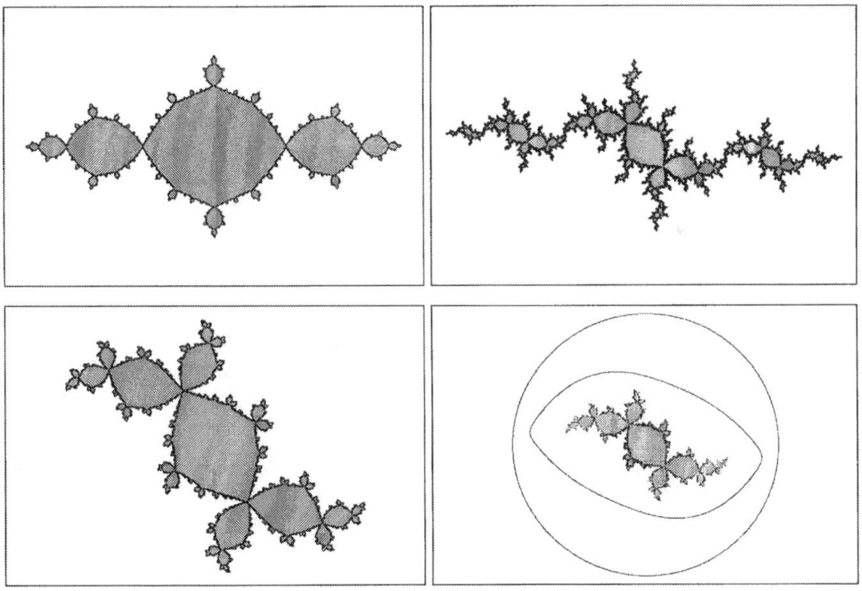

Figure 9. Renormalization.

We can now state two recent results towards the classification of quadratic polynomials.

Theorem 6.3 (Yoccoz) *Suppose $c \in M$ and $f(z) = z^2 + c$ has no indifferent cycle. Then either:*

- *f is combinatorially rigid, $J(f)$ is locally connected and M is locally connected at c, or*
- *f is infinitely renormalizable.*

Theorem 6.4 (Lyubich) *Combinatorial rigidity holds for uncountably many infinitely renormalizable quadratic polynomials $f(z) = z^2 + c$ with explicitly described combinatorial classes.*

Carrying Yoccoz's argument further, Lyubich and Shishikura proved independently:

Theorem 6.5 *If $f(z) = z^2 + c$ has no indifferent cycle and is only finite renormalizable, then $J(f)$ has measure zero.*

Renormalization operators. Renormalization is closely related to *universality* in dynamics.

To explain this phenomenon, for $c \in M \cap \mathbb{R} = [-2, 1/4]$ let $A_c \subset \mathbb{R}$ denote the set of accumulation points of $P_c^n(0)$. The set A_c often plays the role of an attractor for P_c over the reals; for example, it coincides with the attracting cycle if P_c has one.

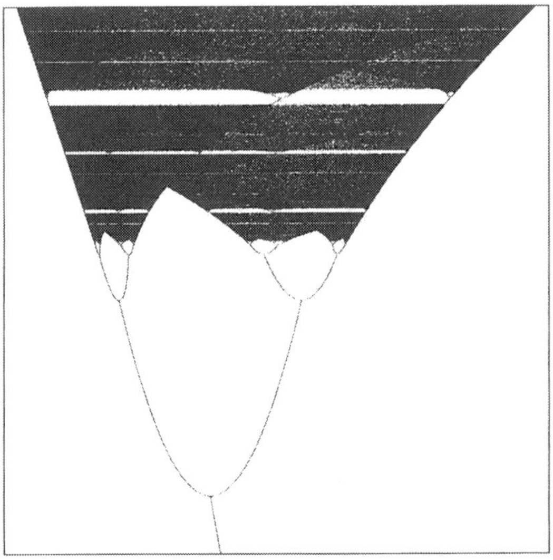

Figure 10. Bifurcation diagram.

A classical experiment is to plot A_c as a function of c. Figure 10 renders the set $\{(x, c) : x \in A_c, -0.6 < c < -2.0\}$. Initially A_c is a single attracting fixed point; as c decreases (moving vertically in the figure), the attractor undergoes a sequence of period doublings, accumulating to a Cantor set at the parameter $c_F \approx -1.401155189$. In physical terms c_F marks a ***phase transition*** from periodicity to chaos in this family. The ***Feigenbaum polynomial*** $f(z) = z^2 + c_F$ is infinitely renormalizable; moreover the renormalization of f^2 is topologically conjugate to f itself.

In other families of real dynamical systems one finds cascades of period doublings remarkably similar to the quadratic family: for example the Hausdorff dimension of A_c and the pace of period doubling are independent of the particular family examined, and thus ***universal.***

In the late 1970s Feigenbaum and Coullet-Tresser offered an explanation of this universality in terms of a fixed point for renormalization. Sullivan has recently given a rigorous proof of many aspects of this explanation; here is a precise statement in the language of quadratic-like mappings that follows from his work.

A quadratic-like map is *real* if $f(\overline{z}) = \overline{f(z)}$. Let $\mathcal{G}$ denote the set of real quadratic-like maps f with connected Julia sets, up to affine conjugacy near $K(f)$. We say $[f_n] \to [f]$ in $\mathcal{G}$ if there are representatives converging uniformly on a neighborhood of $K(f)$. Let $\mathcal{G}^{(p)} \subset \mathcal{G}$ denote the set of f such that f^p is renormalizable. The *renormalization operator*

$$\mathcal{R}_p : \mathcal{G}^{(p)} \to \mathcal{G}$$

sends f to the germ of its quadratic-like restriction $f^p : U_p \to V_p$. We may now state:

Theorem 6.6 (Real fixed points of renormalization) *Suppose the critical point $z = 0$ has period p under $z \mapsto z^2 + s$, where $s \in \mathbb{R}$. Then:*

1. *The limit $s^\infty = \lim s^{*n}$ exists, and is the unique real fixed point of tuning by s;*
2. *if $f(z) = z^2 + s^\infty$, then $\mathcal{R}_p^n(f) \to F$, where F is a fixed-point of renormalization; and*
3. *$\mathcal{R}_p^n(g) \to F$ for any $g \in \mathcal{G}$ which is topologically conjugate to f on $K(g) \cap \mathbb{R}$.*

Corollary 6.7 *The real fixed points of $\mathcal{R}_p$ are classified by the real superstable points $s \in M$ of period p.*

Example. For the Feigenbaum polynomial $f(z) = z^2 + c_F$, we have $c_F = (-1)^\infty$, and after rescaling the renormalizations of f^{2^n} converge to a solution F of the functional equation $F \circ F(z) = \alpha F(\alpha^{-1} z)$.

3-manifolds which fiber over the circle. There is an intriguing analogy between the renormalization operators and the mapping class group of a surface. Consider a closed surface S of genus 2 or more, and let $\psi : S \to S$ be a pseudo-Anosov mapping class. Then ψ acts on $AH(\pi_1(S))$, and a fixed point corresponds to a discrete faithful representation satisfying the functional equation

$$\rho \circ \psi_*(\gamma) = \alpha^{-1}\rho(\gamma)\alpha$$

for some $\alpha \in \mathrm{Aut}(\widehat{\mathbb{C}})$. Although $\Gamma = \rho(\pi_1(S))$ is geometrically infinite, the group generated by Γ and α is *expanding*, *cocompact*, and gives a hyperbolic structure on the closed 3-manifold T_ψ which fibers over the circle with monodromy ψ.

Similarly, a fixed point for the renormalization operator $\mathcal{R}_p$ gives a solution to the functional equation

$$F^p(z) = \alpha^{-1}F(\alpha(z)),$$

and while the critical point of F is recurrent, the dynamical system generated by F and $z \mapsto \alpha z$ has good expansion properties.

To complete the classification of quadratic polynomials, establish local connectivity of the Mandelbrot and prove the density of expanding dynamics, it remains only to prove combinatorial rigidity in the infinitely renormalizable case. Although Theorem 6.4 handles many cases, even the following is still open:

Question 6.8 *Is the Feigenbaum polynomial combinatorially rigid?*

Understanding the expansion implicit in renormalization may be useful in resolving these questions.

7 Notes and references.

§1 Introduction. Classical references for conformal dynamical systems as seen by Poincaré, Fatou and Julia include [Poin], [Julia], and [Fatou]. The dictionary between rational maps and Kleinian groups was introduced in [Sul4] and is discussed at length in [Sul2]. The scope of conformal dynamics is broader than our focus here; it includes real one-dimensional dynamics and codimension-one holomorphic foliations, as well iterated entire functions, correspondences, and higher-dimensional Kleinian groups.

§2 Finiteness theorems and Teichmüller space. Ahlfors' finiteness theorem and the Ahlfors measure zero problem appear in [Ah]; the number of triply-punctured spheres is bounded in [Gr]. Sullivan's no wandering domains theorem appears in [Sul4] (see also [Bers3]), and the Teichmüller space of a holomorphic dynamical system is developed in [McS]. Bers' area theorem and Shishikura's bounds are in [Bers1] and [Shi1].

§3 Expanding dynamics. A candidate 3-dimensional object for a rational maps is constructed in [LM]. A prototype for the conjecture on the density of convex cocompact groups is [Bers2, Conjecture II].

The relation of hyperbolicity to structural stability is discussed for Kleinian groups in [Sul5], and for rational maps in [MSS], [McS] and [Mc4]. Theorem 3.6 and its Corollary can be found in [McS].

Density of hyperbolicity has been announced for the family of entire functions $f_\lambda(z) = \lambda \tan z$ [KK].

§4 Topology of hyperbolic 3-manifolds. Marden's work on the topology of geometrically finite 3-manifolds appears in [Mar]. The basic reference for the ending lamination, pleated surfaces, topological tameness and their relation to the Ahlfors conjecture is [Th1]. Important further developments appear in [Bon], [Oh], [Min1], [Can] and [Min2]. The Whitehead continuum (whose complement gives a manifold with a wild end) is discussed in [Rol, §3I].

Thurston's geometrization theorem for Haken manifolds is discussed in [Mor1], [Th2] and [Th3]; a new approach to the the proof is surveyed in [Mc3].

The Hausdorff dimension of limit sets is studied in [Sul1], [Sul3], [Tu] and [BJ].

The local connectivity of geometrically finite limit sets (Theorem 4.18) is assembled from [AM], [Mask2, Theorem 6], [Mask1, Theorem 2] and [Ab, Theorem 3]; see also [Fl]. A more detailed proof is given in [AM]. Equivariant Peano curves are discussed in [Th2] and [CT]; [Min1] contains more general local connectivity results.

§5 Topology of rational maps. Thurston's characterization of critically finite rational maps appears in [Th4], [DH2]; it is compared to the geometrization of 3-manifolds in [Mc2]. The problem of enumerating critically finite *polynomials* is addressed in [BFH] and [Po]; [Rees2] studies the combinatorics of rational maps of degree two.

Sierpiński Julia sets and components of the space of expanding maps are discussed in [Mc1, §4] (which includes a version of Question 5.3), [Mil2], [Pil], [Mak] and [Pet]. Sierpiński's curve first appears in [Sie]; its uniqueness is proved in [Why]. For the role of trees in conformal dynamics, see [Mor2], [Shi2].

Tuning, mating and other surgery constructions can be found in [Dou1] and [Tan]. Combinatorics near a degree two Lattès example is developed in [Ber].

The landing of external rays is discussed in [Mil3, §18], [Hub] and [Mc4, §6]. A proof that C_d *cellular*, and therefore connected, appears in [Lav].

A polynomial Julia set of positive measure is constructed in [NS], and an example which is not ergodic on its Julia set appears in [St]. Positive measure sets of ergodic maps are constructed in [Rees1].

For local connectivity of geometrically finite Julia sets, see [TY].

§6 Renormalization. External angles for Julia sets and the Mandelbrot set are studied in [DH1], which includes a proof that local connectivity of M implies the density of expanding polynomials. A more recent treatment appears in [Dou2]. See [Mil1] for the tuning operators and self-similarity of the Mandelbrot set.

Background on complex renormalization can be found in [Mc4]; for a historical collection including experimental work on physical systems, see [Cvi]. Work of Yoccoz on the local connectivity of $J(f)$ and M is presented in [Yoc], [Mil4] and [Hub]. These breakthroughs were foreshadowed by work of Branner and Hubbard on cubic polynomials with disconnected Julia sets [BH]. Theorem 6.5 on the measure of Julia sets appears in [Lyu2]. For Lyubich's results on infinitely renormalizable quadratic polynomials, see [Lyu1].

Sullivan's work on renormalization appears in [Sul6], and is expanded upon in [MeSt]. A parallel approach to renormalization and 3-manifolds which fiber over the circle is the subject of [Mc5]. Theorem 6.6 is implicit in Sullivan's work and stated explicitly in [Mc5, Thm 7.16].

The illustrations. Figure 2 and the left frame of Figure 5 were drawn using a computer program written by David Wright [MMW]. The remaining pictures were produced by the author. In Figure 6, the Kleinian group is generated by reflections through the sides of a regular stunted ideal tetrahedron with dihedral angles $2\pi/7$; the rational map is $f(z) \approx -0.695620(z-1)^2(z+2)/(3z-2)$. The second frame of Figure 8 is a blowup around $c \approx -0.160 + 1.034i$, a superstable point of period 4. The parameters for Figure 9 are $s_1 = -1$, $s_2 \approx -0.22561 + 0.744862i$ and $s_1 * s_2 \approx -1.13000 + 0.24033i$.

Bibliography

[Ab] W. Abikoff. On boundaries of Teichmüller spaces and on Kleinian groups: III. *Acta Math.* **134**(1975), 211–237.

[AM] W. Abikoff and B. Maskit. Geometric decompositions of Kleinian groups. *Amer. J. Math* **99**(1977), 687–698.

[Ah] L. Ahlfors. Finitely generated Kleinian groups. *Amer. J. of Math.* **86**(1964), 413–429.

[AM] J. Anderson and B. Maskit. On the local connectivity of limit sets of Kleinian groups. *Preprint, 1995.*

[Ber] J. Bernard. *Dynamique des perturbations d'un example de Lattès.* Thesis, Univ. Paris-Sud Orsay, 1994.

[Bers1] L. Bers. Inequalities for finitely generated Kleinian groups. *J. d'Analyse Math.* **18**(1967), 23–41.

[Bers2] L. Bers. On boundaries of Teichmüller spaces and on kleinian groups: I. *Annals of Math.* **91**(1970), 570–600.

[Bers3] L. Bers. On Sullivan's proof of the finiteness theorem and the eventual periodicity theorem. *Am. J. Math.* **109**(1987), 833–852.

[BFH] Bielefeld, Y. Fisher, and J. Hubbard. The classification of critically preperiodic polynomials as dynamical systems. *J. AMS* **5**(1992), 721–762.

[BJ] C. J. Bishop and P. W. Jones. Hausdorff dimension and Kleinian groups. *Preprint.*

[Bon] F. Bonahon. Bouts des variétés hyperboliques de dimension 3. *Annals of Math.* **124**(1986), 71–158.

[BH] B. Branner and J. H. Hubbard. The iteration of cubic polynomials, Part II: Patterns and parapatterns. *Acta Mathematica* **169**(1992), 229–325.

[Can] R. D. Canary. Ends of hyperbolic 3-manifolds. *J. of the AMS* **6**(1993), 1–35.

[CT] J. W. Cannon and W. P. Thurston. Group invariant Peano curves. *Preprint.*

[Cvi] P. Cvitanović. *Universality in Chaos.* Adam Hilger Ltd, 1984.

[Dou1] A. Douady. Chirurgie sur les applications holomorphes. In *Proceedings of the International Congress of Mathematicians*, pages 724–738. American Math. Soc., 1986.

[Dou2] A. Douady. Descriptions of compact sets in C. In L. R. Goldberg and A. V. Phillips, editors, *Topological Methods in Modern Mathematics*, pages 429–465. Publish or Perish, Inc., 1993.

[DH1] A. Douady and J. Hubbard. *Étude dynamique des polynômes complexes.* Pub. Math. d'Orsay, 1984.

[DH2] A. Douady and J. Hubbard. A proof of Thurston's topological characterization of rational maps. *Acta Math.* **171**(1993), 263–297.

[Fatou] P. Fatou. Sur les équations fonctionnelles. *Bull. Sci. Math. France* **47-48**(1919-20), 161–271, 33–94, 208–314.

[Fl] W. J. Floyd. Group completions and limit sets of Kleinian groups. *Invent. math.* **57**(1980), 205–218.

[Gr] L. Greenberg. On a theorem of Ahlfors and conjugate subgroups of Kleinian groups. *Amer. J. Math* **89**(1967), 56–68.

[Hub] J. H. Hubbard. Local connectivity of Julia sets and bifurcation loci: three theorems of J.-C. Yoccoz. In L. R. Goldberg and A. V. Phillips, editors, *Topological Methods in Modern Mathematics*, pages 467–511. Publish or Perish, Inc., 1993.

[Julia] G. Julia. Mémoire sur l'itération des applications fonctionnelles. *J. Math. Pures et Appl.* **8**(1918), 47–245.

[KK] L. Keen and J. Kotus. Dynamics of the family $\lambda \tan z$. *Preprint 1994.*

[Lav] P. Lavaurs. *Systèmes dynamiques holomorphes: explosion de points périodiques paraboliques.* Thesis, Univ. Paris-Sud Orsay, 1989.

[Lyu1] M. Lyubich. Geometry of quadratic polynomials: Moduli, rigidity, and local connectivity. *Stony Brook IMS Preprint 1993/9.*

[Lyu2] M. Lyubich. On the Lebesgue measure of the Julia set of a quadratic polynomial. *Stony Brook IMS Preprint 1991/10.*

[LM] M. Lyubich and Y. Minsky. Laminations in holomorphic dynamics. *Stony Brook IMS Preprint 1994/20.*

[Mak] P. Makienko. Unbounded components in parameter space of rational maps. *ICTP Trieste Preprint IC/93-84.*

[MSS] R. Mañé, P. Sad, and D. Sullivan. On the dynamics of rational maps. *Ann. Sci. Éc. Norm. Sup.* **16**(1983), 193–217.

[Mar] A. Marden. The geometry of finitely generated kleinian groups. *Annals of Math.* **99**(1974), 383–462.

[Mask1] B. Maskit. On boundaries of Teichmüller spaces and on kleinian groups: II. *Annals of Math.* **91**(1970), 607–639.

[Mask2] B. Maskit. Intersection of component subgroups of Kleinian groups. In *Discontinuous Groups and Riemann Surfaces*, pages 349–367. Princeton Annals of Math Studies 79, 1974.

[Mc1] C. McMullen. Automorphisms of rational maps. In *Holomorphic Functions and Moduli I*, pages 31–60. Springer-Verlag, 1988.

[Mc2] C. McMullen. Rational maps and Kleinian groups. In *Proceedings of the International Congress of Mathematicians Kyoto 1990*, pages 889–900. Springer-Verlag, 1991.

[Mc3] C. McMullen. Riemann surfaces and the geometrization of 3-manifolds. *Bull. AMS* **27**(1992), 207–216.

[Mc4] C. McMullen. *Complex Dynamics and Renormalization.* Annals of Math Studies 135, Princeton University Press, 1994.

[Mc5] C. McMullen. *Renormalization and 3-Manifolds which Fiber over the Circle.* To appear, Annals of Math. Studies, Princeton University Press.

[McS] C. McMullen and D. Sullivan. Quasiconformal homeomorphisms and dynamics III: The Teichmüller space of a holomorphic dynamical system. *Preprint.*

[MeSt] W. de Melo and S. van Strien. *One-Dimensional Dynamics.* Springer-Verlag, 1993.

[Mil1] J. Milnor. Self-similarity and hairiness in the Mandelbrot set. In M. C. Tangora, editor, *Computers in Geometry and Topology*, pages 211–259. Lect. Notes Pure Appl. Math., Dekker, 1989.

[Mil2] J. Milnor. Geometry and dynamics of quadratic rational maps. *Experimental Math.* **2**(1993), 37–83.

[Mil3] J. Milnor. Dynamics in one complex variable: Introductory lectures. *Stony Brook IMS Preprint 1990/5.*

[Mil4] J. Milnor. Local connectivity of Julia sets: Expository lectures. *Stony Brook IMS Preprint 1992/11.*

[Min1] Y. Minsky. On rigidity, limit sets, and end invariants of hyperbolic 3-manifolds. *J. of the AMS* **7**(1994), 539–588.

[Min2] Y. Minsky. The classification of punctured torus groups. *Preprint.*

[Mor1] J. Morgan. On Thurston's uniformization theorem for three dimensional manifolds. In *The Smith Conjecture*, pages 37–125. Academic Press, 1984.

[Mor2] J. Morgan. Trees and hyperbolic geometry. In *Proceedings of the International Congress of Mathematicians Berkeley 1986*, pages 590–597. American Math. Soc., 1987.

[MMW] D. Mumford, C. McMullen, and D. Wright. Limit sets of free two-generator kleinian groups. *Preprint.*

[NS] T. Nowicki and S. van Strien. Polynomial maps with a Julia set of positive Lebesgue measure: Fibonacci maps. *Preprint, 1993.*

[Oh] K. Ohshika. Ending laminations and boundaries for deformation spaces of Kleinian groups. *J. London Math. Soc.* **42**(1990), 111–121.

[Pet] C. L. Petersen. No elliptic limits for quadratic maps. *Preprint, MAT-Report No. 1994-45.*

[Pil] K. Pilgrim. *Cylinders for Iterated Rational Maps.* Thesis, University of California, Berkeley, 1994.

[Poin] H. Poincaré. Mémoire sur les groupes kleinéens. *Acta Math.* **3**(1883), 49–92.

[Po] A. Poirier. On post critically finite polynomials, Part two: Hubbard trees. *Preprint Stony Brook IMS 1993/5.*

[Rees1] M. Rees. Positive measure sets of ergodic rational maps. *Ann. scient. Éc. Norm. Sup.* **19**(1986), 383–407.

[Rees2] M. Rees. A partial description of parameter space of rational maps of degree two: Part I. *Acta Math.* **168**(1992), 11–87.

[Rol] D. Rolfsen. *Knots and Links.* Publish or Perish, 1976.

[Shi1] M. Shishikura. On the quasiconformal surgery of rational functions. *Ann. Sci. Éc. Norm. Sup.* **20**(1987), 1–30.

[Shi2] M. Shishikura. Trees associated with the configuration of Herman rings. *Ergod. Th. & Dynam. Sys* **9**(1989), 543–560.

[Sie] W. Sierpiński. Sur une courbe cantorienne qui contient une image biunivoque et continue de toute courbe donnée. In *Oeuvres Choisies, t. II*, pages 107–119. PWN – Éditions sci. de Pologne, 1975.

[St] S. van Strien. Transitive rational maps which are not ergodic. *To appear, Ergod. Th. & Dynam. Sys.*

[Sul1] D. Sullivan. Growth of positive harmonic functions and Kleinian group limit sets of zero planar measure and Hausdorff dimension two. In *Geometry at Utrecht*, pages 127–144. Springer-Verlag Lecture Notes in Math. 894, 1981.

[Sul2] D. Sullivan. Conformal dynamical systems. In *Geometric Dynamics*, pages 725–752. Springer-Verlag Lecture Notes No. 1007, 1983.

[Sul3] D. Sullivan. Entropy, Hausdorff measures new and old and limit sets of geometrically finite Kleinian groups. *Acta Math.* **153**(1984), 259–277.

[Sul4] D. Sullivan. Quasiconformal homeomorphisms and dynamics I: Solution of the Fatou-Julia problem on wandering domains. *Annals of Math.* **122**(1985), 401–418.

[Sul5] D. Sullivan. Quasiconformal homeomorphisms and dynamics II: Structural stability implies hyperbolicity for Kleinian groups. *Acta Math.* **155**(1985), 243–260.

[Sul6] D. Sullivan. Bounds, quadratic differentials and renormalization conjectures. In F. Browder, editor, *Mathematics into the Twenty-first Century: 1988 Centennial Symposium, August 8-12*, pages 417–466. Amer. Math. Soc., 1992.

[Tan] Tan L. Matings of quadratic polynomials. *Ergod. Th. & Dynam. Sys* **12**(1992), 589–620.

[TY] Tan L. and Yin Y. Local connectivity of the Julia set for geometrically finite rational maps. *ENS Lyon Preprint UMPA-94-121.*

[Th1] W. P. Thurston. *Geometry and Topology of Three-Manifolds.* Princeton lecture notes, 1979.

[Th2] W. P. Thurston. Three-dimensional manifolds, Kleinian groups and hyperbolic geometry. *Bull. AMS* **6**(1982), 357–381.

[Th3] W. P. Thurston. Hyperbolic structures on 3-manifolds I: Deformations of acylindrical manifolds. *Annals of Math.* **124**(1986), 203–246.

[Th4] W. P. Thurston. On the combinatorics and dynamics of iterated rational maps. *Preprint.*

[Tu] P. Tukia. The Hausdorff dimension of the limit set of a geometrically finite Kleinian group. *Acta Math.* **152**(1984), 127–140.

[Why] G. T. Whyburn. Topological characterization of the Sierpiński curve. *Fund. Math.* **45**(1958), 320–324.

[Yoc] J.-C. Yoccoz. Sur la connexité locale des ensembles de Julia et du lieu de connexité des polynômes. *In preparation.*

Quantum cohomology and its associativity

Gang Tian[1]
MASSACHUSETTS INSTITUTE OF TECHNOLOGY
CAMBRIDGE, MA 02139

1 Introduction

The notion of quantum cohomology, which was first proposed in Witten's study of two dimensional nonlinear sigma models [W1], also see [Va], plays a fundamental role in understanding the phenomenon of mirror symmetry for Calabi-Yau manifolds. This phenomenon was first observed by physicists, who were motivated by their knowledge of topological field theory.

A topological sigma model theory starts with correlation functions. These are supposed to be computable as intersection numbers of cycles in the moduli space of holomorphic maps from Riemann surfaces to arbitrary manifolds. The key axiom of a topological field theory is the composition law for correlation functions. Basically, the composition law governs how correlation functions change under surgery of Riemann surfaces, such as gluing several Riemann surfaces together. One of its striking consequences, pointed out by Witten, is the associativity equation for quantum multiplications.

In last few years, there has been rapid progress on the mathematical aspects of topological field theory. A mathematical theory of quantum cohomology has been established. Solutions of the associativity (WDVV) equation have been constructed by using solutions of certain Cauchy-Riemann equations. Many applications of the associativity equation have been found in the study of integrable systems, algebraic geometry, etc.. In these notes, we will expose some of these recent developments, with an emphasis on the geometric aspects of quantum cohomology.

[1]Partially supported by a NSF grant and a Sloan fellowship.

The first step in the mathematical approach to topological field theory, is to define appropriate symplectic invariants which are the mathematical counterparts of correlation functions in Witten's topological sigma model (cf. [W1], [W2]). In the next section, we will review the construction of these invariants. Our method of defining these invariants is to show that for certain almost complex structure on a symplectic manifold V, an intersection theory on moduli spaces of generalized holomorphic maps can be built up.

To define the correlation functions, Witten constructed an appropriate topological sigma model Lagrangian and studied corresponding Feynman path integrals. At the moment, the latter do not exist in the rigorous mathematical sense. The goal of his physical discussions is to construct quantum field theory, at least formally. Witten also gave a formal reduction of the computation of the Feynman path integral to intersection theory on moduli spaces which are compact and smooth.

However, as Witten pointed out in [W2], his arguments for the reduction break down when the behavior of the moduli spaces is bad. In fact, moduli spaces of holomorphic maps are often noncompact and there are indeed contributions from the infinity, to the symplectic invariants defined in [RT1], [RT2] of the moduli spaces of holomorphic maps. Moreover, it is unclear how the symplectic structure of V is used in physical arguments. Since a sequence of holomorphic maps may not have any limit map on a general almost complex manifold, the role of symplectic structure seems to be essential. From the mathematical standpoint, it is still a very interesting problem how to define any useful invariants for general almost complex manifolds.

An algebraic geometer may be tempted to define the invariants directly by counting genuine genus g holomorphic curves in an algebraic manifold. As we will see, invariants defined in this way do not satisfy the composition law predicted by the physicists. Therefore, they are not the same as the correlation functions used by Witten. It was shown for the first time in [RT1], that the right invariants can be defined in terms of solutions of inhomogeneous Cauchy-Riemann equations. These invariants do indeed satisfy the composition law. The construction in [RT1] also provides ample examples of topological sigma models.

There are also purely mathematical motivations for a theory of symplectic invariants and quantum cohomology. Symplectic invariants (i.e., Gromov-Witten invariants) can be used to distinguish symplectic topology on differential manifolds. The problem of whether two symplectic manifolds of different shape are symplectically equivalent was first studied by Gromov in his seminal paper [Gr], and subsequently by Mcduff, Ruan, et. al. ([MS], [R1]). More recently, thanks to the new Seiberg-Witten invariants and the work of C. Taubes, many important results have been obtained on the structure of 4-dimensional symplectic manifolds. In particular, Taubes has proved that CP^2 has a unique symplectic structure.

Floer cohomology was introduced by A. Floer, in the course of his investigation on fixed points of symplectic diffeomorphisms or periodic orbits of

Hamiltonian systems. By the work of Floer, Hofer and Salamon, Ono, on any weakly monotone symplectic manifold, the number of fixed points for a nondegenerate symplectic diffeomorphism is not less than the total Betti number (cf. [F], [HS], [On]). This solves the Arnold conjecture for nondegenerate symplectic diffeomorphisms on weakly monotone manifolds.

For degenerate symplectic diffeomorphisms, the Arnold conjecture is largely open. It is expected that multiplicative structures on the Floer cohomology can be used to study the degenerate case (cf. [Ho], [OV]). As we will see, it turns out that the ring structure on the Floer cohomology is the same as the quantum ring (cf. section 9).

On algebraic manifolds, the GW-invariants generalize classical Schubert invariants. The composition law makes it possible to enumerate many Schubert invariants ([KM], [RT1]). Moreover, solutions of the associativity equation yield new Hodge bundles, and also produce a family of quantum ring structures. New geometric information concerning variation of Hodge structure can be obtained, via quantum cohomology and Mirror symmetry. For instance, the work of Candelas, et. al. ([COGP]) indicates that the GW-invariants on a Calabi-Yau can be computed via the variation of Hodge structures on its mirror manifold.

The organization of these notes is as follows:

In Section 3, we compute GW-invariants in several important cases. We give an application of the symplectic invariants to a problem on symplectic deformation equivalence, which was first studied by Ruan. We will also discuss briefly a very recent and remarkable result of C. Taubes.

In Section 4, we prove that our symplectic invariants satisfy the fundamental axiom – composition law.

The quantum cohomology ring structure is constructed on symplectic manifolds in Section 5. A few important examples are discussed and a quantized version of the Hard Lefschetz Theorem is formulated.

In Section 6, we discuss the WDVV equation and its relation to isomonodromic deformation, or equivalently Riemann-Hilbert problem. Formal series solutions of the WDVV equation are constructed on symplectic manifolds by using rational holomorphic maps.

Section 7 contains some applications of the associativity to enumerative geometry. In particular, we are able to show that the formal solutions of the WDVV equation constructed in Section 6 are indeed convergent in some region for a large class of Fano manifolds.

In section 8, we formulate a conjecture of Witten in terms of GW-invariants.

In section 9, we discuss the equivalence of the quantum cohomology ring and the exterior multiplication on the symplectic Floer cohomology.

The proof of the composition law is based on the technique of degeneration of domains for holomorphic maps. It is also natural to ask what would happen if we degenerate the target. It turns out that in this case, one can also obtain nice degeneration formulas, once one understands the combinatorial structure of maps in symplectic varieties with normal crossings. These are discussed in Section 10. There, we also give an application of such a degeneration formula to the convergence of formal soultions of the WDVV on Calabi-Yau manifolds.

In Section 11, we discuss the Mirror Symmetry Conjecture.

The author would like to thank Dr. B. Siebert for many useful conversations during the preparation of these notes.

The first version of this paper was submitted to the Proceeding of the first conference on current developments in mathematics, Boston, at the end of March, 1995.

Contents

2 The GW-invariants

In this section, we define the symplectic invariants of [RT1] and [RT2]. We will call them GW-invariants. Special cases of these invariants were first studied by Ruan in [R1], [R2] (also see [Gr], [MS]). More precisely, he defined two very important cases of these invariants when either the domain is S^2, or the target is a 4-dimensional symplectic manifold. The physical counterparts of the invariants are Witten's sigma model correlation functions (cf. [W1], [W2]). We will also discuss a few basic properties of the latter invariants and compute them in some cases.

Let (V, ω) be a compact symplectic manifold, $A \in H_2(V, Z)$ and J be an almost complex structure on V satisfying: $\omega(v, Jv) > 0$ for any nonzero tangent vector v, i.e., J is ω-tamed according to Gromov. As usual, $\mathcal{M}_{g,k}$ denotes the moduli space of genus g Riemann surfaces with k-marked points $(2g + k \geq 3)$ and $\overline{\mathcal{M}}_{g,k}$ denotes its Deligne-Mumford compactification. The universal family of curves $\pi : \overline{\mathcal{U}}_{g,k} \to \overline{\mathcal{M}}_{g,k}$ is a projective algebraic variety. We put $\mathcal{U}_{g,k} = \pi^{-1}(\mathcal{M}_{g,k})$. We fix an embedding $\overline{\mathcal{U}}_{g,k} \to \mathcal{P}^N$. Roughly speaking, an inhomongeneous term ν is an anti-J-linear section of $Hom(\pi_1^* T\mathcal{P}^N, \pi_2^* TV)$ over $\mathcal{P}^N \times V$, where π_i is the projection from $\mathcal{P}^N \times V$ to its i-th factor.

A marked Riemann surface in $\mathcal{M}_{g,k}$ will be denoted by $(\Sigma, x_1, \cdots, x_k)$, or simply $(\Sigma, \{x_i\})$ when there is no ambiquity. A (J,ν)-map from Σ is a continuous map $f : \Sigma \to V$, which is smooth on the regular part $\mathrm{Reg}(\Sigma)$ of Σ, satisfying the following inhomogeneous Cauchy-Riemann equation:

$$J \cdot df(x) + df \cdot j_\Sigma(x) = \nu(\phi(x), f(x)),\ x \in \mathrm{Reg}(\Sigma) \qquad (2.1)$$

where j_Σ is the complex structure on Σ. In particular, if $\nu = 0$, we say that f is J-holomorphic . We put

$$\mathcal{M}_A(g,k,J,\nu) = \{(\Sigma, \{x_i\}, f) \mid (\Sigma, \{x_i\}) \in \mathrm{Reg}(\mathcal{M}_{g,k}),\ f : \Sigma \mapsto V \text{ is a } (J,v)\text{-map},\ f_*[\Sigma] = A\} \qquad (2.2)$$

where $\mathrm{Reg}(\mathcal{M}_{g,k})$ denotes the regular part of $\mathcal{M}_{g,k}$.

Proposition 2.1. (cf. [RT2]) *For a generic pair (J,ν), the moduli space $\mathcal{M}_A(g,k,J,\nu)$ is an oriented smooth manifold of dimension $2c_1(V)(A)+2(3-n)(g-1)+2k$.*

The proof of this is standard and follows from an application of the Sard-Smale Transversality Theorem. Such a technique was used before by many people, notably, S. Donaldson, A. Floer, D. Mcduff, C. Taubes, K. Uhlenbeck, et. al..

By the Gromov-Uhlenbeck compactness theorem ([PW], [Ye] and [RT1, Proposition 3.1]), for any sequence $\{(\Sigma_i, x_{i1}, \cdots, x_{ik}, f_i)\}$, by taking a subsequence if necessary, we may assume that $(\Sigma_i, x_{i1}, \cdots, x_{ik})$ converges to some $(\Sigma_\infty, \cdots)$ in $\overline{\mathcal{M}}_{g,k}$, and f_i converges to a map f_∞ into V satisfying: (1) the domain Σ'_∞ of f_∞ is connected, and a union of Σ_∞ with some bubble trees of S^2's; (2) $f_\infty|_{\Sigma_\infty}$ is a (J,ν)-map and f_∞ restricts to J-holomorphic maps on bubble trees. Let $\{x_{\infty 1}, \cdots, x_{\infty k}\}$ be the limit in Σ'_∞ of the sequence $\{x_{i1}, \cdots, x_{ik}\}$. The limiting points are distinct. Some of them may lie in bubble trees instead of Σ_∞. Let G be the subgroup consisting of the automorphisms of the bubble trees which fix bubble components, all marked points $x_{\infty 1}, \cdots, x_{\infty k}$ and all singular points in Σ'_∞. It is easy to see that G is trivial if there are no bubbles. We say that two limits $(\Sigma'_\infty, x_{\infty 1}, \cdots, x_{\infty k}, f_\infty)$ and $(\tilde{\Sigma}'_\infty, \tilde{x}_{\infty 1}, \cdots, \tilde{x}_{\infty k}, \tilde{f}_\infty)$ are equivalent, if (1) $(\Sigma'_\infty, x_{\infty 1}, \cdots, x_{\infty k})$ is equal to $(\tilde{\Sigma}'_\infty, \tilde{x}_{\infty 1}, \cdots, \tilde{x}_{\infty k})$; (2) there is a $\sigma \in G$ such that $\tilde{f}_\infty \cdot \sigma = f_\infty$. Let $\overline{\mathcal{M}}_A(g,k,J,\nu)$ be the set of the limits of sequences in $\mathcal{M}_A(g,k,J,\nu)$ modulo the above equivalence. Then $\overline{\mathcal{M}}_A(g,k,J,\nu)$ is compact.

Consider the natural evaluation map

$$e : \mathcal{M}_A(g,k,J,\nu) \to \mathcal{M}_{g,k} \times V^k \qquad (2.3)$$
$$e((\Sigma, x_1, \cdots, x_k, f)) = (\Sigma, x_1, \cdots, x_k) \times (f(x_1), \cdots, f(x_k))$$

This can be easily extended to a map, still denoted by e, from the compactification $\overline{\mathcal{M}}_A(g,k,J,\nu)$ into $\overline{\mathcal{M}}_{g,k} \times V^k$. To define our invariants, we need to prove that $e\left(\overline{\mathcal{M}}_A(g,k,J,\nu)\backslash\mathcal{M}_A(g,k,J,\nu)\right)$ is of smaller dimension in V^k.

Proposition 2.2. (cf. [RT2]) *Assume that (V,ω) is semi-positive. Then for a generic pair (J,ν), the image $e\left(\overline{\mathcal{M}}_A(g,k,J,\nu)\backslash\mathcal{M}_A(g,k,J,\nu)\right)$ is a set in $\overline{\mathcal{M}}_{g,k} \times V^k$ of dimension no more than $2c_1(V)(A)+2(3-n)(g-1)+2k-2$.*

We say that (V,ω) is semi-positive if for a ω-tamed almost complex strcture J on V, there is no J-holomorphic map $f : S^2 \mapsto V$ satisfying: $3-n \leq \int_{S^2} f^*c_1(V) < 0$, where $c_1(V)$ denotes the first Chern class of (V,ω). Clearly, this condition of semi-positivity holds for any symplectic manifolds of complex dimension ≤ 3. Furthermore, any algebraic manifolds with numerically positive anticanonical bundle, for instance, Calabi-Yau manifolds, are semi-positive.

As a matter of fact, the semi-positivity in Proposition 2.2 can be removed. This can be accomplished by generalizing a theorem of E.Bishop to symplectic manifolds. The Bishop theorem says that an analytic set in a projective space can be compactified by adding a set of codimension at least two, if the induced metric has finite volume.

Proof of Proposition 2.2: The basic idea of its proof is to stratify

$$\overline{\mathcal{M}}_A(g,k,J,\nu)\backslash\mathcal{M}_A(g,k,J,\nu)$$

according to the topology of the domain of limiting maps and the configuration of marked points. Let $\mathcal{M}_A(D,J,\nu)$ be one stratum. Then we first drop the multiplicity of bubbles and obtain a moduli space $\mathcal{M}_{A'}(D,J,\nu)$. By semi-positivity, we have $c_1(V)(A') \leq c_1(V)(A)$. For a generic (J,ν), by using the transversality arguments, one can show that $\dim(\mathcal{M}_{A'}(D,J,\nu))$ is no more than $2c_1(V)(A)+2(3-n)(g-1)+2k$, in fact, strictly smaller if the domain is in $\overline{\mathcal{M}}_{g,k}\backslash\mathcal{M}_{g,k}$. However, if the domain is not in $\overline{\mathcal{M}}_{g,k}\backslash\mathcal{M}_{g,k}$, then the parametrization group G is of dimension at least two, so the image $e(\mathcal{M}_{A'}(D,J,\nu))$ has codimension at least two. The proposition follows.

Now we can define the GW-invariant

$$\Psi^V_{(A,g,k)} : H_*(\overline{\mathcal{M}}_{g,k},\boldsymbol{Q}) \times H_*(V,\boldsymbol{Z})^k \mapsto \boldsymbol{Q}$$

as follows: let $\alpha_1,\cdots,\alpha_k$ be any homology classes in $H_*(V,\boldsymbol{Z})$ and $[K]$ be a rational homology class in $H_*(\overline{\mathcal{M}}_{g,k},\boldsymbol{Q})$. If

$$\operatorname{codim}([K]) + \sum_i \operatorname{codim}(\alpha_i) = 2c_1(V)(A) + 2(3-n)(g-1) + k, \tag{2.4}$$

we choose a pseudo-submanifold representative, denoted by K, of $[K]$ such that its top strata lie in $\operatorname{Reg}(\mathcal{M}_{g,k})$. By (2.4) and Proposition 2.2, for generic pseudo-submanifolds $S_i : P_i \mapsto V$ representing α_i, $K \times \prod_i S_i(P_i)$

intersects $e(\overline{M}_A(g,k,J,\nu))$ transversally at finitely many smooth points of $\mathcal{M}_A(g,k,J,\nu)$. By Proposition 2.1, we can assign a sign to each of these intersection points. We define $\Psi^V_{(A,g,k)}([K];\alpha_1,\cdots\alpha_k)$ to be the algebraic sum of these points. If (2.4) does not hold, we simply define

$$\Psi^V_{(A,g,k)}([K];\alpha_1,\cdots,\alpha_k)=0.$$

For simplicity, we sometimes abbreviate this as $\Psi^V_{(A,g,k)}([K];\{\alpha_i\})$. Also we often write $\Psi^V_{(A,g,k)}(\{\alpha_i\})$ for $\Psi^V_{(A,g,k)}([K];\{\alpha_i\})$ in the case that K is the whole space $\overline{\mathcal{M}}_{g,k}$.

Theorem 2.3. ([RT2]) *Assume that (V,ω) is a compact semi-positive symplectic manifold. Then $\Psi^V_{(A,g,k)}$ is a symplectic invariant. In particular, it is independent of choices of (J,ν).*

This can be proved by constructing cobordisms. The basic techniques in the proof are versions of Proposition 2.1, 2.2 for cobordisms. For instance, let (J',ν') be another pair with properties stated in Proposition 2.2, 2.3. Then by using identical arguments in the proof of Proposition 2.2, 2.3, for a generic path $\{(J_t,\nu_t)\}$ from (J,ν) to (J',ν'), one can show the following:

(1) the set $\bigcup_t e(M_A(g,k,J_t,\nu_t))$ is a smooth submanifold in V^k with boundary

$$e(M_A(g,k,J,\nu))\cup(-e(M_A(g,k,J',\nu')));$$

(2) $\bigcup_t e(\overline{M}_A(g,k,J_t,\nu_t))$ is a compactification of $\bigcup_t e(M_A(g,k,J_t,\nu_t))$;

(3) $\bigcup_t e((\overline{M}_A(g,k,J_t,\nu_t)\backslash M_A(g,k,J_t,\nu_t))$ is of codimension at least two.

Hence, $K\times\prod_i S_i(P_i)$ has the same intersection number with both $e(\overline{M}_A(g,k,J,\nu))$ and $e(\overline{M}_A(g,k,J',\nu'))$ in V^k. It follows that

$$\Psi^V_{(A,g,k)}([K];\{\alpha_i\})$$

is independent of choices of (J,ν).

More recently, J. Li and the author can construct the GW-invariants for general symplectic manifolds.

It is possible to define the invariant $\Psi^V_{(A,g,k)}$ by using differential forms. The difficulty is to prove that the compactification $\overline{\mathcal{M}}_A(g,k,J,\nu)$ carries a fundamental class. This requires understanding how the strata in $\overline{\mathcal{M}}_A(g,k,J,\nu)$ are attached to each other. Analytically, it amounts to a delicate deformation theory near singular curves. It is proved in [RT3] that $\mathcal{M}_A(0,k,J,\nu)$ carries a fundamental class. Presumably, the proof can be extended to arbitary genus. Assuming that this is done, we should be able to prove

$$\begin{aligned}&\Psi^V_{(A,g,k)}([K];\alpha_1,\cdots,\alpha_k)\\&=\int_{\overline{\mathcal{M}}_A(g,k,J,\nu)}e^*(\pi_1^*PD([K])\wedge\pi_2^*\alpha_1^*\wedge\cdots\wedge\pi_{k+1}^*\alpha_k^*),\end{aligned}\tag{2.5}$$

where $PD([K])$ (resp. α_i^*) denotes the Poincare dual of $[K]$ (resp. α_i) in $\overline{\mathcal{M}}_{g,k}$ (resp. V) as usual, and π_j is the projection from $\overline{\mathcal{M}}_{g,k} \times V^k$ onto its jth-factor.

Here are some basic properties of Ψ:

(1) $\Psi^V_{(A,g,k)} = 0$, *if either* $\omega(A) < 0$ *or* $c_1(V)(A) + (n-3)(1-g) < 0$, *in particular,* $\Psi_{(0,g,k)} = 0$ *for any* $g \geq 2$ *and* $n \geq 4$.

(2) $\Psi^V_{(A,g,k)}$ *is multilinear and supersymmetric on* $H_*(V, Z)^k$ *with respect to the* Z_2*-grading by even and odd degrees.*

(3) $\Psi^V_{(A,g,k)}$ *takes integer classes in* $H_*(\overline{\mathcal{M}}_{g,k}, Z) \times H_*(V, Z)^k$ *to integers, where the integer classes are those represented by cycles whose top strata lie in* $\text{Reg}(\mathcal{M}_{g,k})$.

There is a natural map $\pi_k : \overline{\mathcal{M}}_{g,k} \mapsto \overline{\mathcal{M}}_{g,k-1}$: for $C = (\Sigma, x_1, \cdots, x_k)$ in $\overline{\mathcal{M}}_{g,k}$, if x_k is not in any rational components of Σ which contain only three distinguished points, then we define

$$\pi_k(C) = (\Sigma, x_1, \cdots, x_{k-1})$$

Note that a distinguished point of Σ is either a singular point or a marked point. If x_k is in one of the above rational components, we contract this component and obtain a stable curve C' in $\overline{\mathcal{M}}_{g,k-1}$, and define $\pi_k(C) = C'$. Clearly, π_k is continuous.

Then it follows easily from (2.5)

(4) *For any* $\alpha_1, \cdots, \alpha_{k-1}$ *in* $H_*(V, Z)$, *we have*

$$\begin{aligned} &\Psi^V_{(A,g,k)}([K]; \alpha_1, \cdots, \alpha_{k-1}, [V]) \\ =&\Psi^V_{(A,g,k-1)}(\pi_{k*}([K]); \alpha_1, \cdots, \alpha_{k-1}) \end{aligned} \tag{2.6}$$

(5) *Let* α_n *be in* $H_{2n-2}(V, Z)$ *and* $[K]$ *in* $H_*(\overline{\mathcal{M}}_{g,k-1}, Q)$, *then*

$$\begin{aligned} &\Psi^V_{(A,g,k)}\left(\left[\pi_k^{-1}(K)\right]; \alpha_1, \cdots, \alpha_{k-1}, \alpha_k\right) \\ =&\hat{\alpha}_k(A)\Psi^V_{(A,g,k-1)}([K]; \alpha_1, \cdots, \alpha_{k-1}) \end{aligned} \tag{2.7}$$

where $\hat{\alpha}_k$ *is the Poincare dual of* α_k.

It follows from (4) that $\Psi^V_{(A,0,3)}(\alpha_1, \alpha_2, [V])$ is equal to zero if $A \neq 0$ and the cap product $\alpha_1 \cap \alpha_2$ if $A = 0$.

These properties are formulated in [KM] as axioms. They (except (3)) are essentially stated in [W2] for the correlation functions in Witten's topological sigma model theory. Our proof of these properties is inspired by Witten's arguments in [W2].

Remark 2.4: The mixed invariant $\Phi_{(A,\omega,g)}$ in [RT1] can be identified with certain Ψ by choosing appropriate cycles $[K]$. More precisely, for any $k, l \geq 0$

with $2g + k \geq 3$, put $[K_{k,l}]$ to be the closure of the cycle $K^0_{k,l}$ in $\overline{\mathcal{M}}_{g,k+l}$, where $K^0_{k,l}$ is the set of all $(\Sigma, x_1, \cdots, x_{k+l})$ in $\mathcal{M}_{g,k+l}$ with $(\Sigma, x_1, \cdots, x_k)$ being a fixed point in $\mathcal{M}_{g,k}$. Then

$$\Psi^V_{(A,g,k+l)}([K_{k,l}]; \alpha_1, \cdots, \alpha_{k+l}) = \Phi_{(A,\omega,g)}(\alpha_1, \cdots, \alpha_k | \alpha_{k+1}, \cdots, \alpha_{k+l})$$

It follows from (4): $\Phi_{(A,\omega,g)}(\alpha_1, \cdots | \cdots, \alpha_{k+l})$ is 0 if $\dim(\alpha_{k+l}) > 2n - 2$ (cf. [RT1]).

To compute the GW-invariant, it is often more convenient to use holomorphic maps with respect to a particular complex structure, instead of (J, ν)-maps for a generic (J, ν) as in the definition. For instance, in case V is algebraic, a natural question is if Ψ can be defined in terms of holomorphic maps, i.e., Ψ is a classical enumerative invariant. The following theorem provides a partial answer to this question.

Theorem 2.5. *If V is a complex homogeneous manifold, then the genus zero GW-invariant $\Psi^V_{(A,0,k)}$ is enumerative, i.e., it can be defined to be the number of rational curves through given subvarieties in general position.*

This is proved in [RT1] for Grassmannian manifolds (also see [Ber]), and in [LT1] for general homogeneous manifolds. Also see [CF] for defining the invariants algebraically for flag manifolds. Since the tangent bundle is numerically positive for homogeneous manifolds, the moduli space of rational curves has expected dimension and its compactification is reasonably good. So one can use holomorphic maps to define $\Psi_{(A,0,k)}$. It was also stated in [KM] that the GW-invariants can be defined algebraically if the tangent bundle is positive in the suitable sense. In general, the genus zero GW-invariant $\Psi^V_{(A,0,k)}$ should be equal to the number of rational curves with homology class A, whenever the corresponding moduli space of rational curves has expected dimension. The latter should be true for any Fano manifold V and sufficiently large $c_1(V)(A)$.

It follows from Theorem 2.5 that if $V = CP^n$, $\Psi_{([\ell],0,3)}(H^{l_1}, H^{l_2}, H^{l_3}) = 1$ whenever $l_1 + l_2 + l_3 = 2n + 1$, where ℓ is a line in CP^n and H denotes the hyperplane class. Moreover, if $V = CP^2$, $\Psi^V_{(d[\ell],0,3d-1)}(H^2, \cdots, H^2)$ is the number of rational curves through $3d - 1$ points in general position.

Very little is known for higher genus invariants even in case of homogeneous manifolds. However, in [BDW], Bertram, Daskalopoulos and Wentworth defined invariants of higher genus on Grassmannian manifolds in terms of quotient sheaves. Later, Bertram proved that those invariants satisfy trace formulas [Ber]. It turns out that these trace formulas are identical to the Intrilligate-Vafa formulas for the GW-invariants, which were proved by B. Siebert and myself in [ST1]. Therefore, their invariants are the same as the GW-invariants $\Psi^V_{(A,g,k)}([pt]; \cdots)$, where pt denotes a point in $\mathcal{M}_{g,k}$.

Given a general algebraic manifold V, the moduli space $\mathcal{M}_A(g, k, J, 0)$ may not have expected dimension and its compactification may behave very badly. Therefore, one needs to find a subvariety (virtual moduli cycle) in $\overline{\mathcal{M}}_A(g,k,J,0)$, which plays the role of the moduli space for a generic (J,ν). A naive approach can be described as follows: let $\overline{\mathcal{U}}_{g,k}$ be the universal family of stable curves, and $\overline{\mathcal{U}}_A(g,k,J,0)$ be the fiber product of $\overline{\mathcal{M}}_A(g,k,J,0)$ and $\overline{\mathcal{U}}_{g,k}$ over $\overline{\mathcal{M}}_{g,k}$. Then there is a natural evaluation map e from $\overline{\mathcal{U}}_A(g,k,J,0)$ into V. Put $E = e^*TV$ and take its first direct image $\mathcal{R}^1\pi_{1*}E$ over $\overline{\mathcal{M}}_A(g,k,J,0)$, whose fiber over $(\Sigma,\cdots;f)$ is $H^1(\Sigma,f^*TV)$. The following is expected.

Assume that $\overline{\mathcal{M}}_A(g,k,J,0)$ is a d-dimensional Kähler orbifold, and $\mathcal{R}^1\pi_{1*}E$ is an orbifold bundle of rank r outside a subvariety of codimension greater than $r = d - c_1(V)(A) - (3-n)(g-1) - k$. Then

$$\Psi^V_{(A,g,k)}([K];\alpha_1,\cdots,\alpha_k) = \int_{\overline{\mathcal{M}}_A(g,k,J,\nu)} c_r(\mathcal{R}^1\pi_{1*}E)\wedge e^*(\pi_1^*PD([K])\wedge\pi_2^*\alpha_1^*\wedge\cdots\wedge\pi_{k+1}^*\alpha_k^*) \quad (2.8)$$

where e, π_j are as in (2.5), and $c_r(\mathcal{R}^1\pi_{1*}E)$ is the orbifold Chern class.

This means that $c_r(\mathcal{R}^1\pi_{1*}E)$ is Poincare dual to a subvariety in $\overline{\mathcal{M}}_A(g,k,J,0)$, which plays the role of the moduli space for a generic almost complex structure. In general, there are serious difficulties with this approach because $\overline{\mathcal{M}}_A(g,k,J,0)$ may have very bad singularities and the push-forward $\mathcal{R}^1\pi_{1*}E$ may not be a bundle. For instance, the rank of $\mathcal{R}^1\pi_{1*}E$ may be greater than r, the difference between $\dim\mathcal{M}_A(g,k,J,0)$ and the expected dimension. However, it is expected that an analog of intersection theory using normal cones (cf. [Fu]) can be developed to define the GW-invariant purely algebraically. Recently, Jun Li and I have constructed GW-invariants for any algebaric manifolds ([LT2]).

It follows from (2.8) that $\Psi^V_{(0,1,1)}([pt];[V])$ is equal to the Euler number of V. In a similar way, one can also compute $\Psi^V_{(0,1,1)}([\overline{\mathcal{M}}_{1,1}];\cdot)$ and $\Psi^V_{(0,g,0)}$ for $g\geq 2$ and $n\leq 3$.

3 More Examples and Applications

In this section, we compute the GW-invariants for some manifolds and give an application to a deformation equivalence problem.

Let $p: W\mapsto Z$ be a family of symplectic manifolds (V_z,ω_z), where Z is connected and $V_z = p^{-1}(z)$. Then the fundamental group $\pi_1(Z)$ induces a monodromy group Γ which acts on $H_*(V,\mathbf{Z})$, where $V = V_z$ for some z. The following can be easily proved by using the definition.

Theorem 3.1. *For any $\sigma\in\Gamma$, we have*

$$\Psi^V_{(\sigma(A),g,k)}([K];\sigma(\alpha_1),\cdots,\sigma(\alpha_k)) = \Psi^V_{(A,g,k)}([K];\alpha_1,\cdots,\alpha_k) \quad (3.1)$$

There is a useful application of this to Lefschetz pencils of algebraic varieties. Let $W\subset CP^N$ be an algebraic submanifold and V be a generic hyperplane section in W. The second Lefschetz theorem (cf. [AF]) says that the kernel Ker $(H_n(V,\mathbf{Z}) \mapsto H_n(W,\mathbf{Z}))$, where $n = \dim V$, is generated by vanishing cycles e_i. One may arrange these vanishing cycles such that

(1) $e_i \cdot e_j = (-1)^{\frac{n(n-1)}{2}}(1+(-1)^n)\delta_{ij}$. Note that each e_i is represented by a real n-sphere, so $1+(-1)^n$ is the Euler number of this sphere.

(2) For each i, there is a σ_i in the monodromy group of this pencil satisfying:

$$\sigma_i(e_j) = e_j + (-1)^{\frac{(n+1)(n+2)}{2}}(e_j \cdot e_i)e_i,$$

in particular, if n is even, $\sigma(e_j) = e_j$ for $j \neq i$ and $\sigma_i(e_i) = -e_i$.

One can deduce from these and Theorem 3.1

Corollary 3.2. *Let V, W, e_i, etc. be as above. We further assume that A is invariant under the monodromy group. Then $\Psi^V_{(A,g,k)}$ vanishes on $e_i^{2m-1} \times H_*(V,\mathbf{Z})_i$ for each i and $m > 0$, where $H_*(V,\mathbf{Z})_i$ is the σ_i-invariant part of $H_*(V,\mathbf{Z})$.*

In particular, if V is a generic hypersurface in CP^{n+1}, then by the Lefschetz Theorem, $H_*(V,\mathbf{Z})$ is isomorphic to

$$\{[H]^i | 0 \leq i \leq n\} \oplus H_n(V,\mathbf{Z})_0$$

where H is a hyperplane in CP^{n+1} and $H_n(V,\mathbf{Z})_0$ is the primitive part, which is generated by vanishing cycles. Therefore, for any $n \geq 3$ and $\alpha_{l+1},\cdots,\alpha_k$ in $H_n(V,\mathbf{Z})_0$, we have

$$\Psi^V_{(A,g,k)}(\cdot\,; H^{i_1},\cdots,H^{i_l},\alpha_{l+1},\cdots,\alpha_k) = 0 \tag{3.2}$$

whenever $k-l$ is odd.

For a general algebraic manifold V, we denote by $H_{*,\mathrm{inv}}(V,\mathbf{Z})$ the subring of $H_*(V,\mathbf{Z})$ generated by those classes in $H_{2n-2}(V,\mathbf{Z})$, which can be represented by algebraic cycles, in other words, $H_{*,\mathrm{inv}}(V,\mathbf{Z})$ is generated by $H_{n-1,n-1}(V,\mathbf{Z})$ through cap products. Let $H_*(V,\mathbf{Q})_0$ be its orthogonal complement. The following is expected

For any $\alpha_1,\cdots,\alpha_k$ in $H_{*,\mathrm{inv}}(V,\mathbf{Z})$ and γ in $H_*(V,\mathbf{Q})_0$, we have

$$\Psi^V_{(A,g,k+1)}([K];\alpha_1,\cdots,\alpha_k,\gamma) = 0 \tag{3.3}$$

whenever K is an analytic cycle.

We have already seen that this conjecture is true in many cases, such as complete intersections in any projective spaces.

Let us outline a heuristic proof as follows: For simplicity, we assume that $K = \overline{\mathcal{M}}_{o,k+1}$ and $H_2(V,\mathbf{Z}) = \mathbf{Z}$. Without loss of generality, we further assume that each α_i is represented by a subvariety, still denoted by α_i, in general position. Let Z be the moduli space of rational curves with homology class A, which intersect subvarieties α_i $(1 \leq i \leq k)$. Through the evaluation map, Z can be regarded as an effective holomorphic cycle in V, and $\Psi^V_{(A,o,k+1)}(\alpha_1, \cdots, \alpha_k, \gamma)$ can be defined as the intersection number of Z with γ. By the Lefschetz Theorem (cf. [GH]), γ is of the form $\tilde{\gamma} \cap H^l$, where $\tilde{\gamma}$ is a primitive cycle in V of real dimension $n + l'$ and H^l is an intersection of l ample divisors $(l \leq l')$. Note that $\dim_{\boldsymbol{R}} Z = n + 2l - l'$ is even, so $\dim_{\boldsymbol{R}}(\tilde{\gamma}) \leq 2n - 2$. Since $\tilde{\gamma}$ is primitive, for any smooth intersection S of $l' - l + 1$ ample divisors, γ can be represented by a cycle lying in $V \backslash S$.

On the other hand, since α_i are generic, Z should be contained in a smooth intersection of ample divisors in V. Using the fact that $\dim_{\boldsymbol{R}} \tilde{\gamma} = n + l' \leq 2n - 2$, one can embed Z as a subvariety in an intersection S of $l' - l + 1$ ample divisors. Therefore, $\gamma \cap Z = \emptyset$. This implies (3.3).

The only technical problem with these arguments is to construct Z rigorously. It can be overcome in view of my joint work with Jun Li in [LT2], where we construct the quantum cohomology on algebraic manifolds by purely algebraic method. One can also construct this cycle Z by using the results in [RT2].

Next we give an application to deformation equivalence of symplectic manifolds. Two symplectic manifolds (V, ω) and (V', ω') are deformation equivalent if and only if there is a diffeomorphism $\phi : V \mapsto V'$ and a family of symplectic forms ω_t such that $\omega_0 = \omega$ and $\omega_1 = \phi^* \omega'$. The classical Darboux theorem implies that all symplectic manifolds of the same dimension are locally isomorphic. Therefore, deformation equivalence is a global problem.

An important step was taken by J. Moser. He proved that (V, ω) and (V, ω') are equivalent if the cohomology class $[\omega]$ can be connected to $[\omega']$ in $H^2(V, \boldsymbol{R})$ by symplectic forms.

Usually, it is rather difficult to study such a deformation problem. For instance, in the complex surface theory, a long-standing problem is whether or not the moduli space of complex structures is always connnected on a given differential 4-manifold. Since any deformation of a Kähler surface is still Kähler, this can be related to the symplectic deformation problem. Nevertheless, Theorem 2.4 provides a tool for studying such a deformation problem, namely, two compact symplectic manifolds (V, ω) and (V', ω') are deformation equivalent only if the invariants Ψ^V and $\Psi^{V'}$ are the same. Ruan gave first two non-deformation equivalent, 6-dimensional symplectic manifolds, which are diffeomorphic to each other (cf. [R2], [MS]). Let Y be the Barlow surface, which is homeomorphic to the blow-up X of CP^2 at eight points. They have the same homology $H_*(X, \boldsymbol{Z}) = H_*(Y, \boldsymbol{Z})$, but they are not diffeomorphic. A result of C.T.C.Wall implies that $X' = X \times CP^1$ and $Y' = Y \times CP^1$ are diffeomorphic to each other. Ruan showed that $\Psi^{X'}_{(A,0,3)} \neq 0$ and $\Psi^{Y'}_{(A,0,3)} = 0$, where A represents the class of one of the blown-up points in X. Therefore, X' and Y' are not deformation equivalent. Very recently, M. Gross and Ruan can even construct two non-deformation equivalent Calabi-Yau 3-folds.

We recall a result of C.T.C. Wall, saying that two 4-manifolds X, Y are homeomorphic if and only if the stabilized manifolds $X' = X \times CP^1$, $Y' = Y \times CP^1$ are diffeomorphic. An analogy of this in symplectic topology would be ([R2]): given two homeomorphic symplectic 4-manifolds X and Y with $c_1(X) = c_1(Y)$, X is diffeomorphic to Y if and only if X' can be symplectically deformed to Y'. Note that X', Y' carry natural symplectic structures and $c_1(X) = c_1(Y)$ is necessary for X', Y' being deformation equivalent. Ruan's symplectic manifolds give the first indication that this might be true. Heuristically, the moduli space of horizontal rational curves in X' (resp. Y') is just X (resp. Y). If X' and Y' are deformation equivalent, then there is a cobordism from X to Y through moduli spaces of rational curves. It is not unreasonable to expect that such a cobordism is nice and imposes a strong relation between X and Y.

Using the GW-invariant Ψ, we will be able to prove this for simply-connected elliptic surfaces $E^n_{p,q}$ (p, q coprime). Here, E^1 is the blow-up of CP^2 at nine generic points, E^n denotes the fiber connected sum of n copies of E^1, and $E^n_{p,q}$ is obtained from E^n by logarithmic transformations along two smooth fibers with multiplicity p and q. Let A_p, A_q be the two multiple fibers in $E^n_{p,q}$. It is computed in [RT2]

$$\Psi^{E^n_{p,q}\times CP^1}_{(A_p,1,1)}(\overline{\mathcal{M}}_{1,1};\alpha) = 2pA\cdot\alpha,\ \Psi^{E^n_{p,q}\times CP^1}_{(A_q,1,1)}(\overline{\mathcal{M}}_{1,1};\alpha) = 2qA\cdot\alpha \tag{3.4}$$

where $A = pA_p = qA_q$ is the fiber class and α is in $H_2(E^n_{p,q}, \boldsymbol{Z})$. This implies that the 3-folds $E^n_{p,q} \times CP^1$ are not deformation equivalent to each other. On the other hand, using the Donaldson invariant, Friedman and Morgan proved (cf. [FM]) that the surfaces $E^n_{p,q}$ are not diffeomorphic to each other, but they are homeomorphic for each fixed n.

Finally, we would like to mention a truely remarkable theorem, announced recently by C. Taubes ([Ta]). Taubes' theorem relates the GW invariant to the Seiberg-Witten invariant. Given a $Spin_{\boldsymbol{C}}$ structure (denoted by $\mathcal{L}$) on V, we denote by $S^+_{\mathcal{L}}$ its positive spinor bundle. The Clifford multiplication induces a natural map $\tau : \mathrm{EndS}^+_{\mathcal{L}} \mapsto \Lambda_+ \otimes \boldsymbol{C}$, which maps a self-adjoint endomorphism into an imaginary valued form. Then the Seiberg-Witten equations are

$$D_A\psi = 0 \text{ and } F^+_A = \frac{1}{4}\tau(\psi\otimes\psi^*) \tag{3.5}$$

where A is a connection on $L = \det(S^+_{\mathcal{L}})$, F^+_A is the self-dual part of its curvature, and ψ is a section of $S^+_{\mathcal{L}}$. The Seiberg-Witten invariant $SW(\mathcal{L})$ is defined in terms of the solutions of (3.5). For instance, if the solutions of (3.5) are isolated, then $SW(\mathcal{L})$ just counts the number of solutions with appropriate sign (cf. [Ta], [W4]).

Theorem 3.3. ([Ta]) *Let V be a compact, oriented, symplectic 4-manifold with $b_2^+ \geq 3$, and $\mathcal{L}$ be a $Spin_{\boldsymbol{C}}$-structure with $d = a\cdot a + c_1(V)(a) \geq 0$. Note that d is always even. Put $a \neq 0$ to be the Poincare dual to $\frac{1}{2}(c_1(L) + c_1(V))$*

and $g = 1 + d/2$. Then the SW-invariant $SW(\mathcal{L})$ can be identified with the GW-invariant $\Psi^V_{(a,g,d/2)}(\overline{\mathcal{M}}_{g,d/2}; [p], \cdots, [p])$, where p is a point in V.

This implies in particular that the GW-invariants in (3.6) are differential invariants. The Seiberg-Witten invariant is ± 1 for $L = K_V$, which is proved in [W4] for Kähler surfaces and in [Ta] for symplectic 4-manifolds, $\Psi^V_{(K_V, K_V^2+1,0)} \neq 0$. This implies existence of a holomorphic curve of genus $K_V^2 + 1$. Moreover, it is observed in [W4] that all Kähler surfaces of general type satisfy: $SW(\mathcal{L}) = 0$ unless $L = K_V^{-1}$ or K_V, so we have $\Psi^V_{(A,1+d/2,d/2)} = 0$ for any A dual to $mc_1(K_V)$ with $m > 1$ and d, g as in Theorem 3.4. Similar conclusions can be drawn for symplectic manifolds, too (cf. [Ta]).

4 The Composition Law

The composition law relates certain GW-invariants to counting degenerate holomorphic maps. Its classical cousin in enumerative algebraic geometry is the degeneration formula, which was only derived individually in very special cases. It was never developed into a systematic procedure as the composition law provides. One reason might be that classical counting of holomorphic curves, particularly of higher genus, can not have any simple composition laws! The other reason might be that it is very difficult to have a good deformation theory in algebraic geometry. Since our Ψ^V is a symplectic invariant, we only require a deformation theory in symplectic geometry. We will be able to establish this.

Recall that in the definition of $\Psi^V_{(A,g,k)}([K]; \cdots)$, we required that the top strata of K intersect with $\mathcal{M}_{g,k}$. Roughly speaking, the composition law means that this constrain on K can be dropped, i.e., the top strata of k can lie in $\overline{\mathcal{M}}_{g,k} \backslash \mathcal{M}_{g,k}$. Let us describe it in more details.

Assume $g = g_1 + g_2$ and $k = k_1 + k_2$ with $2g_i + k_i \geq 2$. Fix a decomposition $S = S_1 \cup S_2$ of $\{1, \cdots, k\}$ with $|S_i| = k_i$. Then there is a canonical embedding $\theta_S : \overline{\mathcal{M}}_{g_1,k_1+1} \times \overline{\mathcal{M}}_{g_2,k_2+1} \mapsto \overline{\mathcal{M}}_{g,k}$, which assigns to marked curves $(\Sigma_i; x_1^i, \cdots, x_{k_1+1}^i)$ $(i = 1, 2)$, their union $\Sigma_1 \cup \Sigma_2$ with $x_{k_1+1}^1$ identified to x_1^2 and remaining points renumbered by $\{1, \cdots, k\}$ according to S.

There is another natural map $\mu : \overline{\mathcal{M}}_{g-1,k+2} \mapsto \overline{\mathcal{M}}_{g,k}$ by gluing together the last two marked points.

Choose a homogeneous basis $\{\beta_b\}_{1 \leq b \leq L}$ of $H_*(V, \mathbf{Z})$ modulo torsion. Let (η_{ab}) be its intersection matrix. Note that $\eta_{ab} = \beta_a \cdot \beta_b = 0$ if the dimensions of β_a and β_b are not complementary to each other. Put (η^{ab}) to be the inverse of (η_{ab}). Now we can state the composition law, which consists of two formulas.

Theorem 4.1. ([RT2]) *Let $[K_i] \in H_*(\overline{\mathcal{M}}_{g_i,k_i+1}, \mathbf{Q})$ $(i = 1, 2)$ and $[K_0] \in$*

$H_*(\overline{\mathcal{M}}_{g-1,k+2}, \boldsymbol{Q})$. *For any* $\alpha_1, \cdots, \alpha_k$ *in* $H_*(V, \boldsymbol{Z})$. *Then we have*

$$\begin{aligned}\Psi^V_{(A,g,k)} &(\theta_{S*}[K_1 \times K_2]; \{\alpha_i\}) \\ &= \sum_{A=A_1+A_2} \sum_{a,b} \Psi^V_{(A_1,g_1,k_1+1)}([K_1]; \{\alpha_i\}_{i\leq k}, \beta_a)\eta^{ab} \\ &\qquad \times \Psi^V_{(A_2,g_2,k_2+1)}([K_2]; \beta_b, \{\alpha_j\}_{j>k})\end{aligned} \tag{4.1}$$

$$\begin{aligned}\Psi^V_{(A,g,k)}&(\mu_*[K_0]; \alpha_1, \cdots, \alpha_k) \\ &= \sum_{a,b} \Psi^V_{(A,g-1,k+2)}([K_0]; \alpha_1, \cdots, \alpha_k, \beta_a, \beta_b)\eta^{ab}\end{aligned} \tag{4.2}$$

Let us sketch the proof of (4.1). The other case is similar. The proof is based on the Gromov-Uhlenbeck compactness theorem and a gluing theorem ([RT1]). We regard $K_1 \times K_2$ as a cycle in $\overline{\mathcal{M}}_{g,k}$. Let E_i be a sequence of cycles converging to $E_\infty = K_1 \times K_2$ such that the top strata of E_i lie in $\mathrm{Reg}(\mathcal{M}_{g,k})$. Put

$$\begin{aligned}\mathcal{N}_{A,i} = \{&(\Sigma, x_1, \cdots, x_k, f) \in \mathcal{M}_A(g, k, J, \nu) \\ &\mid (\Sigma, x_1, \cdots, x_k) \in E_i, f(x_i) \in C_i\}\end{aligned} \tag{4.3}$$

where C_i denotes a generic cycle representing α_i. Using Proposition 2.1, 2.2, one can show that for a generic pair (J, ν), each $\mathcal{N}_{A,i}$ consists of finitely many smooth points in $\mathcal{M}_A(g, k, J, \nu)$, consequently,

$$\Psi^V_{(A,g,k)}(\theta_*[K_1 \times K_2]; \alpha_1, \cdots, \alpha_k)$$

counts the number of elements in $\mathcal{N}_{A,i}$ with sign for each $i < \infty$. Note that $\theta_*[K_1 \times K_2] = [E_i]$.

Next we show that $\mathcal{N}_{A,i}$ converges to $\mathcal{N}_{A,\infty}$. Proposition 2.2 implies that $\lim \mathcal{N}_{A,i} \subset \mathcal{N}_{A,\infty}$. The following proposition is just Corollary 6.1 in [RT1] in slightly different form.

Proposition 4.2. *Let* $(\Sigma_\infty, x_{1\infty}, \cdots, x_{k\infty}, f_\infty)$ *be in* $\mathcal{N}_{A,\infty}$. *Assume that* f_∞ *is regular in the sense: the linearization* L_{f_∞} *of* (2.1) *at* f_∞ *has trivial cokernel* (cf. [RT1], p. 51-52 for details, and this assumption is true for a generic (J, ν)). *Then for i sufficiently large, there is a unique* $(\Sigma_i, x_{1i}, \cdots, x_{ki}, f_i)$ *in* $\mathcal{N}_{A,i}$ *which is close to* $(\Sigma_\infty, x_{1\infty}, \cdots, x_{k\infty}, f_\infty)$. *Moreover, orientations of the corresponding moduli spaces are the same at* $(\Sigma_i, x_{1i}, \cdots, x_{ki}, f_i)$ *and* $(\Sigma_\infty, x_{1\infty}, \cdots, x_{k\infty}, f_\infty)$.

This shows that $\lim \mathcal{N}_{A,i} = \mathcal{N}_{A,\infty}$, so $\Psi^V_{(A,g,k)}(\theta_{S*}[K_1 \times K_2]; \{\alpha_i\})$ is equal to the number of elements in $\mathcal{N}_{A,\infty}$, counted with sign. Each f_∞ in $\mathcal{N}_{A,\infty}$ consists of two (J, ν)-maps $f_{\infty 1}, f_{\infty 2}$ from two components of Σ_∞. Moreover, the image of $(f_{\infty 1}, f_{\infty 2})$ in $V \times V$ intersects with the diagonal class Δ. So the left side of (4.1) is the number of pairs of (J,ν)-maps $(f_{\infty 1}, f_{\infty 2})$ which intersect Δ, counted with sign. However, by constructing a cobordism as in the proof of Theorem 2.4, one can show that this number depends only the homology class $[\Delta]$ of Δ. On the other hand, by the Künneth formula,

$$[\Delta] = \sum_{a,b} \eta^{ab} \beta_a \otimes \beta_b$$

So (4.1) follows.

Corollary 4.3. *Let $\Phi_{(A,\omega,g)}$ be the mixed invariant in* [RT1] (see Remark 2.4). *Then for any $g = g_1 + g_2$ with $2g_1 + m \geq 3$ and $2g_2 + k - m \geq 3$, we have*

$$\begin{aligned} &\Phi_{(A,\omega,g)}(\{\alpha_i\}_{1\le i\le k} | \{\alpha_{k+j}\}_{1\le j\le l}) \\ &= \sum_{A=A_1+A_2} \sum_{a,b} \sum_{S=S_1\cup S_2} \epsilon(S) \Phi_{(A_1,\omega,g_1)}(\{\alpha_i\}_{1\le i\le m} | \{\alpha_j\}_{j\in S_1}) \\ &\qquad \times \Phi_{(A_2,\omega,g)}(\{\alpha_i\}_{m+1\le i\le k} | \{\alpha_j\}_{j\in S_2}) \end{aligned} \tag{4.4}$$

where $S = \{k+1, \cdots, k+l\}$, $S_1 \cap S_2 = \emptyset$, $\epsilon(S)$ denotes the sign of the permutation induced by S on those α_i with odd degree. In particular, all mixed invariants can be computed in terms of $\Psi^V_{(A,0,k)}(\overline{\mathcal{M}}_{0,k}; \cdots)$, i.e., primitive GW-invariants of genus zero.

Roughly speaking, such a primitive invariant counts the number of rational curves in V through certain cycles in general position.

Mcduff and Salamon in [MS], Liu Gang in [Li] gave an alternative proof of Corollary 4.3 in case $g = 0$ and $k = 4$.

For each $k \geq 4$, there is a fibration $\pi : \overline{\mathcal{M}}_{0,k} \mapsto \overline{\mathcal{M}}_{0,k-1}$ such that all fibers over $\mathcal{M}_{0,k-1}$ are CP^1. Then one can show that any $[K]$ in $H_*(\overline{\mathcal{M}}_{0,k}, Q)$ can be represented by a cycle in $\overline{\mathcal{M}}_{0,k-1}$, as long as $[K]$ is not of top dimension. Therefore, one can deduce from the composition law

Corollary 4.4. *All genus zero GW-invariants $\Psi^V_{(A,0,k)}$ can be reduced to primitive GW-invariants.*

Later we will see that all genus zero GW-invariants can be actually reduced to finitely many primitive GW-invariants for a large class of Fano manifolds.

5 Quantum Cohomology

Let (V,ω) be a semi-positive, compact symplectic manifold. The quantum cohomology on V is the cohomology $H^*(V, \boldsymbol{Z}\{H_2(V)\})$ with a ring structure defined by GW-invariants. The Novikov ring $\boldsymbol{Z}\{H_2(V)\}$, which first appeared in Novikov's study of Morse theory for multivalued functions (cf. [No]), can be described as follows (cf. [HS], [Pu], [RT1]): choose a basis $q_1, \cdots, q_s$ of $H_2(V, \boldsymbol{Z})$, we identify the monomial $q^d = q_1^{d_1} \cdots q_s^{d_s}$ with the sum $\sum_{i=1}^s d_i q_i$. This turns $H_2(V)$ into a multiplicative ring, i.e., $q^d \cdot q^{d'} = q^{d+d'}$. This multiplicative ring has a natural grading defined by $\deg(q^d) = 2c_1(V)(\sum d_i q_i)$. Then $\boldsymbol{Z}\{H_2(V)\}$ is the graded homogeneous ring generated by all formal power series $\sum_{d=(d_1,\cdots,d_s)} n_d q^d$ satisfying: $n_d \in \boldsymbol{Z}$, all q^d with $n_d \neq 0$ have the same degree and the number of n_d with $\omega(\sum d_i q_i) \leq c$ is finite for any $c > 0$. If V is a Fano manifold or a monotone symplectic manifold, then $\boldsymbol{Z}\{H_2(V)\}$ is just a group ring.

Now we can define a ring structure on $H^*(V, \boldsymbol{Z}\{H_2(V)\})$. For any α^*, β^* in $H^*(V, \boldsymbol{Z})$, we define the quantum multiplication $\alpha^* \bullet \beta^*$ by

$$\alpha^* \bullet \beta^*(\gamma) = \sum_{A \in H_2(V,\boldsymbol{Z})} \Psi^V_{(A,0,3)}(\alpha, \beta, \gamma) q^A \tag{5.1}$$

where $\gamma \in H_*(V, \boldsymbol{Z})$. Equivalently, if $\{\beta_a\}$ is a basis as in Theorem 4.1,

$$\alpha^* \bullet \beta^* = \sum_A \sum_{a,b} \Psi^V_{(A,0,3)}(\alpha, \beta, \beta_a)\, \eta^{ab}\, \beta_b^*\, q^A \tag{5.2}$$

Note that if $A = \sum a_i q_i$, we identify A with $(a_1, \cdots, a_s)$. In general, any α^*, β^* in $H^*(V, \boldsymbol{Z}\{H_2(V)\})$ can be written:

$$\alpha^* = \sum_d \alpha_d^* q^d, \ \beta^* = \sum \beta_{d'}^* q^{d'}$$

where α_d^*, $\beta_{d'}^*$ are in $H^*(V, \boldsymbol{Z})$. We define

$$\alpha^* \bullet \beta^* = \sum_{d,d'} \alpha_d^* \bullet \beta_{d'}^*\, q^{d+d'} \tag{5.3}$$

Recall that the degree of $\alpha_d^* q^d$ is $\deg(\alpha^*) + \deg(q^d)$. It follows that the multiplication preserves the degree. However, it is not clear at all if the multiplication is associative. Given α^*, β^*, γ^* in $H_*(V, \boldsymbol{Z})$ and any δ in $H_*(V, \boldsymbol{Z})$, we have

$$((\alpha^* \bullet \beta^*) \bullet \gamma^*)(\delta) = \sum_{A,B} \sum_{a,b} \Psi^V_{(A,0,3)}(\alpha, \beta, \beta_a) \eta^{ab} \Psi^V_{(A,0,3)}(\beta_b, \gamma, \delta)$$

$$(\alpha^* \bullet (\beta^* \bullet \gamma^*))(\delta) = \sum_{A,B} \sum_{a,b} \Psi^V_{(A,0,3)}(\alpha, \beta_a, \delta) \eta^{ab} \Psi^V_{(B,0,3)}(\beta, \gamma, \beta_b)$$

So the associativity means that for any fixed A in $H_2(V, \boldsymbol{Z})$,

$$\sum_{A_1+A_2=A} \sum_{a,b} \Psi^V_{(A_1,0,3)}(\alpha, \beta, \beta_a) \eta^{ab} \Psi^V_{(A_2,0,3)}(\beta_b, \gamma, \delta)$$

$$= \sum_{A_1+A_2=A} \sum_{a,b} \Psi^V_{(A_1,0,3)}(\alpha, \beta_a, \delta) \eta^{ab} \Psi^V_{(A_2,0,3)}(\beta, \gamma, \beta_b) \tag{5.4}$$

But by either Theorem 4.1 or Corollary 4.3, both sides of (5.4) are equal to $\Psi^V_{(A,0,4)}([p];\alpha,\beta,\gamma,\delta)$. Therefore, we have

Theorem 5.1. ([RT1, Theorem 8.4]) *The quantum multiplication $\bullet$ is associative, consequently, there is an associative, supercommutative, graded ring structure, i.e., quantum ring structure, on $H_*(V, \mathbf{Z}\{H_2(V)\})$.*

In physics and sometimes mathematical literatures, one substitutes q by $(e^{-t\omega(q_1)}, \cdots, e^{-t\omega(q_s)})$, so the quantum product becomes

$$\alpha^* \cdot \beta^* = \sum_A \Psi^V_{(A,0,3)}(\alpha,\beta,\beta_a)\eta^{ab}\beta^*_b e^{-t\omega(A)} \tag{5.5}$$

In particular, this converges to the classical cup product as $t \to \infty$. If $c_1(V) > 0$, then

$$\alpha^* \cdot \beta^* = \alpha^* \cup \beta^* + \sum_{c_1(V)(A)>0} \phi_A$$

where ϕ_A is of degree $\deg(\alpha^* \cup \beta^*) - 2c_1(V)(A)$.

Examples:
(1) ([GK]) Let F_{n+1} be the Flag manifold of all sequences of subspaces

$$E_1 \subset E_2 \subset \cdots \subset E_n$$

of $\mathbf{C}^{n+1}$ with $\dim_{\mathbf{C}} E_j = j$. Its cohomology ring is generated by the first Chern classes u_j of the universal line bundles $L_j \mapsto F_{n+1}$ with fiber E_{j+1}/E_j. Note that $\sum u_j = 0$. Givental and Kim observed that there is a natural isomorphism

$$H^*(F_{n+1}, \mathbf{Z}\{H_2(F_{n+1})\}) = \frac{\mathbf{Z}[u_0, \cdots, u_n, q_0, \cdots, q_n]}{\mathcal{I}}$$

where $\mathcal{I}$ denotes the ideal generated by coefficients c_j of the characteristic polynomial

$$\det(A_n + \lambda) = \lambda^n + c_1\lambda^{n-1} + \cdots + c_{n+1},$$

and A_n is defined by induction: $A_0 = u_0$, $A_1 = u_0u_1 + q_1$, $\cdots$, $A_k = u_kA_{k-1} + q_kA_{k-2}, \cdots$. Note that these c_j are the Poisson commuting integrals of the Toda lattice studied by J. Moser (cf. [GK], [MS]). The quantum cohomology has been computed (in a heuristic way) for partial Flag manifolds by Astashkevich and Sadov [AS]. Recently, these computations have been rigorously verified (cf. [CF], [Lu]).

Before we discuss further examples, we give a simple lemma on the general structure of quantum cohomology of Fano manifolds. This lemma is proved in [ST1], but we will state it slightly differently.

Lemma 5.4. *Let (V,ω) be a compact, semipositive symplectic manifold, such as Fano manifolds. If its ordinary cohomology ring has a presentation*

$Z[x_1, \cdots, x_L]/\mathcal{I}$, where $\mathcal{I}$ is an ideal $\{f_1, \cdots, f_m\}$ generated by homogeous polynomials $f_1, \cdots, f_m$, then the quantum ring is isomorphic to

$$\frac{Z[x_1, \cdots, x_L, q]}{\{F_1(x, q), \cdots, F_m(x, q)\}}$$

where $F_i(x, q)$ are homogeneous polynomial in x, q such that $F(x, 0) = f(x)$.

(2) The quantum cohomology of the Grassmannian $G(r, n)$ has been computed in [ST1], [W3], [Pu]. Let S be the tautological bundle over $G(r, n)$ of complex k-planes in C^n. It is known that $H^*(G(r, n), Z)$ is given by

$$\frac{Z[x_1, \cdots, x_r]}{\{s_{n-r+1}, \cdots, s_n\}}$$

where s_j are Segre classes, defined inductively by $s_j = -x_1 s_{j-1} - \cdots - x_{j-1}s_1 - x_j$. In fact, x_i corresponds to the i-th Chern class $c_i(S)$ ($i = 1, \cdots, r$). Using Lemma 5.4, one can prove (see [ST1])

$$H^*(G(r, n), Z\{H_2(G(r, n))\}) = \frac{Z[x_1, \cdots, x_r, q]}{\{s_{n-r+1}, \cdots, s_{n-1}, s_n + (-1)^r q\}}.$$

(3) Batyrev has computed the quantum cohomology of toric Fano varieties (cf. [Ba]). More recently, Morrison and Plesser were able to generalize Batyrev's computations to singular toric varieties (cf. [MP]).

(4) Let V be a hypersurface in CP^{n+1} ($n \geq 3$) of degree $d \leq n + 1$. It is a Fano manifold. Its rational cohomology is generated by the hyperplane class H and the primitive cohomology $H^n(V, Q)_0$, with the relations:

$$H^{n+1} = 0, \ H \cup \alpha = 0, \ \alpha \cup \beta = \frac{1}{d}\left(\int_V \alpha \wedge \beta\right) H^n$$

for $\alpha, \beta \in H^n(V, Q)_0$. Since all primitive classes are Poincare dual to vanishing cycles, using the discussion in section 3, one can show

$$H_{n+1} = \sum_{i=1}^{[\frac{n+1}{r}]} a_i H_{n+1-ir} q^i, \ H \bullet \alpha = \max\{0, d-n\}\tilde{\alpha}\, q,$$

$$\alpha \bullet \beta = \frac{1}{d}\int_V \alpha \wedge \beta \left(H_n - \sum_{i=1}^{[\frac{n}{r}]} b_i H_{n-ir} q^i\right)$$

where H_l denotes the quantum product of l classes H, $r = n + 2 - d$ is the index and $\alpha, \tilde{\alpha}, \beta \in H^n(V, Q)_0$.

If $r \geq 2$, then $H \bullet \alpha = 0$. By the associativity,

$$0 = (H \bullet \alpha) \bullet \beta = \frac{1}{d}\int_V \alpha \wedge \beta \left(H_{n+1} - \sum_{i=1}^{[\frac{n}{r}]} b_i H_{n+1-ir} q^i\right)$$

It follows that $a_{n+1} = 0$, $a_i = b_i$ for $1 \leq i \leq n$, whenever $H^*(V, \boldsymbol{Q})_0$ is non-trivial. The coefficient $a_1 = b_1$ encodes the contribution of lines in V. By a standard computation, one can deduce from the above a result of Beauville [Be1]: in case $n \geq 2d - 3$, the quantum cohomology of V over rational numbers is generated by H and $H^n(V, \boldsymbol{Q})_0$ with relations

$$H_{n+1} = d^d H_{d-1} q, \; H \bullet \alpha = 0, \; \alpha \bullet \beta = \frac{1}{d} \int_V \alpha \wedge \beta \left(H_n - d^d H_{d-2} q\right)$$

for $\alpha, \beta \in H^n(V, \boldsymbol{Q})_0$.

If $r = 1$, i.e., $d = n + 1$, then for any β in $H^n(V, \boldsymbol{Q})$,

$$\tilde{\alpha} \cup \beta([V]) = \Psi^V_{(1,0,3)}(H, \alpha, \beta).$$

It is shown in [Be1] that

$$\Psi^V_{(1,0,3)}(H, \alpha, \beta) = -(n+1)^n \int_V \alpha \wedge \beta.$$

Therefore, for any primitive class α, we have

$$\left(H + (n+1)^{(n+1)}\right) \bullet \alpha = 0.$$

In the preliminary version of this paper, I dropped the term $(n+1)^{(n+1)}$ in the above equation. This error was pointed to me by A. Collino.

Using the above equations, one can deduce

$$\begin{aligned} a_i &= b_i + (n+1)^{(n+1)} b_{i-1} \text{ for } 2 \leq i \leq n+1, \\ A_1 &= b_1 - (n+1)^{(n+1)}. \end{aligned}$$

These discussions are valid for complete intersections in $\boldsymbol{C}P^{n+l}$.

In general, let V be any compact Kähler manifold and $\omega \in H^2(V, \boldsymbol{R})$ be a Kähler class. The Hard Lefschetz Theorem states that $H^*(V, \boldsymbol{R})$ has the following decomposition

$$H^*(V, \boldsymbol{R}) = \oplus_{k=1}^{[\frac{n}{2}]} \oplus_{i=0}^{n-k} \omega^i \cup P^k(V), \tag{5.6}$$

where $P^k(V)$ is the kth-primitive part of $H^k(V, \boldsymbol{R})$ with respect to the Kähler class ω. In view of (3.3) and its generalization, we propose

Conjecture 5.5. *Let V be a compact Kähler manifold. For each k, we denote by $\{H^2(V, \boldsymbol{R}), P^k(V)\}$ the subspace of $H^*(V, \boldsymbol{R})$ spanned by classes of the form $\alpha^i \cup \beta$, where $\alpha \in H^2(V, \boldsymbol{R})$ and $\beta \in P^k(V)$. Then*

$$H^2(V, \boldsymbol{R}) \bullet \{H^2(V, \boldsymbol{R}), P^k(V)\} \subset \{H^2(V, \boldsymbol{R}), P^k(V)\}. \tag{5.7}$$

This can be thought as a quantized version of the Hard Lefschetz Theorem. If this is true, then the quantum multiplication on $H^*(V, \boldsymbol{Z}\{H_2(V)\})$ induces a subring structure on $H^*_{\text{inv}}(V, \boldsymbol{Z}\{H_2(V)\})$, which is just

$$H^*_{\text{inv}}(V, \boldsymbol{R}) \otimes \boldsymbol{Z}\{H_2(V)\}.$$

If V is a Fano manifold with $H^2(V, \mathbf{Z}) = \mathbf{Z}$, then the conjecture means that for any $k \geq 3$,

$$\oplus_{i=0}^{n-k} c_1(V)_i \bullet P^k(V) \;=\; \oplus_{i=0}^{n-k} c_1(V)^i \cup P^k(V). \tag{5.8}$$

In this case, one may further expect that $c_1(V) \bullet \alpha = c_1(V) \cup \alpha$ for any $\alpha \in P^k(V)$.

Finally, we would like to mention that Crauder and Miranda [CM] computed quantum cohomology on rational surfaces, such as cubic surfaces in $\mathbf{C}P^3$. Recently, Siebert and the author proved a recursion formula for the quantum cohomology of the moduli spaces of stable bundles over Riemann surfaces Σ_g (cf. [ST2]). In view of [Do], [DS], this computes certain product structure on the instanton Floer cohomology on 3-manifolds $S^1 \times \Sigma_g$.

6 The WDVV Equation

In fact, there is a family of quantum multiplications, containing $\bullet$ of the last section as a special case. These quantum multiplications make use of all the primitive GW-invariants, while $\bullet$ uses only the invariant $\Psi^V_{(A,0,3)}$.

Let $\{\beta_a\}_{1\leq a\leq L}$ be an integral basis of $H_*(V, \mathbf{Z})$ modulo torsion. Any $w \in H^*(V, \mathbf{C})$ can be written as $\sum t_a \beta_a^*$. Clearly, $w \in H^*(V, \mathbf{Z})$ if all t_a are integers. We define the quantum multiplication $\bullet_w$ by

$$\begin{aligned} &\alpha^* \bullet_w \beta^*(\gamma) \\ =&\sum_A \sum_{k\geq 0} \frac{\epsilon(\{a_i\})}{k!} \Psi^V_{(A,0,k+3)}(\alpha, \beta, \gamma, \beta_{a_1}, \cdots, \beta_{a_k}) t_{a_1} \cdots t_{a_k} q^A \end{aligned} \tag{6.1}$$

where $\alpha, \beta, \gamma \in H_*(V, \mathbf{Z})$, and $\epsilon(\{a_i\})$ is the sign of the induced permutation on odd dimensional β_a. Obviously, this multiplication reduces to $\bullet$ at $w = 0$. As we argued in the last section, the associativity of $\bullet_w$ is equivalent to

$$\begin{aligned} &\sum_{A=A_1+A_2} \sum_{a,b} \sum_{\{1,\cdots,k\}=S_1\cup S_2} \epsilon(S_1, S_2) \Psi^V_{(A_1,0,|S_1|+3)}(\alpha, \beta, \beta_a, \{\beta_{a_i}\}_{a_i\in S_1}) \eta^{ab} \\ &\qquad \times \Psi^V_{(A_2,0,|S_2|+3)}(\beta_b, \gamma, \delta, \{\beta_{a_i}\}_{a_i\in S_2}) \\ =& \sum_{A=A_1+A_2} \sum_{a,b} \sum_{\{1,\cdots,k\}=S_1\cup S_2} (-1)^{\deg(\beta^*\cup\gamma^*)} \epsilon(S_1, S_2) \\ &\qquad \times \Psi^V_{(A_1,0,|S_1|+3)}(\alpha, \gamma, \beta_a, \{\beta_{a_i}\}_{a_i\in S_1}) \eta^{ab} \\ &\qquad \times \Psi^V_{(A_2,0,|S_2|+3)}(\beta_b, \beta, \delta, \{\beta_{a_i}\}_{a_i\in S_2}) \end{aligned} \tag{6.2}$$

for all $\alpha, \beta, \gamma, \delta$, where $\epsilon(S_1, S_2)$ is the sign of the permutation induced by S_1, S_2 on odd dimensional β_a. However, (6.2) follows easily from Corollary 4.3 or Theorem 4.1. Therefore, we have

Theorem 6.1. *The quantum multiplications $\bullet_w$ are associative.*

The associativity (6.2) can be restated in terms of a solution of WDVV equations. To avoid additional technicalities in using superstructures, we confine ourselves to the case $w \in W = H^{\mathrm{even}}(V, \boldsymbol{C})$. The general case can be treated similarly (cf. [KM]).

Following Witten, we define a generating function (prepotential in physical literature)

$$\Phi^V(w,q) = \sum_{A \in H_2(V,\boldsymbol{Z})} \sum_{k \geq 3} \frac{1}{k!} \Psi^V_{(A,0,k)}(\overline{\mathcal{M}}_{0,k}; w, \cdots, w)\, q^A \tag{6.3}$$

This function is a formal power series in w, q. In fact, the generating function in [W2] is equal to $\Phi^V(w,(1,\cdots,1))$, i.e., Φ^V specified at $q=(1,\cdots,1)$. Its convergence remains to be a difficult problem in general (cf. section 7). By inserting q, we gain the advantage of avoiding the convergence problem. We may arrange the basis $\{\beta_a\}$ such that β_a is an even class if and only if $a \leq N$. Then for any $a, b, c \leq N$,

$$(\beta_a \bullet_w \beta_b)(\beta_c) = \frac{\partial^3 \Phi^V}{\partial t_a \partial t_b \partial t_c}(w) \tag{6.4}$$

Then (6.2) implies

Theorem 6.2. ([RT1]) *Let V be a semi-positive symplectic manifold. Then for any specified number q, Φ^V_w satisfies the associativity equation:*

$$\sum_{a,b} \frac{\partial^3 \Phi^V}{\partial t_i \partial t_j \partial t_a} \eta^{ab} \frac{\partial^3 \Phi^V}{\partial t_k \partial t_l \partial t_b} = \sum_{a,b} \frac{\partial^3 \Phi^V}{\partial t_i \partial t_k \partial t_a} \eta^{ab} \frac{\partial^3 \Phi^V}{\partial t_j \partial t_l \partial t_b} \tag{6.5}$$

where $1 \leq i, j, k, l \leq N$.

According to Dubrovin (cf. [Du], [RT1], [MS]), a solution of (6.4) gives rise to a family of flat connections. If we define $\Gamma = \{\Gamma^c_{ab}\}$ by

$$\Gamma^c_{ab} = \sum_e \eta^{ce} \frac{\partial^3 \Phi^V}{\partial t_a \partial t_b \partial t_e}$$

then $\nabla^\epsilon = d + \epsilon\Gamma$ defines a family of connections on the tangent bundle TW over W, where d denotes the standard differentiation on W. Then (6.5) is the same as flatness of ∇^ϵ. In fact, Dubrovin observed

Theorem 6.3. *Each ∇^ϵ is a torsion-free, flat connection satisfying:*

$$d\langle \alpha, \beta \rangle = \langle \nabla^\epsilon \alpha, \beta \rangle$$

where $\langle \cdot, \cdot \rangle$ is the inner product on W induced by (η_{ab}). Moreover, $\nabla^\epsilon \mathbf{1} = \sqrt{-1}$ for the identity $\mathbf{1}$ on W.

The generating function Φ^V has the following scaling property: let $\sigma(\lambda)$ ($\lambda \in \boldsymbol{C}^*$) be the scaling map:

$$\sigma(\lambda)(t_1, \cdots, t_N) = (\lambda^{\deg(\beta_1^*)-2} t_1, \cdots, \lambda^{\deg(\beta_N^*)-2} t_N) \tag{6.6}$$

Then

$$\Phi^V(\sigma(\lambda)(w), \lambda^{-2c_1(V)} q) = \lambda^{2(\dim_{\boldsymbol{C}} V-3)} \Phi^V(w, q) \tag{6.7}$$

Note that $\lambda^{-2c_1(V)} q = (\lambda^{-2c_1(V)(q_1)} q_1, \cdots, \lambda^{-2c_1(V)(q_s)} q_s)$. Consequently, the Dubrovin connection satisfies:

$$\Gamma_{ab}^c(\sigma(\lambda)w, \lambda^{-2c_1(V)} q) = \lambda^{\deg(\beta_c^*)-\deg(\beta_a^*)-\deg(\beta_b^*)} \Gamma_{ab}^c \tag{6.8}$$

These scaling properties follow simply from the definition of Ψ^V. Infinitesimally, (6.8) can be reformulated ([Du]) as

$$\mathcal{L}_X \Phi^V = 2(\dim_{\boldsymbol{C}} V - 3)\Phi^V \tag{6.9}$$

where $\mathcal{L}_X$ denotes the Lie derivative, and X is the vector field

$$\sum_{a \le N} (\deg(\beta_a^*) - 2) t_a \partial_a - \sum_s 2c_1(V)(q_s) \partial_{q_s} \tag{6.10}$$

The Witten-Dijkgraaf-Verlinde-Verlinde (WDVV) equation is just the associativity equation (6.4) together with the scaling property (6.9). It follows from Theorem 6.3 and (6.9) that quantum multiplications $\bullet_w$ induce the structure of a so-called Frobenius algebra on W (see [Du], [KM]).

In order to further investigate the prepotential, we arrange the basis $\{\beta_a\}$ such that $\deg(\beta_N^*) = 2n$ and $\deg(\beta_a^*) = 2n - 2$ whenever $N - s \le a < N$. Then using Remark 2.4, one can deduce

$$\begin{aligned} &\frac{\partial^3 \Phi^V}{\partial t_a \partial t_b \partial t_c}(t_1, \cdots, t_N; q) \\ = &\frac{\partial^3 \Phi^V}{\partial t_a \partial t_b \partial t_c}(t_1, \cdots, t_{N-s-1}, 0, \cdots, 0; e^{t_{N-s}} q_1, \cdots, e^{t_{N-1}} q_s) \end{aligned} \tag{6.11}$$

It follows that

$$\Phi^V(t_1, \cdots, t_N; q) - \Phi^V(t_1, \cdots, t_{N-s-1}, 0, \cdots, 0; e^{t_{N-s}} q_1, \cdots, e^{t_{N-1}} q_s)$$

is a quadratic polynomial.

Examples: (1) $V = \boldsymbol{CP}^n$, choose $\beta_a = H^{n+1-a}$ $(1 \le a \le n+1)$, where H is the hyperplane class. For $n = 2$, we have

$$\Phi^{CP^2}(t_1, t_2, t_3; 1) = \frac{1}{2}(t_1 t_3^2 + t_3 t_2^2) + \sum_{k \ge 1} \frac{n_k t_1^{3k-1}}{(3k-1)!} e^{k t_2} \tag{6.12}$$

where n_k is the number of rational curves of degree k in CP^2 through $3d-1$ points in general position.

In general, let $S_{n,k}$ be the set of $(n-1)$-tuples $\{k_a\}_{1\le a\le n-1}$ with $k_i \ge 0$ and $\sum_{i=1}^{n-1} ik_{n-i} = (n+1)k + n - 3$. Then we have

$$\Phi^{CP^n}(t_1,\cdots,t_{n+1};1) = \frac{1}{6}\sum_{a+b+c=2n+3} t_a t_b t_c + \sum_{k\ge 1}\sum_{\{k_a\}\in S_{n,k}} \frac{\sigma_k^{CP^n}(k_1,\cdots,k_{n-1})t_1^{k_1}\cdots t_{n-1}^{k_{n-1}}}{k_1!\cdots k_{n-1}!}e^{kt_n} \tag{6.13}$$

where $\sigma_k^V(k_1,\cdots,k_{n-1})$ denotes the number of degree k rational curves in V through k_1 points, ... k_{n-1} subspaces of dimension $n-2$ in general position.

(2) Let V be a hypersurface in CP^{n+1} of degree $d \le n+2$. Then the hyperplane section $H \cap V$ generates a subring $H_{\text{inv}}(V, Z)$ of $H_*(V, Z)$, which is invariant under the monodromy group. Let Φ_{inv}^V be the restriction of Φ^V to $H_{\text{inv}}(V, C)$. Then by the discussions in section 3, Φ_{inv}^V also satisfies the WDVV equation. Furthermore, if we put $S_{V,k}$ to be the set of $n-1$-tuples $\{k_a\}_{1\le a\le n-1}$ with $k_i \ge 0$ and $\sum_{i=1}^{n-1} ik_{n-i} = (n+2-d)k + n - 3 > 0$, then

$$\begin{aligned}&\Phi_{\text{inv}}^V(t_1,\cdots,t_{n+1};1)\\ =&\frac{d}{6}\sum_{a+b+c=2n+3} t_a t_b t_c + \sum_{k\ge 1}\sum_{\{k_a\}\in S_{V,k}} \frac{\sigma_k^V(k_1,\cdots,k_{n-1})t_1^{k_1}\cdots t_{n-1}^{k_{n-1}}}{k_1!\cdots k_{n-1}!}e^{kt_n}\end{aligned} \tag{6.14}$$

If V is a Calabi-Yau 3-fold, i.e., $d = 5, n = 3$, it has been proved (cf. [AM], [Vo]) that

$$\frac{\partial^3\Phi^V(t_1,t_2,t_3,t_4)}{\partial t_3^3} = 5 + \sum_{k\ge 1}\frac{k^3 n_k e^{kt_3}}{1-e^{kt_3}} \tag{6.15}$$

$$\frac{\partial^3\Phi^V(t_1,t_2,t_3,t_4)}{\partial t_i\partial t_j\partial t_k} = \frac{5}{2}\frac{\partial^3}{\partial t_i\partial t_j\partial t_k}(t_1t_4^2 + 2t_2t_3t_4),\ (i,j,k)\ne(3,3,3),$$

where n_k denotes the number of irreducible rational curves in V.

In the following, we will take $q = (1,\cdots,1)$.

Consider the open subset M in $H^*(V, C)$ where Φ^V is well-defined and the quantum multiplication by $X(w)$ on $H^*(V, C)$ has no multiple spectrum, i.e., semi-simple according to B. Dubrovin. Therefore, we can find new local coordinates $u_1,\cdots,u_N$ of M, which are simply eigenvalues of the quantum multiplication $X(w)\bullet_w$ on $H^*(V, C)$. Dubrovin proved that in these new coordinates $u_1,\cdots,u_N$, the prepotential Φ^V can be extended to be a meromorphic function on $(CP^1)^N$. When the action $X(w)\bullet_w$ is nilpotent, like in the case of Calabi-Yau manifolds, it is unclear how to find such new coordinates that Φ^V can be extended.

To find his extension, Dubrovin identifies the WDVV equation with a certain Riemann-Hilbert problem ([Du]).

Denote by ∂_a the differentiation $\frac{\partial}{\partial t_a}$ in the direction of β_a. Then flatness of the Dubrovin connection ∇^ϵ is equivalent to the compatibility of the linear system

$$\partial_a \xi^b \;=\; -\,\epsilon \Gamma^b_{ac} \xi^c \tag{6.16}$$

This gives a "Lax pair" for the associativity equation (6.5), where ϵ plays the role of the spectral parameter. For any ϵ , this system has N linearly independent solutions, which give rise to a flat coordinate system for ∇^ϵ. In fact, the solutions $\{\xi^a\}$ of (6.16) correspond one-to-one to flat sections $\xi = \xi^a \partial_a$ of ∇^ϵ.

Next, the standard machinery of scaling reduction of integrable systems suggests to add a differential equation in the spectral parameter ϵ for the section $\xi = \xi(t, \epsilon)$ (cf. Flaschka and Newell [FN], Jimbo et. al. [JMS], Its and Novokshenov [IN]).

Lemma 6.4. ([Du, Proposition 3.1]) *The WDVV equation is equivalent to compatibility of the system* (6.16) *with the equation*

$$2\epsilon \frac{\partial \xi^a}{\partial \epsilon} = \epsilon X^p \Gamma^a_{bp} \xi^b + \partial_b X^a \xi^b \tag{6.17}$$

where X is defined in (6.10), *consequently,* $\partial_b X^a = (\deg(\beta_a) - 2)\delta_{ab}$.

Proof: Introduce differential operators D_b and D_ϵ as follows:

$$\begin{aligned} D_b \xi^a &= \partial_b \xi^a + \epsilon \Gamma^a_{bc} \xi^c \\ D_\epsilon \xi^a &= \frac{\partial \xi^a}{\partial \epsilon} - \frac{1}{2}\left(X^b \Gamma^a_{bc} \xi^c + \frac{1}{\epsilon} \partial_b X^a \xi^b \right) \end{aligned} \tag{6.18}$$

Then compatibility simply means that $[D_b, D_\epsilon] = 0$, namely, they commute with each other. A straightforward computation shows

$$\begin{aligned} &(D_b D_\epsilon - D_\epsilon D_b)\xi^a \\ =& -\frac{1}{2}\left(\partial_b X^p \Gamma^a_{pc} \xi^c + \partial_c X^p \Gamma^a_{pb} \xi^c - \partial_c X^a \Gamma^c_{pb} \xi^p + X^p \partial_b \Gamma^a_{pc} \xi^c\right) \\ & - \Gamma^a_{bc}\xi^c - \frac{\epsilon}{2}(\Gamma^a_{bc}\Gamma^c_{pe} - \Gamma^c_{be}\Gamma^a_{pc})\xi^e X^p \qquad (6.19) \\ =& -\frac{1}{2}\left((\deg(\beta_b) + \deg(\beta_c) - \deg(\beta_a))\Gamma^a_{bc} + X^p \partial_p \Gamma^a_{bc}\right)\xi^c \qquad \text{(associativity)} \\ =&0 \qquad \text{(homogeneity of } \Gamma^a_{bc}) \end{aligned}$$

The lemma is proved.

The compatibility of (6.17) can be reformulated (cf. [KM]) as vanishing of the curvature of ∇ on $H^*(V, \boldsymbol{C}) \times \boldsymbol{C}$, where ∇ is the connection such that D_b, D_ϵ are its induced differentiations on components ξ^a of any section $\xi = \xi^a \partial_a$.

The basic idea (also see [FN], [IN], [JMS]) is to parametrize the solutions of the WDVV equation by the monodromy data of the operator (6.11). The problem is to find the monodromy. This was done by Dubrovin under a certain semi-simplicity assumption on the quantum product $\bullet_t$.

Consider the operator

$$\Lambda = 2\frac{d}{d\epsilon} - U - \frac{1}{\epsilon}V, \tag{6.20}$$

where U and V are two complex $N \times N$ ϵ-independent matrices, U is a diagonal matrix with distinct eigenvalues, and V is skew-symmetric. The solutions of $\Lambda\varphi = 0$ are analytic multivalued functions in $C\backslash\{0\}$. The monodromy of these solutions is called monodromy of Λ.

Assume that for a given w in $H^*(V, C)$, the multiplication $X\bullet_w$, where X is as in (6.10), induces a semi-simple $N \times N$ matrix. Then the eigenvalues $u_1, \cdots, u_N$ of this matrix give rise to new coordinates near w in $H^*(V, C)$. One can show that the intersection form (flat metric) η is diagonalized under the new coordinates $u_1, \cdots, u_N$, i.e.,

$$\eta(\frac{\partial}{\partial u_i}, \frac{\partial}{\partial u_j}) = \eta_{ii}\delta_{ij}. \tag{6.21}$$

Put $\tilde{\partial}_i = \frac{1}{\sqrt{\eta_{ii}}}\frac{\partial}{\partial u_i}$, then $\eta(\tilde{\partial}_i, \tilde{\partial}_j) = \delta_{ij}$. Moreover, we have

$$\tilde{\partial}_i X = (u_j - u_i)\gamma_{ij}\tilde{\partial}_j, \ \gamma_{ij} = \gamma_{ji} = \frac{1}{\sqrt{\eta_{jj}}}\frac{\partial\sqrt{\eta_{ii}}}{\partial u_j}. \tag{6.22}$$

These γ_{ij} are called the rotation coefficients.

For any section $\xi = \xi^a\partial_a = \zeta^i\tilde{\partial}_i$, $D_\epsilon\xi^a = 0$ for all a if and only if for all i,

$$\Lambda_u\zeta^i = (2\frac{d}{d\epsilon} - U - \frac{1}{\epsilon}V)\zeta^i = 0, \tag{6.23}$$

where $U = \mathrm{diag}(u_1, \cdots, u_N)$ and $V = ((u_j - u_i)\gamma_{ij})_{1\le i,j\le N}$ (cf. [Du]). Since Λ_u is just the restriction of the flat connection ∇ to $\{u\} \times C$, the monodromy of Λ_u is independent of u. Thus solutions of the WDVV equation can be embedded into the space of isomonodromy deformations of operators Λ of the form (6.20). Each operator Λ has two singularities in the ϵ-sphere $C \cup \infty$: the regular singularity at $\epsilon = 0$ and the irregular one at $\epsilon = \infty$. The monodromy of Λ at $\epsilon = 0$ is determined by the matrix V. The monodromy at $\epsilon = \infty$ is specified by a $N \times N$ Stokes matrix (cf. [Du]). By analyzing the Stokes matrix at ∞ and using a result of T. Miwa, Dubrovin showed

Theorem 6.5 ([Du]). *Let Φ be any solution of the WDVV equation such that the induced multiplication by X is semi-simple. Then the corresponding matrix-valued function V, in coordinates $u_1, \cdots, u_N$, extends to a meromorphic function on the universal covering of $CP^{N-1}\backslash\{\text{diagonals}\}$.*

One can also identify the WDVV with the equation of isomonodromy deformations of an operator with regular singularities (cf. [Du]). This can be done by using the Laplace transformation. We will only carry out formal arguments here. These arguments can be made rigorous.

We denote by $L(f)$ the Laplace transform of f, namely,

$$L(f)(\lambda) = \int e^{-\lambda\epsilon} f(\epsilon) d\epsilon \tag{6.24}$$

Put $\zeta = L(\xi) = L(\xi^a)\partial_a$. Then $D_\epsilon \xi = 0$ implies

$$(2\lambda\delta_c^a - X^p\Gamma_{pc}^a)\frac{\partial\zeta^c}{\partial\lambda} - \deg(\beta_a)\zeta^a = 0, \tag{6.25}$$

in coordinates $u_1, \cdots, u_N$, this is the same as

$$\frac{\partial\zeta}{\partial\lambda} + (2\lambda - U)^{-1}(V - 2I)\zeta = 0, \tag{6.26}$$

The points $\lambda = u_1, \cdots, u_N, \infty$ are regular singularities of the coefficients. The monodromy preserving deformations of (6.26) were described by Schlesinger at the begining of this century. In modern language, they form a moduli space of flat holomorphic bundles, over punctured spheres, with fixed holonomy at each puncture. Therefore, the solution Φ^V of the WDVV equation gives rise to a meromorphic flat connection on the moduli space, which can be extended to be quasi-projective.

Finally, we briefly discuss the semi-simplicity assumption. We say that V is semi-simple in the sense of Dubrovin, if for a generic w, the quntum multiplication $X\bullet_w$ on $H^*(V, C)$ has only simple eigenvalues.

If V is a Calabi-Yau manifold, then for dimensional reason, the GW-invariant

$$\Psi^V_{(A,0,k+3)}(\alpha, \beta, \gamma, \{\beta_{a_i}\})$$

in (6.1) vanishes unless $\deg\alpha^* + \deg\beta^* + \deg\gamma^* \leq \dim_C V$. This implies that $X\bullet_w$ corresponds to certain lower triangular matrix with respect to a basis $\{\beta_a\}$ of $H^*(V, C)$. So the multiplication $X\bullet_w$ always has only one eigenvalue. Therefore, no Calabi-Yau manifold is semi-simple.

However, we expect that the semi-simplicity condition holds for Fano manifolds. If so, the above discussion indicates that counting rational curves in a Fano manifold can be identified with a deformation problem of flat bundles over a punctured sphere. In a sense, this can be thought as the Mirror symmetry phenomenon for Fano manifolds.

Let V be an n-dimensional Fano manifold. We put $\omega = \frac{1}{r}c_1(V)$, where r is the index. By the Hard Lefschetz Theorem, we have the decomposition

$$H^*(V, C) = \oplus_{k=0}^{[\frac{n}{2}]} \oplus_{i=0}^{n-k} \omega^i \cup P^k(V, C),$$

where $P^k(V,C) = \{\alpha \in H^*(V,C) \mid \omega^{n-k+1} \cup \alpha = 0\}$ is the kth-primitive part of $H^*(V,C)$. For each k, we choose a basis $\{\beta_{ik}\}_{1\le i\le N_k}$ of $P^k(V)$. Clearly, we have

$$N_0 = 1, \ N = \sum_{k=0}^{[\frac{n}{2}]}(n-k)N_k.$$

As before, we denote by $\omega_l = \omega \bullet \cdots \bullet \omega$ the quantum product of ω by l times. Since

$$\omega_l \bullet \beta_{ik} = \omega^l \cup \beta_{ik} + \text{ terms of lower degree}$$

$\{\omega_l \bullet \beta_{ik}\}$ forms a basis of $H^*(V,C)$.

Let U_w be the $N \times N$-matrix representing the quantum multiplication $X(w)\bullet_w$ in the basis $\{\omega_l \bullet \beta_{ik}\}$, and let h_w be its characteristic polynomial, i.e., $h_w(\lambda) = \det(\lambda I - U_w)$. Obviously, V is semi-simple if and only if h_w is minimal for a generic w, namely, h_w has only simple roots.

If $V = CP^n$, then $N = n+1$ and $h_0(\lambda) = \lambda^{n+1} - (-2(n+1))^{n+1}$. The roots of h_0 are $-2(n+1)e^{\frac{2\pi k\sqrt{-1}}{n+1}}$, where $k = 0, 1, \cdots, n$. So CP^n is semi-simple.

Next we take V to be the n-dimensional quadratic hypersurface in CP^{n+1} with $n \ge 3$. Then $N = n + 2 - \frac{1}{2}(1-(-1)^n)$. If n is odd, $N = n+1$ and consequently, $H^*(V,Q)$ is generated by $H^2(V,Z)$. A simple computation shows that $h_0(\lambda) = \lambda^{n+1} - 4(-2n)^n\lambda$. Its roots are 0, $-2n(4e^{2\pi k\sqrt{-1}})^{\frac{1}{n}}$ ($k = 0, 1, \cdots, n-1$), so V is semi-simple. If n is even, $H^*(V,C) = H^*_{\text{inv}}(V,C) \oplus H^*(V,C)_0$. Using that $c_1(V) \bullet \alpha = 0$ for $\alpha \in H^*(V,C)_0$, one can easily show that $h_0(\lambda) = \lambda^{n+2} - 4(-2n)^n\lambda^2$. So h_0 has only one multiple root 0 with multiplicity two. However, V is still semi-simple in the sense of Dubrovin. To see this, we denote by ω the positive generator of $H^2(V,Z)$. Note that $c_1(V) = n\omega$. Put $w = t\omega^n$. Then $X(w) = (2n-2)t\omega^n - 2c_1(V)$. An easy computation shows that h_w has simple roots for t sufficiently small. So V is semi-simple.

Conjecture 6.6. *Any Fano manifold is semi-simple in the sense of Dubrovin.*

Using the discussion in section 3, 5, one may be able to check this for Fano hypersurfaces in projective spaces.

There is a weaker version of the above conjecture. Let Φ^V_{inv} be the restriction of the prepotential to $H^*_{\text{inv}}(V,C)$. Assuming (3.3), we can easily deduce that Φ^V_{inv} also satisfies the WDVV equation with the scaling field X_{inv} on $H^*_{\text{inv}}(V,C)$, which is induced by X. Then one can ask if $X_{\text{inv}}\bullet_w$, which acts on $H^*_{\text{inv}}(V,C)$, is semi-simple for a generic w in $H^*_{\text{inv}}(V,C)$, where V is any Fano manifold. It does not seem to be very hard to check this for hypersurfaces, in view of Corollary 3.2 and Example (4) in section 5. For instance, if V is a cubic hypersurface in CP^4, one can easily show that the quantum multiplication by X_{inv} has only simple eigenvalues for a generic w.

7 Applications to Enumerative Geometry

The associativity law has some direct and interesting applications in counting holomorphic curves in algebraic varieties. In this section, we will count the number of rational curves on Fano manifolds, such as hypersurfaces, assuming that $\Psi^V_{(A,0,k)}$ actually counts the number of rational curves in those algebraic varieties. It is believed that this assumption holds at least for rational curves of large degree on Fano manifolds. We will always denote by $H^*_{\rm inv}(V, \boldsymbol{Z})$ the ring generated by $H^2(V, \boldsymbol{Z})$ modulo torsion.

For simplicity, we will start with a Fano manifold V with $H_2(V, \boldsymbol{Z}) = \boldsymbol{Z}$ such that the associativity law holds for the restriction of Φ^V to $H^*_{\rm inv}(V, \boldsymbol{C})$, for example, complete intersections in $\boldsymbol{C}P^{n+1}$ ($n \geq 3$) or Fano manifolds satisfying (3.3). We will abbreviate $\Psi^V_{(A,0,k)}(H^{j_1}, \cdots, H^{j_k})$ by $\Psi^V_A(j_1, \cdots, j_k)$, where H is the Poincare dual of the positive generator in $H^2(V, \boldsymbol{Z})$. Roughly speaking, $\Psi^V_A(j_1, \cdots, j_k)$ counts the number of rational curves in V with homology class A through subvarieties H^{j_1},...,H^{j_k}.

Theorem 7.1. *Let V be as above. Then for $j_1 \geq \cdots \geq j_k \geq 2$,*

$$
\begin{aligned}
&\Psi^V_d(j_1, j_2, j_3, j_4, \cdots, j_k) - \Psi^V_d(j_1, j_2+1, j_3-1, j_4, \cdots, j_k) \\
&= d\left(\Psi^V_d(j_1+j_3-1, j_2, j_4, \cdots, j_k) - \Psi^V_d(j_1+j_2, j_3-1, j_4, \cdots, j_k)\right) \qquad (7.1)\\
&+ \sum_{m=1}^{d-1}\sum_{i=0}^{n}\sum_{\{4,\cdots,k\}=S_1\cup S_2} m\Big(\Psi^V_{d-m}(j_1, j_2, i, \{j_a\}_{a\in S_1})\cdot\Psi^V_m(j_3-1, n-i, \{j_a\}_{a\in S_2}) \\
&\qquad -\Psi^V_{d-m}(j_1, j_3-1, i, \{j_a\}_{a\in S_1})\cdot\Psi^V_m(j_2, n-i, \{j_a\}_{a\in S_1})\Big)
\end{aligned}
$$

In particular, there is a recursion formula for computing σ^V_d.

Proof: The proof follows directly from the associativity law. Special cases of (7.1) have been derived before. In case $V = \boldsymbol{C}P^n$, (7.1) is derived in [KM] and [RT1], in case V is a hypersurface of odd dimension, (7.1) is derived in [RT1]. By Property (1) in section 2, we have $\sigma^V_0(j_1, \cdots, j_l) = \deg(V)$ or 0 according to whether or not $l = 3$ and $j_1 + j_2 + j_3 = n$. Moreover, $H^i \cap H^j = \deg(V)$ if $i + j = n$, $= 0$ if $i + j \neq 0$. Then using (2.6) to eliminate H, one can easily deduce (7.1) from (6.2) with $\alpha = H^{j_1}$, $\beta = H^{j_2}$, $\gamma = H^{j_3-1}$, $\delta = H$ and $\{\beta_{a_i}\} = \{H^{j_i}\}_{4\leq i\leq k}$.

An easy corollary of this theorem is that all primitive GW-invariants $\Psi^V_A(j_1, \cdots, j_k)$ can be computed in terms of 3-point invariants $\Psi^V_d(H^i, H^j, H^k)$ with $i + j + k = 2n - rd \geq 0$, where r is the index of V, i.e., $c_1(V) = rH^*$. In particular, if V is a hypersurface in $\boldsymbol{C}P^{n+1}$ of degree $\leq \frac{n+3}{2}$, then all primitive GW-invariants of genus zero can be explicitely computed (cf. section 5, example 4). In case $V = CP^2$, (7.1) becomes

$$n_d = \frac{1}{2} \sum_{\substack{d_1+d_2=d \\ d_1,d_2>0}} \frac{(d_1 d_2(3dd_1 d_2 - 2d^2 + 6d_1 d_2)(3d-4)!}{(3d_1-1)!(3d_2-1)!} n_{d_1} n_{d_2} \tag{7.2}$$

where n_d are given in last section. This formula was first observed by Kontsevich (cf. [KM]).

Theorem 7.1 can be used to derive an estimate on growth of degrees of moduli spaces of degree d rational curves, or equivalently, growth of numbers of rational curves through subvarieties as d tends to infinity. More precisely, for each n, there are positive constants $\lambda_1, \cdots, \lambda_{n-1}$ such that

$$\frac{\sigma_d^V(k_1, \cdots, k_{n-1})}{k_1! \cdots k_{n-1}!} \leq \lambda_1^{k_1} \cdots \lambda_{n-1}^{k_{n-1}} \tag{7.3}$$

It follows immediately from this that the formal series $\Phi_{\rm inv}^V(t_1, \cdots, t_{n+1}, 1)$ is convergent whenever $|t_1 e^{\frac{t_n}{n+2-d}}| < \lambda_1$, ..., $|t_{n-1} e^{\frac{(n-1)t_n}{n+2-d}}| < \lambda_{n-1}$. It seems that the generating function Φ^V is also convergent in some region of $H^*(V, C)$, since all other cohomology classes are dominated by powers of the Poincare dual of the hyperplane section. It is not hard to give λ_i explicitly. However, it seems unclear what is exactly the asymptotic growth of σ_d^V.

To generalize Theorem 7.1 to any Fano manifolds, we need further information on the GW-invariants Ψ^V. Assuming that $H^2(V, Z)$ generates the cohomology ring and using the associativity law, Kontsevich and Manin showed (see [KM]) that $\Psi^V_{(A,0,k)}(\{\alpha_i\})$ can be computed in terms of 3-point invariants for rational curves. However, the assumption is too strong for most Fano manifolds. A natural question is whether or not one can compute the restriction of Ψ^V to the part $H^*_{\rm inv}(V, Z)$, which is generated by $H^2(V, Z)$. The key question is if the restriction $\Psi^V_{\rm inv}$ satisfies the WDVV equation, which would follow from (3.3) for rational curves. Therefore, one can show

Theorem 7.2. *Let V be a Fano manifold such that* (3.3) *is true for rational curves. Then any invariant $\Psi^V_{(A,0,k)}(\{\alpha_i\})$, where $\{\alpha_i\} \subset H^*_{\rm inv}(V, Z)$, can be computed in terms of 3-point invariants for rational curves. In particular, the number of rational curves with sufficiently large degree can be computed.*

Finally, let us say a few words on higher genus cases. The composition law implies that all GW-invariants with fixed conformal structure can be computed by taking the "trace" on $\Psi^V_{(A,0,k)}$. However, it is still unclear how $\Psi^V_{(A,g,k)}([p]; \cdots)$ can be used to count the number of irreducible genus g algebraic curves in V with fixed conformal structure.

8 The Generalized Witten Conjecture

In this section, using the GW-invariant Ψ^V, we put a conjecture of Witten ([W2]) on a rigorous mathematical footing. We will also verify this conjecture mathematically for small genus. Most arguments in this section are due to Witten.

As before, we denote by $\overline{\mathcal{U}}_{g,k}$ the universal family of curves over $\overline{\mathcal{M}}_{g,k}$. Then each marked point x_i gives rise to a section, still denoted by x_i, of the fibration $\overline{\mathcal{U}}_{g,k} \mapsto \overline{\mathcal{M}}_{g,k}$. If $\mathcal{K}_{\mathcal{U}|\mathcal{M}}$ denotes the cotangent bundle to fibers of this fibration, we put $\mathcal{L}_{(i)} = x_i^*(\mathcal{K}_{\mathcal{U}|\mathcal{M}})$. Following Witten, we put

$$\langle \tau_{d_1,\alpha_1}\tau_{d_2,\alpha_2}\cdots\tau_{d_k,\alpha_k}\rangle_g(q) \;=\; \sum_{A\in H_2(V,\boldsymbol{Z})} \Psi^V_{(A,g,k)}([K_{d_1,\cdots,d_k}];\{\alpha_i\})\,q^A \quad (8.1)$$

where $\alpha_i \in H_*(V,\boldsymbol{Q})$ and $[K_{d_1,\cdots,d_k}]$ is the Poincare dual of $c_1(\mathcal{L}_{(1)})^{d_1}\cup\cdots\cup c_1(\mathcal{L}_{(k)})^{d_k}$. Symbolically, $\tau_{d,\alpha}$'s denote "quantum field theory operators". As before, choose a basis $\{\beta_a\}_{1\le a\le N}$ of $H_{*,\mathrm{even}}(V,\boldsymbol{Z})$ modulo torsion. We introduce formal variables $t_{r,a}$, where $r=0,1,2,\cdots$ and $1\le a\le N$. Witten's generating function (cf. [W2]) is now simply defined to be

$$F^V(t_{r,a};q) = \left\langle e^{\sum_{r,a} t_{r,a}\tau_{r,\beta_a}}\right\rangle = \sum_{n_{r,a}}\prod_{r,a}\frac{(t_{r,a})^{n_{r,a}}}{n_{r,a}!}\left\langle\prod_{r,a}\tau_{r,\beta_a}^{n_{r,a}}\right\rangle \quad (8.2)$$

where the $n_{r,a}$ are arbitrary collections of nonnegative integers, almost all zero, labeled by r,a. The summation in (8.2) is over all values of the genus g and all homotopy classes A of (J,ν)-maps. Sometimes, we write F_g^V to be the part of F^V involving only GW-invariants of genus g. It is clear that F^V is a generalization of Φ^V in the last section. Indeed this generalized function contains more information on the underlying manifold, for instance, using Taubes' theorem (section 3), one can show that for an algebraic surface V of general type,

$$F^V(t_{r,a};q) \;=\; F^V(t_{r,a};0) + q^{K_V}e^{\tau_{0,0}} + \cdots, \quad (8.3)$$

while Φ^V depends only on the intersection form of V.

One of Witten's goals is to find out the equations which F^V satisfies. The case that V is a point corresponds to the 2-dimensional topological gravity, where F^V is governed by the KdV hierarchy, conjectured by Witten ([W2]) and verified by Kontsevich ([Ko]). In general, it is not clear what equations F^V would solve, though there are partial results for $V = \boldsymbol{CP}^1$ (see [Ho]). However, in [W2], Witten made a conjecture on F^V, which we will describe now.

Assume that $\beta_1 = [V]$. Following Witten's arguments in [W2], one can deduce from (2.6) that F^V satisfies the generalized string equation:

$$\frac{\partial F^V}{\partial t_{0,1}} = \frac{1}{2}\eta_{ab}t_{0,a}t_{0,b} + \sum_{i=0}^{\infty}\sum_a t_{i+1,a}\frac{\partial F^V}{\partial t_{i,a}} \quad (8.4)$$

Following Witten, one can introduce

$$U = \frac{\partial^2 F^V}{\partial t_{0,1}\partial t_{0,\sigma}},\quad U' = \frac{\partial^3 F^V}{\partial t_{0,1}^2\partial t_{0,\sigma}},\quad \cdots,\quad U_\sigma^{(l)} = \frac{\partial^{l+2}F^V}{\partial t_{0,1}^{l+1}\partial t_{0,\sigma}},\quad \text{for } l\ge 0 \quad (8.5)$$

We will regard $U^{(l)}$ to be of degree l. By a differential function of degree k we mean a function $G(U, U', U'', \cdots)$ of degree k in that sense. In particular, any function of the form $G(U)$ is of degree zero, and $(U')^2$ has degree two.

Witten's Conjecture: *For every $g \geq 0$, there are differential functions $G_{m,a,n,b}(U_\sigma, U'_\sigma, \cdots)$ of degree $2g$ such that*

$$\frac{\partial^2 F_g}{\partial t_{m,a} \partial t_{n,b}} = G_{m,a,n,b}(U_\sigma, U'_\sigma, \cdots) \tag{8.6}$$

up to and including terms of genus g.

To explain the significance of this conjecture, we quote from Witten ([W2]): if V is a point, it is a consequence of the KdV hierarchy. Indeed, the KdV hierarchy has a stronger property — $G_{m,a,n,b}$ is a differential function of degree at most $2(m+n)$ even for g tending to infinity.

It was pointed out by Witten that the composition law implies

$$\frac{\partial^3 F_0}{\partial t_{d_1,a_1} \partial t_{d_2,a_2} \partial t_{d_3,a_3}} = \sum_{a,b} \frac{\partial^2 F_0}{\partial t_{d_1-1,a_1} \partial t_{0,a}} \eta^{ab} \frac{\partial^3 F_0}{\partial t_{0,b} \partial t_{d_2,a_2} \partial t_{d_3,a_3}} \tag{8.7}$$

and consequently, the conjecture for $g = 0$.

In [RT2], we showed a weaker version of the conjecture for $g \leq 2$. The special case $g = 1$ and $V = CP^1$ was checked in [W2].

9 Quantum Cohomology and Symplectic Floer Cohomology

The symplectic Floer cohomology was defined for a hamiltonian symplectomorphism whose fixed points are nondegenerate. A crucial step to a solution of the Arnold conjecture is to show that the Floer cohomology is the same as the ordinary cohomology. Furthermore, Floer also introduced a multiplication on the Floer cohomology, *extrinsic multiplication*, by the ordinary cohomology. This multiplication is very useful for the solution of the Arnold conjecture in the case of degenerate fixed points [F], [OV]. On the other hand, there is a natural way to define an intrinsic multiplication on the Floer cohomology by using perturbed J-holomorphic maps from a pair of pants to the symplectic manifold. In fact, such an intrinsic multiplication has been formally defined by Betz and Rade [BR]. It has been conjectured that all these multiplications (intrinsic and extrinsic) are the same as the quantum multiplication. For the extrinsic multiplication, a physical argument was given by Sadov [Sa]. In [Pu], Piunikhin outlined a different approach towards the proof of the equivalence of the extrinsic multiplication and the quantum multiplication by using intersection theory. Recently, this conjecture was solved independently by Piunikhin, Salamon, Schwarz [PSS] and Ruan, Tian [RT3].

The two proofs are completely different. In [RT3], the authors established the symplectic Floer cohomology for Bott-type hamiltonian symplectomorphisms and then the equivalence follows as a corollary, while the method in [PSS] is more direct.

Given any function H on $V \times S^1$, we can associate a vector field X_H as follows:

$$\omega(X_H(z,t), v) = v(H)(z,t), \text{ for any } v \in T_z V \qquad (9.1)$$

We call H a periodic hamiltonian and X_H the hamiltonian vector field associated to H. Let $\phi_t(H)$ be the integral flow of the hamiltonian vector field X_H. Then $\phi_1(H)$ is a hamiltonian symplectomorphism.

Definition 9.1. *A periodic hamiltonian H is of Bott-type if and only if the fixed-point set $F(\phi_1(H))$ of $\phi_1(H)$ is nondegenerate in the sense of Bott, i.e., $F(\phi_1(H))$ consists of smooth submanifolds in V and $\phi_1(H)$ is nondegenerate in the normal directions of these submanifolds.*

From now on, we fix a Bott-type periodic hamiltonian function H. Let $\mathcal{L}(V)$ be the space of contractible maps (sometimes called contractible loops) from S^1 into V and $\tilde{\mathcal{L}}(V)$ be the universal covering of $\mathcal{L}(V)$, namely, $\tilde{\mathcal{L}}(V)$ is as follows:

$$\tilde{\mathcal{L}}(V) = \{(u,x) | u \in \mathcal{L}(V), x : D^2 \to V \text{ such that } u = x|_{\partial D^2}\}/\sim, \qquad (9.2)$$

where the equivalence relation $\sim$ is the homotopic equivalence of x. The covering group of $\tilde{\mathcal{L}}$ over $\mathcal{L}$ is $\pi_2(V)$. We can define a symplectic action functional on $\tilde{\mathcal{L}}(V)$,

$$a_H((u,x)) = -\int_{D^2} x^*\omega + \int_0^{2\pi} H(t, u(t))dt \qquad (9.3)$$

It follows from the closedness of ω that a_H descends to the quotient space by $\sim$. The Euler-Lagrange equation of a_H is

$$\dot{u} + X_H(t, u(t)) = 0 \qquad (9.4)$$

Let $\boldsymbol{R}(H)$ be the critical point set of a_H, i.e., the set of smooth contractible loops satisfying the Euler-Lagrange equation.

We can assign a Conley-Zehnder index to each component of $\boldsymbol{R}(H)$. So

$$\boldsymbol{R}(H) = \cup_i \boldsymbol{R}_i(H),$$

where $\boldsymbol{R}_i(H)$ consists of critical points in $\boldsymbol{R}(H)$ with the Conley-Zehnder index i.

Consider the space of trajectories

$$\tilde{\mathcal{N}} = \bigcup \tilde{\mathcal{N}}(i,j),$$

where

$$\tilde{\mathcal{N}}(i,j) = \{u \in \tilde{\mathcal{N}};\ \lim_{s\to-\infty} u_s \in \boldsymbol{R}_i(H),\ \lim_{s\to\infty} u_s \in \boldsymbol{R}_j(H)\}.$$

Clearly, $\boldsymbol{R}^1$ acts on $\tilde{\mathcal{N}}$ by translations in time. Let $\mathcal{N} = \tilde{\mathcal{N}}/\boldsymbol{R}^1$ and $\mathcal{N}(i,j) = \tilde{\mathcal{N}}(i,j)/\boldsymbol{R}^1$. For a generic J, $\mathcal{N}(i,j)$ is a smooth, oriented manifold of dimension $i-j-1+\dim \boldsymbol{R}_i(H)$. Furthermore, it admits a compactification $\overline{\mathcal{N}}(i,j)$, which is an abstract geometric chain (cf. [RT3, Theorem 2.4]). An abstract geometric chain of dimension n is a finite simplicial complex P such that its singular set P_s is a subcomplex and $\dim(P_s) \le n-2$, $\dim(P_s \cap \partial P) \le n-3$ (cf. [Fut]). A geometric chain is a pair (P,f), where P is an abstract geometric chain and f is a map from P.

One defines the boundary maps

$$\partial_-^i : \overline{\mathcal{N}}(i,j) \to \boldsymbol{R}_i(H),\ \partial_+^j : \overline{\mathcal{N}}(i,j) \to \boldsymbol{R}_j(H).$$

For a generic J, the boundary maps $\partial_\pm^i$ are smooth and transverse to each other whenever they have the same target space.

The group $\Gamma_0 = \{A \in \pi_2(V); c_1(V)(A) = 0\}$ acts on $\tilde{\mathcal{L}}(V)$ and generates (possibly) infinitely many components in $\boldsymbol{R}_i$. We denote by $\boldsymbol{R}_i^s(H)$ the components of $\boldsymbol{R}_i(H)$, so $\boldsymbol{R}_i(H) = \bigcup \boldsymbol{R}_i^s(H)$. For each $R_i^s(H)$, let $C_m(R_i^s(H), \boldsymbol{Z})$ be the space of geometric chains in $R_i^s(H)$ transverse to all the boundary maps into $R_i^s(H)$. Then $H_*(C_*(R_i^s(H), \boldsymbol{Z}), \partial) = H_*(R_i^s(H), \boldsymbol{Z})$. Now we define a boundary map:

$$\begin{aligned}\partial_k : C_m(R_i(H), \boldsymbol{Z}) = \oplus C_m(R_i^s(H), \boldsymbol{Z}) &\mapsto C_{m+k-1}(R_{i-k}, \boldsymbol{Z})\\ \partial_k((P,f)) &= (P \times_{R_i(H)} \overline{\mathcal{N}}(i, i-k), \partial_+^{i-k} \cdot \pi_2)\end{aligned}$$

where π_2 is the natural projection from the fiber product $P \times_{R_i(H)} \overline{\mathcal{N}}(i, i-k)$ onto the second factor. We define the cochain complex $C^*(R_i(H), \boldsymbol{Z})$ of cochains $f = \sum_s f_s$, where $f_s \in \mathrm{Hom}(C_j(R_i^s(H), \boldsymbol{Z}), \boldsymbol{Z})$ satisfying:

$$\#\{s | f_s \ne 0, a_H(R_i^s) \ge c\} < \infty$$

for any c. Define the coboundary map δ_k by $\delta_k(\alpha) = \alpha\partial_k$. Furthermore, we define $\partial_0 = (-1)^{n+j}\partial$ and δ_0 similarly. Put $\delta^H = \oplus\delta_k$, then $\delta^H\delta^H = 0$. We define $HF^*(V,H)$ to be the cohomology of $(C^*(R_*(H), \boldsymbol{Z}), \delta^H)$. This is just the Floer cohomology in case H is nondegenerate. It is proved in [RT3] that $HF^*(V,H)$ is independent of particular choices of Bott-type Hamiltonian functions. Therefore, we may often write $HF^*(V)$ for $HF^*(V,H)$. Choosing $H = 0$, one can show that $HF^*(V) = H^*(V, \boldsymbol{Z}\{H_2(V)\})$.

One can introduce a multiplicative structure on $HF^*(V)$ in two ways: extrinsic and intrinsic. Here we just discuss the extrinsic multiplication. We refer the readers to [RT3] for the intrinsic one. The extrinsic product is an action of $H^*(V, \boldsymbol{Z})$ on $HF^*(V)$. Let A be a generic cocycle representing $\alpha^* \in H^k(V, \boldsymbol{Z})$ and $\gamma \in C^{m-i}(R_i(H), \boldsymbol{Z})$, then we define $A(\gamma) \in C^{m+k}(V, H)$

as follows: let $A(\gamma) = \sum A^j(\gamma)$, where $A^j(\gamma) \in C^{m+k-j}(R_j, \boldsymbol{Z})$. Consider the evaluation map

$$e : \tilde{\mathcal{N}}(j, i) \to V$$

by $e(u) = u(0,0)$. One can compactify $\tilde{\mathcal{N}}(j,i)$ by adding boundary components

$$\mathcal{N}(j, k_1) \times \cdots \times \mathcal{N}(k_{p-1}, k_p) \times \tilde{\mathcal{N}}(k_p, k_{p+1}) \times \mathcal{N}(k_{p+1}, k_{p+2}) \times \cdots \times \mathcal{N}(k_s, i).$$

We denote this compactification by $\overline{\mathcal{N}}^*(j,i)$. Then $\overline{\mathcal{N}}^*(j,i)$ is an abstract geometric chain (cf. [RT3, Lemma 3.1]) and e extends over $\overline{\mathcal{N}}^*(j,i)$.

For any geometric chain $(P, f) \in C_{m+k-j}(R_j^s(H), \boldsymbol{Z})$, the fiber product

$$P \times_{R_j^s} \overline{\mathcal{N}}^*(j,i)$$

is an abstract geometric chain of dimension $m + k$. Together with the map

$$\partial^+ \times e : P \times_{R_j^s} \overline{\mathcal{N}}^*(j,i) \to R_i(H) \times V,$$

it gives a geometric chain of $R_i(H) \times V$. We define

$$A(\gamma)((P, f)) = \gamma \times A((P \times_{R_j^s} \overline{\mathcal{N}}^*(j,i), \partial^+ \times e))$$

One can show that $\delta A \cdot - A \cdot \delta = 0$, i.e., $A\cdot$ is a cochain map. Therefore, it induces an action of α^* on $H^*(V, \boldsymbol{Z})$, the extrinsic product, on the Bott-type Floer cohomology $HF^*(V, H)$. Moreover, this action is independent of H. Choosing a nondegenerate H, one can easily show that this extrinsic multiplication is the same as Floer's. On the other hand, when $H = 0$, the extrinsic multiplication is just the quantum multiplication, if we view α^* as an element of the quantum cohomology. Therefore, we have

Theorem 9.2. *The extrinsic multiplication is equivalent to the quantum multiplication.*

This allows us to compute the multiplicative structure on the Floer cohomology.

10 Degenerating Families

A symplectic degeneration is a fibration $\pi : \mathcal{V} \mapsto \Delta \subset \boldsymbol{C}$ satisfying: $\mathcal{V}$ is a symplectic manifold and $V_t = \pi^{-1}(t)$ is a symplectic submanifold if $t \neq 0$, a union of symplectic submanifolds with simple normal crossings if $t = 0$. It is not hard to find an ω-tamed almost complex structure J on $\mathcal{V}$, such that π is J-holomorphic. Degenerations of algebraic manifolds are examples of such a degeneration. In this section, we will loosely discuss how GW-invariants on a smooth fiber are related to those on the singular fiber in a symplectic degeneration $\pi : \mathcal{V} \mapsto \Delta$. For simplicity, we assume that the central fiber V_0

has no triple points, i.e., no three components intersect. In fact, any degeneration can be decomposed into finitely many such simple degenerations, so the general case can be reduced to special cases by induction. For simplicity, we will only consider the case of rational curves here.

First we introduce a notion of marked rational V_0-trees, or simply V_0-trees. Let $V_0 = \bigcup_{i=1}^m V_{0i}$, where V_{0i} are irreducible. A V_0-tree T consists of an ordinary tree with branches T_0, T_j $(1 \le j \le l)$ with a marking $(i(j), a_j)$ attached to each T_j, satisfying: (1) $1 \le i(j) \le m$, $a_j \in H_2(V_{0i(j)}, \mathbf{Z})$; (2) If $T_j \cap T_{j'} \ne \emptyset$, then $i(j) \ne i(j')$; (3) For any j, the number of j' with fixed $i' = i(j')$ and $T_{j'} \cap T_j \ne \emptyset$ is the same as the intersection number $a_j \cdot V_{0i'}$ in $\mathcal{V}$. As usual, such a tree is rational if it does not contain any cycles, or equivalently, l is equal to the number of intersection points in T. Two V_0-trees T and T' are isomorphic if there is a tree isomorphism $\phi : \{T_j\} \mapsto \{T'_j\}$ such that $\phi(T_0) = T'_0$ and $(i(j), a_j) = (i'(j'), a'_{j'})$ whenever $\phi(T_j) = T'_{j'}$. Denote by G_T the group of automorphisms of T.

Rational V_0-trees arise naturally as configurations of limiting holomorphic maps. Fix $A \in H_2(\mathcal{V}, \mathbf{Z})$ and a generic inhomogeneous term ν on $\mathcal{V}$ compatible with the fibration. Denote by $\mathcal{M}_{A,t}$ the moduli space of (J, ν)-maps f from S^2 into V_t with homology class A when viewed as a map to $\mathcal{V}$. Let $f_t \in \mathcal{M}_{A,t}$, by the Gromov-Uhlenbeck compactness theorem and taking a subsequence if necessary, we may assume that f_t converges to f_0 in V_0. The map f_0 consists of a (J, ν)-map f_{00} and J-holomorphic maps f_{0j} from S^2, i.e., bubbles, where $1 \le j \le l$. Suppose that the sequence $\{f_t\}$ and (J, ν) are generic. Then images of f_{0j} intersect each other transversally. Now we associate f_0 with a rational tree $T(f_0)$ as follows: branches of $T(f_0)$ correspond to components of f_0 with $T(f_0)_0$ associated to f_{00}, $T(f_0)_j$ intersects with $T(f_0)_{j'}$ if and only if $Im(f_{0j})$ intersects $Im(f_{0j'})$, for each branch $T(f_0)_j$, the marking $(i(j), a_j)$ is given by the V_i in which f_{0j} lies and the homology class of f_{0j} in V_i. Obviously, $T(f_0)$ is rational. It can be proved (cf. [Ti1]) that $T(f_0)$ is actually a V_0-tree. This follows from that f_0 is the limit of smooth maps from S^2 and the total space $\mathcal{V}$ is smooth.

Let $\mathcal{T}_A$ be the set of equivalence classes of rational V_0-trees T with $A = \sum_{j=0}^{|T|-1} a_j$, where $|T|$ is the number of branches in T. For any $T \in \mathcal{T}_A$, put

$$\mathcal{M}_{T,0} = \{f_0 = \cup_j f_{0j} \mid f_{00} : \text{ a } (J, \nu) - \text{map}\}.$$

Note that $-\sum_{j=0}^{|T|-1} a_j \cdot V_{0i(j)}$ is twice the number $s(T)$ of intersection points in T. The rationality implies that $s(T) = |T| - 1$. Then one can show that the "virtual" dimension of $\mathcal{M}_{T,0}$ is given by

$$\begin{aligned}
&\dim \mathcal{M}_{T,0} \\
&= \sum_{j=0}^{|T|} \left(2c_1(V_{i(j)})(a_j) + 2n - (n-1)a_j \cdot (V_0 - V_{0i(j)})\right) - 6(|T| - 1) \qquad (10.1) \\
&= 2c_1(\mathcal{V})(A) + 2n + 2(n-3)(|T| - s(T) - 1) = \dim \mathcal{M}_{A,t}
\end{aligned}$$

In fact, using the gluing theorem in [RT1], one can show

Lemma 10.1. *Assume that $\mathcal{V}$ is semi-positive. We denote by* $\lim \mathcal{M}_{A,t}$ *the set of limits of maps in* $\mathcal{M}_{A,t}$. *Then for a generic* (J,ν), $\lim \mathcal{M}_{A,t}$ *coincides with* $\bigcup_{T\in\mathcal{T}_A} \mathcal{M}_{T,0}$ *outside finitely many subsets of codimension at least two.*

Following the arguments in defining the GW-invariant, one can define an invariant $\Psi^{V_0}_{(T,k)}$ on $H_*(\mathcal{V},Z)^{\otimes k}$ for each $T \in \mathcal{T}_A$. Roughly speaking, if $\alpha_1,\cdots,\alpha_k$ are in $H_*(\mathcal{V},Z)$, $\Psi^{V_0}_{(T,k)}(\{\alpha_i\})$ counts with sign the number of $(f;x_1,\cdots,x_k)$ satisfying: (1) $f \in \mathcal{M}_{T,0}$; (2) x_i are smooth points in the domain of f; (3) $f(x_i) \in \alpha_i$. It follows from Lemma 10.1,

Theorem 10.2. *Let $\mathcal{V}$ be a fibration as above. Then we have the following degeneration formula:*

$$\Psi^{V_t}_{(A,0,k)} = \sum_{T\in\mathcal{T}_A} \Psi^{V_0}_{(T,k)} \qquad (10.2)$$

where $t \neq 0$.

This theorem has many applications. For instance, one can apply this to certain Calabi-Yau 3-folds V, such as quintics in CP^4 and prove that the GW-potential Φ^V is convergent for $Re(t_i)$ sufficiently large, or equivalently, the number of rational curves in V of degree d grows exponentially with d. The idea is simple: let V be a quintic, then it has a semistable degeneration into a quartic plus a hyperplane. Formula (10.2) tells us that the number of rational curves in V can be bounded in terms of the number of rational curves in a quartic and a hyperplane. The last two manifolds are all Fano, so we can apply results in section 7.

11 Mirror Symmetry and Calabi-Yau Manifolds

A two-dimensional conformal topological field theory consists of a linear superspace with a nondegenerate inner product and a family of correlations satisfying certain axioms. According to physicists, each Calabi-Yau manifold V gives rise to two conformal topological field theories, called A- and B-models (cf. [W5]). The correlation functions of the A-model are determinated by counting holomorphic maps from a Riemann surface into V, while the correlation functions of the B-model can be computed by calculating periods of classical differential forms. Mirror Symmetry gives a correspondence between those two models. One remarkable consequence of Mirror Symmetry is the prediction of Candelas, de la Ossa, Green and Parkes on the number of rational curves in a generic quintic hypersurface V in CP^4 ([COGP]). They predicted that the generating function Φ^V in section 6 can be found by studying the 4-th order linear differential operators satisfied by some hypergeometric functions, namely,

$$\sum_{k=0}^{\infty} \frac{(5k)!\, q^k}{(k!)^5}$$

consequently, they computed the number n_d of rational curves in V of degree d for all d. Only the first three numbers have been computed rigorously by mathematicians ([Kz], [ES]). They do coincide with the prediction of Candelas, etc.. However, Kontsevich has made some progress on the number of higher degree [Ko2]. Since 1991, more explicit Calabi-Yau manifolds have been studied and the number of rational curves on them have been computed based on the Mirror Symmetry Hypothesis (cf. [CKM], [HKTY]). Also certain Hodge theoretic explanation of B-models has been provided based on work of Griffiths, etc. (cf. [Mo]). However, Mirror Symmetry remains largely a mystery. Mathematical aspects of the A-model are essentially formal besides some special computations in concrete examples. The most interesting part on rational curves and geometry of Calabi-Yau manifolds still needs basic mathematical understanding. In this section, we will give a brief discussion on the Mirror Symmetry Conjecture in mathematical terms.

Let V be a Calabi-Yau manifold, i.e., a compact Kähler manifold with vanishing real first Chern class. It is known (cf. [Be2]) that a finite covering of V admits a decomposition $T^k_{\boldsymbol{C}} \times V_1 \times V_2$, where V_1 is a hyperKähler manifold and $h^{2,0}(V, \boldsymbol{C}) = 0$.

Now we assume that $\dim_{\boldsymbol{C}} V = 3$, then there is no factor V_1 in the above decomposition, so without loss of generality, we assume that V is simply connected, which implies that $h^{1,0}(V, \boldsymbol{C}) = 0$ and $h^{3,0}(V, \boldsymbol{C}) = 1$. Let $\mathcal{M}_V$ be the moduli space of complex structures on V. By the Bogomolov-Tian-Todorov theorem, $T_V \mathcal{M}_V$ is naturally isomorphic to $H^1(V, T_V)$, the infinitesimal deformation space. To avoid technicalities, we assume that there is a universal family $\pi : \mathcal{U}_V \mapsto \mathcal{M}_V$. Then the direct image $\mathcal{F}^0 = \mathcal{R}^3 \pi_* \boldsymbol{C}_{\mathcal{U}_V}$ is a holomorphic bundle over $\mathcal{M}_V$. There is a canonical flat connection ∇^B, namely, the Gauss-Manin connection, on $\mathcal{F}^0$ such that all integral cohomology classes are parallel sections. The bundle $\mathcal{F}^0$ has a natural filtration

$$\mathcal{F}^3 \subset \mathcal{F}^2 \subset \mathcal{F}^1 \subset \mathcal{F}^0$$

where $\mathcal{F}^k$ is the subbundle with fibers $\mathcal{F}^k|_V = \oplus_{p \geq k} H^{p,3-p}(V, \boldsymbol{C})$. The connection ∇^B is compatible with this filtration in the sense (Griffiths transversality): ∇^B can be written as $\nabla^{LC} + \theta + \overline{\theta}$, where ∇^{LC} is a hermitian connection and θ is holomorphic such that for all $k > 0$,

$$\theta : \mathcal{F}^k \mapsto \mathcal{F}^{k-1} \otimes \Omega^1_{\mathcal{U}_V|\mathcal{M}_V}, \ \theta^2 = 0$$

Note that $T_{\mathcal{U}_V|\mathcal{M}_V}$ is naturally isomorphic to $(\mathcal{F}^2/\mathcal{F}^3) \otimes (\mathcal{F}^3)^*$. It follows that θ is locally determinated by a holomorphic section, still denoted by θ, in the symmetric product $S^3 \Omega^1_{\mathcal{U}_V|\mathcal{M}_V}$.

On the other hand, the prepotential Φ^V in section 6 induces a flat connection ∇^A, at least formally, on $\mathcal{H}^0 = \oplus_{0\leq k\leq 3} H^k(V,\Omega^k)$ over $H^1(V,\Omega^1)$. If we write $\nabla^A = \nabla^{LC} + \Gamma$, then

$$\eta_{ae}\Gamma^e_{bc}(t) = \frac{\partial^3 \phi^V}{\partial t_a \partial t_b \partial t_c}(t)$$
$$= \psi^V_{(0,0,3)}(\beta_a,\beta_b,\beta_c) + \sum_{A\neq 0} \frac{\beta^*_a(A)\beta^*_b(A)\beta^*_c(A)\, n_A e^{t^*(A)}}{1-e^{t^*(A)}}$$

where n_A is the number of irreducible rational curves in V with homology class A.

As pointed out in the last section, if V is any complete intersection in CP^N, such as a quintic hypersurface, the connection ∇^A is actually well-defined in the region $\{\omega \in H^1(V,\Omega^1) | Re(\omega) << 0\}$. It is flat and is compatible with the obvious filtration $\mathcal{H}^p = \oplus_{0\leq k\leq 3-p} H^k(V,\Omega^k)$. The following conjecture is folklore.

Conjecture 11.1: *∇^A can be always extended to be a flat meromorphic connection on a quasi-projective manifold, moreover, it is often the same as ∇^B of another Calabi-Yau manifold V' via variation of Hodge structures.*

This is one important part of the Mirror Symmetry Conjecture. The other part is to show that the natural coordinates in the definition of ∇^A are the same as those flat coordinates, which were predicted by physicists (cf. [COGP]) and mathematically reformulated in [Mo]. In particular, the Mirror Symmetry Conjecture implies that the prepotential F^A on V can be identified with certain F^B on V' after an appropriate change of variables (cf. [COGP], [Mo]), where the Calabi-Yau manifold V' is given in Conjecture 11.1 and called the mirror of V. The prepotential F^B is much easier to calculate. For instance, in [COGP], Candelas et. al. computed the number of rational curves in a quintic hypersurface by calculating F^B on the mirror manifold (assuming that it exists).

It is obvious that the part of the conjecture on the extension of ∇^A is more tractable. Let us provide an evidence for this conjecture as follows.

Let V be a Calabi-Yau 3-fold and C_i $(1 \leq i \leq k)$ be smooth rational curves in V with $H^1(C_i, N_{C_i|V}) = 0$, where $N_{C_i|V}$ denotes the normal bundle of C_i in V. Contracting these rational curves, one obtains a 3-fold V_0 with k ordinary double points. A smoothing of V_0 is a smooth 4-fold $\pi : \mathcal{V} \mapsto \Delta \subset C$ such that its generic fiber $\pi^{-1}(t)$ is smooth and $V_0 = \pi^{-1}(0)$. Put V_t to be the generic fiber. This process from V to V_t is actually an analytic surgery. It provides a method of finding new complex manifolds, which is due to Clemens (see [Fr]). It is shown by Friedman, Kawamata, Tian ([Fr], [Ka], [Ti2]) that V_0 has a smoothing if and only if there are nonzero λ_i satisfying: $\sum_i \lambda_i [C_i] = 0$ in $H_2(V, C)$. Such V and V_t are said to be adjacent to each other. In general, V_t may not be projective. We say that two Calabi-Yau 3-folds V and V' are united if there are finitely many Calabi-Yau 3-folds V_α

($\alpha = 1, \cdots, L$) such that $V_1 = V$, $V_L = V'$ and V_α, $V_{\alpha+1}$ are adjacent to each other for all α. In fact, R. Friedman, M. Reid speculated independently that any two Calabi-Yau 3-folds are united.

Now suppose that two Calabi-Yau 3-folds V, V' are united. We will see that the GW-invariants Ψ^V and $\Psi^{V'}$ are related in a simple way. It is clear that we only need to show this in case V and V' are adjacent. Without loss of generality, assume that $V' = V_t$ as above. There is a natural map $p: H_2(V, \mathbf{Z}) \mapsto H_2(V', \mathbf{Z})$. One can show that the kernel of p is $\sum_{i=1}^k \mathbf{Z}[C_i]$, i.e.,

$$H_2(V', \mathbf{Z}) = H_2(V, \mathbf{Z}) / \sum_{i=1}^k \mathbf{Z}[C_i]$$

For simplicity, we may assume that $H_2(V', \mathbf{Z})$ is torsion-free.

By perturbing the complex structure of V outside C_i, one may arrange all holomorphic rational curves other than C_i to lie entirely in $V \backslash \bigcup_i C_i$. These curves will survive during the blow-down and smoothing, so they are in 1-1 correspondence with rational curves in V'. Then for any nonzero A' in $H_2(V, \mathbf{Z})$, we have

$$n_{A'} = \sum_{p(A)=A'} n_A,$$

where $n_{A'}$ (resp. n_A) is the number of irreducible rational curves in V' (resp. V) with homology class A' (resp. A).

On the other hand, if $p(A) = 0$, then $n_A = \#\{C_i | [C_i] = A\}$, in particular, $n_{[mC_i]} = 0$ if there is no C_j with $[C_j] = m[C_i]$. Here we used the fact that V' is projective.

For any t with Poincare dual t^* in $H^2(V, \mathbf{C})$, we have

$$\begin{aligned}
\frac{\partial^3 \Phi^V}{\partial t_a \partial t_b \partial t_c(t)} &= \Psi^V_{(0,0,3)}(\beta_a, \beta_b, \beta_c) + \sum_{p(A)\neq 0} \frac{\beta_a^*(A)\beta_b^*(A)\beta_c^*(A)\, n_A e^{t^*(A)}}{1 - e^{t^*(A)}} \\
&\quad + \sum_{A=[C_i]} \frac{\beta_a^*(A)\beta_b^*(A)\beta_c^*(A)\, n_A e^{t^*(A)}}{1 - e^{t^*(A)}} \\
&= \Phi_1^V + \Phi_2^V
\end{aligned}$$

We may identify any t' in $H_{n-2}(V', \mathbf{C})$ with a point in $H_{n-2}(V, \mathbf{C})$. Then $t'^*([C_i]) = 0$. It follows from the above that as t tends to t', $\Phi_1^V(t)$ converges to $\Phi^{V'}(t')$. On the other hand, $\Phi_2^V(t)$ has a simple pole at t'.

This shows how the connection ∇^A changes, when one Calabi-Yau manifold V moves to another Calabi-Yau V'. Symmetrically, when V' moves to V through degeneration and then small resolution, the connection ∇^B changes in a similar way as the above. This makes us wonder if the above motion from one Calabi-Yau to another is preserved under Mirror Symmetry, more precisely, if V moves to V' through blow-down and smoothing, then the mirror of V goes to the mirror of V' through degeneration and small resolution.

It seems to be even more interesting to see what would happen when one of united Calabi-Yau 3-folds, say V', is not projective, for example, $H_2(V', Z) = 0$. In this case, V' has to be diffeomorphic to the connected sum of finitely many copies of $S^3 \times S^3$.

Nevertheless, for the first part of Conjecture 11.1, the central problem is to find the monodromy group of ∇^A. The above method of studying adjacent Calabi-Yau manifolds may provide an insight into this problem.

Bibliography

[AF] A. Andreotti and T. Frankel, The second Lefschetz theorem on hyperplane sections, Global Analysis, edited by Spenser and S. Iyanaga, Princeton Univ. Press.

[AJ] M. Atiyah and L. Jeffrey, Topological Lagrangians and cohomology, preprint, 1990.

[AM] Aspinwall and D. Morrison, Topological field theory and rational curves, Comm. in Math. Phys., vol., 151 (1993), 245-262.

[AS] A. Astashkevich and V. Sadov, Quantum cohomology of partial flag manifolds $F_{n_1,\cdots,n_k}$, preprint, 1994.

[Ba] V. Batyrev, Dual polyhedra and mirror symmetry for Calabi-Yau hypersurfaces in toric varieties, preprint, Essen Univ., 1992.

[Be1] A. Beauville, Quantum cohomology of complete intersections, preprint, 1995.

[Be2] A. Beauville, Variétés Kähleriennes dont la lère classe de Chern est nulle. J. Diff. Geom., 18 (1983), 755-782.

[Ber] A. Bertram, Modular Schubert calculus, I, II, preprint.

[BDW] A. Bertram, G. Daskalopoulos and R. Wentworth, Gromov invariants for holomorphic maps from Riemann surfaces to Grassmannians, preprint.

[BR] M. Betz and J. Rade, Products and relations in symplectic Floer homology, preprint, 1995.

[CF] I. Ciocan-Fontanine, Quantum cohomology of Flag varieties, preprint, 1995.

[CKM] P. Candelas, A. Font, S. Katz and D. Morrison, Mirror Symmetry for Two Parameter Models - II, Nucl. Phys. B429 (1994) 626-674.

[CM] B. Crauder and R. Miranda, Quantum cohomology of rational surfaces, preprint, 1994.

[COGP] P. Candelas, X.C. de la Ossa, P. Green and L. Parkes, A pair of Calabi-Yau manifolds as an exactly soluble superconformal theory, Essays on Mirror manifolds, edited by Yau, International Press, Hong Kong, 1992.

[Do] S. Donaldson, Floer homology and algebraic geomrty, preprint.

[DS] S. Dostoglou and D. Salamon, Self-dual instantons and holomorphic curves, Ann. of Math., 1994.

[Du] B. Dubrovin, Integrable systems in topological field theory, Nucl. Phys. B379 (1992), 627-689. Also Geometry of 2d topological field theory, preprint, 1994.

[ES] G. Ellingsrud and S. Stromme, The number of twisted cubic curves on the generic quintic threefold, Essays on Mirror manifolds, edited by Yau, International Press, Hong Kong, 1992.

[F] A. Floer, Symplectic fixed points and holomorphic spheres, Comm. Math. Phys, 120 (1989), 575-611.

[Fk] K. Fukaya, Floer homology for connected sum, preprint.

[FM] R. Friedman and J. Morgan, On the diffeomorphism types of certain algebraic surfaces I, II, J. Diff. Geom., 27 (1988).

[FN] H. Flaschka and A. Newell, Comm. Math. Phys., 76 (1980).

[Fr] R. Friedman, On threefolds with trivial canonical bundle, Complex Geometry and Lie Group, Proc. of Symp. in Pure Math., vol. 53 (1991), AMS.

[Fu] W. Fulton, Intersection theory. Ergeb. Math. Grenzgb., 3 Folge, vol. 2, Springer-Verlag, 1983.

[Fut] K. Fukaya, Morse homotopoy, A_∞ category and Floer cohomologies, preprint.

[GH] P. Griffiths and J. Harris, Principles of Algebraic Geometry, John Wiley & Sons, New York.

[GK] A. Givental and B. Kim, Quantum cohomology of flag manifolds and Toda lattices, preprint, 1993.

[Gr] M. Gromov, Pseudo holomorphic curves in symplectic manifolds, Invent. math., 82 (1985), 307-347.

[HKTY] S. Hosono, A. Klemm, S. Theisen and S.T. Yau, Mirror symmetry, mirror map and applications to complete intersection Calabi-Yau spaces, Nucl. Phys., B 344 (1995), 501-554.

[Ho] H. Hori, Constraints for topological strings in $D \geq 1$, preprint.

[HS] H. Hofer and D. Salamon, Floer homology and Novikov rings, preprint.

[IN] A.R. Its and V. Yu. Novokshenov, The Isomonodromic Deformation Method in the Theory of Painlevé Equations, Lec. Notes in Mathematics, 1191.

[JMS] A. Jimbo, T. Miwa and M. Sato, Publ. RIMS, 14 (1978), 223; 15 (1979), 201, 577, 871; 16 (1980), 531.

[Ka] Y. Kawamata, preprint.

[KM] M. Kontsevich and Y. Manin, GW classes, Quantum cohomology and enumerative geometry, Comm.Math.Phys., 164 (1994), 525-562.

[Ko] M. Kontsevich, Intersection theory on the moduli space of curves and the matrix airy function, Comm. Math. Phys., 147 (1992).

[Ko2] M. Kontsevich, Enumeration of rational curves via torus actions, "The Moduli Space of Curves", edited by R.Dijkgraaf, C.Faber, G. van der Geer, series Progress in Mathematics vol.129, Birkhauser 1995.

[Kz] S. Katz, On the finiteness of rational curves on quintic hypersurfaces, Compositio Math., 60 (1986).

[Li] G. Liu, Associativity of quantum multiplication, preprint, October, 1994.

[LT1] J. Li and G. Tian, Quantum cohomology of homogeneous varieties, to appear in J. Alg. Geom.

[LT2] J. Li and G. Tian, Virtual moduli cycles and GW-invariants, in preparation.

[Lu] P. Lu, A rigorous definition of fiberwise quantum cohomology and equivariant quantum cohomology, preprint, 1995.

[Mi] T. Miwa, Painlevé property of monodromy preserving equations and the analyticity of τ-functions, Publ. RIMS, 17 (1981), 703-721.

[MP] D. Morrison and R. Plesser, preprint, 1994.

[MS] D. McDuff and D. Salamon, J-holomorphic curves and quantum cohomology, University Lec. Series, vol. 6, AMS.

[Mo] D. Morrison, Picard-Fuchs equations and mirror maps for hypersurfaces, Essays on Mirror manifolds, edited by Yau, International Press, Hong Kong, 1992.

[No] S. Novikov, Multivalued functions and functionals - analogue of the Morse theory, Soviet Math. Dokl. 24 (1981).

[On] K. Ono, On the Arnold conjecture for weakly monotone symplectic manifolds, Inv. Math., 1994.

[OV] K. Ono and L. Van, Cup-length estimate for symplectic fixed points, preprint, 1993.

[PSS] S. Piunikhin, D. Salamon and M. Schwarz, preprint, 1995.

[Pu] S. Piunikhin, Quantum and Floer cohomology have the same ring structure, preprint, 1994.

[PW] T. Parker and J. Woflson, A compactness theorem for Gromov's moduli space, J. Geom. Analysis, 3 (1993), 63-98.

[R1] Y. Ruan, Topological Sigma model and Donaldson type invariants in Gromov theory, preprint.

[R2] Y. Ruan, Symplectic topology on algebraic 3-folds, Jour. Diff. Geom., 39 (1994), 215-227.

[RT1] Y. Ruan and G. Tian, A mathematical theory of quantum cohomology, To appear in J. Diff. Geom., 1995; announcement, Math. Res. Let., vol 1, no 1 (1994), 269-278.

[RT2] Y. Ruan and G. Tian, Higher genus symplectic invariants and sigma model coupled with gravity, in preparation.

[RT3] Y. Ruan and G. Tian, Bott-type symplectic Floer cohomology and its multiplication structures, preprint, 1995.

[Sa] V. Sadov, On the equivalence of Floer's and quantum cohomology, preprint, 1993.

[ST1] B. Siebert and G. Tian, On quantum cohomology rings of Fano manifolds and a formula of Vafa and Intriligator, preprint, 1994.

[ST2] B. Siebert and G. Tian, Quantum cohomology of moduli spaces of stable bundles, in preparation.

[Ta] C. Taubes, The Seiberg-Witten and the Gromov invariants, preprint, 1995.

[Ti1] G. Tian, Rational curves on degenerating families, in preparation.

[Ti2] G. Tian, Smoothing 3-folds with trivial canonical bundle and ordinary double points, Essays on Mirror manifolds, edited by Yau, International Press, Hong Kong, 1992.

[Va] C. Vafa, Topology mirrors and quantum rings, Essays on mirror manifolds, edited by S. T. Yau, International Press, Hong Kong 1992.

[Vo] C. Voisin, A mathematical proof of Aspinwall-Morrison formula, preprint, 1995.

[W1] E. Witten, Topological sigma models, Comm. Math. Phys., 118 (1988).

[W2] E. Witten, Two dimensional gravity and intersection theory on moduli space, Surveys in Diff. Geom., 1 (1991), 243-310.

[W3] E. Witten, The Verlinde algebra and the cohomology of the Grassmannian, preprint, December, 1993.

[W4] E. Witten, Monopoles and 4-manifolds, Math. Res. Letters vol. 1 (1994), 769-796.

[W5] E. Witten, Mirror manifolds and topological field theory, Essays on Mirror manifolds, edited by Yau, International Press, Hong Kong, 1992.

[Ye] R. Ye, Gromov's compactness theorem for pseudo-holomorphic curves, Trans. Amer. math. Soc., 1994.

Made in United States
Orlando, FL
05 November 2026

90229887R00227